YOUMARES 8 – Oceans Across Boundaries: Learning from each other

Icebergs at the mouth of Scoresby Sund (Kangertittivaq), East Greenland. Maria S. Merian Expedition MSM 56, July 2016. (Photo: Boris Koch, AWI)

Simon Jungblut • Viola Liebich • Maya Bode
Editors

YOUMARES 8 – Oceans Across Boundaries: Learning from each other

Proceedings of the 2017 conference for YOUng MARine RESearchers in Kiel, Germany

Editors

Simon Jungblut
BreMarE – Bremen Marine Ecology, Marine Zoology
University of Bremen
Bremen, Germany

Alfred Wegener Institute
Helmholtz Centre for Polar and Marine Research
Bremerhaven, Germany

Maya Bode
BreMarE – Bremen Marine Ecology, Marine Zoology
University of Bremen
Bremen, Germany

Deutsche Gesellschaft für Meeresforschung
(DGM) e.V., Biozentrum Klein Flottbek
Hamburg, Germany

Viola Liebich
Deutsche Gesellschaft für Meeresforschung
(DGM) e.V., Biozentrum Klein Flottbek
Hamburg, Germany

To all Young Marine Researchers

Foreword

YOUMARES is a bottom-up conference, which has been organized for 8 years now by highly engaged young people that are enthusiastic about marine sciences. It was initiated by the working group on studies and education of the German Society for Marine Research (DGM) with the aim of building a network for young marine researchers.

From my perspective, part of the success of the YOUMARES conferences is the bottom-up concept that generates a multitude of new and creative ideas, presentation formats, and communication approaches. Another unique feature is that YOUMARES is also open for pupils and young university students interested in marine sciences. As a wonderful example, the organizers of a previous YOUMARES conference contacted local schools and convinced an English teacher to introduce the topic of fisheries biology in her class. As a result, the entire school class later attended the YOUMARES fisheries biology session.

The challenge of the bottom-up concept is the natural fluctuation within the organizing committee and it requires highly engaged people with good organizing skills to sustain YOUMARES. In my view, on the other hand, the fundamental benefit for team members and participants is a substantial gain of soft skills, long-lasting contacts and friendships, and the build-up of personal networks.

Over 180 participants from 23 nations attended YOUMARES 8 and it was, again, highly inspiring to see the creative ideas developed by organizers and participants. Apart from the science itself, many sessions at YOUMARES 8 addressed gender aspects in science, compatibility of research and family, or proposal writing aspects, which are not always part of the classical education in an early scientist's career.

The publication of these proceedings is unique and faced several challenges: Is the science sound and does the effort interfere with ongoing tasks, for example, in the authors' PhD project? How can the proceedings be financed without having a research institution in the background? Who takes care of organizing submissions, peer-review process, and revisions? All of these aspects were dealt with by the organizers with enormous creativity and momentum. Their effort included proposal writing, acquisition of funding and supporters who helped organizing contributions and reviews.

This peer-reviewed publication documents the YOUMARES effort and, at the same time, supports the future careers of the contributors. Several chapters inherently express young marine researchers' concerns toward the fundamental environmental and societal challenges in the marine realm, such as climate change, littering, or human pressure on coasts. Meeting these challenges requires multidisciplinary, international, and cross-generation interplay, and to cite one chapter: "The Static, Boundary-Based Norm of Scientific Thinking Must Be Overcome."

I congratulate the organizers and contributors for their effort and recommend reading these proceedings – the laurels of highly engaged marine researchers, who will shape marine science in future.

Boris Koch

Alfred Wegener Institute, Helmholtz Centre for Polar and Marine Research, Bremerhaven, Germany
MARUM Center for Marine Environmental Sciences, Bremen, Germany
University of Applied Sciences, Bremerhaven, Germany
March 28, 2018

Preface

This proceedings volume is the final product of the YOUMARES 8 conference, held from 13 to 15 September 2018 in Kiel, Germany. YOUMARES is a conference series organized by and for YOUng MARine RESearchers under the auspices of the German Society for Marine Research (Deutsche Gesellschaft für Meeresforschung e.V. – DGM). Especially bachelor, master, and PhD students from all fields of marine sciences are asked to contribute to the conference. Their presentations represent current issues of marine research and are organized in thematic sessions, which are hosted mostly by PhD students or young post-docs. In addition to organizing and moderating their session, the session hosts are given the opportunity to write a literature review of a session-related topic of their choice. These literature reviews, together with all conference abstracts, are compiled in this book. The articles, i.e., peer-reviewed chapters of this book, represent the current state of knowledge of their specific topic, while the corresponding abstracts represent ongoing research projects.

The 2017 edition of the YOUMARES series was hosted by the Kiel University and the GEOMAR Helmholtz Centre for Ocean Research in Kiel. Over 180 young researchers contributed over 90 talks and 27 poster presentations. Including all helpers, this eighth edition of YOUMARES was the biggest YOUMARES conference so far. The icebreaker event took place in the foyer of the east shore building of GEOMAR, whereas presentations, talks, and workshops were held in a seminar building of the Kiel University. Keynote talks were given by Prof. Dr. Mojib Latif (GEOMAR Helmholtz Centre for Ocean Research, Kiel) on "The Role of the Oceans in Climate Change," and by Dr. Claudia Hanfland (Alfred Wegener Institute, Helmholtz Centre for Polar and Marine Research, Bremerhaven) on "Career Planning – Advice from the Cheshire Cat."

We hope that these articles and abstracts are a source of knowledge and inspiration for the conference participants, authors, and all interested people. We hope that this book will provide the conference participants with sustainable memories about the conference in Kiel and that it also encourages interested people to join the YOUMARES network.

Bremen, Germany — Simon Jungblut
Hamburg, Germany — Viola Liebich
Bremen, Germany — Maya Bode
March 2018

Acknowledgments

We would like to thank many helpers for taking over smaller and bigger tasks during all phases of the preparation and realization of the conference in Kiel. Without the strong support of these volunteers, organizing a conference of such a size would be impossible. We would like to thank all of them: Jan Brüwer, Andreas Eich, Joeline Ezekiel, Thea Hamm, Lena Heel, Lisa Hentschel, Dorothee Hohensee, Elham Kamyab, Maral Khosravi, Veloisa Mascarenhas, Ola Nour, Olga Sazonova, Timothy Tompson, and Mara Weidung.

We are very grateful to the Kiel University and the GEOMAR Helmholtz Centre for Ocean Research Kiel for providing the space and rooms for the conference and icebreaker venues. Special thanks go to Wiebke Basse (Integrated School of Ocean Sciences ISOS of the Kiel Cluster of Excellence "The Future Ocean") and Michael Mattern for their organizational and technical support prior to and during the conference.

The keynote lectures of Mojib Latif (GEOMAR Helmholtz Centre for Ocean Research Kiel) and Claudia Hanfland (Alfred Wegener Institute, Helmholtz Centre for Polar and Marine Research, Bremerhaven) received much attention. We are grateful to both for presenting interesting and stimulating plenary talks.

The workshops during the conference were organized by several people to which we are all grateful: Francisco Barboza, Hanna Campen, Markus Franz, Jonas Geburzi, Lydia Gustavs, Daniel Hartmann, Marie Heidenreich, Lisa Hentschel, Maysa Ito, Veit Klimpel, Frank Schweikert, and Martin Visbeck.

Several partners financially supported the conference: Norddeutsche Stiftung für Umwelt und Entwicklung, SubCtech, DFG-Schwerpunktprogramm Antarktisforschung, Bornhöft Meerestechnik, Aida, and Deutsche See.

Springer Nature provided book vouchers to award the best oral and poster presentations.

The Zoological Museum Kiel is thanked for offering free entrance to their exhibitions for all conference participants.

The Staats- und Universitätsbibliothek Bremen and the Alfred Wegener Institute, Helmholtz Centre for Polar and Marine Research supported the publication of this proceedings book.

Thanks go to Alexandrine Chernonet, Judith Terpos, and Springer for their support during the editing and publishing process of this book.

All chapters of this book have been peer-reviewed by internationally renowned scientists. The reviews contributed significantly to the quality of the chapters. We would like to thank all reviewers for their time and their excellent work: Tina Dohna, Erik Duemichen, Tor Eldevik, Lucy Gwen Gillis, Gustaaf Hallegraeff, Charlotte Havermans, Ferenc Jordán, Trevor McIntyre, Paul Myers, Ingo Richter, Paris Vasilakopoulos, Aurore Voldoire, Jan Marcin Węsławski, Christian Wild, Argyro Zenetos, and further anonymous reviewers.

We editors are most grateful to all participants, session hosts, and presenters of the conference and to the contributing authors of this book. You all did a great job in presenting and representing your (fields of) research. Without you, YOUMARES would not be worth to organize.

Contents

Contributors

Mona Andskog Faculty of Biology and Chemistry, University of Bremen, Bremen, Germany

Leibniz Centre for Tropical Marine Research (ZMT), Bremen, Germany

Francisco R. Barboza GEOMAR Helmholtz Centre for Ocean Research, Kiel, Germany

Maya Bode BreMarE – Bremen Marine Ecology, Marine Zoology, University of Bremen, Bremen, Germany

Deutsche Gesellschaft für Meeresforschung (DGM) e.V., Biozentrum Klein Flottbek, Hamburg, Germany

Jan David Brüwer Red Sea Research Center, Division of Biological and Environmental Science and Engineering (BESE), King Abdullah University of Science and Technology (KAUST), Thuwal, Saudi Arabia

Faculty of Biology and Chemistry, University of Bremen, Bremen, Germany

Max Planck Institute for Marine Microbiology, Bremen, Germany

Hagen Buck-Wiese Faculty of Biology and Chemistry, University of Bremen, Bremen, Germany

Max Planck Institute for Marine Microbiology, Bremen, Germany

Camila Campos Alfred Wegener Institute (AWI), Helmholtz Centre for Polar and Marine Research, Bremerhaven, Germany

Xochitl Cormon Institute for Marine Ecosystem and Fishery Science, Centre for Earth System Research and Sustainability (CEN), University of Hamburg, Hamburg, Germany

Maha J. Cziesielski Red Sea Research Centre, King Abdullah University of Science and Technology, Thuwal, Kingdom of Saudi Arabia

Tina Dippe GEOMAR Helmholtz Centre for Ocean Research Kiel, Kiel, Germany

Hannah S. Earp Faculty of Biology and Chemistry, University of Bremen, Bremen, Germany

Leibniz Centre for Tropical Marine Research (ZMT), Bremen, Germany

School of Ocean Sciences, Bangor University, Menai Bridge, Wales, UK

Markus Franz GEOMAR Helmholtz Centre for Ocean Research, Kiel, Germany

Jonas C. Geburzi Zoological Institute and Museum, Kiel University, Kiel, Germany

Alfred Wegener Institute, Helmholtz Centre for Polar and Marine Research, Wadden Sea Station, List/Sylt, Germany

Jana K. Geuer Alfred Wegener Institute (AWI), Helmholtz Centre for Polar and Marine Research, Bremerhaven, Germany

Thea Hamm GEOMAR Helmholtz Center for Ocean Research, Kiel, Germany

Jan Harlaß GEOMAR Helmholtz Centre for Ocean Research Kiel, Kiel, Germany

Brigitte C. Heylen Behavioural Ecology and Ecophysiology, University of Antwerp, Antwerp, Belgium

Terrestrial Ecology Unit, Ghent University, Ghent, Belgium

Myriel Horn Alfred Wegener Institute (AWI), Helmholtz Centre for Polar and Marine Research, Bremerhaven, Germany

Maysa Ito GEOMAR Helmholtz Centre for Ocean Research, Kiel, Germany

Simon Jungblut BreMarE – Bremen Marine Ecology, Marine Zoology, University of Bremen, Bremen, Germany

Alfred Wegener Institute, Helmholtz Centre for Polar and Marine Research, Bremerhaven, Germany

Laura Käse Alfred Wegener Institute (AWI), Helmholtz Centre for Polar and Marine Research, Biologische Anstalt Helgoland, Helgoland, Germany

Therese Keck Institute for Space Sciences, Freie Universität Berlin, Berlin, Germany

Martin Krebs GEOMAR Helmholtz Centre for Ocean Research Kiel, Kiel, Germany

Viola Liebich Deutsche Gesellschaft für Meeresforschung (DGM) e.V., Biozentrum Klein Flottbek, Hamburg, Germany

Claudia Lorenz Alfred Wegener Institute (AWI), Helmholtz Centre for Polar and Marine Research, Biologische Anstalt Helgoland, Helgoland, Germany

Joke F. Lübbecke GEOMAR Helmholtz Centre for Ocean Research Kiel, Kiel, Germany

Faculty of Mathematics and Natural Sciences, Christian Albrechts University, Kiel, Germany

Maciej K. Mańko Department of Marine Plankton Research, Institute of Oceanography, University of Gdańsk, Gdynia, Poland

Veloisa Mascarenhas Institut für Chemie und Biologie des Meeres, Universität Oldenburg, Wilhelmshaven, Germany

Morgan L. McCarthy School of Biological Sciences, The University of Queensland, St. Lucia, QLD, Australia

Marine Biology, Vrije Universiteit Brussel (VUB), Brussels, Belgium

Dominik A. Nachtsheim Institute for Terrestrial and Aquatic Wildlife Research, University of Veterinary Medicine Hannover, Büsum, Germany

BreMarE – Bremen Marine Ecology, Marine Zoology, University of Bremen, Bremen, Germany

Sarah Piehl Department of Animal Ecology I and BayCEER, University of Bayreuth, Bayreuth, Germany

Natalie Prinz Faculty of Biology and Chemistry, University of Bremen, Bremen, Germany

Leibniz Centre for Tropical Marine Research (ZMT), Bremen, Germany

Camilla Sguotti Institute for Marine Ecosystem and Fishery Science, Centre for Earth System Research and Sustainability (CEN), University of Hamburg, Hamburg, Germany

Katarzyna S. Walczyńska Department of Marine Plankton Research, Institute of Oceanography, University of Gdańsk, Gdynia, Poland

Agata Weydmann Department of Marine Plankton Research, Institute of Oceanography, University of Gdańsk, Gdynia, Poland

About the Editors

Dr. Simon Jungblut is a marine ecologist and zoologist. After completing a bachelor's degree in biology and chemistry at the University of Bremen, Germany, he studied the international program Erasmus Mundus Master of Science in Marine Biodiversity and Conservation at the University of Bremen; the University of Oviedo, Spain; and Ghent University, Belgium. Afterwards, he completed a PhD project entitled: "Ecology and Ecophysiology on Invasive and Native Decapod Crabs in the Southern North Sea" at the University of Bremen in cooperation with the Alfred Wegener Institute, Helmholtz Centre for Polar and Marine Research in Bremerhaven and was awarded the doctoral title in natural sciences at the University of Bremen in December 2017.

Since 2015, Simon is actively contributing to the YOUMARES conference series. After hosting some conference sessions, he is the main organizer of the scientific program since 2017.

Dr. Viola Liebich is a biologist from Berlin, who worked on invasive tunicates for her diploma thesis at the Wadden Sea Station Sylt of the Alfred Wegener Institute. With a PhD scholarship by the International Max Planck Research School for Maritime Affairs, Hamburg, and after her thesis work at the Institute for Hydrobiology and Fisheries Science, Hamburg, and the Royal Netherlands Institute for Sea Research, Texel, the Netherlands, she finished her thesis "Invasive Plankton: Implications of and for Ballast Water Management" in 2013.

For three years, until 2015, Viola Liebich worked for a project on sustainable brown shrimp fishery and stakeholder communication at the WWF Center for Marine Conservation and started her voluntary YOUMARES work one year later. In 2017, she also became elected member of the DGM steering group. She is currently working as a self-employed consultant on marine and maritime management (envio maritime).

Dr. Maya Bode is a marine biologist, who accomplished her Bachelor of Science in biology at the University of Göttingen, Germany, and her Master of Science in marine biology at the University of Bremen, Germany. Thereafter, she completed her PhD thesis entitled "Pelagic Biodiversity and Ecophysiology of Copepods in the Eastern Atlantic Ocean: Latitudinal and Bathymetric Aspects" at the University of Bremen in cooperation with the Alfred Wegener Institute, Helmholtz Centre for Polar and Marine Research in Bremerhaven and the German Center for Marine Biodiversity Research (DZMB) at the Senckenberg am Meer in Wilhelmshaven. She received her doctorate in natural sciences at the University of Bremen in March 2016.

Since 2016, Maya is a board member of the German Society for Marine Research (DGM) and actively contributes to the YOUMARES conference series as organizer of the scientific program.

YOUMARES – A Conference from and for YOUng MARine RESearchers

Viola Liebich, Maya Bode, and Simon Jungblut

Abstract

YOUMARES is an annual early-career scientist conference series. It is an initiative of the German Society for Marine Research (DGM) and takes place in changing cities of northern Germany. The conference series is organized in a bottom-up structure: from and for YOUng MARine RESearchers. In this chapter, we describe the concept of YOUMARES together with its historical development from a single-person initiative to a conference venue of about 200 participants. Furthermore, the three authors added some personals experiences and insights, what YOUMARES means to them.

Concept and Structure of YOUMARES

Education is the central key component for the progression of societies. As such, it is the basis to cope with the *challeng*es of globalization. At the same time, the oceans are the biggest and most important ecosystem, securing the survival capabilities of mankind on earth. It is, therefore, of pivotal interest that young researchers commit themselves to shape the future of this ecosystem in a sustainable way. To jointly develop the most important future topics, a vibrant and interdisciplinary network of research, economy, and society is necessary. As such, YOUMARES is much more than a regular annual research conference. It is a platform which aims to establish a network especially for early career scientists (Einsporn 2011). It thereby promotes the research and communication activities of High School, Bachelor, Master, and PhD students. Similar to regular conferences, the participants have the possibility to present their research in oral or poster presentations. Additionally, different kinds of workshops, plenary discussions and social events enable the participants to extensively exchange with each other at eye level. Providing an exchange platform should ultimately lead to a young researcher network and to the enhancement of individual and collective competence (Fig. 1).

YOUMARES is an initiative of the working group "Studies and Education" of the German Society for Marine Research (Deutsche Gesellschaft für Meeresforschung e.V. – DGM). Right from the beginning in 2010 on, an essential part of the idea was to drive the organization of the conference bottom-up (Einsporn 2011). The whole conference is organized by early career scientists. In each winter a core organization team publishes a "Call for Sessions", which encourages young marine researchers from all kinds of scientific fields to apply alone or in pairs for hosting one of the scientific sessions at the upcoming conference. The applications contain the CVs, a motivation letter and most importantly a "Call for Abstracts" for the proposed session. If two or more applicant groups propose similar sessions, the core organization team brings them into contact and encourages them to organize a joint session. Once the applications are reviewed and the sessions are being set, the different "Calls for Abstracts" are published. The session hosts have several responsibilities. They handle the abstracts of their sessions and organize, structure, and moderate their session at the actual conference. Additionally, they are asked to write a literature review of the field of research (or one aspect of it)

V. Liebich (✉)
Deutsche Gesellschaft für Meeresforschung (DGM) e.V., Biozentrum Klein Flottbek, Hamburg, Germany
e-mail: enviomaritime@gmail.com

M. Bode
BreMarE – Bremen Marine Ecology, Marine Zoology, University of Bremen, Bremen, Germany

Deutsche Gesellschaft für Meeresforschung (DGM) e.V., Biozentrum Klein Flottbek, Hamburg, Germany
e-mail: mabode@uni-bremen.de

S. Jungblut
BreMarE – Bremen Marine Ecology, Marine Zoology, University of Bremen, Bremen, Germany

Alfred Wegener Institute, Helmholtz Centre for Polar and Marine Research, Bremerhaven, Germany
e-mail: jungblut@uni-bremen.de

S. Jungblut et al. (eds.), *YOUMARES 8 – Oceans Across Boundaries: Learning from each other*,
https://doi.org/10.1007/978-3-319-93284-2_1

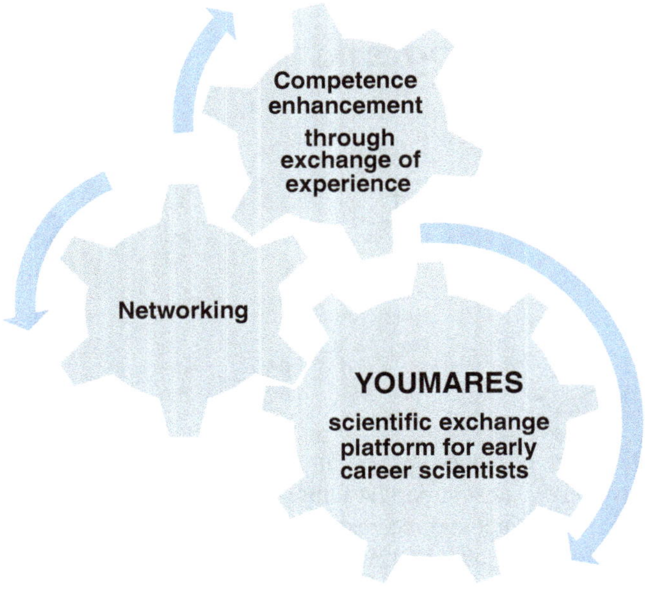

Fig. 1 The interplay between the provision of an exchange platform for early career scientists, networking efforts, and the enhancement of competence

they cover with their session. The product of all these efforts of the session hosts is the book at hand. It summarizes the literature reviews of most sessions and all presenter abstracts of the latest edition of the conference series, YOUMARES 8, held from 13 to 15 September 2017 in Kiel, Germany.

A Brief History of Getting Larger

YOUMARES was established by the initiative of a single person – Marc Einsporn. Marc came up with the idea of a platform where especially the young generation of scientists would be able to exchange and to present their research to an audience of a similar career stage. Starting off as a national conference, the first YOUMARES took place under a different name ("Netzwerktreffen junge Meeresforschung") in Hamburg in June 2010 with less than 50 participants (Table 1, Einsporn 2010). Already 1 year later, the name "YOUMARES" was established and it took place with about 130 participants over 3 days in September (Einsporn 2011). From then on, the conference acquired an international reputation and was held each September in different cities in northern Germany. By 2017, eight editions of YOUMARES took place; so far in seven different cities (Table 1). Already in 2012, participants came from more than ten different countries, in 2013 from more than 15 different countries (Wiedling and Einsporn 2012, Einsporn et al. 2013). Over the years, YOUMARES has expanded into the largest meeting of young marine scientists in Germany. The most recent edition, YOUMARES 8, had about 195 participants and 95 oral presentations (Table 1). Organizing an event of this size obviously requires a large team of organizers and helpers.

The topical sessions of each YOUMARES edition offer an interesting insight into the spectrum and the diversity of research early career scientists are conducting in the marine field (Table 2). In few cases, the same people applied for hosting a session in subsequent years. However, some topics are reoccurring relatively often over the years as for instance aquaculture, plastic pollution, invasive species, coral reefs and polar regions.

How to Get in Contact: Personal Experiences as a Young Researcher

YOUMARES – Science Works Best When Being Shared

Viola Liebich

I had joined YOUMARES as a participant some years ago when I was still a PhD student. When I first heard about this conference I didn't realize just how special it was, to be honest. Being on-site, I liked the atmosphere and noticed the rather young audience. However, it was only later in my PhD career that I joined 'big' and 'professional' meetings in an international set-up. The topic of my PhD was the introduction of invasive species via ballast water and I took a turn joining an EU project 'with application'. Applied science still has a bit of a stale taste to it for many researchers. The different worlds seem to collide on ballast water management conferences when biologists meet vessel fleet managers, government representatives, lawyers, engineers, and project managers – the guys in suits as they were called in my old institute. Dinners often were five courses served with wine you had to fight off to be not re-filled all the time. Now, was that an inspiring and relaxing atmosphere? No, I enjoyed the nice food but didn't feel comfortable talking to people I didn't know and went home with a missed chance to enlarge my network. But YOUMARES had showed me that we are as scientists not alone with our topics, ideas, questions, and problems. I learned from my first supervisor that science works best when being shared and that is, in my opinion, what YOUMARES also stands for.

Thus, when I got the chance to organize this year's YOUMARES as head of the team, I recalled that feeling. Above all, I wanted to create that easy atmosphere with people of similar minds – as if we would all meet up in a student house kitchen. At the same time, we had the expectation to offer a professional conference. The bottom-up approach done by young volunteers when organizing it should not be an excuse that the conference and everything around it doesn't provide you the best options. Although it was often challenging to find the time to call after sponsors, facility

Table 1 Key data of YOUMARES conferences until 2017

Year	Dates	Place	Motto	No. participants	No. sessions	No. talks	No. posters	Reference
2010	12 June	Hamburg	Netzwerktreffen Junge Meeresforschung – Young marine research: Diversity and similarities	46	4	17	?	Einsporn (2010)
2011	07–09 September	Bremerhaven	YOUMARES 2 – Oceans amidst science, innovation and society	130	6	31	33	Einsporn (2011)
2012	12–14 September	Lübeck	YOUMARES 3 – Between space and seafloor – Aqua vita est	130	10	60	50	Wiedling and Einsporn (2012)
2013	11–13 September	Oldenburg	YOUMARES 4 – From coast to deep sea: Multiscale approaches to marine sciences	150	15	53	35	Einsporn et al. (2013)
2014	10–12 September	Stralsund	YOUMARES 5 – Opportunities and solutions – Research for changing oceans	100	10	35	16	Jessen and Golz (2014)
2015	16–18 September	Bremen	YOUMARES 6 – A journey into the blue – Ocean research and innovation	126	14	47	27	Jessen et al. (2015)
2016	11–13 September	Hamburg	YOUMARES 7 – People and the 7 seas – Interaction and innovation	110	11	42	29	Bode et al. (2016)
2017	13–15 September	Kiel	YOUMARES 8 – Oceans across boundaries: Learning from each other	195	15	95	27	This contribution

Table 2 Session topics of YOUMARES conferences until 2017

Year	Session number and session	Reference
2010	(1) Biologie und Chemie (Biology and chemistry)	Einsporn (2010)
	(2) Fernerkundung (Remote sensing)	
	(3) Mikro- und Molekularbiologie (Micro- and molecular biology)	
	(4) Aquakultur (Aquaculture)	
2011	(1) Human impacts on the oceans and subsequent environmental responses	Einsporn (2011)
	(2) Remote sensing: Higher orbits for deeper understanding	
	(3) Aquaculture: Main research priorities to fulfill our need for sustainable seafood	
	(4) Living with the Sea: Coastal livelihoods and management	
	(5) Marine technologies – The art of engineering in synergy with natural sciences	
	(6) Ocean of diversity: From micro scales to macro results	
2012	(1) Aliens from inner space: Where do they come from, what do they do and how can we stop them?	Wiedling and Einsporn (2012)
	(2) Between sea and Anthroposphere: Marine socio-economics in an era of global change	
	(3) Environmental changes in the pelagic: Consequences and acclimatization strategies – From plankton to fish	
	(4) Integrated aquaculture – Polyculture of plants, invertebrates and finfish	
	(5) Ocean modelling: Theory & concepts	
	(6) Physical oceanography – Between measuring and modelling	
	(7) Reefs from shallow to deep – Environmental constraints and perspective	
	(8) The aquatic climate archive: Tracking the rise and fall of ancient civilizations.	
	(9) Lessons from the past, for the present and the future?	
	(10) Water resources in coastal areas – Scarcity and management implications	
2013	(1) Dissolved Organic Matter (DOM) – small in size but large in impact: Basis of life in the world's ocean	Einsporn et al. (2013)
	(2) Aquatic microorganisms: Between producers, consumers and pathogens	
	(3) Marine plastic pollution: From sources to solutions	
	(4) Importance of coral reefs for coastal zones: Services, threats, protection strategies	
	(5) Fluctuations in cephalopod and jellyfish abundances: Reasons and potential impacts on marine ecosystems	
	(6) Responses of marine fish to environmental stressors	
	(7) The ecosystem approach and beyond: Multidisciplinary science for sustainability in fisheries	
	(8) Aquaculture: Fish feeds the world – but how?	

(continued)

Table 2 (continued)

Year	Session number and session	Reference
	(9) How to integrate blue biotechnology in food industry and medicine	
	(10) Marine measurement technologies: Science and engineering	
	(11) Operational oceanography	
	(12) Methods and applications of ocean remote sensing	
	(13) Coping with uncertainties in marine science – From crisis management to the new risk approaches in the Baltic Sea chemicals management	
	(14) Marine habitat mapping: Stretching the blue marble on a map	
	(15) What's up with coral reefs?	
2014	(1) Small-scale fisheries research – Towards sustainable fisheries using a multi-entry perspective	Jessen and Golz (2014)
	(2) Individual engagement in environmental change	
	(3) Aquaculture in a changing ocean	
	(4) Coral reef ecology, management and conservation in a rapidly changing ocean environment	
	(5) Tools and methods supporting an ecosystem based approach to marine spatial management	
	(6) Measurement and control engineering – The clockwork in marine science	
	(7) Aquatic plastic pollution – Tackling environmental impacts with new solutions	
	(8) Mangrove forests – An endangered ecological and economic transition zone between ocean and land	
	(9) Effects of global climate change on emerging infectious diseases of marine fish	
	(10) Cold water research – From high latitude coasts to deep sea trenches	
2015	(1) Frame works for sustainable management of water resources	Jessen et al. (2015)
	(2) Population genetics as a powerful tool for the management and sustainability of natural resources	
	(3) Cephalopods and society: Scientific applications using cephalopods as models	
	(4) Challenges and innovative solutions for monitoring pollution and restoration of coastal areas	
	(5) ScienceTainment	
	(6) From invasive species to novel ecosystems	
	(7) From outer space to the deep-sea: Remote sensing in the twenty-first century	
	(8) No living without the ocean: Social-ecological systems in the marine realm	
	(9) How our behavior can make the difference in ocean conservation	
	(10) Recent approaches in coral reef research: Traditional and novel applications towards building resilience	
	(11) Latest developments in land-based aquaculture	
	(12) Active study in times of Bologna	
	(13) Multispecies and ecosystem models for fisheries management and marine conservation	
	(14) Aquatic plastic pollution	
2016	(1) From egg to juvenile: Advances and novel applications to study the early life history stages of fishes	Bode et al. (2016)
	(2) Dissolved organic matter in aquatic systems: Assessment and applications	
	(3) Fighting eutrophication in shallow coastal waters	
	(4) Deep, dark and cold – Frontiers in polar and deep sea research	
	(5) Going global: Invasive and range-expanding species	
	(6) How do communities adapt?	
	(7) Marine species interactions and ecosystem dynamics: Implications for management and conservation	
	(8) Coastal and marine pollution in the Anthropocene: Identifying contaminants and processes	
	(9) Social dimensions of environmental change in the coastal marine realm	
	(10) Phytoplankton: Are we all looking at it differently? Diverse methods and approaches to the study of marine phytoplankton	
	(11) Coral reefs and people in changing times	
2017	(1) Sentinels of the sea: Ecology and conservation of marine top predators	This contribution
	(2) Reading the book of life – -omics as a universal tool across disciplines	
	(3) Physical processes in the tropical and subtropical oceans: Variability, impacts, and connections to other components of the climate system	
	(4) Cephalopods: Life histories of evolution and adaptations	
	(5) Ecosystems dynamics in a changing world: Regime shifts and resilience in marine communities	
	(6) The interplay between marine biodiversity and ecosystems functioning: Patterns and mechanisms in a changing world	

(continued)

Table 2 (continued)

Year	Session number and session	Reference
	(7) Ocean optics and ocean color remote sensing	
	(8) Polar ecosystems in the age of climate change	
	(9) The physics of the Arctic and subarctic oceans in a changing climate	
	(10) Phytoplankton in a changing environment – Adaptation mechanisms and ecological surveys	
	(11) How do they do it? – Understanding the success of marine invasive species	
	(12) Coastal ecosystem restoration – Innovations for a better tomorrow	
	(13) Microplastics in aquatic habitats – Environmental concentrations and consequences	
	(14) Tropical aquatic ecosystems across time, space and disciplines	
	(15) Open session	

details, caterers, accommodation offers, and of course all the scientific input, we put our mind to it. And I am very proud of this year's team. We achieved all we could have hoped for and managed to make YOUMARES 8 the biggest one so far!

YOUMARES and the DGM – Interlinking the Young and the Experienced

Maya Bode

My first contact to YOUMARES was from a different point of view: When I was in the final stage of my PhD thesis, I participated in the DGM-Meeresforum in Bremen in 2015 where marine researchers met politicians and climate scientists. Discussions about hot topics such as the plastic problem, geoengineering and deep sea diversity, and the limits and responsibility of human actions were indeed inspiring. Especially the interdisciplinary exchange between young and experienced researchers was extremely motivating: that we, as young marine researchers, really have the possibility to change what is going on in the world, if we efficiently use our resources, as such work together, constantly update ourselves about recent research findings and interlink various disciplines of marine sciences, engineering, social sciences, politics, and economics. As vast as the ocean may appear, we know and experience these days that resources and ecosystem's carrying capacities are limited and already overexploited in many regions of the world ocean. Efficient science with the ultimate aim to serve nature and society needs creativity and constant interdisciplinary exchange of knowledge. During the last decades, the society of marine scientists has grown and together with new technologies and sophisticated networking, we have the opportunity – better than ever before – to exchange new findings, bring our knowledge into the world and enhance interdisciplinary research, partnerships, and cooperation. YOUMARES serves as such a platform and has the potential to make marine research more efficient in the future.

To help to aim this goal, I became a member of the DGM in 2015 and helped organizing the YOUMARES 7 as scientific coordinator. Then, in 2016, I became a board member of the DGM with the main motivation to enhance the exchange of experienced and young marine researchers. Since 2015, the DGM-Meeresforum takes place each year, 1 day before the YOUMARES, bringing together young and experienced scientists, in the afternoon by inspiring talks and discussions and later in the evening by getting together at the icebreaker party of the YOUMARES. The DGM was founded in 1980 as a platform for exchange of information and views on all kinds of marine topics, having around 400 members nowadays. For the future, I would like to be part of the DGM growing larger and achieving a new standing and reputation among marine researchers and political institutions. With the experience of the DGM members and potential new young members, together with the DGM-Meeresforum and YOUMARES as an annual meeting and conference, we create a large and sustainable network all around the world.

YOUMARES – A Conference for the Future

Simon Jungblut

My first contact with YOUMARES was back in 2013. The conference was about to be held at the University of Oldenburg and was obviously growing bigger in the last editions. During this time, I was a student in the "Erasmus Mundus Master of Science in Marine Biodiversity and Conservation" and based at the University of Bremen. At some point, I read about YOUMARES online and shortly thereafter some posters appeared on the black-boards in our faculty building. The posters advertised YOUMARES as "convention for young scientists and engineers". That raised my interest. I identified myself with being a "young scientist" and decided to participate in the conference as a listener. The whole conference was interesting and amazing. I spoke to a lot of other participants and learned about their study programs and institutes. In addition, the talks and posters were interesting and informative. Right from the beginning I liked the concept of giving young students and scientists a relaxed and open platform to present and discuss their first research projects. After hosting sessions in the years 2015 and 2016, I took over the scientific organization of

YOUMARES 8 in 2017. I was responsible for the scientific program of the conference. This included collecting and first review of session applications and later abstract applications, the arrangement of the time schedule and the on-site coordination of hosts, conference participants and plenary speakers.

Being a part of the organization team was a totally new aspect for me. I liked to connect people and to bring them together to discuss and to network. The bases for shaping the networking experiences of young researchers are, to my experience, the shared research interests of the participants but also that the conference provides useful interdisciplinary workshops and other socializing activities. Thus, I see the future of YOUMARES in promoting such workshops and activities, side by side with the scientific presentations. Participants should be able to present their research to a broad, young audience and to participate workshops providing skills, which are useful for their future scientific life. Additionally, there should be enough room and time to effectively connect to other young scientists. Connecting young researchers might be a key component to help them establish collaborations. In this sense, a conference like YOUMARES helps to make research more efficient and more interdisciplinary, which ultimately might be a step towards a more efficient battle against the big problems the world ocean is facing right now.

References

Bode M et al (2016) People and the 7 seas – interaction and innovation. In: Conference proceedings of the YOUMARES 7 conference, Hamburg. Available at: https://www.youmares.org/past-conferences/youmares-7/

Einsporn M (2010) Young marine research: diversity and similarities. Group photo collage with participants of the Netzwerktreffen Junge Meeresforschung, Hamburg. Available at: https://www.youmares.org/past-conferences/youmares-1/

Einsporn M (2011) Oceans amidst science, innovation and society. In: Proceedings of the YOUMARES 2 conference, Bremerhaven. Available at: https://www.youmares.org/past-conferences/youmares-2/

Einsporn M et al (2013) Recent impulses to marine science and engineering – from coast to deep sea: multiscale approaches to marine sciences. In: Proceedings of the YOUMARES 4 conference, Oldenburg. Available at: https://www.youmares.org/past-conferences/youmares-4/

Jessen C, Golz V (2014) Opportunities and solutions – research for our changing oceans. Book of Abstracts of the YOUMARES 5 conference, Stralsund. Available at: https://www.youmares.org/past-conferences/youmares-5/

Jessen C et al (2015) A journey into the blue – ocean research and innovation. In: Conference book of the YOUMARES 6 conference, Bremen, 2015. Available at: https://www.youmares.org/past-conferences/youmares-6/

Wiedling J, Einsporn M (2012) Recent impulses to marine science and engineering. Between space and seafloor – aqua vita est. In: Proceedings of the YOUMARES 3 conference, Lübeck. Available at: https://www.youmares.org/past-conferences/youmares-3/

Can Climate Models Simulate the Observed Strong Summer Surface Cooling in the Equatorial Atlantic?

Tina Dippe, Martin Krebs, Jan Harlaß, and Joke F. Lübbecke

Abstract

Variability in the tropical Atlantic Ocean is dominated by the seasonal cycle. A defining feature is the migration of the inter-tropical convergence zone into the northern hemisphere and the formation of a so-called cold tongue in sea surface temperatures (SSTs) in late boreal spring. Between April and August, cooling leads to a drop in SSTs of approximately 5°. The pronounced seasonal cycle in the equatorial Atlantic affects surrounding continents, and even minor deviations from it can have striking consequences for local agricultures.

Here, we report how state-of-the-art coupled global climate models (CGCMs) still struggle to simulate the observed seasonal cycle in the equatorial Atlantic, focusing on the formation of the cold tongue. We review the basic processes that establish the observed seasonal cycle in the tropical Atlantic, highlight common biases and their potential origins, and discuss how they relate to the dynamics of the real world. We also briefly discuss the implications of the equatorial Atlantic warm bias for CGCM-based reliable, socio-economically relevant seasonal predictions in the region.

T. Dippe (✉) · M. Krebs · J. Harlaß
GEOMAR Helmholtz Centre for Ocean Research Kiel, Kiel, Germany
e-mail: tdippe@geomar.de; jharlass@geomar.de

J. F. Lübbecke
GEOMAR Helmholtz Centre for Ocean Research Kiel, Kiel, Germany

Faculty of Mathematics and Natural Sciences, Christian Albrechts University, Kiel, Germany
e-mail: jluebbecke@geomar.de

The Equatorial Atlantic: A Climate Hot Spot

The tropical oceans are a crucial element of the global climate system. Defined here as the ocean area between 15°N and 15°S, they occupy only about 13% of the earth's surface, but receive approximately 30% of the global net surface insolation.[1] Processes both in the ocean and the atmosphere redistribute surplus heat from low to higher latitudes. Without these mechanisms, the tropics would get steadily warmer, while the polar regions would radiate away more heat than they receive and hence continue to cool. The oceans help to establish the overall radiative equilibrium that is responsible for our relatively stable climate (Trenberth and Caron 2001).

Apart from the energy surplus, another defining feature of an equatorial ocean is that the effect of the earth's rotation vanishes at the equator, giving rise to a physical framework that is subtly different from its higher-latitude counterpart. The effect of the earth's rotation manifests in a pseudo-force that is called the Coriolis force. It deflects large-scale motion towards the right of the movement on the northern hemisphere and towards the left on the southern hemisphere. It provides rotation to large weather systems and explains why large-scale movement curves or even becomes circular. An exception is the equator, where the Coriolis force vanishes and movement can be straightforward. Additionally, the non-existent Coriolis force at the equator acts as a barrier for the transmission of information within the ocean, for example

[1] Based on data by Trenberth et al. (2009).

S. Jungblut et al. (eds.), *YOUMARES 8 – Oceans Across Boundaries: Learning from each other*,
https://doi.org/10.1007/978-3-319-93284-2_2

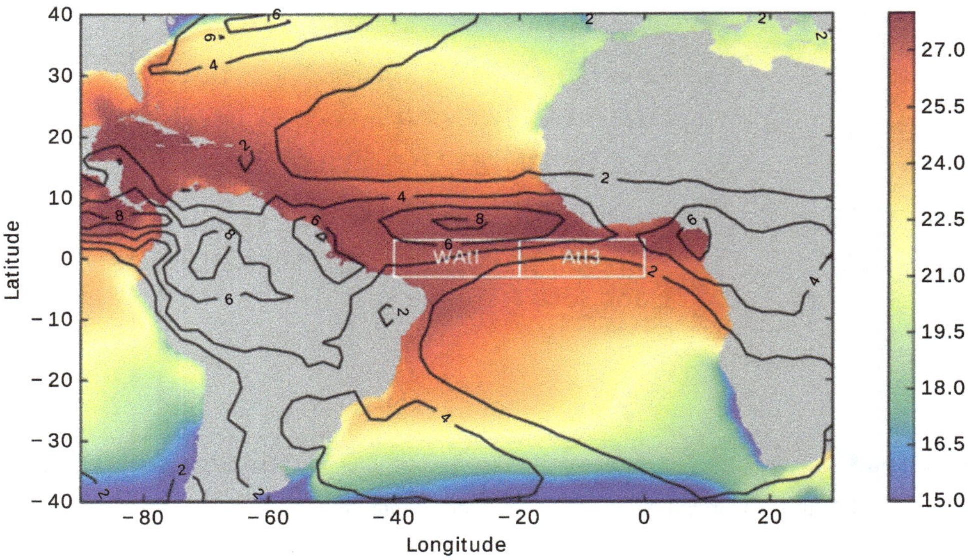

Fig. 1 The observed tropical Atlantic mean state sea surface temperature (SST) and precipitation: Annual mean sea surface temperatures are shown as shading, precipitation in contours. White boxes indicate the Atl3 and WAtl region in the eastern and western tropical Atlantic, respectively. The used datasets are the NOAA Optimum Interpolated SST dataset (OISST, Reynolds et al. 2007; Banzon et al. 2016), and the NOAA Climate Prediction Center (CPC) Merged Analysis of Precipitation dataset. (CMAP, Xie and Arkin 1997)

via equatorial Kelvin waves. Communicating information from the southern to the northern hemisphere and vice versa is hence a non-trivial enterprise in the ocean.

While the basic set-up of the marine tropical climate system is identical in all three tropical oceans, details differ between basins. The Pacific Ocean has the largest extent and is characterized by a relatively simple land-ocean geometry; it behaves much like a perfect theoretical ocean. The tropical Atlantic, in contrast, is much narrower and the surrounding continents interact with the ocean in complex ways. For example, the tropical Atlantic appears to be more susceptible to extra-equatorial influences (e.g., Foltz and McPhaden 2010; Richter et al. 2013; Lübbecke et al. 2014; Nnamchi et al. 2016), and variability is due to a number of interacting mechanisms on overlapping time scales (Sutton et al. 2000; Xie and Carton 2004). Therefore, the tropical Atlantic is less readily understood than the tropical Pacific, and still poses substantial challenges to the scientific community.

The mean state of the tropical Atlantic is characterized by a complex interplay of atmospheric and oceanic features. These are i) the trade wind systems of both the northern and southern hemispheres, ii) a system of alternating shallow zonal[2] currents in the ocean, and iii) a zonal gradient in upper-ocean heat content that is also reflected in a pronounced zonal gradient in sea surface temperatures (SSTs), with warm temperatures in the west and cooler surface waters in the east. Figure 1 illustrates the mean state of SST and precipitation.

The trade winds are part of the climate system's hemispheric response to the strong temperature gradient between the polar and the equatorial regions. Intense (solar) surface heating at the equator produces warm and humid, ascending air masses. During the ascend, part of the air moisture condensates and releases latent heat, which further accelerates the rising motion. The upward flow moves mass from the surface layer towards the top of the troposphere, effectively decreasing surface pressure and forming a low-pressure trough. At the surface, a compensation flow towards the low-pressure trough is established. Due to the rotation of the earth, however, the flow veers to the west and creates the surface trade winds. The northeasterly and southeasterly trade winds of the northern and southern hemispheres, respectively, converge in the inter-tropical convergence zone (ITCZ), a zonal band of intense precipitation and almost vanishing horizontal winds (Fig. 1). Because the ITCZ is located to the north of the equator in the Atlantic, the equatorial Atlantic is not dominated by the ITCZ itself, but by the trade wind system of the southern hemisphere that provides relatively steady easterly winds on the equator. (See below for why the ITCZ is, on average, not residing on the equator in the tropical Atlantic.)

A consequence of the easterly wind forcing at the ocean surface and the vanishing Coriolis force at the equator is that the wind pushes the warm surface waters westward. Water piles up to the east of Brazil in the Atlantic warm pool, providing water temperatures of approximately 28 °C at the surface. Conversely, the surface layer of warm water in the eastern tropical Atlantic is thinned out considerably – the eastern part of the basin stores much less heat in the upper ocean than the western part. A pronounced zonal gradient in upper-ocean heat content is established. Figure 8a illustrates this mean state.

The pressure below the ocean surface is not uniform across the basin either. At the equator, the bulk of warm water in the western ocean basin adds pressure to the water

[2] "Zonal" refers to an east-west orientation, i.e. one that is parallel to the equator. A north-south orientation is called "meridional".

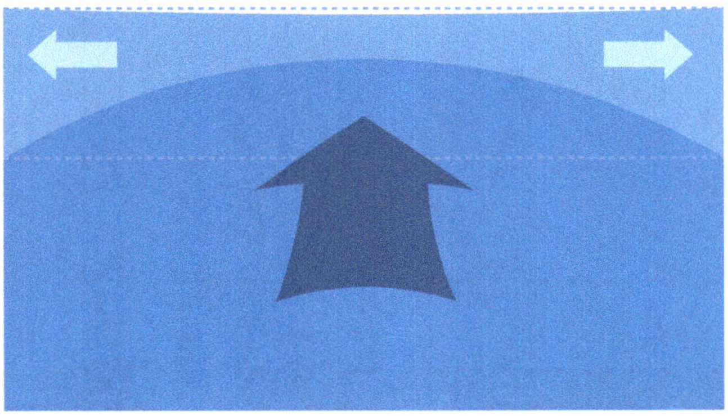

Fig. 2 Upwelling driven by horizontal divergence. Consider an ocean in a state of rest. In a simple model, a layer of warm water is sitting on top of a layer of colder water. Both the interfaces between the warm surface layer and the atmosphere, and between the colder subsurface water and surface layer are approximately even (horizontal dashed blue lines). When a divergence is created in the upper layer, mass is transported away from the divergence (light blue arrows in the surface layer). Because water is approximately incompressible, mass must be conserved. A vertical flow from the subsurface layer compensates the horizontal divergence (dark blue, upward arrow). In reality, this domes the interface between the surface and the subsurface layers. The sea surface adapts to the doming interface by decreasing in a similar fashion, albeit with a much smaller amplitude

column, while eastern ocean pressure is reduced. The resulting east-west pressure gradient is balanced by a strong eastward current right below the surface – the equatorial undercurrent (EUC) (Cromwell 1953; Cromwell et al. 1954). At the surface, on the other hand, the direct wind forcing and meridional pressure gradients produce a complex system of alternating zonal current bands (e.g., Schott et al. 2003; Brandt et al. 2006, 2008).

The three-dimensional flow of the upper equatorial oceans directly below the well-mixed surface layer is characterized by a slow but steady upward motion of, at best, a few meters per day (Rhein et al. 2010). This so-called "upwelling" is maintained by two processes. First, the Coriolis force deflects the off-equatorial components of the wind-induced westward displacement of surface water masses into opposite directions. On the northern hemisphere, westward flow veers north, while the Coriolis force directs it south on the southern hemisphere. Zonal wind-driven upper ocean mass transports diverge; they effectively transport mass away from the equator. However, because mass is conserved, sea level sags imperceptibly, and upwelling transports colder, subsurface water closer to the surface by creating a "dome" in the interface between the warm surface water and cooler subsurface water. The ratio between the surface and subsurface layer thicknesses changes in response to the surface divergence. Figure 2 illustrates how divergent flow in the surface layer creates upwelling and changes the geometry of the involved interfaces between both the atmosphere and the ocean, and the ocean surface and subsurface layers.

Second, a small meridional contribution to the equatorial wind field contributes to maintaining equatorial upwelling. These meridional contributions are illustrated in Fig. 7b by the equatorial wind vectors that do not point straight to the west but rather to the northwest, as they are part of the southern hemisphere trade wind regime crossing the equator into the northern hemisphere for most of the year. In the ocean, they induce meridional surface mass transports slightly off the equator (Philander and Pacanowski 1981). Again, the Coriolis force redirects these meridional motions into zonal mass transports of opposite signs, which contribute to the upper ocean horizontal divergence.

Over the course of the year, the set-up of this basic state varies. Due to the tilted rotational axis of the earth, the latitude of maximum insolation shifts into the northern hemisphere in boreal – i.e. northern hemispheric – summer, and into the southern hemisphere in boreal winter. The ITCZ, accompanied by the trade wind systems of both hemispheres, migrates in a similar fashion. However, the ITCZ does not oscillate around the equator but stays north of it for most of the year (Hastenrath 1991; Mitchell and Wallace 1992). Xie (2004) reviewed the "riddle" of the asymmetric ITCZ and concluded that it is, contrary to intuition, not so much the overall distribution of landmasses and oceans that anchors the Atlantic ITCZ to the northern hemisphere, but a combination of air-sea coupling and the shape of the West-African shoreline. More recently, Frierson et al. (2013) also demonstrated how the meridional temperature gradient between the warm northern hemisphere and the relatively colder southern hemisphere impacts the ITCZ behavior. All factors combine to pull the trade wind system of the southern hemisphere across the equator and establish the highest SSTs to the north of the equator.

Driven by the changing trade wind systems, the zonal surface current systems vary in strength and location. The intensity of the Equatorial Undercurrent, while firmly pinned to the equator, varies as well (Johns et al. 2014). Variations in the wind forcing lead to seasonally recurring intensifications of the zonal heat content gradient.

One of the most striking elements of the tropical Atlantic seasonal cycle is the formation of the Atlantic cold tongue in the eastern equatorial Atlantic during boreal summer. The cold tongue is characterized by an intense cooling of the upper ocean. Figure 3a shows that SSTs in the Atl3 region (3°S–3°N, 20°W–0°E) drop from 28 °C to about 23 °C between April and August, forming a distinct, tongue-shaped pattern of relatively cool surface water that stretches from the West African coast into the central equatorial Atlantic (Figs. 3b, c). The observed temperature difference between April and August in the upper 50 m of the Atl3 region alone corresponds to a change in thermal energy of

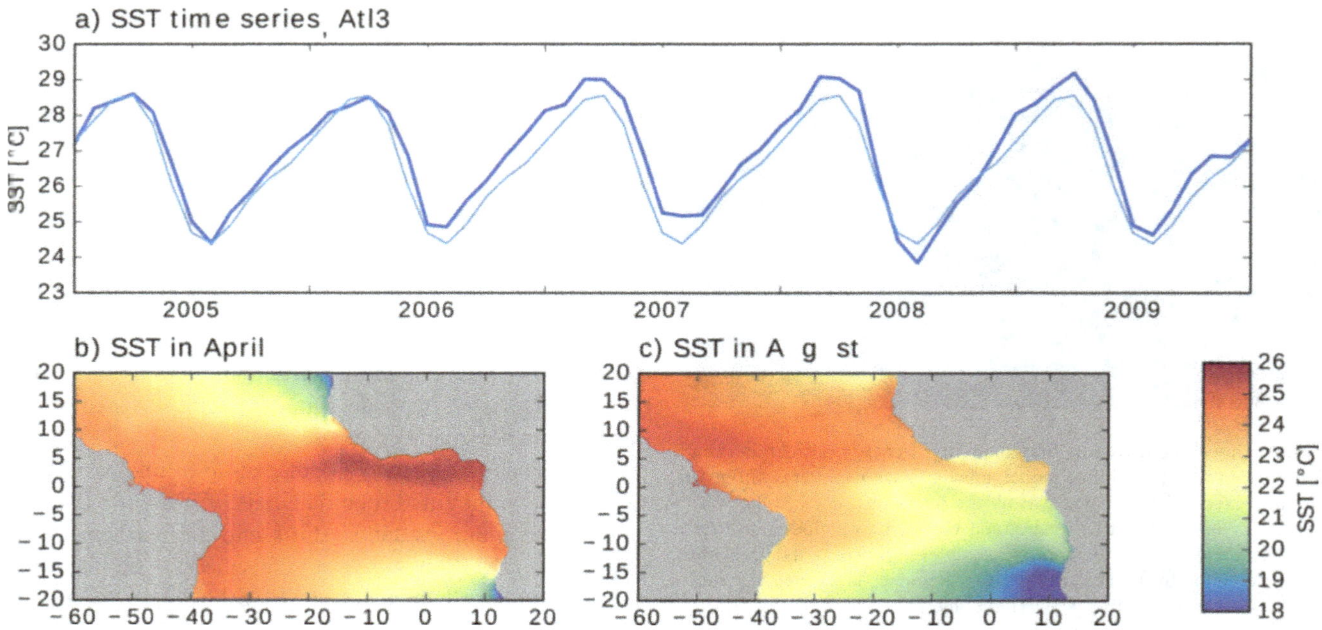

Fig. 3 Observed cold tongue based on the NOAA Optimum Interpolated SST dataset (OISST). (**a**) Exemplary time series of monthly mean Atl3 sea surface temperature (SST, dark blue) and the climatological seasonal cycle (light blue). For the seasonal cycle, monthly mean data has been averaged for each calendar month for the period 1981–2012. (**b**) and (**c**) Climatological SST fields for April and August, illustrating the climatological conditions when SSTs reach their maximum just before the onset of the cold tongue, and when the cold tongue is fully developed, respectively

351.16 EJ.[3] That is 13 times the US-American energy consumption of 2014, or 2.6 times the total global energy consumption of 2011.

The formation of the cold tongue co-occurs with seasonal changes in the atmospheric circulation. An important and well-known aspect of this is the strong co-variability between the onset of the cold tongue and the onset of the West African monsoon (e.g., Okumura and Xie 2004; Brandt et al. 2011a; Caniaux et al. 2011), a key element of large-scale precipitation in western Africa and hence a crucial factor of agriculture. Understanding the complex processes that shape the coupled atmosphere-ocean-land climate system of the equatorial Atlantic is a task of high societal relevance.

In concert with accurate and long-term observations, climate models are an essential tool to investigate the equatorial Atlantic. Here we address the question of how well state-of-the-art climate models are able to reproduce the observed seasonal cycle of the equatorial Atlantic. The section "Climate models: A crash course" gives an overview on coupled climate models and introduces the concept of model biases. The section "Can climate models reproduce the observed seasonality of the equatorial atlantic climate system?" reports common biases in the tropical Atlantic and how they relate to the formation of the modeled cold tongue. An outlook in the last section addresses the usefulness of climate models for studies of cold tongue variability, a crucial source of tropical Atlantic climate variability that strongly affects the surrounding continents.

Climate Models: A Crash Course

Climate models numerically solve the Navier-Stokes equations for a set of specified assumptions. The Navier-Stokes equations are a system of non-linear partial differential equations that describe the behavior of fluids, from a drop of water that hits the surface of a puddle, to global circulation systems such as the trade wind systems. They are highly complex and can only be solved numerically when they are approximated to focus on a specific class of fluid processes. For climate models, these processes are mostly related to the large-scale global circulation, synoptic phenomena, and possibly mesoscale phenomena[4] such as ocean eddies. The approximated Navier-Stokes equations that are used in current climate models are called the primitive equations.

Climate models consist of a number of "building blocks". The two core building blocks are an atmosphere and an ocean general circulation model (GCM). Given appropriate surface and boundary forcing, both GCM types can be run

[3] Based on thermal data from the World Ocean Atlas (WOA2013v2, Locarnini et al. 2013).

[4] Size on the order of 10–50 km.

independently. Phillips (1956) demonstrated this by designing the first successful atmospheric GCM. To allow the oceanic and atmospheric blocks to interact with each other, a coupling module exchanges information at the air-sea interface. A coupler and the atmospheric and oceanic GCMs together form the simplest coupled GCM (CGCM). Such a basic CGCM lacks a number of relevant processes, relating for example to the land and sea ice components of the climate system or the impact of vegetation. To introduce these important aspects into the model, CGCMs are "upgraded" with additional building blocks to form earth system models. If a basic CGCM is a simple brick house of only one room, a full-fledged earth system model is a mansion with specialized rooms for different tasks. Important additional building blocks for an earth system model are modules that simulate the behavior of sea ice, ice sheets and snow cover on land, vegetation and other surface processes such as river runoff into the ocean, atmospheric chemistry, biogeochemistry in the ocean or even geological processes of varying complexity.

In order to solve the model equations numerically, CGCMs need to discretize the real world into finite spatial and temporal units. The basis for such a discretization is a three-dimensional grid of grid boxes that each contain a single value of a given variable. The CGCM applies the model equations to the grid boxes and integrates them forward in time. Essentially, each grid box is a mini-model that is, however, exchanging information with neighboring grid boxes.

An important characteristic of a model grid is its resolution, i.e. the size of its grid boxes.[5] It defines, among other things, which processes can be resolved. As an example, consider the development of cumulus clouds. While cumulus clouds have a horizontal scale of less than 10 km, state-of-the-art models use a resolution of about 100 km. On such a grid CGCMs cannot simulate cumulus clouds directly. Consequently, the climatic impacts of such clouds have to be parameterized, i.e. their effect must be captured by the model in a simpler way that is supported by observations. For convective[6] and mixing processes alone – important aspects of cumulus clouds -, a number of parameterization schemes exist that subtly alter the behavior of large-scale processes in the models.

In addition to horizontal processes, models must be able to capture vertical motions in the climate system. Cumulus clouds, for example, extent vertically throughout varying portions of the troposphere, and vertical movement within clouds is a key factor of precipitation. On a larger scale, ascending air masses within the ITCZ define an important aspect of the tropical climate system (cf. Section "The equatorial atlantic: A climate hot spot"). Models need to be able to reproduce these vertical movements. They require vertical layering, giving rise to the three-dimensional structure of a model grid. A common feature of all models is that their vertical levels are unevenly distributed. Because properties usually change drastically close to the air-sea interface, resolving these strong gradients requires a high vertical resolution. Conversely, the thickest levels are farthest away from the air-sea interface. In the ocean, the last model level usually ends at the sea floor; the atmosphere, however, is not bounded that clearly. Some models only resolve the troposphere, our "weather" sphere that reaches up to approximately 15 km, while a number of recent atmosphere models incorporate the stratosphere as well (up to 80 km).

Figure 4 illustrates schematically how the different "building blocks" of a CGCM work together and how the real world must be discretized into grid boxes to allow a numerical solution of the primitive equations.

CGCMs are initialized either from a state of rest – i.e. the ocean and atmosphere are without motion and only establish their general circulation patterns during the first stage of the simulation, the so-called "spin-up" – or from a more specific state that is generally derived from observations. In both cases, the model needs time to smooth out initial imbalances and establish an equilibrium. Additionally, climatically relevant forcing parameters must be prescribed to the model in the form of boundary conditions. A prominent example of such a boundary condition is the strength and variability of the solar forcing, our energy source on earth, or the atmospheric CO_2 concentration.

Climate models are used to address a host of research questions. They aid scientists in interpreting observations, infer mechanisms, or provide information on how the climate system might evolve in the future. All of these tasks, however, require that CGCMs are able to produce a realistic climate. Due to various limitations, this is not always the case. A common manifestation of the shortcomings of a climate model is the formation of biases.

A bias is a systematic difference between the modeled and the observed climate. This difference can occur in any statistical property of any model variable. While standard biases are routinely monitored during the development and application of climate models, non-obvious biases may be present in simulations that look fine otherwise. Consider, for example, SST in a given location. While routine bias controls may have found a realistic mean SST, closer inspection could reveal that SST anomalies tend to be too high. Because positive and negative anomalies cancel each other out on

[5] Note that, usually, not all grid boxes of a GCM have the same size, neither in terms of absolute surface area, nor in terms of longitudinal and latitudinal extent. A common practice in ocean models, for example, is to refine the latitudinal resolution towards the equator to better resolve the fine structures of the equatorial oceans. In a similar fashion, Sein et al. (2016) recently discussed grid layouts for ocean models that increase their spatial resolution in certain target areas.

[6] Convection: upward motion in the atmosphere.

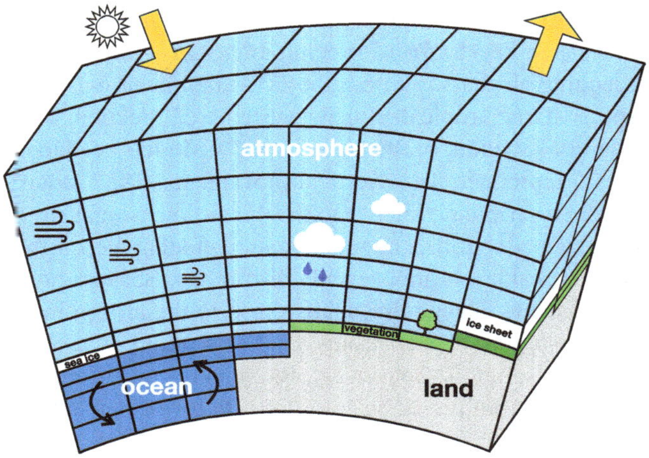

Fig. 4 Schematic of a Coupled General Circulation Model (CGCM). On the most basic level, the earth is a closed system that receives energy from the sun and radiates away thermal energy (yellow arrows at the "top of the atmosphere"). A CGCM tries to simulate the processes within this system. It consists of a number of modules that interact with each other. Important modules in state-of-the-art CGCMs are the ocean-and-sea-ice module, the atmospheric module, and additional modules that simulate, for example, land surface processes or vegetation. These "building blocks" of the CGCM exchange information with each other via an additional "coupling module". Coupling is a computationally expensive operation that can account for up to a third of the total required computational resources of a CGCM. A CGCM solves an approximation to the Navier-Stokes equations numerically. These are a set of non-linear partial differential equations that describe the motion of fluids. To solve them, the model must discretize the real world into finite spatial and temporal units. In the three-dimensional space domain, this discretization results in a layered grid. Each grid box contains a single value for each model variable. Processes acting on spatial scales that are smaller than the extent of the grid box must be parameterized. Prominent examples of these "sub-grid" processes are, for example, the formation of clouds and precipitation

average, this biased variance would not be obvious. In a similar manner, positive and negative SST anomalies might not be distributed realistically, with the model perhaps producing a few very strong positive anomalies and many weak negative anomalies that still form the expected average. In this case, the modeled SST distribution is skewed with respect to observations.

An additional limitation on the hunt for biases is that a bias can only be diagnosed in comparison to a reliable observational benchmark. Many parameters of the real climate system, however, are hard to observe or have only been observed for a short time. In general, large-scale patterns on the earth's surface and throughout the atmosphere can be observed relatively easily with satellite-borne remote sensing instruments. SST, for example, has been carefully monitored by a number of satellite missions since the 1980s. Processes below the ocean surface, however, can usually not be monitored from space. Instead, observational data have to be obtained by measurements from ships, moored instruments and autonomous vehicles such as gliders and floats. For the tropical oceans, the TAO/TRITON mooring array in the Pacific (McPhaden 1995), the PIRATA array in the Atlantic (Bourlès et al. 2008), and the RAMA array in the Indian Ocean (McPhaden et al. 2009) provide, among others, information on temperature, salinity, current velocities and air-sea fluxes. Additionally, an increasing number of hydrographic observations have become available over the last decade due to the Argo program (Roemmich et al. 2009). While all of these measurements provide invaluable information about the state of the tropical oceans, they are not spatially continuous and have only been operational for the last few decades. Obtaining information about the evolution of the climate system in the past remains a core challenge of climate research.

Although no climate model is exactly like the other, some biases are shared by a wide range of state-of-the-art CGCMs. Figure 5 shows the global pattern of the annual mean SST bias for the average of 33 CGCMs and an experiment with the Kiel Climate Model (KCM, Park et al. 2009). Positive values indicate that modeled SST is warmer than in observations and vice versa. We validated the performance of these CGCMs and the KCM in terms of SST against the satellite derived Optimum Interpolated SST dataset (OISST, Reynolds et al. 2007; Banzon et al. 2016). Figure 5 shows that while the KCM is a unique model that has individual flaws and strengths, the characteristics of its equatorial Atlantic SST bias are well comparable to other current CGCMs (examples of other models are shown, among others, in Wahl et al. 2011; Xu et al. 2014; Ding et al. 2015; Harlaß et al. 2017).

The KCM is a state-of-the-art CGCM that was integrated with radiative forcing for the period 1981–2012 in rather coarse resolution. The ocean-sea ice model NEMO (Madec 2008) was run with 31 vertical levels and a horizontal resolution of 2° that is refined to 0.5° in the equatorial region. The atmospheric model ECHAM5 (Roeckner et al. 2003) is run with 19 vertical levels and a global horizontal resolution of approximately 3.75°. Results from KCM simulations are selected here for consistency reasons. We stress again that while the KCM differs wildly from other CGCMs in some aspects, its simulation of the tropical Atlantic is representative for most current-generation CGCMs.

Can Climate Models Reproduce the Observed Seasonality of the Equatorial Atlantic Climate System?

The Equatorial Atlantic Warm Bias: Symptoms

The annual mean SST bias varies considerably between different regions of the ocean (Fig. 5). Striking features of the global SST bias pattern are the pronounced warm biases

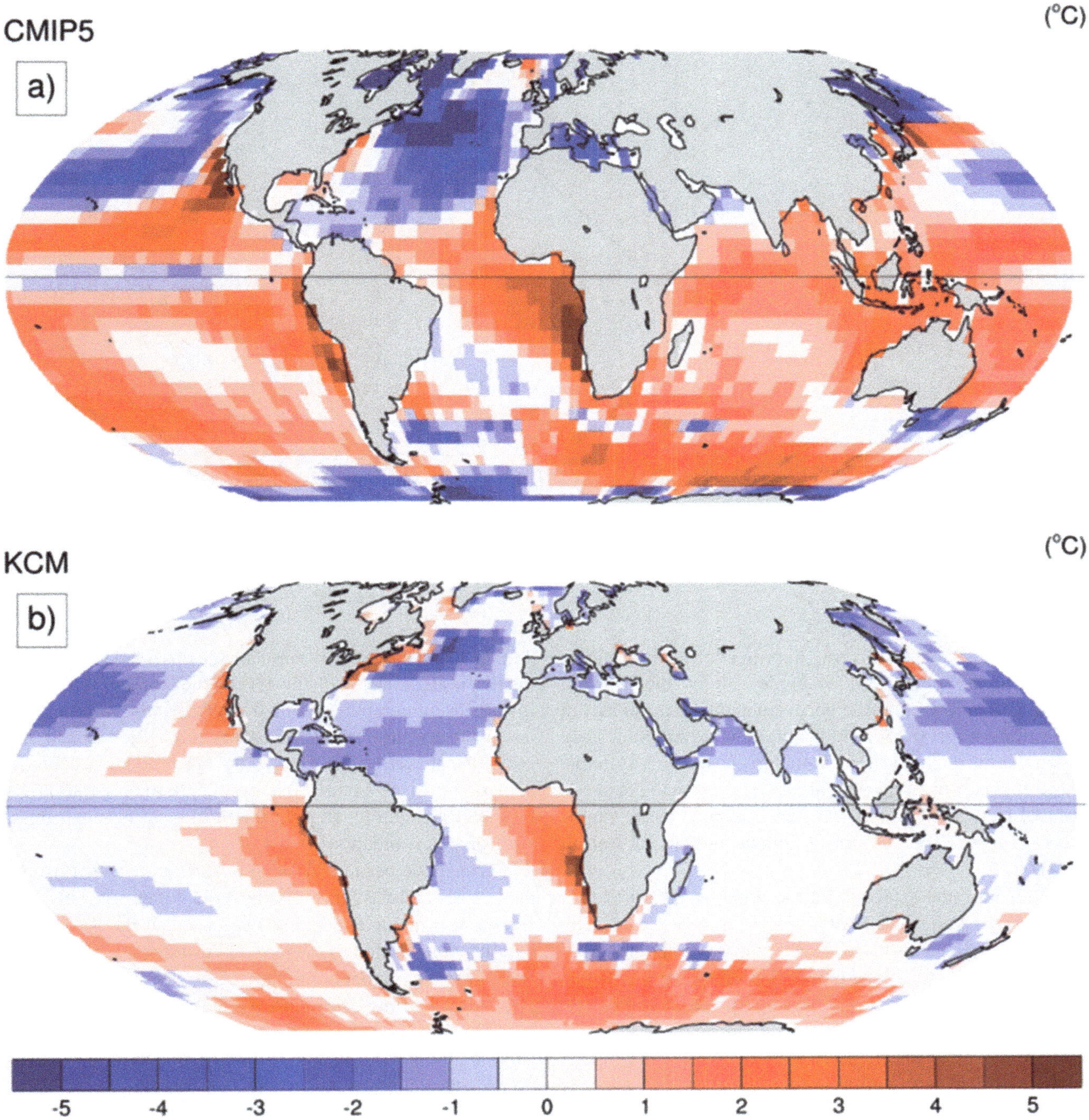

Fig. 5 Annual mean global sea surface temperature (SST) bias in (**a**) the ensemble mean of 33 Coupled General Circulation Models (CGCMs) contributing to the Coupled Model Intercomparison Project, Phase 5 (CMIP5, Taylor et al. 2012) and (**b**) one integration of the Kiel Climate Model (KCM). For CMIP5, the chosen experiments were "historical" experiments that were forced by the observed changes in atmospheric composition. The KCM was run with an atmospheric horizontal resolution of approximately 3.75° and with 19 vertical levels. The ocean model had a horizontal resolution of 2° that was refined to 0.5° towards the equator, and 31 vertical levels. The annual mean SST bias was diagnosed with respect to the NOAA Optimum Interpolated SST dataset (OISST) for the period 1982–2009. Using an ensemble mean of three ensembles instead of a single integration to diagnose the KCM SST bias changed the results only negligibly. This demonstrates how robust a feature the annual mean SST bias pattern is in the KCM

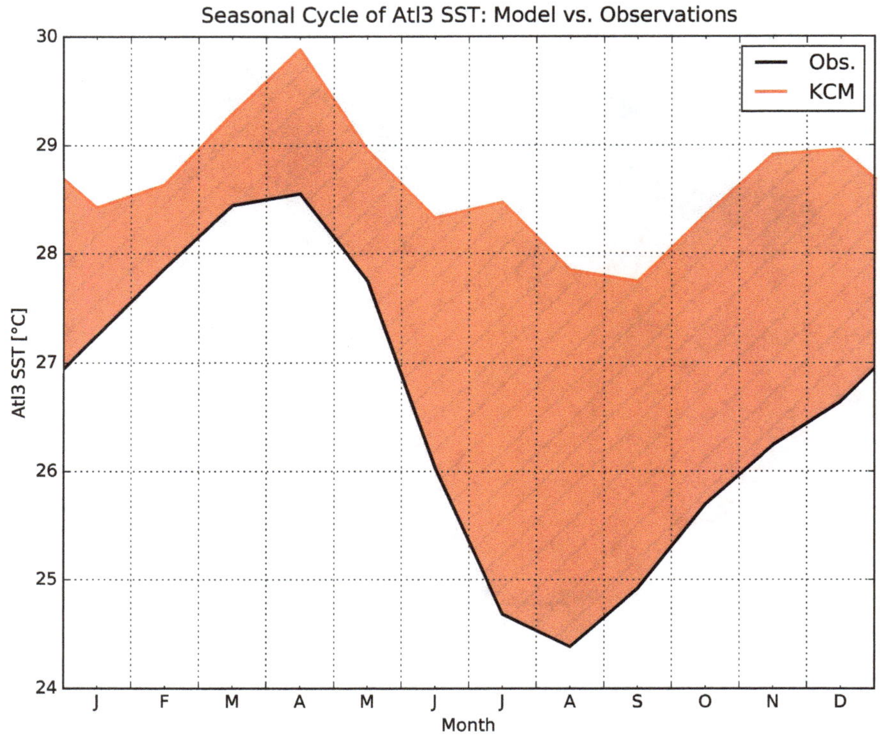

Fig. 6 Seasonal cycle of Atl3 sea surface temperature (SST) in observations (NOAA Optimum Interpolated SST dataset, black), and the Kiel Climate Model (KCM, red). Red shading illustrates the bias magnitude for each month

along the subtropical western shorelines of all continents. These biases appear, for example, along the western US-American as well as the Peruvian and Chilean coasts in the Pacific, or off Angola and Namibia in the Atlantic. They are anchored to the eastern boundary upwelling systems, where cold subsurface waters are brought close to the ocean surface. Here, SST biases can reach annual mean amplitudes of up to 7 °C in current climate models (Xu et al. 2014).

In this section, we focus on the pronounced warm bias that covers the equatorial Atlantic cold tongue region. The annual mean SST bias in the Atl3 region has a magnitude of approximately 2 °C.[7] In the upper 50 m of Atl3 in the KCM, this corresponds to a heat surplus of approximately 380 EJ, an amount of energy that could melt 47 times the ice volume of the Antarctic ice sheet.[8]

An important aspect of the equatorial Atlantic SST bias is that it varies over the course of the year. Figure 6 shows that the SST bias of the KCM is smallest in boreal winter, with a value of less than 1 °C in February. During the cold tongue formation, it rapidly increases to almost 4 °C until July. For the rest of the year, it slowly decreases again. This implies that the KCM struggles to simulate the observed strong cooling that is associated with the development of the cold tongue in boreal summer. Indeed, Fig. 6 shows that the KCM – similar to most state-of-the-art CGCMs (e.g., Richter and Xie 2008; Richter et al. 2014b) – does not produce a coherent cold tongue that is comparable in strength to observations. A key process of the equatorial Atlantic climate system is missing from the simulations.

Because the ocean and the atmosphere are strongly coupled in the tropics, the missing cold tongue is only one symptom of a fundamentally biased equatorial Atlantic in current climate models. Figure 7 illustrates the bias of the zonal wind component in the KCM. During spring, the KCM strongly underestimates the magnitude of zonal wind in the western tropical Atlantic (Fig. 7a). While the absolute value of zonal wind is higher in the KCM than in observations, especially during spring, the magnitude is much smaller. Instead of the generally easterly winds (negative values), associated with the trade winds, the KCM simulates very weak westerly winds (positive values). This "westerly wind bias" – so-called because the simulated zonal winds are much too westerly compared to the observed trade winds – is another typical bias pattern in state-of-the-art GCMs. It agrees with an ITCZ that is displaced too far to the south, a feature that is common to both coupled and atmosphere-only GCMs (e.g., Doi et al. 2012; Richter et al. 2012; Siongco et al. 2015).

An important question is: How do the different bias symptoms relate to each other dynamically, and how do these

[7] Note, however, that by no means *all* climate models develop such a strong equatorial Atlantic warm bias. Some models are capable of simulating a more realistic tropical Atlantic, but these models represent but a tiny minority of all current CGCMs.

[8] We used the thermal data from WOA2013v2 to compare our model results with. The Antarctic ice volume is based on the Bedmap2 dataset (Fretwell et al. 2013).

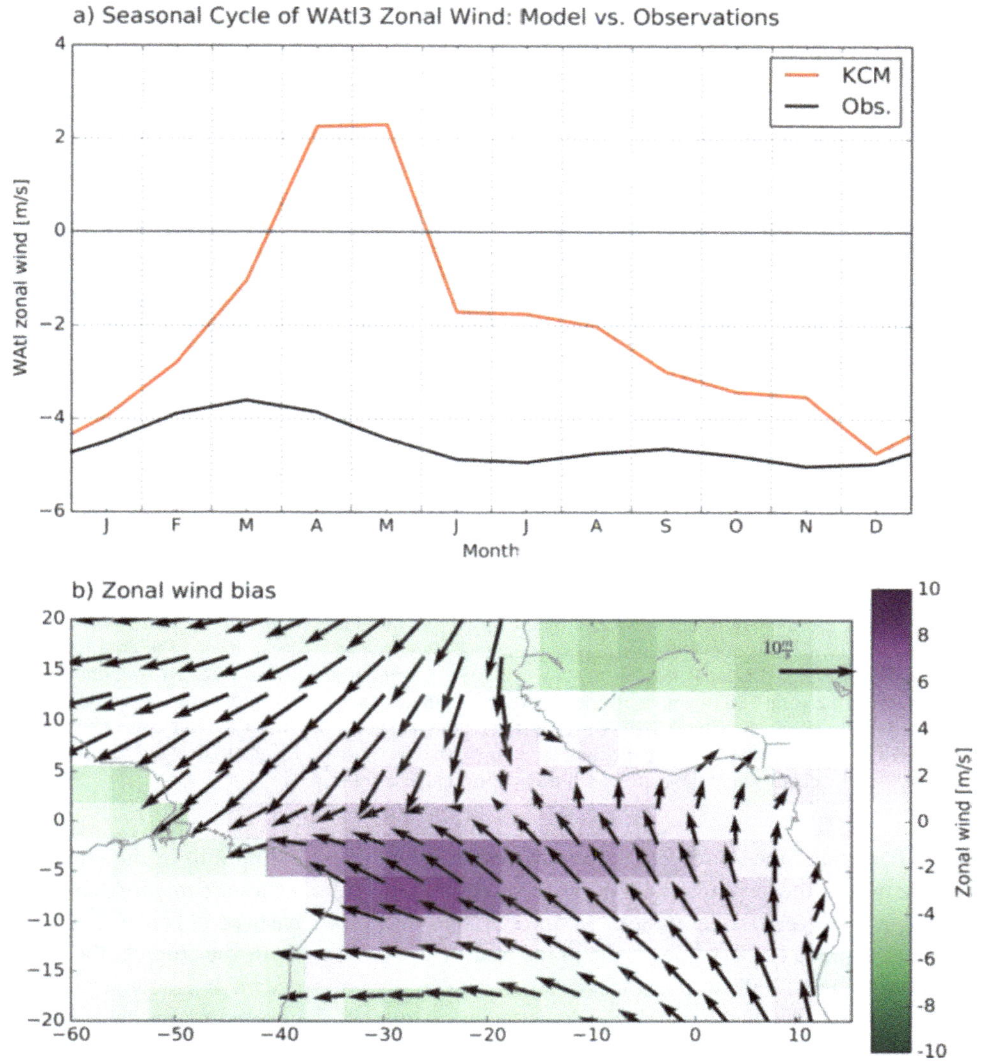

Fig. 7 Tropical Atlantic near-surface winds and zonal wind bias in spring. (**a**) Same as Fig. 6, but for the zonal component of 10 m wind in WAtl. (**b**) Climatological mean of observed 10 m wind (arrows) and the Kiel Climate Model (KCM) zonal wind bias in February–April (shading) in the equatorial Atlantic. The wind climatology is based on the Scatterometer Climatology of Ocean Winds (SCOW, Risien and Chelton (2008)). Arrows combine the zonal and meridional components of the climatological 10 m wind, while shading only refers to the zonal component of the wind

dynamics compare to the observed processes that shape the tropical Atlantic climate system? In the next subsection, we first review the basic processes that establish the observed seasonal cycle in the tropical Atlantic, and then compare the observations with what is happening in state-of-the-art climate models.

Which Processes Produce the Equatorial Atlantic Warm Bias?

A good first assumption about the seasonal cycle is that it is driven by the seasonal movement of the sun. Such a seasonal cycle should be symmetric. In the tropical Atlantic, however, it is clearly asymmetric. Figure 6 shows that the cooling period between April and August is much shorter – or, equivalently, more intense – than the subsequent period of gradual warming that lasts until the following April. Processes other than the seasonal forcing of solar insolation must contribute to the fast growth of the summer cold tongue.

Recent studies of the tropical Atlantic suggest that the rapid formation of the cold tongue involves a coupled, positive feedback (Keenlyside and Latif 2007; Burls et al. 2011; Richter et al. 2016). A feedback establishes a relationship between two or more variables. In a negative feedback small perturbations in one variable are compensated by changes in the other such that the system returns to its original, stable state. The opposite is true for a positive feedback. Here, a perturbation – even a small one – in one variable provokes changes in the other variables that reinforce the original perturbation. The system continues to diverge from its initial state. The perturbation grows until the feedback is disrupted.

The dominant positive feedback in the equatorial oceans is the Bjerknes feedback (Bjerknes 1969). It relates three key properties of an equatorial ocean basin to each other: SST in the eastern ocean basin; zonal wind variability in the western

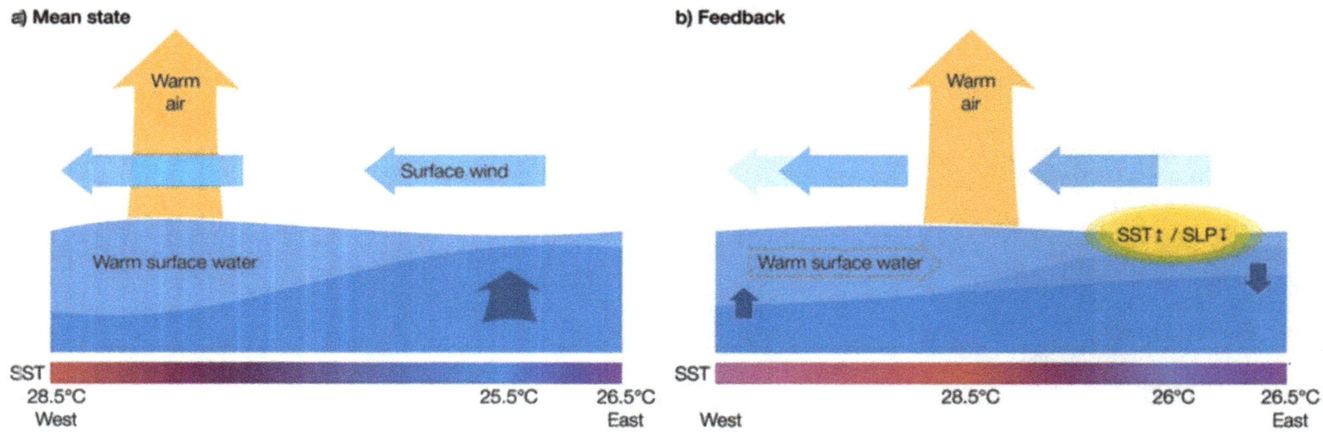

Fig. 8 The Bjerknes feedback. (**a**) Mean state. Along the equator, the surface wind field is dominated by the trade winds of the southern hemisphere. Both the zonal and meridional components of the trade winds contribute to surface divergences close to the equator, producing equatorial upwelling (thick blue arrow). Steady equatorial easterly wind forcing (blue arrows) pushes warm surface waters (light blue layer) towards the western ocean basin and builds up the warm pool. Warm and moist air rises above the warm pool (orange arrow). In contrast, the surface mixed layer is thin in the eastern basin, upwelling is more efficient there, and SSTs are, on average, cooler than in the warm pool (approximately 25.5 °C and 28.5 °C, respectively; the equatorial SST distribution is sketched in the bar below the figure). (**b**) The positive Bjerknes feedback alters the state of the tropical ocean. The trade winds weaken, and zonal surface winds in the western ocean basin decrease. The balance between the subsurface pressure gradient and wind stress forcing is disrupted, and part of the warm pool "sloshes back" into the central ocean basin, redistributing warm surface water more evenly across the ocean basin. The tilt in the interface between the surface and subsurface waters decreases, and upwelling is less efficient in providing cold subsurface water to the surface layer in the eastern ocean basin. The cold tongue region warms (orange ovals). Sea level pressure (SLP) over the warm anomalies decreases, and convection shifts towards the central ocean basin. The surface wind response to this shift in surface convection and the zonal SLP distribution further weakens the trade wind regimes and closes the feedback

ocean basin; and the zonal distribution of upper ocean heat content along the equator, with large heat reservoirs and thick surface layers in the western warm pool, and thin surface layers in the cold tongue region in the east.

Figures 8a and b illustrate, respectively, the mean state of an equatorial ocean and how the Bjerknes feedback alters it. Consider a weakening of the easterly trade winds in the western ocean basin (or equivalently a decrease in easterly zonal wind stress at the ocean surface). The balance between the wind stress and the piled-up warm water in the western ocean basin temporarily fades, and the piled-up warm pool "sloshes back" into the eastern ocean basin, redistributing the upper ocean heat content more evenly across the equatorial basin.[9] The zonal gradient in heat content is leveled out, and the additional heat in the eastern ocean basin helps to establish a positive SST anomaly. This process can last several months in the equatorial Pacific and approximately one month in the equatorial Atlantic. (These different time scales are mainly due to the different east-west extents of the basins and hence signal propagation speeds.)

In the tropics, the atmosphere is closely coupled to the ocean. It reacts strongly to underlying SST variability by developing an anomalous wind field that converges over a warmer-than-usual patch of water (Gill 1980). The local changes in the wind field co-occur with changes in the zonal pressure gradient along the equator. The altered zonal pressure gradient in turn induces further weakening of the easterly trade winds in the western ocean basin, closing the feedback loop. An equivalent process with opposite signs takes place when the trade winds intensify in the western ocean basin.

The Bjerknes feedback is restricted to the equatorial ocean basins. While the ingredients of the feedback – wind, upper ocean heat content and SST variability – are present in every region of the ocean and usually interact with each other in one way or the other, the fully coupled Bjerknes feedback requires that information is zonally transmitted across almost the entire zonal extent of the basin, both in the atmosphere and the ocean. This is only possible when the Coriolis force vanishes or is negligibly small, since it would otherwise deflect the involved physical motions into curved movements. A direct, zonal exchange between the eastern and western ocean basins would not be possible in the presence of the Coriolis force.

[9] In the framework of this explanation, an interesting observation is that the Bjerknes feedback can only operate as long as the reservoir of warm water in the western warm pool is not empty. Once this is the case, the feedback breaks down, the SST anomaly stops to grow and the warm pool fills up again. A negative feedback has replaced the positive feedback. For the tropical Pacific, this sequence of alternating feedbacks has been described by Jin (1997) in the framework of the *recharge oscillator*. The name relates to the idea that the equatorial ocean is "charged" with warm water in the warm pool region – or, equivalently, heat – that is then discharged to the atmosphere during a warm event.

In the tropical Atlantic, a number of seasonal processes in the coupled atmosphere-ocean system produce a climate state that allows the Bjerknes feedback to operate during early boreal summer. Although we explain the processes in a sequential manner below, note that clear causalities are hard to establish in a coupled system. Different aspects of the phenomenon – here: the northward movement of the ITCZ and the development of the Atlantic cold tongue – cannot be disentangled from each other. Neither does the ITCZ move north *because* of the cold tongue development, nor does the cold tongue develop *because* the ITCZ moves north. Rather, both phenomena co-occur as manifestations of the same coupled phenomenon.

One key ingredient of the equatorial Atlantic seasonal cycle is the northward migration of the marine ITCZ (Xie and Philander 1994). In boreal spring, the ITCZ is in its southernmost position. The trade wind regimes of both hemispheres converge close to the equator and produce weak equatorial surface winds. When the ITCZ moves north in late boreal spring, the southern hemisphere trade winds cross the equator. Starting in March–April, surface winds intensify (illustrated by an increase in magnitude in Fig. 7a) and contribute to enhanced equatorial upwelling.

The spring strengthening of western equatorial *zonal* surface winds enhances the zonal gradient in upper ocean heat content. Strong easterly winds push the surface waters more efficiently towards the western warm pool, thinning out the warm surface layer in the eastern ocean basin and transporting the cooling signal westward. As a result, cold subsurface water lodges closer to the ocean surface. This background state requires very little subsurface water to be mixed into the surface layer to produce a substantial cooling. The western equatorial zonal spring winds "precondition" the eastern equatorial Atlantic for the formation of the cold tongue (e.g., Merle 1980; Okumura and Xie 2006; Grodsky et al. 2008; Hormann and Brandt 2009; Marin et al. 2009).

In concert with the development of the first seasonal cooling signals in May and June, the West African monsoon sets in (e.g., Okumura and Xie 2004; Brandt et al. 2011b; Caniaux et al. 2011; Giannini et al. 2003). From an atmospheric perspective, the monsoon onset is characterized by accelerating southeasterly surface winds in the Gulf of Guinea in late boreal spring. The strengthening *meridional* component of these winds enhances upwelling slightly to the south of the equator, and downwelling slightly to the north. The intensified upwelling provides additional initial cooling to the eastern equatorial region by mixing colder subsurface water into the warm surface layer. From the ocean perspective, on the other hand, cooling SSTs in the eastern equatorial Atlantic intensify the southerly winds in the Gulf of Guinea, which in turn contributes to the northward migration of convection and rainfall associated with the West African monsoon (Okumura and Xie 2004).

Lastly, oceanic processes contribute to the formation of the cold tongue. A number of studies found that vertical mixing at the base of the surface layer – where temperature gradients are strongest – seasonally varies in strength (e.g., Hazeleger and Haarsma 2005; Jouanno et al. 2011; Hummels et al. 2013, 2014). A likely explanation for this is that the intensities of the westward surface current and the eastward equatorial undercurrent vary over the course of the year. When the relative velocities of the two currents are strong, the vertical velocity shear at their boundary increases,[10] and frictional processes mix colder subsurface water into the warm surface layer. Figure 9 illustrates both the spring state of the tropical Atlantic and the basic processes that produce the first cooling signals in early boreal summer.

The net effect of these interacting processes – the northward migration of the ITCZ and the associated strengthening of the southern hemisphere trade winds on the equator, the thinning of the of eastern equatorial surface layer, the enhanced upwelling along the equator and especially in the cold tongue region, and the increased mixing at the base of the surface mixed layer – is that the first cold anomalies develop in the eastern equatorial Atlantic in late April. The atmosphere in turn reacts to the cold anomalies, and the Bjerknes feedback sets in. Starting in May, it lends additional growth to the cold tongue (Burls et al. 2011). In August, the seasonally active Bjerknes feedback loop breaks down (Dippe et al. 2017) and a more moderate warming sets in. In the absence of the Bjerknes feedback the cold tongue can no longer be maintained and dissolves, due to mixing processes in the ocean and surface heat exchange with the atmosphere.

Many models struggle to simulate a seasonally active Bjerknes feedback that is comparable to observations in both strength and seasonality. Richter and Xie (2008) pointed out that model performance with respect to the Atlantic Bjerknes feedback is quite diverse between models that participated in the Coupled Model Intercomparison Project, Phase 3 (CMIP3, Meehl et al. 2007). Likewise, Deppenmeier et al. (2016) found systematic weaknesses in the CMIP5 models. For example, many models displace the Atlantic warm pool towards the central equatorial Atlantic (Chang et al. 2007; Richter and Xie 2008; Liu et al. 2013). This displacement is a consequence of the westerly wind bias in the western equatorial Atlantic (Wahl et al. 2011; Richter et al. 2012, 2014b). Figure 7 illustrates for the KCM that the spring winds are much weaker in the model than in observations. Consequently, the surface wind stress is not sufficient to pile up warm sur-

[10]"Velocity shear" is a different term for "velocity gradient". A flow is sheared when different layers of the flow have different velocities. Depending on the magnitude of the shear and the viscosity of the fluid, the shear produces local turbulence and mixing due to frictional processes within the fluid. If no turbulence occurs, the flow is called "laminar".

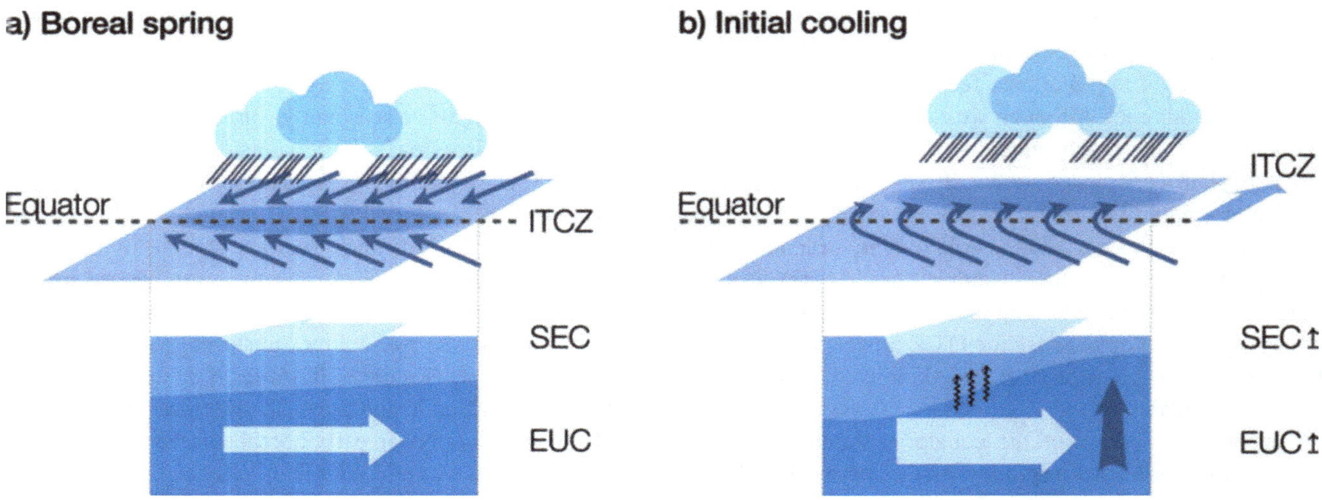

Fig. 9 Initial cold tongue cooling in the tropical Atlantic. (**a**) Spring conditions. The highest sea surface temperatures (SSTs) and the lowest sea level pressures (SLPs) are found approximately on the equator (dashed black line), forming the equatorial low pressure trough (dark-blue shading). The trade wind systems of both the northern and the southern hemispheres (dark blue arrows) converge in the trough and anchor the Inter-Tropical Convergence Zone (ITCZ, clouds and strong precipitation) to the equator. Zonal surface wind forcing is relaxed during spring, warm surface waters are distributed more evenly across the basin. At the ocean surface, the South Equatorial Current (SEC) transports water towards the west. Below the surface, close to the interface between the surface layer and the subsurface, the Equatorial Under-Current (EUC) transports water towards the east. (**b**) Initial cold tongue cooling: In early boreal summer, the ITCZ migrates away from the equator into the northern hemisphere. The trade winds of the southern hemisphere follow the low pressure trough and cross the equator. In the western ocean basin, zonal surface winds increase and push the warm surface water more efficiently towards the west. The warm pool deepens in the west, while the surface layer thins in the east. Additionally, both the meridional and zonal components of the wind field in the eastern ocean basin strengthen and contribute to a local surface divergence that is compensated by enhanced upwelling (thick, dark-blue arrow). Lastly, both the SEC and EUC increase in strength. Enhanced vertical velocity gradients in the vicinity of the interface between the surface and the subsurface water layers produce shear instabilities (black squiggly lines) that mix the cold subsurface water efficiently into the surface layer

face waters in the western ocean basin in a manner comparable to observations. Heat content is distributed more evenly across the equatorial ocean basin and supplies additional heat to the eastern surface layer. Even if the model produced wind variability that could serve as a valid initial perturbation to trigger the Bjerknes feedback,[11] the biased background state of the ocean could not support the feedback. The cold tongue fails to establish.

An interesting equivalent of this mechanism has been observed in the real ocean by Marin et al. (2009). The study compares the Atlantic cold tongue in two years with grossly different wind variability and finds that in the year with relatively weak spring winds in the western equatorial Atlantic – this compares well to the climatological, biased state in many CGCMs -, the zonal heat content gradient in the upper ocean does not develop. The winds fail to precondition the tropical Atlantic for the growth of the cold tongue.

Studies with current atmospheric GCMs have found the westerly wind bias in boreal spring to be an intrinsic feature of (uncoupled) atmospheric GCMs (Richter et al. 2012, 2014b; Harlaß et al. 2017). Coupling an already biased atmospheric GCM to an ocean GCM induces positive feedbacks that amplify the wind and SST biases in the equatorial Atlantic. Additionally, Grodsky et al. (2012) showed that an ocean GCM, too, is intrinsically biased in the tropical Atlantic, although the magnitude of this bias is much smaller than the warm bias in a coupled model.

The atmospheric westerly wind bias has been linked to a seesaw pattern in rainfall biases over South America and Africa (Chang et al. 2007; Richter et al. 2012, 2014b; Patricola et al. 2012). The proposed physical mechanism that links precipitation to the wind is the following: Tropical rainfall is tied to strong convection. Ascending moist and warm air masses create a local negative pressure anomaly at the surface that alters the zonal gradient in surface pressure along the equator. Surface winds, in turn, are dynamically related to surface pressure gradients.[12]

A current hypothesis of what prevents climate models from developing a cold tongue comparable to observations in

[11] This is by no means a given. As shown below and hinted at above, the equatorial Atlantic bias also manifests in the atmosphere and may well prevent the model from establishing the link between eastern ocean SST and western ocean wind variability that is necessary to close the Bjerknes feedback loop.

[12] Wind compensates pressure gradients. That is why large-scale storm systems are organized around low core pressures: The storm winds try to flow into the low pressure at the "heart" of the storm and eliminate the strong pressure gradient between the storm center and the storm environment. The Coriolis force provides rotation to storm systems by deflecting the pressure compensation flow.

boreal summer thus is: Opposing rainfall biases in South America and Africa produce a zonal surface pressure gradient along the equator that is weaker than in observations. The resulting winds in the equatorial western Atlantic are too weak in magnitude and cannot reproduce the observed distribution of upper ocean heat content. Consequently, the seasonally induced equatorial upwelling in early boreal summer is not sufficient to produce the observed cooling that finally triggers the Bjerknes feedback.

In agreement with these mechanisms, a number of studies have found that a physically sound way to reduce the equatorial Atlantic warm bias is to improve the atmospheric models. Tozuka et al. (2011) showed that tweaking the convection scheme can project strongly on the ability of the models to simulate the correct distribution of climatological SSTs in the equatorial Atlantic. Harlaß et al. (2015) conducted a number of experiments with the KCM that varied both the horizontal and vertical resolution of the atmospheric GCM, while keeping a constant coarse resolution for the ocean GCM. For sufficiently high atmospheric resolutions, the western equatorial wind bias strongly decreased and the equatorial Atlantic warm bias nearly vanished. The seasonal cycle as a whole greatly improved. In a follow-up study, Harlaß et al. (2017) found that sea level pressure and precipitation gradients along the equator are not sensitive to the atmospheric resolution. Nevertheless, the wind bias in their study decreased significantly. To explain this, they propose that the position of maximum precipitation and zonal momentum transport play an important role in giving rise to the zonal wind bias. Zonal momentum can be either transported by mixing it from the free troposphere into the boundary layer or by meridional advection into the western equatorial Atlantic (Zermeño-Diaz and Zhang 2013; Richter et al. 2014b, 2017). These findings agree with the study of Richter et al. (2014a), who found that zonal wind variability in the western equatorial Atlantic is strongly related to vertical momentum transports in the overlying atmosphere. Further studies by Voldoire et al. (2014), Wahl et al. (2011), and DeWitt (2005) confirm the importance of the atmospheric component of a CGCM to properly simulate the complex tropical Atlantic climate system.

Outlook: Implications for the Usability of CGCMs in the Equatorial Atlantic

Using the KCM, a CGCM that simulates the tropical Atlantic in a manner very similar to a wide range of state-of-the-art CGCMs, we have shown exemplary that coupled global climate models currently struggle to simulate a realistic equatorial Atlantic climate system. The dominant feature of this problem is that CGCMs struggle to simulate the defining feature of the seasonal cycle – the formation of the Atlantic cold tongue in early boreal summer. An important cause of this bias is a strong and seasonally varying westerly wind bias in equatorial zonal wind in atmospheric models that is present even in the absence of atmosphere-ocean coupling. While much progress has been made in understanding and reducing the equatorial Atlantic warm bias, many models still produce a profoundly unrealistic seasonal cycle in the equatorial Atlantic. How does this shortcoming affect the usefulness of coupled models in the equatorial Atlantic?

A key task of climate models is to forecast deviations from the expected climate state. For seasonal predictions, the expected climate state is the climatological seasonal cycle. Some of these deviations are generated randomly and are, by definition, unpredictable. Others are the product of – sometimes potentially predictable – climate variability.

In the tropical Atlantic, the dominant mode of year-to-year SST variability is the Atlantic Niño[13] (Zebiak 1993). The Atlantic Niño is essentially a modulation of the seasonal formation of the cold tongue (Burls et al. 2012). This modulation can manifest in a range of different cold tongue measures. For example, cold tongue growth might set in earlier (or later), the cold tongue might cool more strongly, or it might, in its mature phase, occupy a larger area in the tropical Atlantic than usual. Caniaux et al. (2011) argued that all of these measures reveal an aspect of cold tongue variability, but that they do not vary consistently with each other.

Still, the Atlantic Niño is generally described in terms of Atl3 summer SSTs. While the seasonal cycle of Atl3 SSTs spans a range of roughly 5 °C, interannual variations of Atl3 SST between May and July rarely exceed amplitudes of 1 °C (Fig. 10a). The seasonal cycle of the tropical Atlantic is by far the dominant signal in Atl3 SSTs (Fig. 10b). It is the background against which the interannual variability of the Atlantic Niño plays out.

Even though the Atlantic Niño constitutes only a relatively small deviation from the seasonal cycle, its effects on adjacent rainfall patterns can be substantial (e.g., Giannini et al. 2003; García-Serrano et al. 2008; Polo et al. 2008; Rodríguez-Fonseca et al. 2011). A key demand of African countries, where food security heavily relies on agriculture, is hence to be able to reliably predict the amplitude of the Atlantic Niño a few months, ideally even more than a season, ahead. Only such relatively long-ranged forecasts would allow African farmers to adapt their farming strategy

[13] The name "Atlantic Niño" refers to the Pacific El Niño, because the pattern of Atlantic Niño SST anomalies is similar to the Pacific El Niño. Apart from this, a number of differences exist between the two phenomena (discussed for example in Keenlyside and Latif (2007), Burls et al. (2011), Lübbecke and McPhaden (2012), Richter et al. (2013), and Lübbecke and McPhaden (2017). Nnamchi et al. (2015, 2016) argued that the Atlantic Niño might not be dynamical in nature, but a product of atmospheric noise forcing. Alternative names for the Atlantic Niño are Atlantic Zonal Mode or Atlantic Cold Tongue Mode.

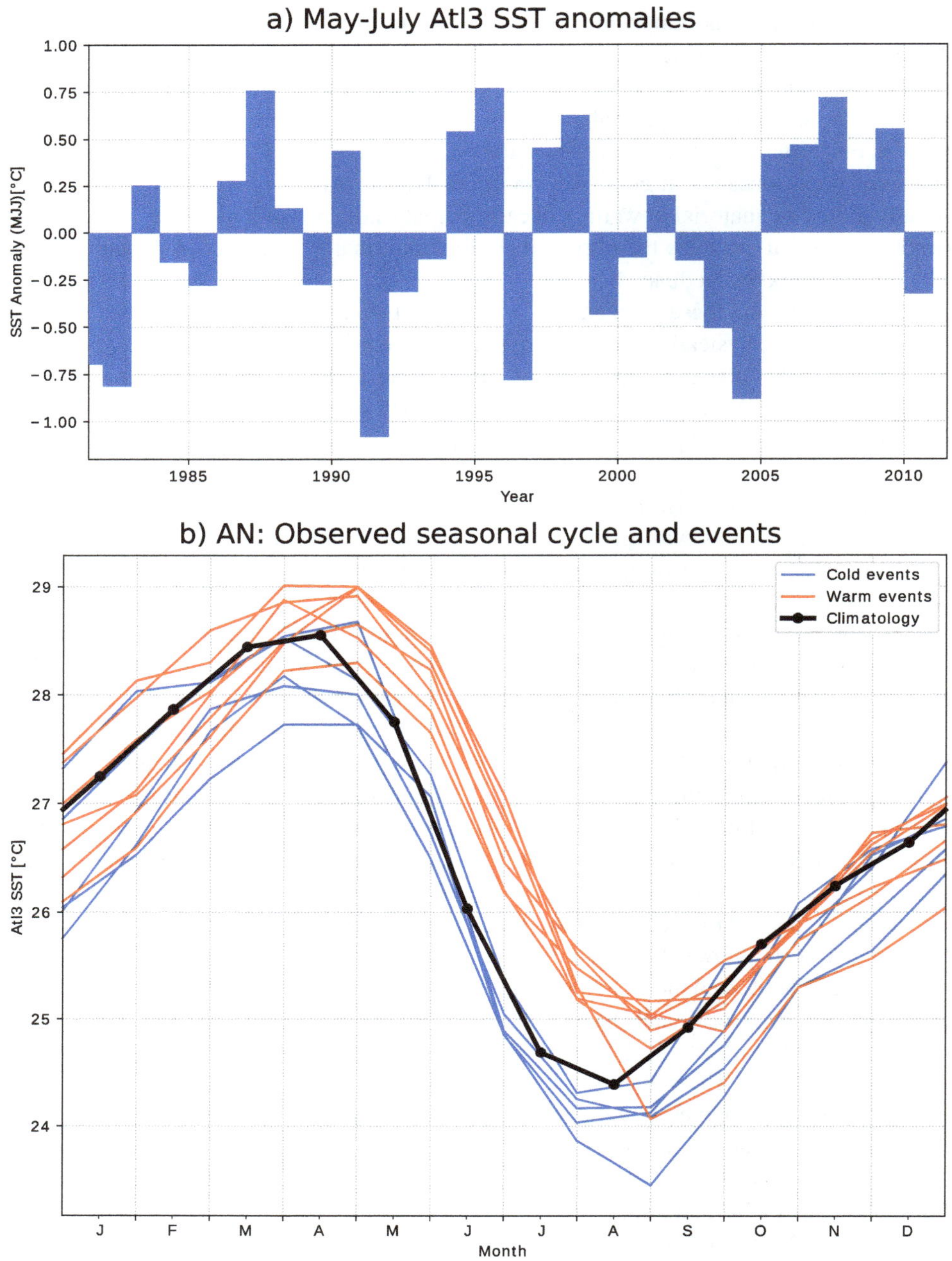

Fig. 10 The observed Atlantic Niño, based on the NOAA Optimum Interpolated SST dataset (OISST). (**a**) Time series of May–June-July (MJJ) Atl3 sea surface temperature (SST) anomalies. (Anomalies of a time series that, for each year, averaged MJJ monthly means together. Positive values indicate that the observed Atl3 region was warmer in MJJ of that year than on average.) (**b**) Observed seasonal cycle of Atl3 SST (black) and SST trajectories for individual years that produced warm (red) and cold (blue) Atlantic Niño events

for the upcoming season. Unfortunately, most models perform very poorly with respect to the Atlantic Niño and can provide hardly any predictive skill (Stockdale et al. 2006; Richter et al. 2017).

One reason for these shortcomings is that a prerequisite to simulate the variability of Atlantic cold tongue growth is a model that produces a realistic cold tongue. Indeed, Ding et al. (2015) showed that even a symptomatic – as opposed to a dynamically motivated and hence more process-oriented – reduction of the equatorial Atlantic SST bias in the KCM greatly improves the ability of the model to track the observed Atlantic Niño variability. This serves as an example of how the mean state interacts with climate variability. How the bias influences the predictive skill of the KCM for tropical Atlantic SST and whether the real climate system actually provides the potential to produce reliable forecasts of Atlantic Niño variability a few months in advance are the subjects of current research.

In general, the equatorial Atlantic warm bias has been an important issue since the earliest attempts of coupled global climate modeling (Davey et al. 2002) and continues to challenge the scientific community. It serves as an important reminder that model output should not always be taken at face value. Rather, models can struggle to represent observed physical processes, even though their physical basis in the form of the approximated Navier-Stokes equations is sound.

In the equatorial Atlantic, the entire coupled system is off-key in coupled global climate models due to the misrepresentation of crucial physical processes. However, alternative ways exist to study the tropical Atlantic with the help of models. Akin to early modeling studies of the El Niño-Southern Oscillation, statistical models can provide some insight into the equatorial Atlantic (e.g., Wang and Chang 2008; Chang et al. 2004). Simulations with ocean-only GCMs help to understand the oceanic response to atmospheric processes (e.g., Lübbecke et al. 2010). Additionally, regional climate models of the equatorial Atlantic have been employed successfully to study different aspects of the region (e.g., Seo et al. 2006; Burls et al. 2011, 2012). Lastly, computational power continues to increase and allows for higher spatial resolution. If the equatorial Atlantic contains predictive potential, future generations of improved CGCMs are likely to unlock it at some point.

The research into various biases, their origins, their dynamics, and, most importantly, possible ways to reduce them, remains a core challenge of the global climate modeling community.

Acknowledgements We would like to thank Richard Greatbatch, Peter Brandt, Mojib Latif, Rebecca Hummels, and Martin Claus for discussing the manuscript with us at an early stage and contributing valuable ideas.

While we finished work on the first submitted draft of the manuscript, our colleague and friend Martin Krebs passed away. We hope this study will serve as a reminder of Martin's outstanding scientific work. TD will fondly remember how she discussed first ideas for the manuscript with Martin and how they came up with a rather wayward way to illustrate how much energy was lost from the Atl3 region during the formation of the cold tongue. Turns out that 2.6 times the total global energy consumption of 2011 is enough to power roughly 70 billion generic 600 W fridges for 4 months.

Appendix

This article is related to the YOUMARES 8 conference session no. 3: "Physical Processes in the Tropical and Subtropical Oceans: Variability, Impacts, and Connections to Other Components of the Climate System". The original Call for Abstracts and the abstracts of the presentations within this session can be found in the appendix "Conference Sessions and Abstracts", chapter "1 Physical Processes in the Tropical and Subtropical Oceans: Variability, Impacts, and Connections to Other Components of the Climate System", of this book.

References

Banzon V, Smith TM, Chin TM et al (2016) A long-term record of blended satellite and in situ sea-surface temperature for climate monitoring, modeling and environmental studies. Earth Syst Sci Data 8(1):165–176. https://doi.org/10.5194/essd-8-165-2016

Bjerknes J (1969) Atmospheric teleconnections from the equatorial Pacific. Mon Weather Rev 97(3):163–172

Bourlès B, Lumpkin R, McPhaden MJ et al (2008) The Pirata program: history, accomplishments, and future directions. Bull Am Meteorol Soc 89(8):1111–1125. https://doi.org/10.1175/2008BAMS2462.1

Brandt P, Schott FA, Provost C et al (2006) Circulation in the central equatorial Atlantic: mean and intraseasonal to seasonal variability. Geophys Res Lett 33(7):L07609. https://doi.org/10.1029/2005GL025498

Brandt P, Hormann V, Bourlès B et al (2008) Oxygen tongues and zonal currents in the equatorial Atlantic. J Geophys Res Oceans 113(4):C04012. https://doi.org/10.1029/2007JC004435

Brandt P, Caniaux G, Bourlès B et al (2011a) Equatorial upper-ocean dynamics and their interaction with the West African monsoon. Atmos Sci Lett 12(1):24–30. https://doi.org/10.1002/asl.287

Brandt P, Funk A, Hormann V et al (2011b) Interannual atmospheric variability forced by the deep equatorial Atlantic ocean. Nature 473(7348):497–500. https://doi.org/10.1038/nature10013

Burls NJ, Reason CJC, Penven P et al (2011) Similarities between the tropical Atlantic seasonal cycle and ENSO: an energetics perspective. J Geophys Res Oceans 116(C11):C11010. https://doi.org/10.1029/2011JC007164

Burls NJ, Reason CJC, Penven P et al (2012) Energetics of the tropical Atlantic zonal mode. J Clim 25(21):7442–7466. https://doi.org/10.1175/JCLI-D-11-00602.1

Caniaux G, Giordani H, Redelsperger JL et al (2011) Coupling between the Atlantic cold tongue and the West African monsoon in boreal spring and summer. J Geophys Res Oceans 116(4):C04003. https://doi.org/10.1029/2010JC006570

Chang P, Saravanan R, Wang F et al (2004) Predictability of linear coupled systems. Part II: an application to a simple model of tropical Atlantic variability. J Clim 17(7):1487–1503. https://doi.org/10.1175/1520-0442(2004)017<1487:POLCSP>2.0.CO;2

Chang CY, Carton JA, Grodsky SA et al (2007) Seasonal climate of the tropical Atlantic sector in the NCAR community climate system model 3: error structure and probable causes of errors. J Clim 20(6):1053–1070. https://doi.org/10.1175/JCLI4047.1

Cromwell T (1953) Circulation in a meridional plane in the central equatorial Pacific. J Mar Res 23:196–213

Cromwell T, Montgomery RB, Stroup ED (1954) Equatorial undercurrent in Pacific ocean revealed by new methods. Science 119(3097):648–649. https://doi.org/10.1126/science.119.3097.648

Davey M, Huddleston M, Sperber K et al (2002) STOIC: a study of coupled model climatology and variability in tropical ocean regions. Clim Dyn 18(5):403–420. https://doi.org/10.1007/s00382-001-0188-6

Deppenmeier AL, Haarsma RJ, Hazeleger W (2016) The Bjerknes feedback in the tropical Atlantic in CMIP5 models. Clim Dyn 47(7–8):2691–2707. https://doi.org/10.1007/s00382-016-2992-z

DeWitt DG (2005) Diagnosis of the tropical Atlantic near-equatorial SST bias in a directly coupled atmosphere-ocean general circulation model. Geophys Res Lett 32(1):L01703. https://doi.org/10.1029/2004GL021707

Ding H, Greatbatch RJ, Latif M et al (2015) The impact of sea surface temperature bias on equatorial Atlantic interannual variability in partially coupled model experiments. Geophys Res Lett 42(13):5540–5546. https://doi.org/10.1002/2015GL064799

Dippe T, Greatbatch RJ, Ding H (2017) On the relationship between Atlantic Niño variability and ocean dynamics. Clim Dyn. https://doi.org/10.1007/s00382-017-3943-z

Doi T, Vecchi GA, Rosati AJ et al (2012) Biases in the Atlantic ITCZ in seasonal-interannual variations for a coarse- and a high-resolution coupled climate model. J Clim 25(16):5494–5511. https://doi.org/10.1175/JCLI-D-11-00360.1

Foltz GR, McPhaden MJ (2010) Interaction between the Atlantic meridional and Niño modes. Geophys Res Lett 37:L18604. https://doi.org/10.1029/2010GL044001

Fretwell P, Pritchard HD, Vaughan DG et al (2013) Bedmap2: improved ice bed, surface and thickness datasets for Antarctica. Cryosphere 7(1):375–393. https://doi.org/10.5194/tc-7-375-2013

Frierson DMW, Hwang YT, Fučkar NS et al (2013) Contribution of ocean overturning circulation to tropical rainfall peak in the northern hemisphere. Nat Geosci 6:940–944. https://doi.org/10.1038/ngeo1987

García-Serrano J, Losada T, Rodríguez-Fonseca B et al (2008) Tropical Atlantic variability modes (1979–2002). Part II: time-evolving atmospheric circulation related to SST-forced tropical convection. J Clim 21(24):6476–6497. https://doi.org/10.1175/2008JCLI2191.1

Giannini A, Saravanan R, Chang P (2003) Oceanic forcing of Sahel rainfall on interannual to interdecadal time scales. Science 302(5647):1027–1030. https://doi.org/10.1126/science.1089357

Gill AE (1980) Some simple solutions for heat-induced tropical circulation. Q J R Meteorol Soc 106(449):447–462. https://doi.org/10.1002/qj.49710644905

Grodsky SA, Carton JA, McClain CR (2008) Variability of upwelling and chlorophyll in the equatorial Atlantic. Geophys Res Lett 35(3):L03610. https://doi.org/10.1029/2007GL032466

Grodsky SA, Carton JA, Nigam S et al (2012) Tropical Atlantic biases in CCSM4. J Clim 25(11):3684–3701. https://doi.org/10.1175/JCLI-D-11-00315.1

Harlaß J, Latif M, Park W (2015) Improving climate model simulation of tropical Atlantic sea surface temperature: the importance of enhanced vertical atmosphere model resolution. Geophys Res Lett 42(7):2401–2408. https://doi.org/10.1002/2015GL063310

Harlaß J, Latif M, Park W (2017) Alleviating tropical Atlantic sector biases in the Kiel climate model by enhancing horizontal and vertical atmosphere model resolution: climatology and interannual variability. Clim Dyn 50:2605. https://doi.org/10.1007/s00382-017-3760-4

Hastenrath S (1991) Climate dynamics of the tropics. Springer, Dordrecht. https://doi.org/10.1007/978-94-011-3156-8

Hazeleger W, Haarsma RJ (2005) Sensitivity of tropical Atlantic climate to mixing in a coupled ocean-atmosphere model. Clim Dyn 25(4):387–399. https://doi.org/10.1007/s00382-005-0047-y

Hormann V, Brandt P (2009) Upper equatorial Atlantic variability during 2002 and 2005 associated with equatorial Kelvin waves. J Geophys Res Oceans 114:C03007. https://doi.org/10.1029/2008JC005101

Hummels R, Dengler M, Bourlès B (2013) Seasonal and regional variability of upper ocean diapycnal heat flux in the Atlantic cold tongue. Prog Oceanogr 111:52–74. https://doi.org/10.1016/j.pocean.2012.11.001

Hummels R, Dengler M, Brandt P et al (2014) Diapycnal heat flux and mixed layer heat budget within the Atlantic cold tongue. Clim Dyn 43(11):3179–3199. https://doi.org/10.1007/s00382-014-2339-6

Jin FF (1997) An equatorial ocean recharge paradigm for ENSO. Part I: conceptual model. J Atmos Sci 54(7):811–829

Johns WE, Brandt P, Bourlès B et al (2014) Zonal structure and seasonal variability of the Atlantic equatorial undercurrent. Clim Dyn 43(11):3047–3069. https://doi.org/10.1007/s00382-014-2136-2

Jouanno J, Marin F, Du Penhoat Y et al (2011) Seasonal heat balance in the upper 100 m of the equatorial Atlantic Ocean. J Geophys Res Oceans 116(9):C09003. https://doi.org/10.1029/2010JC006912

Keenlyside NS, Latif M (2007) Understanding equatorial Atlantic interannual variability. J Clim 20(1):131–142. https://doi.org/10.1175/JCLI3992.1

Liu H, Wang C, Lee SK et al (2013) Atlantic warm pool variability in the CMIP5 simulations. J Clim 26(15):5315–5336. https://doi.org/10.1175/JCLI-D-12-00556.1

Locarnini RA, Mishonov AV, Antonov JI et al (2013) World ocean atlas 2013. Vol. 1: Temperature. Levitus S (ed), Mishonov A (technical ed), NOAA Atlas NESDIS 73

Lübbecke JF, McPhaden MJ (2012) On the inconsistent relationship between Pacific and Atlantic Niños. J Clim 25(12):4294–4303. https://doi.org/10.1175/JCLI-D-11-00553.1

Lübbecke JF, McPhaden MJ (2017) Symmetry of the Atlantic Niño mode. Geophys Res Lett 44(2):965–973. https://doi.org/10.1002/2016GL071829

Lübbecke JF, Böning CW, Keenlyside NS et al (2010) On the connection between Benguela and equatorial Atlantic Niños and the role of the South Atlantic anticyclone. J Geophys Res Oceans 115(C9):C09015. https://doi.org/10.1029/2009JC005964

Lübbecke JF, Burls NJ, Reason CJC et al (2014) Variability in the South Atlantic anticyclone and the Atlantic Niño mode. J Clim 27(21):8135–8150. https://doi.org/10.1175/JCLI-D-14-00202.1

Madec G (2008) NEMO ocean general circulation model reference manuel. Tech. rep., Institut Pierre-Simon Laplace, Paris

Marin F, Caniaux G, Giordani H et al (2009) Why were sea surface temperatures so different in the eastern equatorial Atlantic in June 2005 and 2006? J Phys Oceanogr 39(6):1416–1431. https://doi.org/10.1175/2008JPO4030.1

McPhaden MJ (1995) The tropical atmosphere ocean array is completed. Bull Am Meteorol Soc 76:739–741

McPhaden MJ, Meyers G, Ando K et al (2009) RAMA: the research moored Array for African-Asian-Australian monsoon analysis and prediction. Bull Am Meteorol Soc 90(4):459–480. https://doi.org/10.1175/2008BAMS2608.1

Meehl GA, Covey C, Taylor KE et al (2007) THE WCRP CMIP3 multimodel dataset: a new era in climate change research. Bull Am Meteorol Soc 88(9):1383–1394. https://doi.org/10.1175/BAMS-88-9-1383

Merle J (1980) Seasonal heat budget in the equatorial Atlantic ocean. J Phys Oceanogr 10(3):464–469

Mitchell TP, Wallace JM (1992) The annual cycle in equatorial convection and sea surface temperature. J Clim 5(10):1140–1156

Nnamchi HC, Li J, Kucharski F et al (2015) Thermodynamic controls of the Atlantic Niño. Nat Commun 6:8895. https://doi.org/10.1038/ncomms9895

Nnamchi HC, Li J, Kucharski F et al (2016) An equatorial-extratropical dipole structure of the Atlantic Niño. J Clim 29(20):7295–7311. https://doi.org/10.1175/JCLI-D-15-0894.1

Okumura Y, Xie SP (2004) Interaction of the Atlantic equatorial cold tongue and the African monsoon. J Clim 17(18):3589–3602. https://doi.org/10.1175/1520-0442(2004)017<3589:IOTAEC>2.0.CO;2

Okumura Y, Xie SP (2006) Some overlooked features of tropical Atlantic climate leading to a new Nino-like phenomenon. J Clim 19(22):5859–5874

Park W, Keenlyside N, Latif M et al (2009) Tropical Pacific climate and its response to global warming in the Kiel climate model. J Clim 22(1):71–92. https://doi.org/10.1175/2008JCLI2261.1

Patricola CM, Li M, Xu Z et al (2012) An investigation of tropical Atlantic bias in a high-resolution coupled regional climate model. Clim Dyn 39(9):2443–2463. https://doi.org/10.1007/s00382-012-1320-5

Philander SGH, Pacanowski RC (1981) The oceanic response to cross-equatorial winds (with application to coastal upwelling in low latitudes). Tellus 33(2):201–210. https://doi.org/10.3402/tellusa.v33i2.10708

Phillips NA (1956) The general circulation of the atmosphere: a numerical experiment. Q J R Meteorol Soc 82(352):123–164. https://doi.org/10.1002/qj.49708235202

Polo I, Rodríguez-Fonseca B, Losada T et al (2008) Tropical Atlantic variability modes (1979–2002). Part I: time-evolving SST modes related to West African rainfall. J Clim 21(24):6457–6475. https://doi.org/10.1175/2008JCLI2607.1

Reynolds RW, Smith TM, Liu C et al (2007) Daily high-resolution-blended analyses for sea surface temperature. J Clim 20(22):5473–5496. https://doi.org/10.1175/2007JCLI1824.1

Rhein M, Dengler M, Sültenfuß J et al (2010) Upwelling and associated heat flux in the equatorial Atlantic inferred from helium isotope disequilibrium. J Geophys Res Oceans 115(C8):C08021. https://doi.org/10.1029/2009JC005772

Richter I, Xie SP (2008) On the origin of equatorial Atlantic biases in coupled general circulation models. Clim Dyn 31(5):587–598. https://doi.org/10.1007/s00382-008-0364-z

Richter I, Xie SP, Wittenberg AT et al (2012) Tropical Atlantic biases and their relation to surface wind stress and terrestrial precipitation. Clim Dyn 38(5–6):985–1001. https://doi.org/10.1007/s00382-011-1038-9

Richter I, Behera SK, Masumoto Y et al (2013) Multiple causes of interannual sea surface temperature variability in the equatorial Atlantic Ocean. Nat Geosci 6(1):43–47. https://doi.org/10.1038/ngeo1660

Richter I, Behera SK, Doi T et al (2014a) What controls equatorial Atlantic winds in boreal spring? Clim Dyn 43(11):3091–3104. https://doi.org/10.1007/s00382-014-2170-0

Richter I, Xie SP, Behera SK et al (2014b) Equatorial Atlantic variability and its relation to mean state biases in CMIP5. Clim Dyn 42(1–2):171–188. https://doi.org/10.1007/s00382-012-1624-5

Richter I, Xie SP, Morioka Y et al (2016) Phase locking of equatorial Atlantic variability through the seasonal migration of the ITCZ. Clim Dyn 48(11–12):3615–3629. https://doi.org/10.1007/s00382-016-3289-y

Richter I, Doi T, Behera SK et al (2017) On the link between mean state biases and prediction skill in the tropics: an atmospheric perspective. Clim Dyn 50:3355. https://doi.org/10.1007/s00382-017-3809-4

Risien CM, Chelton DB (2008) A global climatology of surface wind and wind stress fields from eight years of QuikSCAT scatterometer data. J Phys Oceanogr 38(11):2379–2413. https://doi.org/10.1175/2008JPO3881.1

Rodríguez-Fonseca B, Janicot S, Mohino E et al (2011) Interannual and decadal SST-forced responses of the West African monsoon. Atmos Sci Lett 12(1):67–74. https://doi.org/10.1002/asl.308

Roeckner E, Bäuml G, Bonaventura L et al (2003) The atmospheric general circulation model ECHAM5: part 1: model description. MPI Report 349. https://doi.org/10.1029/2010JD014036

Roemmich D, Johnson GC, Riser S et al (2009) The Argo program. Oceanography 22(2):34–43. https://doi.org/10.5670/oceanog.2009.36

Schott FA, Dengler M, Brandt P et al (2003) The zonal currents and transports at 35°W in the tropical Atlantic. Geophys Res Lett 30(7):1349. https://doi.org/10.1029/2002GL016849

Sein DV, Danilov S, Biastoch A et al (2016) Designing variable ocean model resolution based on the observed ocean variability. J Adv Model Earth Syst 8(2):904–916. https://doi.org/10.1002/2016MS000650

Seo H, Jochum M, Murtugudde R et al (2006) Effect of ocean mesoscale variability on the mean state of tropical Atlantic climate. Geophys Res Lett 33(9):L09606. https://doi.org/10.1029/2005GL025651

Siongco AC, Hohenegger C, Stevens B (2015) The Atlantic ITCZ bias in CMIP5 models. Clim Dyn 45(5):1169–1180. https://doi.org/10.1007/s00382-014-2366-3

Stockdale TN, Balmaseda MA, Vidard A (2006) Tropical Atlantic SST prediction with coupled ocean-atmosphere GCMs. J Clim 19(23):6047–6061. https://doi.org/10.1175/JCLI3947.1

Sutton RT, Jewson SP, Rowell DP (2000) The elements of climate variability in the tropical Atlantic region. J Clim 13(18):3261–3284

Taylor KE, Stouffer RJ, Meehl GA (2012) An overview of CMIP5 and the experiment design. Bull Am Meteorol Soc 93(4):485–498. https://doi.org/10.1175/BAMS-D-11-00094.1

Tozuka T, Doi T, Miyasaka T et al (2011) Key factors in simulating the equatorial Atlantic zonal sea surface temperature gradient in a coupled general circulation model. J Geophys Res Oceans 116(6):C06010. https://doi.org/10.1029/2010JC006717

Trenberth KE, Caron JM (2001) Estimates of meridional atmosphere and ocean heat transports. J Clim 14(16):3433–3443

Trenberth KE, Fasullo JT, Kiehl J (2009) Earth's global energy budget. Bull Am Meteorol Soc 90(3) 311–323. https://doi.org/10.1175/2008BAMS2634.1

Voldoire A, Claudon M, Caniaux G et al (2014) Are atmospheric biases responsible for the tropical Atlantic SST biases in the CNRM-CM5 coupled model? Clim Dyn 43(11):2963–2984. https://doi.org/10.1007/s00382-013-2036-x

Wahl S, Latif M, Park W et al (2011) On the tropical Atlantic SST warm bias in the Kiel climate model. Clim Dyn 36(5–6):891–906. https://doi.org/10.1007/s00382-009-0690-9

Wang F, Chang P (2008) Coupled variability and predictability in a stochastic climate model of the tropical Atlantic. J Clim 21(23):6247–6259. https://doi.org/10.1175/2008JCLI2283.1

Xie SP (2004) The shape of continents, air-sea interaction, and the rising branch of the Hadley circulation. In: Diaz HF, Bradley RS (eds) The Hadley circulation: present, past and future. Springer, Dordrecht, pp 121–152

Xie P, Arkin PA (1997) Global precipitation: a 17-year monthly analysis based on gauge observations, satellite estimates, and numerical model outputs. Bull Am Meteorol Soc 78(11):2539–2558. https://doi.org/10.1175/1520-0477(1997)078<2539:GPAYMA>2.0.CO;2

Xie SP, Carton JA (2004) Tropical Atlantic variability: patterns, mechanisms, and impacts. In: Wang C, Xie SP, Carton JA (eds) Earth's climate: the ocean-atmosphere interaction. American Geophysical Union, Washington, DC, pp 121–142. https://doi.org/10.1029/147GM07

Xie SP, Philander SGH (1994) A coupled ocean-atmosphere model of relevance to the ITCZ in the eastern Pacific. Tellus A 46(4):340–350. https://doi.org/10.1034/j.1600-0870.1994.t01-1-00001.x

Xu Z, Chang P, Richter I et al (2014) Diagnosing southeast tropical Atlantic SST and ocean circulation biases in the CMIP5 ensemble. Clim Dyn 43(11):3123–3145. https://doi.org/10.1007/s00382-014-2247-9

Zebiak SE (1993) Air-sea interaction in the equatorial Atlantic region. J Clim 6(8):1567–1586

Zermeño-Diaz DM, Zhang C (2013) Possible root causes of surface westerly biases over the equatorial Atlantic in global climate models. J Clim 26(20):8154–8168. https://doi.org/10.1175/JCLI-D-12-00226.1

The Physical System of the Arctic Ocean and Subarctic Seas in a Changing Climate

Camila Campos and Myriel Horn

Abstract

The Earth's climate is changing and the poles are particularly sensitive to the global warming, with most evident implications over the Arctic. While summer sea ice reduced significantly compared to the previous decades, and the atmospheric warming is amplified over the Arctic, changes in the ocean are less obvious due to its higher inertia. Still, impacts of the changing climate on high-latitude and polar oceans are already observable and expected to further increase. The northern seas are essential regions for the maintenance of the Atlantic Meridional Overturning Circulation, which in turn is a key aspect of the maritime climate. Alterations in heat and freshwater/salinity content in the Arctic Ocean and adjacent seas impact and are closely linked to buoyancy flux distributions, which control the vertical and horizontal motion of water masses, thus impacting the climate system on a longer time scale. In this context, we set our focus on the Arctic Ocean and Atlantic subarctic seas, review some of the contemporary knowledge and speculations on the complex coupling between atmosphere, sea ice, and ocean, and describe the important elements of its physical oceanography. This assessment is an attempt to raise awareness that investigating the pathways and timescales of oceanic responses and contributions is fundamental to better understand the current climate change.

Introduction

The Arctic region (Fig. 1) is a relative small fraction of the globe's surface, but plays a crucial role in determining global climate dynamics due to the intimate and complex couplings between cryosphere, atmosphere, ocean, and land (Serreze et al. 2007). Currently, the Arctic is undergoing remarkable environmental changes and has been in focus of the climate sciences community (Winton 2008; Overland 2016).

The Arctic near surface air temperature is warming twice as fast as the global average (Serreze and Francis 2006). This accelerated response is known as the Arctic amplification (Winton 2008; Serreze and Barry 2011; Cohen et al. 2014), and one of the most dramatic indicators of the Arctic warming has been the decline in the sea ice cover. Satellite observations reveal that the area of the Arctic sea ice during summer has steadily decreased by more than 40% in recent decades (Fig. 2) (Comiso et al. 2008; Pistone et al. 2014). Notwithstanding, observations further show a year-round loss of sea ice extent and thickness (Lindsay and Schweiger 2015; Rothrock et al. 2008), which suggest that from year to year more melt and less recovery is taking place.

The observed rate of sea ice extent reduction during the last three to four decades has occurred faster than anticipated by models participating on the Intergovernmental Panel on Climate Change Fourth Assessment Report: the observed trend for the September sea ice extent was −9.12 ± 1.54% per decade for the period 1979–2006, while the mean decline trend of all the models participating in the report was −4.3 ± 0.3% per decade (Stroeve et al. 2007). The accelerated sea ice decline has likely occurred due to a combination of decadal-scale variability in the coupled ice-ocean-atmosphere-land system and radiative greenhouse gas forcing (e.g., Serreze and Barry 2011; IPCC 2014; Zhang 2015). According to model studies, the Arctic sea ice will continue shrinking and thinning year-round in the course of the twenty-first century as the global mean surface temperature rises, with projections of summer ice free Arctic in the near

Both authors contributed equally.

C. Campos · M. Horn (✉)
Alfred Wegener Institute (AWI), Helmholtz Centre for Polar and Marine Research, Bremerhaven, Germany
e-mail: camila.campos@awi.de; myriel.horn@awi.de

S. Jungblut et al. (eds.), *YOUMARES 8 – Oceans Across Boundaries: Learning from each other*,
https://doi.org/10.1007/978-3-319-93284-2_3

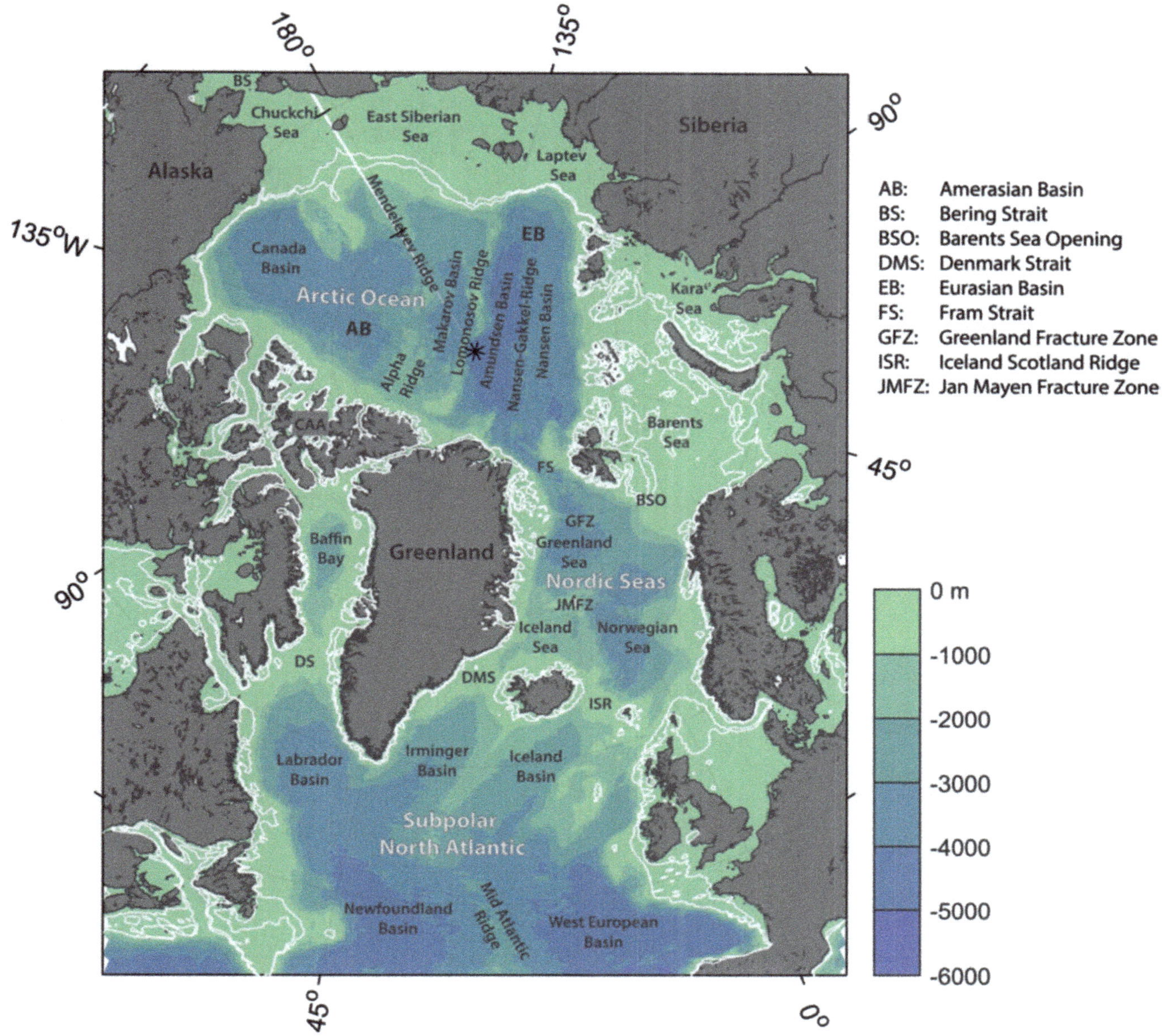

Fig. 1 The northern seas (produced with the help of the colormap from Thyng et al. 2016). Bathymetric and geographical map derived from the 2-min ETOPO2 database

future (Wang and Overland 2009; IPCC 2014). Nevertheless, the impacts of these projections for the weather and climate locally and elsewhere are not sufficiently well understood.

Numerous studies have been published on the relation between Arctic sea ice decline and weather and climate. While some have addressed the question how Arctic sea ice decline impacts climate (Budikova 2009; Vihma 2014; Semmler et al. 2016), Lang et al. (2017) and several others have reviewed the recent decline in Arctic sea ice and the processes responsible for it (Polyakov et al. 2012; Stroeve et al. 2012; Barnes and Screen 2015). By far, the majority of these studies focus on atmospheric pathways and, therefore, our understanding of the mechanisms, pathways, and timescales by which the ocean controls or responds to these changes remains quite limited.

Previous studies addressed how the inflow of the warm Atlantic Water (AW) to the Arctic Ocean contributes to the decline of the sea ice extent and thickness (e.g., Carmack et al. 2015; Onarheim et al. 2014). Itkin et al. (2014) has addressed this problem from the reverse perspective, and showed in idealized experiments that a weaker (i.e., thinner) sea ice cover allows higher momentum transfer into the Arctic Ocean and impacts the surface and intermediate ocean circulation. In other words, there is an intrinsic two-way

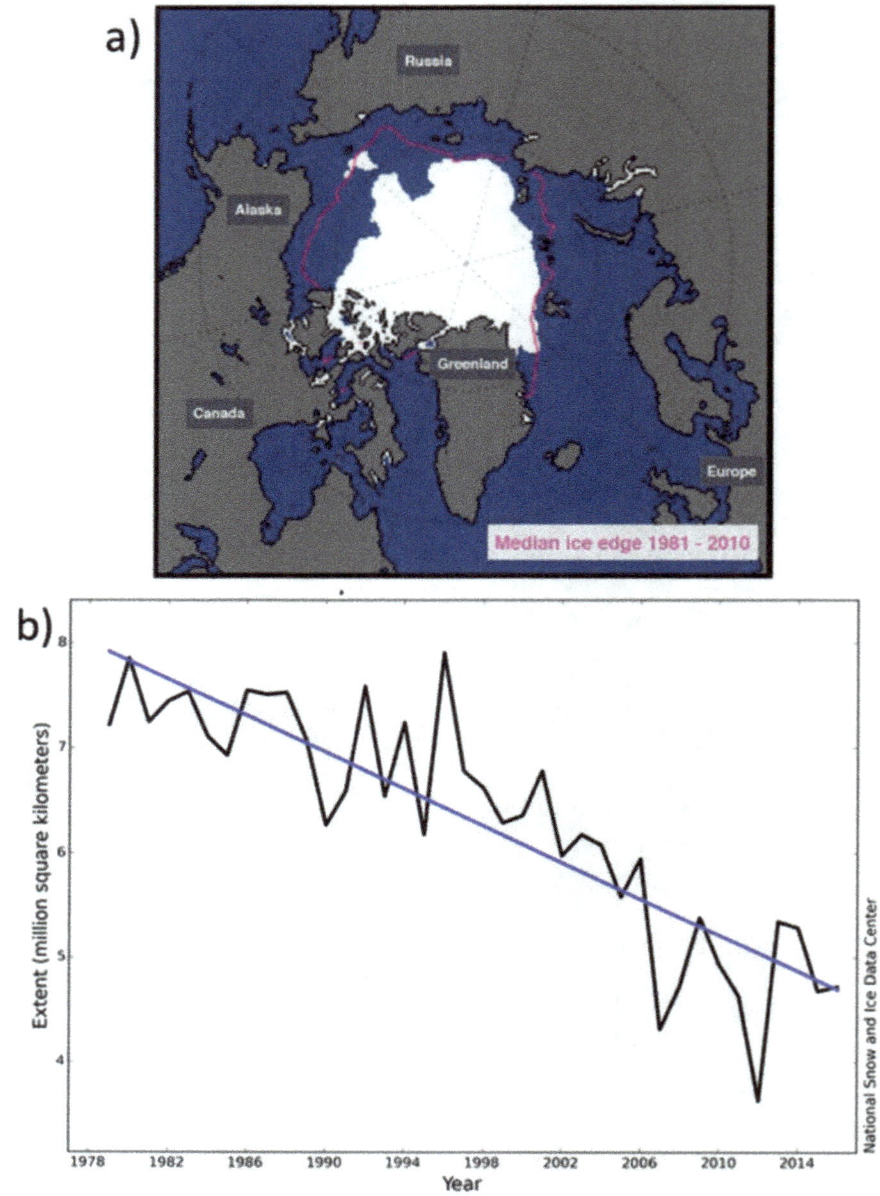

Fig. 2 Arctic summer sea ice decline (Fetterer et al. 2016, provided by the National Snow and Ice Data Center NSIDC, with permission). (**a**) Arctic September (minimum) sea ice extent in 2016 (white area) compared to the median ice edge for the period 1981 to 2010 (fuchsia line) and (**b**) average monthly September sea ice extent for the years 1979–2016, blue line: decline rate of 13.2% per decade relative to the 1981–2010 average

relation between the ocean and the sea ice, and any change in sea ice cover may impact the dynamics and thermodynamics of the ocean. Recent observations suggest that a diminishing sea ice cover to the northeast of Svalbard is responsible for reducing the stratification of the ocean and allowing more upward heat transfer, which preconditions the ice to further melting (Polyakov et al. 2010, 2017).

A significant increase in liquid freshwater content has been observed in the upper Arctic Ocean in the past two decades (Rabe et al. 2011; Giles et al. 2012; Morison et al. 2012; Rabe et al. 2014), while the Arctic sea ice volume has been shrinking significantly (Lindsay and Schweiger 2015). Sea ice and liquid fresh water are important factors for the Arctic Ocean, where they insulate the atmosphere from the warm Atlantic-derived water at intermediate depths, by limiting the upward heat transport, hence influencing the sea ice formation and melting as well as the air temperature.

After a freshening of the subpolar North Atlantic and Nordic Seas from the 1960s to the 1990s, both regions became again more saline thereafter (Curry and Mauritzen 2005; Boyer et al. 2007; Mauritzen et al. 2012). The Nordic Seas and the subpolar North Atlantic are the main regions in the northern hemisphere, where deep water formation takes place and thereby are key regions for global climate (Rhein et al. 2011). Freshwater changes could potentially influence this overturning system and thereby have a profound impact on our climate (Koenigk et al. 2007; Rennermalm et al. 2007).

In this chapter, we provided an introductory overview on the complex interactions of the coupled Arctic system in a changing climate with specific interest in the ways in which Arctic Ocean and adjacent seas may respond and modulate the observed and projected changes over high and mid-latitude. Next, we give an overview of the complex interplay between the dynamics and thermodynamics of the sea ice, atmosphere and ocean. We start by addressing the sea ice cycle, variability, and importance in the climate system (Section "Arctic sea ice"). In Section "Arctic – subarctic atmosphere" we give background information on the Arctic – Subarctic atmosphere (Section "Atmospheric circulation: Why does it matter?") and present the main atmospheric circulation modes (Section "Major modes of atmospheric circulation in the Arctic"). Then, we finally get to discuss the changing climate from the ocean perspective (Section "Ocean"): at first, we describe the main geographical features and the hydrography of the northern seas; subsequently, we address recent research and discussion of the global relevance of the region in a changing world. Final remarks are given in Section "Outlook".

Arctic Sea Ice

Sea Ice Cycle

The sea ice cover has a natural cycle as a consequence of the periodic changes of incident solar radiation over high latitudes. As the cold season arrives, atmospheric temperatures rapidly begin to drop. This leads to a positive thermal gradient from ocean to the surrounding air, resulting in a direct loss of sensible heat from the upper ocean. Dynamical instability in the upper meters of the ocean is generated as a consequence to density changes caused by cooling, and a vertical mixing is maintained until a significant layer of the upper water column approaches homogeneous temperature. Once the ocean freezing temperature of −1.9 °C is achieved, sea ice structures begin to form, and during this process a salt solution (brine) is expelled into the ocean further increasing its density. However, if mixing is deep enough, the surface waters may not reach freezing temperatures due to mixing with the warmer waters at intermediate depths and sea ice formation will not occur.

After initial formation in fall, sea ice continues growing through winter months and increases in vertical and horizontal extent. It can be characterized by highly complex and variable macrostructures, such as ridges, melt ponds, leads and polynyas. By the end of wintertime, the sea ice extent has reached its maximum. During spring, the solar radiation gradually increases thereby initiating the melting phase, which carries on until the next cooling season. If all the sea ice melts away, the area is characterized by the presence of fist year ice. However, if sea ice persists until the end of the warm season a perennial (multiyear) sea ice cover establishes. The fundamental differences between them relate to the vertical growth and surface roughness.

Overall, freezing and melting are controlled by net surface heat energy flux variations during the year, and environmental conditions, e.g., wind and oceanic currents, play a role in determining expansion and thickening. Furthermore, the horizontally confined Arctic Ocean allows for thicker sea ice growth (in comparison to the Southern Ocean), and winter sea ice thickness ranges on average from 3 to 4 m. For more details the reader is referred to Thomas and Dieckmann (2010).

Sea Ice Role in the Climate System

Sea ice is a highly reflective surface, with albedo ranging from 50% to 70%. Albedo is a measure of a surface's reflectivity, and may be even higher if a snow cover is present. A thicker ice pack supports a greater layer of snow and this system can reflect up to 90% of solar energy. Additionally, it acts as an insulator between ocean and atmosphere, and, therefore, restricts heat and momentum fluxes at this interface. If the atmosphere or the ocean warms up (above melting temperatures) sea ice melts and, since the exposed ocean surface has a much lower albedo than sea ice, the overall albedo of polar areas decrease. The low reflectance oceanic surface takes in extra heat, driving major changes in the regional radiative equilibrium and further sea ice melt. The described processes is the so-called ice-albedo feedback mechanism and is accounted as the main reason of nonlinear changes over polar regions (Winton 2008; Serreze and Barry 2011; Vihma 2014). Changes to ocean density caused by the sea ice cycle are important processes for the local oceanic stratification and global oceanic circulation.

A few specific areas of the high latitude oceans are crucial for the production of dense water masses, which contribute to the lower limb of the global oceanic overturning circulation. The upper layers of the ocean are densified through cooling of surface waters and the injection of brine during sea ice formation resulting in vertical mixing and deep convection (Tomczak and Godfrey 1994). In these regions the dense water sinks and is replaced by surface water from other areas and the continuation of this process is one of the drivers of the Meridional Overturning Circulation; the sinking of these waters is compensated by upwelling at other sites (Talley et al. 2011). On the other hand, sea ice constitutes a source of relatively fresh water (with an average salinity ranging from 2 to 7 (Thomas and Dieckmann 2010)) and when it melts it decreases the density of the water directly underneath, creating a stable surface layer. Changes in the water density at the deep convection sites may alter mixing and convection processes. Hence, the presence of sea ice strongly modulates interactions between ocean and atmosphere, namely heat, mass, and momentum transfers.

In addition to all physical aspects, sea ice acts as a key component also for the Arctic ecosystem, it also determines marine transportation and offshore activities, and is of crucial societal importance. A detailed description of these aspects is beyond the scope of the present review, but we refer to the Arctic Climate Impact Assessment – Scientific Report (ACIA 2004) for a more thorough perspective.

Arctic – Subarctic Atmosphere

Atmospheric Circulation: Why Does It Matter?

The polar regions are the world's heat sink: at low latitudes the amount of incoming solar radiation (shortwave) exceeds the emitted infrared radiation (longwave), whereas there is an annual energy deficit at the poles, where more heat is emitted than absorbed. The surplus of energy is then transported from the equatorial region towards the poles in the atmosphere and ocean. In the atmosphere, this manifests as global circulation cells, which, due to turbulent interactions, transfer energy to smaller processes of regional and local importance forcing climate and weather patterns. The latter play a very important role in the coupling with ocean and sea ice, which on the other hand also force changes on the atmospheric circulation. Therefore, global climate and weather are highly dependent on these interactions between the components of the earth system (Taylor 2009).

Though temperatures have been increasing in polar and equatorial regions, it has been amplified at high latitudes, especially over the Arctic (Serreze and Barry 2011). This amplification is attributed to several feedback mechanisms (Taylor et al. 2013) and, even though the ice-albedo feedback is often cited as primary contributor, some studies suggest that other interactions, like the warming of the lower atmosphere might play a bigger role (Pithan and Mauritsen 2014). Serreze and Barry (2011) provide a thorough synthesis of research on Arctic amplification.

The fact that the temperature increase over the Arctic has been happening at a faster rate than the global average, decreases the overall meridional temperature gradient over the globe, which in turn may affect the atmospheric circulation pattern locally as well as remotely (Barnes and Screen 2015). The scientific community has been broadly concerned with possible changes over mid-latitude weather such as, e.g., the occurrence of extreme weather events and the weakening and shifting of the westerly winds (Overland 2016). These winds are strongly coupled to the track and intensity of storm systems travelling at mid-latitudes, hence it is expected that changes in the position and strength of the jet stream leads to noticeable changes in the northern hemispheric daily weather (e.g., Barnes and Screen 2015; Serreze and Barry 2011).

The particular role and responses of the atmosphere in a warming climate are beyond the scope of this work. Thus, for more comprehensive understanding we refer here to several studies which review and investigate responses of large-scale atmospheric circulation to changes in sea ice cover over the Arctic (Budikova 2009; Bader et al. 2011; Vihma 2014; Semmler et al. 2016). Nevertheless, an overview on the background characteristics of the Arctic atmospheric system are given next.

Major Modes of Atmospheric Circulation in the Arctic

As explained above, atmospheric circulation and weather are linked to gradients. The system has an intrinsic seasonal variability upon which these gradients oscillate. To characterize the major atmospheric modes over the Arctic, a brief illustration on its climatology is given in terms of sea level pressure.

The prevailing atmospheric circulation over the Arctic is anticyclonic, which results from an average high-pressure system that spawns winds over the region. Although prevalent, the circulation regime may shift to cyclonic on the time scales of 5–7 years (Proshutinsky et al. 2009). Shifts from one regime to another are forced by changes in the location and intensity of the pressure systems described below. This oscillatory mode is part of the Arctic system's natural variability and may help to explain the significant, basin-scale changes of the Arctic atmosphere-ice-ocean system (Polyakov and Johnson 2000; Proshutinsky et al. 2009, 2015).

The two semi-permanent centers of low pressure, the oceanic Aleutian and Icelandic Lows, and the continental Siberian High, which extends into the Arctic as the Beaufort High, are observed as pronounced features during winter. In summer, the gradients of the polar and subpolar regions are relatively weak, and sea level pressure distribution is dominated by the subtropical, the Azores and the Pacific Highs (McBean et al. 2005). To describe the main states of the atmospheric circulation, indices were created. Based on a surface variable and obtained through statistical analysis, these are used to characterize complex climate processes and explain past variability.

The major mode of variability in the Arctic is the Arctic Oscillation (AO), and is characterized by the relation between the surface pressure anomaly in the Arctic and in mid-latitudes (Thompson and Wallace 1998). When the AO is in its positive phase, surface pressure in the polar region is low. This mode manifests as the strengthening of the zonal westerly winds

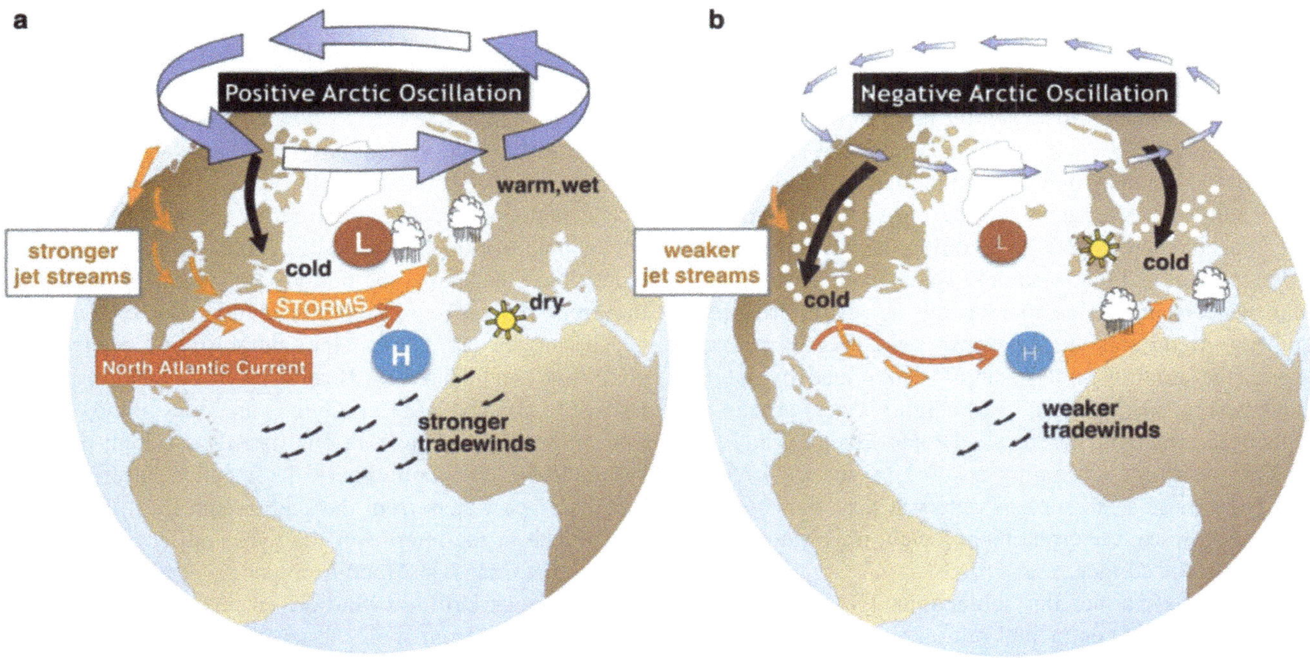

Fig. 3 Schematic of the Arctic Oscillation and its effects (adapted from AMAP 2012, with permission). Positive Arctic Oscillation (**a**) and negative Arctic Oscillation (**b**) Accordingly, the centres of low (red encircled L) and high (blue encircled H) pressure systems over the North Atlantic indicate the corresponding North Atlantic Oscillation phases (**a**: positive, **b**: negative)

which act to confine colder air over the high latitudes. On the other hand, in the negative phase of the AO, surface pressure is high in the Arctic, acting to weaken the atmospheric circulation, and thus, allowing an easier escape of the cold polar air masses towards the mid-latitudes (Fig. 3).

The regional manifestation of the AO in the North Atlantic is the North Atlantic Oscillation (NAO). It is given by the correlation of the main pressure centers in the North Atlantic, namely the Icelandic Low and the Azores High (Fig. 3). Oscillations between positive and negative phases are tied to shifts in storm tracks and associated patterns of precipitation and temperature.

For more detailed information we refer to Serreze and Barry (2014) and Turner and Marshall (2011).

Ocean

Geography of the Arctic Mediterranean

The Arctic Mediterranean consists of two major parts: the Arctic Ocean and the Nordic Seas. The Arctic Ocean is the northernmost part of the Arctic Mediterranean which is enclosed by North America, the Eurasian continent, Svalbard, and Greenland (Fig. 1). The Nordic Seas are enclosed by Svalbard, Norway, Iceland, Scotland and Greenland and include the Greenland Sea, Norwegian Sea, and Iceland Sea (also called the GIN Seas). The Arctic Ocean connects to the Nordic Seas via the Fram Strait (between Greenland and Svalbard, ~2600 m deep) and the Barents Sea Opening (between Svalbard and Norway, ~200 m deep). Other gateways are the narrow channels through the Canadian Arctic Archipelago (Islands North West of Greenland, ~150–230 m deep) and the Bering Strait (~45 m deep and only 50 km wide), which is the only connection to the Pacific Ocean. Towards the Eurasian Continent the Arctic Ocean consists of wide, shallow shelves (<50–300 m deep), which make up almost half of the entire Arctic Ocean and comprise five marginal seas: Barents Sea, Kara Sea, Laptev Sea, East Siberian Sea and Chuckchi Sea. At the coasts of North America and Greenland the shelves are much narrower. The deep basins in the center of the Arctic Ocean are divided into two major parts, the Amerasian Basin and the Eurasian Basin. They are separated by the Lomonosov Ridge, which is approximately 1600 m deep. The Eurasian Basin consists of the Nansen Basin and the Amundsen Basin, which are separated by the Gakkel Ridge. The Mendeleyev Ridge and Alpha Ridge divide the Amerasian Basin into the Makarov Basin and the Canada Basin. With approximately 4500 m depth, the Amundsen Basin is the deepest, while the Canada Basin is by far the largest. The boundary of the Nordic Seas to the North Atlantic is Denmark Strait (~ 500–700 m deep) and the Iceland-Scotland-Ridge (~300–850 m deep). There are two fracture zones in the center of the Nordic Seas: the Greenland Fracture Zone and the Jan Mayen Fracture Zone. The shelves along the Greenland coast are wide and shallow with a steep shelf break.

Arctic Ocean Circulation and Hydrography

Since the Arctic Ocean is a largely enclosed ocean, there are only two water masses that enter the basin from other oceans: Pacific Water (PW) and Atlantic Water (AW). The low-salinity PW is transported through the Bering Strait (Fig. 4) and is mainly advected at the surface into the Amerasian Basin and adjacent shelf regions. In addition to continental runoff and precipitation, the low salinity PW is an important Arctic fresh water source. Due to relatively small differences in temperature throughout the water column of the Arctic Ocean, the stratification is mainly determined by salinity changes (Fig. 5). Thus, the fresh (light) waters stay in the upper ocean and build the so called Polar Mixed Layer. Large parts of the Arctic Ocean are covered by sea ice which is built from these fresh surface waters at near-freezing temperatures. By sea ice formation and melt, freshwater is concentrated at the surface.

The AW is warmer but saltier than PW and Meteoric waters comprising continental run-off and net precipitation. Thereby, it is denser and can be found deeper in the water column. The significant difference in salinity creates a strong halocline between the Polar Mixed Layer and the AW layer establishing a strong permanent stratification in the deep basins (Fig. 5). The halocline, which is defined by high vertical salinity gradients ($32.5 < S < 34.5$), is thickest in the Canada Basin (200–250 m) and thinnest in the Nansen Basin (100–150 m). The temperatures of the halocline remain close to the freezing point. Due to heat loss and mixing with shelf waters there are many modifications of AW at intermediate depths. Just below the halocline, temperatures are highest in the Nansen Basin and decrease towards the Canada Basin (Fig. 5). The densest waters are formed on the shelves of the Barents Sea, where the AW subsequently releases heat to the atmosphere and is mixed with brine rejected from newly formed sea ice before it sinks down the shelf break into the Nansen Basin (Fig. 4). This Arctic bottom water is the densest water of the world ocean, but can only be found in the Arctic region (Tomczak and Godfrey 1994). Only a small part is able to flow over the sill of the Fram Strait into the Nordic Seas balancing the bottom/deep water formation on the Arctic shelves (e.g., Bönisch and Schlosser 1995).

The surface circulation in the central Arctic Ocean mainly comprises two features: the Beaufort Gyre and the Transpolar Drift (Fig. 4). The Beaufort Gyre is an anticyclonic circulation in the Canada Basin that is forced by a high pressure system in the lower atmosphere, the so called Beaufort High. Fresh surface waters from the shelves and from the Pacific accumulate in the interior of the gyre and leave the Arctic through the Canadian Arctic Archipelago or the Fram Strait. The Transpolar Drift is a wind-driven current that directs sea ice and waters from the Siberian shelves and the Bering Strait to the Fram Strait, where they exit the Arctic Ocean into the Nordic Seas and subpolar North Atlantic.

The relatively warm and saline AW, which is the main water source of the entire Arctic Ocean, enters the Nordic Seas from the south. The Norwegian Atlantic Current carries the AW through the Nordic Seas at the surface and splits into two main branches. One branch enters the Arctic Ocean through the Barents Sea Opening, the other, which then is called West Spitsbergen Current, flows through the Fram Strait. Only a part of the West Spitsbergen Current propagates further north into the Arctic Ocean, the other part recirculates close to the Fram Strait. The AW dives underneath

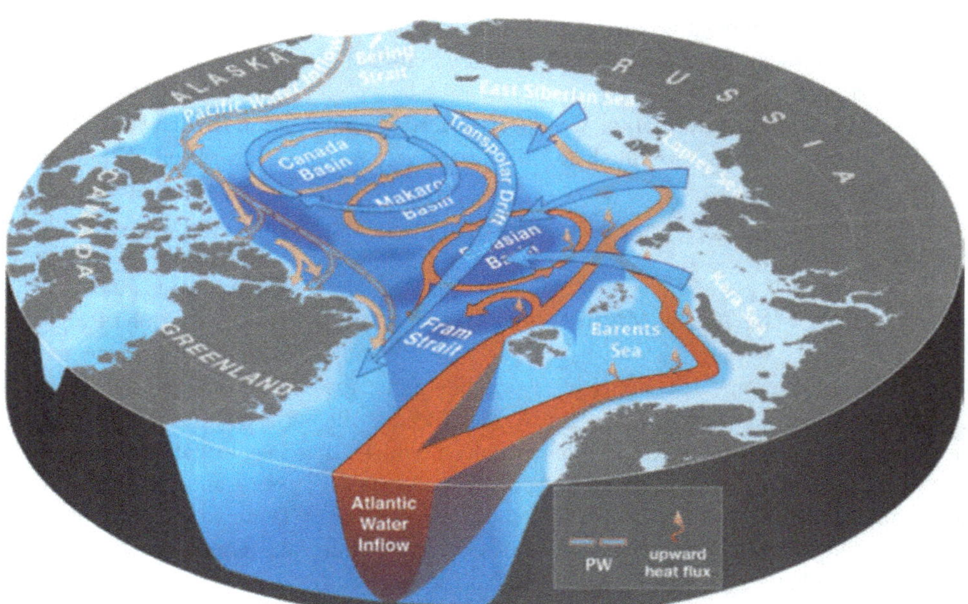

Fig. 4 Schematic of the Arctic Ocean circulation (reproduced from Carmack et al. 2015, American Meteorological Society, used with permission). Blue arrows indicate the surface circulation, pink-blue arrows the main pathways of the Pacific Water at intermediate depths and red arrows show the Atlantic Water circulation. GIN Seas: Greenland-Iceland-Scotland Seas, usually called the "Nordic Seas"

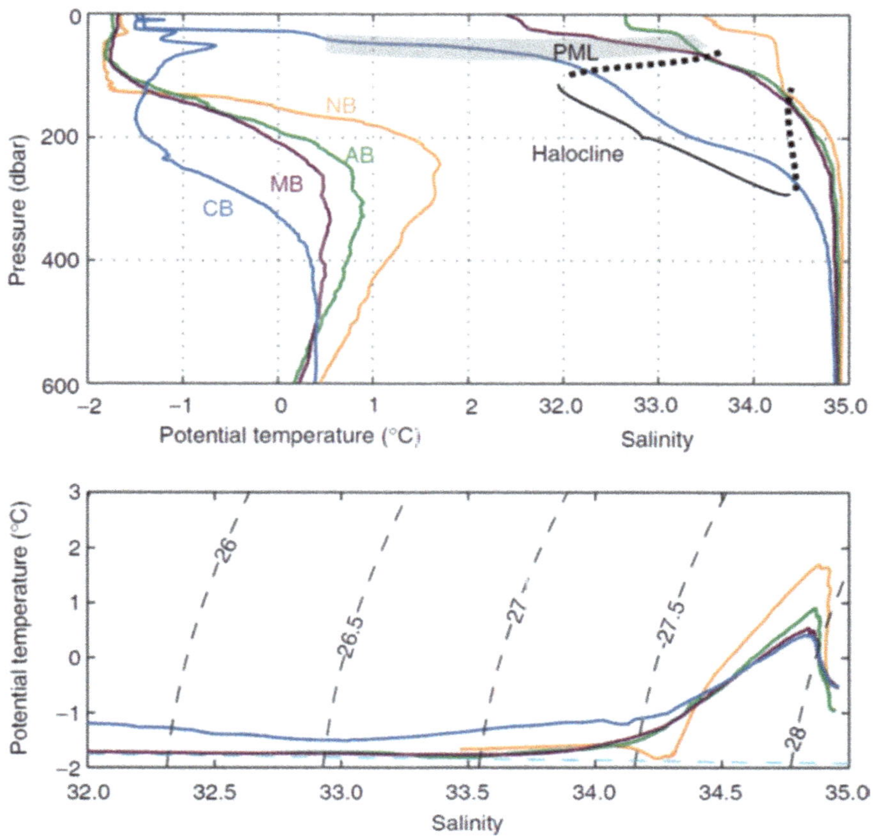

Fig. 5 Upper Arctic Ocean hydrography (reproduced from Rudels 2009, with permission from Elsevier). Potential temperature and salinity of the upper Nansen Basin (NB, orange), Amundsen Basin (AB, green), Makarov Basin (MB, purple), and Canada Basin (CB, blue). *PML* Polar Mixed Layer

the sea ice when reaching it North of Fram Strait and later meets the other cooled AW branch, which flows down the Barents Sea shelf slope into the Nansen Basin. Steered by the topography, the AW flows along the shelf breaks and spreads all over the Arctic Ocean forming counterclockwise circulations in all deep basins (Fig. 4).

For more details on the Arctic Ocean hydrography and circulation see Rudels (2009).

Fresh water

As salt is mainly conserved in the present-day oceans, the only way to change ocean salinity, which determines the stratification in the Arctic Ocean, is by removing or adding fresh water. Therefore, fresh water, both liquid and stored in sea ice, plays a key role in many physical processes in the Arctic. It is of high relevance for local and global climate and the thermohaline circulation, as the fresh surface layer in the Arctic Ocean limits the upward heat transfer from the AW layer to the atmosphere and sea ice and as a freshening of the upper Nordic Seas and subpolar North Atlantic may decrease deep convection due to higher stratification (e.g. Aagard and Carmack 1989; Haak et al. 2003; Häkkinen 1999; Yang et al. 2016).

By freshwater content or oceanic freshwater transport, we understand an equivalent amount of fresh water that is required to dilute water with a reference salinity to obtain the observed salinity. For the Arctic Ocean most studies use a reference salinity of 34.8 as it is approximately the average salinity of the Arctic Ocean (e.g., Aagard and Carmack 1989; Serreze et al. 2006; Holland et al. 2007; Haine et al. 2015). Others choose the reference salinity 35 as it is approximately the salinity of the AW inflow (e.g., Rabe et al. 2011, 2014). With a changing climate and related ocean changes, the estimates of the average Arctic Ocean salinity and AW inflow salinity might need to be adjusted. However, there is a starting discussion amongst scientists on the sensible choice of the reference salinity, challenging the common way to calculate freshwater content and transport (Tsubouchi et al. 2012; Bacon et al. 2015).

More than 100,000 km^3 of fresh water with respect to a reference salinity of 34.8 are stored in the Arctic Ocean (Haine et al. 2015) (Fig. 6). About 95,000 km^3 are stored as liquid freshwater and about 15,000 km^3 in sea ice (Haine et al. 2015). Liquid fresh water is added to the Arctic Ocean by precipitation, continental run-off, glacier/ice sheet/sea ice melt, and PW inflow and is removed by evaporation, sea ice formation and advection to the North Atlantic. All these sources and sinks are affected by the recent climate change

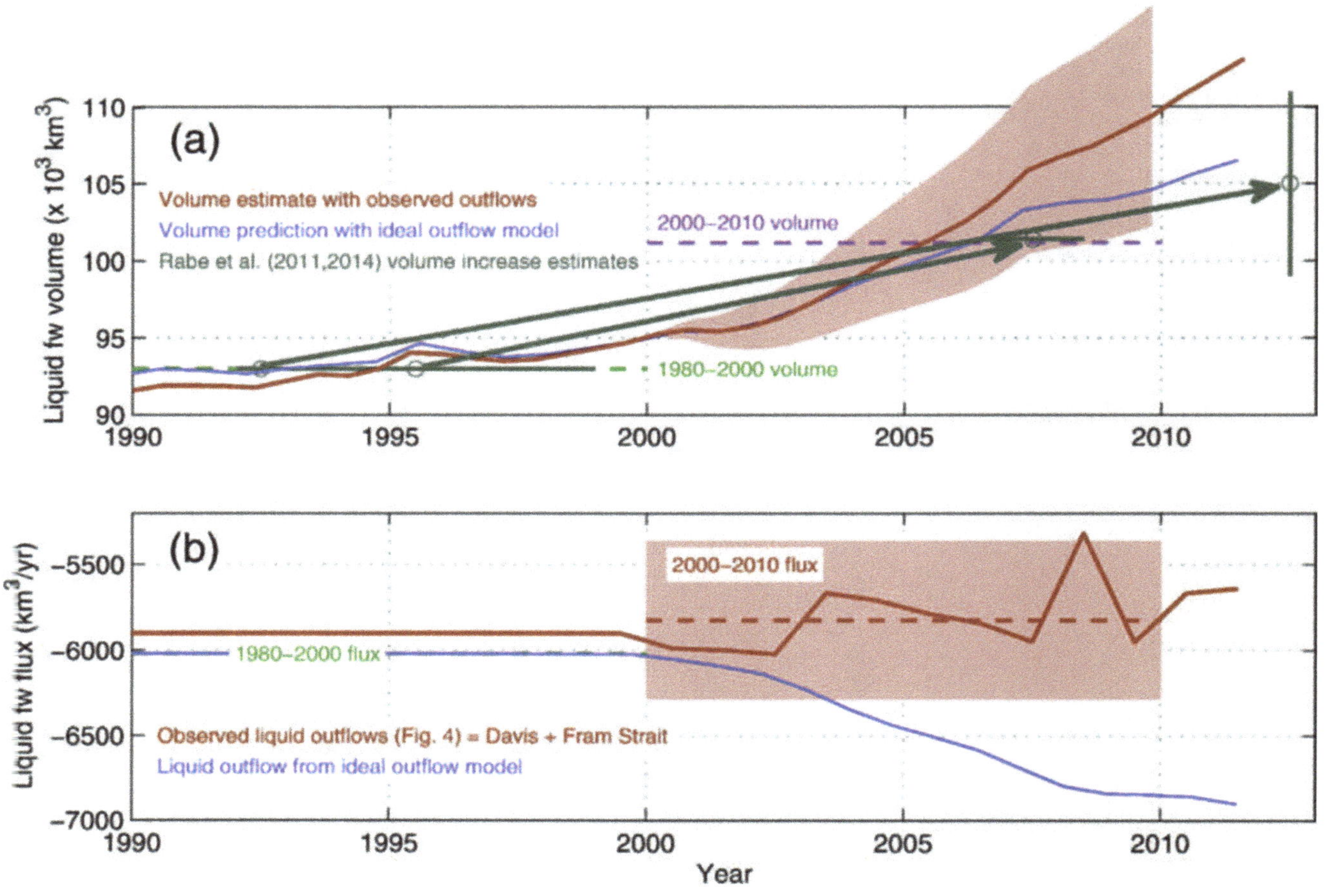

Fig. 6 Arctic freshwater variability (adapted from Haine et al. 2015, with permission from Elsevier). (**a**) Arctic liquid freshwater volume (Arctic Ocean, Canadian Arctic Archipelago and Baffin Bay) and (**b**) export flux (through Davis and Fram Strait) from observations (red) and an ideal outflow model (blue). The arrows indicate estimates from Rabe et al. (2011, 2014)

influencing the freshwater budget of the Arctic Ocean and adjacent seas.

River Runoff and Atmospheric Fluxes

The major source of liquid fresh water in the Arctic Ocean is the continental runoff. The rivers discharge approximately 3300 km^3 year^{-1} fresh water to the upper Arctic Ocean, which accounts for 11% of the total global river discharge (Fichot et al. 2013). Considering that the upper Arctic Ocean only makes up 0.1% of Earth's total ocean volume, this is a remarkable contribution. Net precipitation (precipitation minus evaporation) over the Arctic is estimated to 2200 km^3 year^{-1} (Haine et al. 2015).

Syntheses of Arctic river discharge data revealed an increase of 7–10% in the last 30 to 60 years correlated with the NAO and global mean surface air temperature (Peterson et al. 2002; Overeem and Syvitski 2010). Niederdrenk et al. (2016) showed in a model study that a strong Icelandic low promoting warmer and wetter conditions over Eurasia leads to increased precipitation and thus enhanced river runoff to the Arctic Ocean. This mechanism is proposed to be responsible for the most of the Arctic river runoff variability. Although Déry and Wood (2005) found a 10% decrease in annual river discharge from Canadian rivers into the Arctic Ocean and North Atlantic, the total river discharge into the Arctic Ocean increased by 5.6 km^3 year^{-2} during the second half of the twentieth century (McClelland et al. 2006). There is evidence for an intensification of the global water cycle related to global warming (Huntington 2006) explaining the positive trends in precipitation and continental runoff.

Arctic Glacier and Greenland Ice Sheet Melt

Due to the warming atmosphere and ocean, freshwater fluxes from both, the Greenland Ice Sheet and Arctic glaciers, increased significantly in the last few decades (Yang et al. 2016). The Greenland Ice Sheet mass-loss more than doubled from 2002 to 2009 (Velicogna 2009). Yang et al. (2016) estimated the acceleration of the trend to 20 Gt year^{-2}. Freshwater flux anomalies from surface meltwater and solid ice discharge from 1995 to 2010 sum up to about 3000 km^3 (Bamber et al. 2012). Thereby, the highest freshwater flux with an increase of about 50% was released into the Irminger

Sea and the Labrador Sea (Bamber et al. 2012). This increased freshwater flux is mainly attributed to increased ice discharge from accelerated outlet glaciers in south Greenland (van den Broeke et al. 2009), which might be triggered by the warming of the waters at the glacier ice-ocean interface resulting in increased basal melting (e.g., Holland et al. 2008). Although the freshwater flux is highest in the South, there are also indications in the Northeast of Greenland that warm waters get close to the outlet glaciers and may initiate increasing glacier retreat and associated freshwater fluxes to the ocean (e.g., 79 North Glacier, Schaffer et al. 2017). Also the ice mass loss of glaciers in the Canadian Arctic Archipelago has sharply increased in recent years and almost tripled between 2004 and 2009 (Gardner et al. 2011; Lenaerts et al. 2013).

Oceanic Transport of Sea Ice and Liquid Fresh water

About 2500 km^3 year^{-1} of liquid fresh water (relative to a salinity of 34.8) enter the Arctic Ocean through the Bering Strait, while 3200 km^3 year^{-1} and 2800 km^3 year^{-1} exit the Arctic via the Canadian Arctic Archipelago and Fram Strait, respectively (Serreze et al. 2006; Haine et al. 2015). There are only small amounts of sea ice transported through the Bering Strait (140 km^3 year^{-1}) and the Canadian Arctic Archipelago (160 km^3 year^{-1}), whereas large amounts of sea ice are exported through Fram Strait (1900 km^3 year^{-1} solid freshwater transport) (Serreze et al. 2006; Haine et al. 2015).

Observations presented by Woodgate et al. (2012) showed a slight increase in Bering Strait freshwater flux since 2001 due to increased volume fluxes, which can be explained by changes in the Pacific-Arctic pressure head and local winds. Although the liquid freshwater outflow through the Canadian Arctic Archipelago and the Fram Strait show large interannual variability, there is no significant long-term trend since the beginning of record (Haine et al. 2015). However, a new data record of Fram Strait sea ice area export, which was developed from satellite radar images and surface pressure observations across Fram Strait by Smedsrud et al. (2017), reveals a positive trend of about 5.9% per decade from 1979 to 2014. Ionita et al. (2016) related changes in the simulated Fram Strait sea ice export to atmospheric blocking events over Greenland, which block the winds over the Strait that mainly drive the sea ice transport. These Greenland blocking events are proposed to happen more frequently in recent years due to climate change (e.g., Hanna et al. 2016).

The freshwater export through the various channels of the Canadian Arctic Archipelago varies mainly due to volume flux anomalies governed by variations in the large-scale atmospheric circulation (Jahn et al. 2010a, b; Peterson et al. 2012) or driven by the sea surface height gradient across the strait (e.g., McGeehan and Maslowski 2012; Wekerle et al. 2013). Proshutinsky and Johnson (1997) identified two wind-driven circulation regimes in the Arctic Ocean that either accumulate fresh water in the western Arctic Ocean (anticyclonic) or releases it to the North Atlantic (cyclonic). During anticyclonic circulation regimes, fresh water accumulates in the Beaufort Gyre north of the Canadian Arctic Archipelago due to a wind-driven spin-up as a response to anomalously high sea level pressure over the Arctic (low AO/NAO). During cyclonic regimes (high AO/NAO), the Beaufort Gyre slows down due to cyclonic winds and releases the accumulated fresh water (Proshutinsky et al. 2002; Giles et al. 2012). This fresh water mainly exits the Arctic via the Canadian Arctic Archipelago and partly via the Fram Strait. A tracer study by Jahn et al. (2010a) showed that the main sources of the freshwater export through the Canadian Arctic Archipelago is PW and North American runoff. Although the Arctic Ocean's circulation alternated between the cyclonic and anticyclonic pattern at 5–7 year-intervals in the past, it has remained in an anticyclonic mode for 17 years since 1997 (Proshutinsky et al. 2015). Proshutinsky et al. (2015) speculated that freshwater fluxes from the Greenland Ice Sheet to the North Atlantic interrupted an ocean-atmosphere feedback loop that previously lead to an automatic decadal alternation between cyclonic and anti-cyclonic circulation regimes ("auto-oscillatory system").

The variability of liquid freshwater export through Fram Strait is driven by both, variations in the volume flux and changes in the salinity of the advected waters (e.g., Jahn et al. 2010b). The salinity of the waters exported through Fram Strait depends of the source water, which is mainly Eurasian runoff or PW (Jahn et al. 2010a). During years of an anticyclonic circulation anomaly (low AO) Eurasian runoff is released from the Eurasian Shelf (Jahn et al. 2010a) and directed towards Fram Strait by a strong Transpolar Drift (Morison et al. 2012). During a cyclonic circulation regime (high AO) the Eurasian runoff is kept by a cyclonic circulation in the Eurasian basin (Morison et al. 2012) and PW that is released from the Beaufort Gyre flows along the northern shelf of Greenland and penetrates into Fram Strait (Jahn et al. 2010a). In agreement with this, Karcher et al. (2012) found, from iodine-129 observations and modeling, changing contributions of AW and PW in the Fram Strait outflow to result from changes in the Arctic Ocean circulation as a response to the large-scale atmospheric circulation.

Heat and Volume Fluxes in the Arctic Ocean

It is by now evident that changes in any of the components of the Earth system play a role in determining climate responses over high latitudes and consequent teleconnections. Therefore, it is expected that the interannual variability and recent decline trend of sea ice cover are not

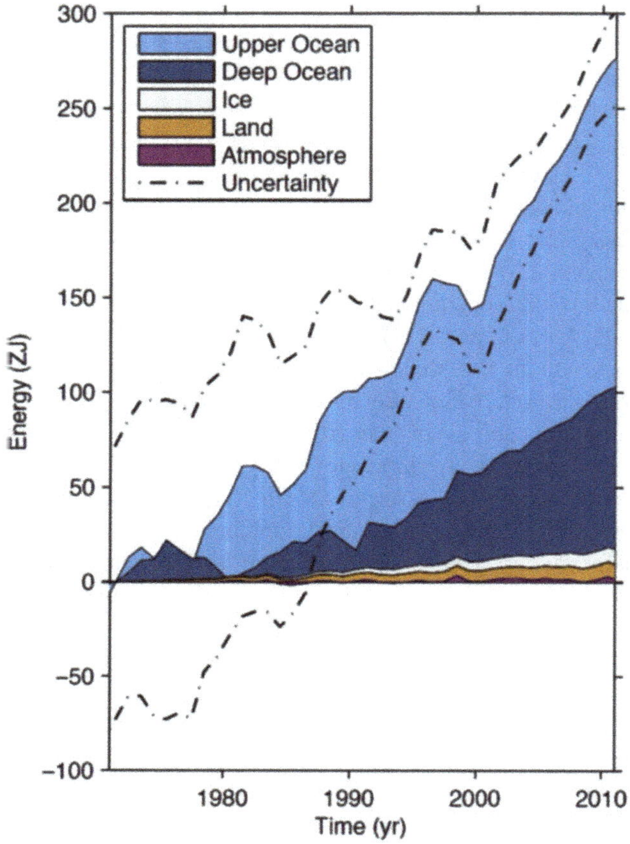

Fig. 7 Stored energy in the Earth climate system (modified from IPCC 2014). Discretisation of energy accumulation change in zeta joules (ZJ) in each component of the climate system relative to 1971 for periods as given. Ocean: upper ocean heat change (from surface to 700 m) in light blue; deep ocean (below 700 m, including below 2000 m estimates starting from 1992) in dark blue. Ice melt: glaciers and ice caps (light grey), Greenland and Antarctic ice sheet estimates starting from 1992, and Arctic sea ice estimate from 1979 to 2008. Land: continental warming (orange). Atmosphere: warming estimate starting from 1979 (purple). Uncertainty in the ocean estimate also dominates the total uncertainty (dot-dashed lines indicate the uncertainty from all five components at 90% confidence intervals)

only controlled by atmospheric heating and cooling, but also largely by heat flux from the underlying ocean. The oceans are the largest thermal reservoir of the Earth's climate system (Fig. 7). According to recent estimates, around 90% of the warming of the Earth's system over the last century has been stored in the oceans (e.g., Levitus et al. 2012; Riser et al. 2016). The biggest share of this amount is kept trapped in the upper ocean, hence being potentially available for atmosphere warming and sea ice melt. Thus, small changes in the pathways and amounts of heat carried and stored by the ocean currents could have a significant effect on present and future changes in Arctic coupled ocean-ice-atmosphere system. For an in-depth review and speculation on this topic see Carmack et al. (2015).

Shortwave radiation and sensible heat fluxes are the sources of net heat input in the upper ocean (Itoh et al. 2011). Most of the heat input to the upper ocean in summer is given off in autumn and winter as longwave radiation and turbulent sensible heat fluxes to the atmosphere cool the open water and the ice/snow surfaces (e.g., Serreze et al. 2009). Still, important observations indicated that a surplus of heat is preserved in the ocean during winter, potentially hindering the seasonal growth of sea ice (e.g., Timmermans 2015).

Another source of heat into the Artic system is the relatively warm AW carried through the main gateways connecting the Arctic Ocean with the Atlantic Ocean: the Fram Strait and Barents Sea. There is also some exchange with the Pacific Ocean, but, according to up-to-date estimates reviewed in Beszczynska-Möller et al. (2012), the net volume and heat fluxes flowing from the Pacific Ocean into the Arctic through the Bering Strait are small, particularly when compared to those through Barents Sea (0.8 Sv and 10* to 20* TW, against 2.0 Sv and 50** to 70*** TW, respectively[1]). In fact, the inflow of AW to the Barents Sea accounts for about half of the northward heat transport to the Arctic Ocean and the Barents Sea combined (Smedsrud et al. 2013). Notwithstanding, the heat carried by the AW into the Barents sea is lost to the atmosphere as latent, sensible, and long wave radiation (Smedsrud et al. 2010).

Atlantic Water

Recent observations report a warming of the North Atlantic Ocean resulting in trackable changes in the Arctic Ocean. Since the 1990's, temperature and salinity have rapidly increased from the eastern North Atlantic subpolar gyre branch to the Fram Strait (Holliday et al. 2008). Furthermore, during the last decade, a net temperature increase of the incoming AW in the Fram Strait of 1 °C has been reported (Schauer et al. 2008; Polyakov et al. 2013). Over the period from 1997 to 2006, an increase of 1 °C was described for AW entering the Barents Sea (Beszczynska-Möller et al. 2012).

The warming of the AW is accompanied by estimates of volume transport increase into the Barents Sea, setting the so-called "Atlantification" of the basin (Årthun et al. 2012; Oziel et al. 2016; Smedsrud et al. 2013). Oziel et al. (2016) suggested that the increased inflow of AW into the Barents Sea would also cause the enhancement of the outflow of the dense modified AW mode (also called Barents Sea Water) into the intermediate layers of the Arctic. The reason to this is that more warm waters could initiate a sea ice melt-freeze

[1]Estimation of heat transport is dependent on a chosen reference temperature. For thorough understanding we refer to Schauer and Beszczynska-Möller (2009). Here:

* referenced to freezing temperature
** referenced to 0 °C
*** heat flux for closed volume budget

loop assuring a constant mixing and sinking of denser water. On the contrary, Rudels et al. (2013) state that under a regime of stronger inflow of the warm AW, no cooling to freezing temperature would occur. In this case, no brine-induced convection would occur, thus modulating the production of less dense Barents Sea Water. In either case, changes in the cooling and mixing of AW in the Barents Sea could impact the ventilation of the intermediate layer inside the Arctic, since 50–80% of the water occupying this layer is influenced by water mass originated on the Barents Sea basin (Schauer et al. 2002).

It has been hypothesized that the inflow of warm AW into the Arctic Ocean has a considerable influence on the decline and variability of sea ice extent and thickness (Årthun et al. 2012; Smedsrud et al. 2013; Carmack et al. 2015; Long and Perrie 2015). Roughly 20% of the total trend in sea ice volume loss since 2004 is related to observed episodes of AW warming (Ivanov et al. 2012).

In their study *"Is weaker sea ice changing the Atlantic Water circulation?"*, Itkin et al. (2014) try to understand and predict the effects of recent loss of sea ice on the overall intermediate circulation. It is presented that a thinner sea ice cover offers less hindrance to momentum transfer to the ocean, thus allowing a spin up of the Arctic circulation. The strengthening of the surface anticyclonic circulation results in the accumulation of water in the interior of the Beaufort Gyre, as a consequence of the enhanced Ekman convergence (Deser et al. 2014; Itkin et al. 2014; Long and Perrie 2015). Later, the stored water adds up to an increased outflow into the Atlantic Ocean, and is compensated by the increased inflow through the Barents Sea.

The Atlantification of the Arctic Ocean has been recently shown to extend even further into the Arctic Ocean as sea ice-ocean-atmosphere interactions change; at the same time the usual vertical stability of the Arctic Ocean is decreasing as the warm waters reach further upward and more heat is made available for inducing further melting of sea ice (Polyakov et al. 2017).

Outlook

Along with pronounced atmospheric warming, the summer Arctic sea ice has been projected to disappear by the second half of the twenty-first century (Overland and Wang 2013). How this will affect the atmosphere-sea ice-ocean interactions and influence the weather and variability in and beyond the Arctic boundaries has been more and more under discussion (Jung et al. 2014; Suo et al. 2017).

Model simulations predict that the hydrological cycle will further intensify and thereby increase the Arctic freshwater content, as well as the liquid freshwater export (e.g., Arzel et al. 2008; Holland et al. 2007). At the same time, the sea ice volume will further shrink, which possibly results in a decreasing sea ice export as projected by Arzel et al. (2008).

Accelerating melt of the Greenland Ice Sheet and Arctic glaciers due to increasing air temperature provides even more freshwater to the North Atlantic with its deep water formation sites (Yang et al. 2016) and contributes to the projected increase in sea level rise (Rahmstorf 2007). Predictions on the impact of the increasing freshwater input to the North Atlantic are very diverse and range from almost no impact to a complete shutdown of the Atlantic Meridional Overturning Circulation (AMOC). Böning et al. (2016) argue from their model simulation that the accumulation of Greenland Ice Sheet melt water has not been large enough yet to have a significant impact on the AMOC. Though, they found that an accumulated freshwater runoff of about 20,000 km^3, that could be reached by 2040 considering the current observed trend in runoff, would slowdown the AMOC by more than 5 Sv. Liu et al. (2017) even projected a complete collapse of the AMOC 300 years after an abrupt doubling of the atmospheric CO_2 concentration from the 1990 level. However, Behrens et al. (2013) showed how sensitive model simulations are to small variations in precipitation and the choice of sea surface salinity restoring to climatological values. From these small variations they found in their model simulation a variation of an accelerated AMOC of ~22 Sv to a nearly-collapsed state of ~6 Sv. Swingedouw et al. (2009) showed with their model experiment that the AMOC response to the freshwater input is not linear and depends on the mean state of the climate.

Changes in ocean characteristics and sea ice cover can have significant influence on the biogeochemical feedbacks and marine ecosystem (Bates and Mathis 2009; Johannessen and Miles 2011). The reduction of the sea ice cover allows for more light to reach and warm the ocean surface, promoting an increase in primary production (Slagstad et al. 2015). This, in turn, is intrinsically related to an expected increase CO_2 uptake by the ocean (Bates and Mathis 2009). Still, some controversy exists with indications that the ocean may soon enough saturate, and loses its CO_2 uptake capacity (Cai et al. 2010).

Indeed, fundamental questions on the pathways and time scales in which the ocean drives and responds to changes in the coupled system still remain unresolved, hence still demanding much effort in order to better understand the role of the northern seas in the context of a changing climate.

Appendix

This article is related to the YOUMARES 8 conference session no. 9: "The Physics of the Arctic and Subarctic Oceans in a Changing Climate". The original Call for Abstracts and the abstracts of the presentations within this session can be found in the appendix "Conference Sessions and Abstracts", chapter "2 The Physics of the Arctic and Subarctic Oceans in a Changing Climate", of this book.

References

Aagard K, Carmack EC (1989) The role of sea ice and other fresh water in the Arctic circulation. J Geophys Res 94:14485–14498. https://doi.org/10.1029/JC094iC10p14485

ACIA (Arctic Climate Impact Assessment) (2004) Impacts of a warming arctic. Cambridge University Press, Cambridge

AMAP (Arctic Monitoring and Assessment Programme) (2012) Changes in Arctic snow, water, ice and permafrost. SWIPA 2011 Overview Report. Arctic Climate Issues 2011

Årthun M, Eldevik T, Smedsrud LH et al (2012) Quantifying the influence of Atlantic heat on Barents Sea ice variability and retreat. J Clim 25:4736–4743. https://doi.org/10.1175/JCLI-D-11-00466.1

Arzel O, Fichefet T, Goosse H et al (2008) Causes and impacts of changes in the Arctic freshwater budget during the twentieth and twenty-first centuries in an AOGCM. Clim Dyn 30:37–58. https://doi.org/10.1007/s00382-007-0258-5

Bacon S, Aksenov Y, Fawcett S et al (2015) Arctic mass, freshwater and heat fluxes: methods and modelled seasonal variability. Phil Trans R Soc A 373:20140169. https://doi.org/10.1098/rsta.2014.0169

Bader J, Mesquita MDS, Hodges KI et al (2011) A review on Northern Hemisphere sea-ice, storminess and the North Atlantic oscillation: observations and projected changes. Atmos Res 101:809–834. https://doi.org/10.1016/j.atmosres.2011.04.007

Bamber J, Van Den Broeke M, Ettema J et al (2012) Recent large increases in freshwater fluxes from Greenland into the North Atlantic. Geophys Res Lett 39:8–11. https://doi.org/10.1029/2012GL052552

Barnes EA, Screen JA (2015) The impact of Arctic warming on the midlatitude jet-stream: can it? has it? Will it? WIREs Clim Chang 6:277–286. https://doi.org/10.1002/wcc.337

Bates NR, Mathis JT (2009) The Arctic Ocean marine carbon cycle: evaluation of air-sea CO_2 exchanges, ocean acidification impacts and potential feedbacks. Biogeosciences 6:2433–2459

Behrens E, Biastoch A, Böning CW (2013) Spurious AMOC trends in global ocean sea-ice models related to subarctic freshwater forcing. Ocean Model 69:39–49. https://doi.org/10.1016/j.ocemod.2013.05.004

Beszczynska-Möller A, Fahrbach E, Schauer U et al (2012) Variability in Atlantic water temperature and transport at the entrance to the Arctic Ocean, 1997–2010. ICES J Mar Sci 69:852–863. https://doi.org/10.1093/icesjms/fss056

Böning CW, Behrens E, Biastoch A et al (2016) Emerging impact of Greenland meltwater on deepwater formation in the North Atlantic Ocean. Nat Geosci 9:523–527. https://doi.org/10.1038/ngeo2740

Bönisch G, Schlosser P (1995) Deep water formation and exchange rates in the Greenland/ Norwegian Seas and the Eurasian Basin of the Arctic Ocean derived from tracer balances. Prog Oceanogr 35:29–52

Boyer T, Levitus S, Antonov J et al (2007) Changes in freshwater content in the North Atlantic Ocean 1955–2006. Geophys Res Lett 34:L16603. https://doi.org/10.1029/2007GL030126

Budikova D (2009) Role of Arctic Sea ice in global atmospheric circulation: a review. Glob Planet Chang 68:149–163. https://doi.org/10.1016/j.gloplacha.2009.04.001

Cai W-J, Chen L, Chen B et al (2010) Decrease in the CO_2 uptake capacity in an ice-free Arctic Ocean Basin. Science 329:556–559. https://doi.org/10.1126/science.1189338

Carmack E, Polyakov IV, Padman L et al (2015) The increaing role of oceanic heat in sea ice loss in the new Arctic. Am Meteorol Soc 96:2079–2106. https://doi.org/10.1175/BAMS-D-13-00177.1

Cohen J, Screen JA, Furtado JC et al (2014) Recent Arctic amplification and extreme mid-latitude weather. Nat Geosci 7:627–637. https://doi.org/10.1038/ngeo2234

Comiso JC, Parkinson CL, Gersten R et al (2008) Accelerated decline in the Arctic Sea ice cover. Geophys Res Lett 35:L01703. https://doi.org/10.1029/2007GL031972

Curry R, Mauritzen C (2005) Dilution of the northern North Atlantic Ocean in recent decades. Science 308:1772–1774. https://doi.org/10.1126/science.1109477

Déry SJ, Wood EF (2005) Decreasing river discharge in northern Canada. Geophys Res Lett 32:L10401. https://doi.org/10.1029/2005GL022845

Deser C, Tomas R, Sun L (2014) The role of ocean – atmosphere coupling in the zonal-mean atmospheric response to Arctic Sea ice loss. J Clim 28:2168–2186. https://doi.org/10.1175/JCLI-D-14-00325.1

Fetterer F, Knowles K, Meier W et al (2016, updated daily) Sea Ice Index, Version 2. Boulder, Colorado USA. NSIDC: National Snow and Ice Data Center. https://doi.org/10.7265/N5736NV7

Fichot CG, Kaiser K, Hooker SB et al (2013) Pan-Arctic distributions of continental runoff in the Arctic Ocean. Sci Rep 3:1053. https://doi.org/10.1038/srep01053

Gardner AS, Moholdt G, Wouters B et al (2011) Sharply increased mass loss from glaciers and ice caps in the Canadian Arctic Archipelago. Nature 473:357–360. https://doi.org/10.1038/nature10089

Giles KA, Laxon SW, Ridout AL et al (2012) Western Arctic Ocean freshwater storage increased by wind-driven spin-up of the Beaufort Gyre. Nat Geosci 5:194–197. https://doi.org/10.1038/ngeo1379

Haak H, Jungclaus J, Mikolajevicz U et al (2003) Formation and propagation of great salinity anomalies. Geophys Res Lett 30:1473. https://doi.org/10.1029/2003GL017065

Haine TWN, Curry B, Gerdes R et al (2015) Arctic freshwater export: status, mechanisms, and prospects. Glob Planet Chang 125:13–35. https://doi.org/10.1016/j.gloplacha.2014.11.013

Häkkinen S (1999) A simulation of Thermohaline effects of a great salinity anomaly. J Clim 12:1781–1795

Hanna E, Cropper TE, Hall RJ et al (2016) Greenland blocking index 1851–2015: a regional climate change signal. Int J Climatol 36:4847–4861. https://doi.org/10.1002/joc.4673

Holland MM, Finnis J, Barrett AP et al (2007) Projected changes in Arctic Ocean freshwater budgets. J Geophys Res 112:G04S55. https://doi.org/10.1029/2006JG000354

Holland PR, Jenkins A, Holland DM (2008) The response of ice shelf basal melting to variations in ocean temperature. J Clim 21:2558–2572. https://doi.org/10.1175/2007JCLI1909.1

Holliday NP, Hughes SL, Bacon S et al (2008) Reversal of the 1960s to 1990s freshening trend in the Northeast North Atlantic and Nordic Seas. Geophys Res Lett 35:L03614. https://doi.org/10.1029/2007GL032675

Huntington TG (2006) Evidence for intensification of the global water cycle: review and synthesis. J Hydrol 319:83–95. https://doi.org/10.1016/j.jhydrol.2005.07.003

Ionita M, Scholz P, Lohmann G et al (2016) Linkages between atmospheric blocking, sea ice export through Fram Strait and the Atlantic Meridional overturning circulation. Sci Rep 6:32881. https://doi.org/10.1038/srep32881

IPCC (2014) Climate change 2014: synthesis report. Contribution of working groups I, II and III to the fifth assessment report of the intergovernmental panel on climate change. IPCC, Geneva

Itkin P, Kracher M, Gerdes R (2014) Is weaker Arctic Sea ice changing the Atlantic water circulation? J Geophys Res Oceans 119:5992–6009. https://doi.org/10.1002/2013JC009633

Itoh M, Inoue J, Shimada K et al (2011) Acceleration of sea-ice melting due to transmission of solar radiation through ponded ice area in the Arctic Ocean: results of in situ observations from icebreakers in 2006 and 2007. Ann Glaciol 52:249–260

Ivanov VV, Alexeev VA, Repina I et al (2012) Tracing Atlantic water signature in the Arctic Sea ice cover East of Svalbard. Adv Meteorol 2012:201818. https://doi.org/10.1155/2012/201818

Jahn A, Tremblay B, Mysak LA et al (2010a) Effect of the large-scale atmospheric circulation on the variability of the Arctic Ocean freshwater export. Clim Dyn 34:201–222. https://doi.org/10.1007/s00382-009-0558-z

Jahn A, Tremblay LB, Newton R et al (2010b) A tracer study of the Arctic Ocean's liquid freshwater export variability. J Geophys Res 115:C07015. https://doi.org/10.1029/2009JC005873

Johannessen OM, Miles MW (2011) Critical vulnerabilities of marine and sea ice – based ecosystems in the high Arctic. Reg Environ Chang 11:S239–S248. https://doi.org/10.1007/s10113-010-0186-5

Jung T, Kasper MA, Semmler T et al (2014) Arctic influence on subseasonal midlatitude prediction. Geophys Res Lett 41:3676–3680. https://doi.org/10.1002/2014GL059961.1

Karcher M, Smith JN, Kauker F et al (2012) Recent changes in Arctic Ocean circulation revealed by iodine-129 observations and modeling. J Geophys Res 117:C08007. https://doi.org/10.1029/2011JC007513

Koenigk T, Mikolajewicz U, Haak H et al (2007) Arctic freshwater export in the 20th and 21st centuries. J Geophys Res 112:G04S41. https://doi.org/10.1029/2006JG000274

Lang A, Yang S, Kaas E (2017) Sea ice thickness and recent Arctic warming. Geophys Res Lett 44:409–418. https://doi.org/10.1002/2016GL071274

Lenaerts JTM, Van Angelen JH, Van Den Broeke MR et al (2013) Irreversible mass loss of Canadian Arctic archipelago glaciers. Geophys Res Lett 40:870–874. https://doi.org/10.1002/grl.50214

Levitus S, Antonov JI, Boyer TP et al (2012) World Ocean heat content and thermosteric sea level change (0–2000 m), 1955–2010. Geophys Res Lett 39:L10603. https://doi.org/10.1029/2012GL051106

Lindsay R, Schweiger A (2015) Arctic Sea ice thickness loss determined using subsurface, aircraft, and satellite observations. Cryosphere 9:269–283. https://doi.org/10.5194/tc-9-269-2015

Liu W, Xie S, Liu Z et al (2017) Overlooked possibility of a collapsed Atlantic Meridional overturning circulation in warming climate. Sci Adv 3:e1601666

Long Z, Perrie W (2015) Scenario changes of Atlantic water in the Arctic Ocean. J Clim 28:5523–5548. https://doi.org/10.1175/JCLI-D-14-00522.1

Mauritzen C, Melsom A, Sutton RT (2012) Importance of density-compensated temperature change for deep North Atlantic Ocean heat uptake. Nat Geosci 5:905–910. https://doi.org/10.1038/ngeo1639

McBean G, Alekseev G, Chen D et al (2005) Arctic climate: past and present. In: Arctic climate impact assessment. Cambridge University Press, Cambridge, pp 22–60

McClelland JW, Déry SJ, Peterson BJ et al (2006) A pan-arctic evaluation of changes in river discharge during the latter half of the 20th century. Geophys Res Lett 33:L06715. https://doi.org/10.1029/2006GL025753

McGeehan T, Maslowski W (2012) Evaluation and control mechanisms of volume and freshwater export through the Canadian Arctic Archipelago in a high-resolution pan-Arctic ice-ocean model. J Geophys Res 117:C00D14. https://doi.org/10.1029/2011JC007261

Morison J, Kwok R, Peralta-Ferriz C et al (2012) Changing Arctic Ocean freshwater pathways. Nature 481:66–70. https://doi.org/10.1038/nature10705

Niederdrenk AL, Sein DV, Mikolajewicz U (2016) Interannual variability of the Arctic freshwater cycle in the second half of the twentieth century in a regionally coupled climate model. Clim Dyn 47:3883–3900. https://doi.org/10.1007/s00382-016-3047-1

Onarheim IH, Smedsrud LH, Ingvaldsen RB et al (2014) Loss of sea ice during winter north of Svalbard. Tellus A 66:23933. https://doi.org/10.3402/tellusa.v66.23933

Overeem I, Syvitski JPM (2010) Shifting discharge peaks in arctic rivers, 1977–2007. Geogr Ann 92A:285–296

Overland JE (2016) A difficult Arctic science issue: midlatitude weather linkages. Pol Sci 10:210–216. https://doi.org/10.1016/j.polar.2016.04.011

Overland JE, Wang M (2013) When will the summer Arctic be nearly sea ice free? Geophys Res Lett 40:2097–2101. https://doi.org/10.1002/grl.50316

Oziel L, Sirven J, Gascard J (2016) The Barents Sea frontal zones and water masses variability (1980–2011). Ocean Sci 12:169–184. https://doi.org/10.5194/os-12-169-2016

Peterson BJ, Peterson BJ, Holmes RM et al (2002) Increasing river discharge to the Arctic Ocean Science 298:2171–2173. https://doi.org/10.1126/science.1077445

Peterson I, Hamilton J, Prinsenberg S et al (2012) Wind-forcing of volume transport through Lancaster sound. J Geophys Res 117:C11018. https://doi.org/10.1029/2012JC008140

Pistone K, Eisenman I, Ramanathan V (2014) Observational determination of albedo decrease caused by vanishing Arctic Sea ice. Proc Natl Acad Sci U S A 111:3322–3326. https://doi.org/10.1073/pnas.1318201111

Pithan F, Mauritsen T (2014) Arctic amplification dominated by temperature feedbacks in contemporary climate models. Nat Geosci 7:181–184. https://doi.org/10.1038/NGEO2071

Polyakov IV, Johnson A (2000) Arctic decadal and interdecadal variability. Geophys Res Lett 27:4097–4100

Polyakov IV, Timokhov LA, Alexeev VA et al (2010) Arctic Ocean warming contributes to reduced polar ice cap. J Phys Oceanogr 40:2743–2756. https://doi.org/10.1175/2010JPO4339.1

Polyakov IV, Walsh JE, Kwok R (2012) Recent changes of Arctic multiyear sea ice coverage and the likely causes. Bull Am Meteorol Soc 93:145–151. https://doi.org/10.1175/BAMS-D-11-00070.1

Polyakov IV, Pnyushkov AV, Rember R et al (2013) Winter convection transports Atlantic water heat to the surface layer in the Eastern Arctic Ocean. J Phys Oceanogr 43:2142–2155. https://doi.org/10.1175/JPO-D-12-0169.1

Polyakov IV, Pnyushkov AV, Alkire MB et al (2017) Greater role for Atlantic inflows on sea-ice loss in the Eurasian Basin of the Arctic Ocean. Science 356:285–291

Proshutinsky AY, Johnson MA (1997) Two circulation regimes of the wind-driven Arctic Ocean between anticyclonic Gudkovich and Nikiforov with a wind-driven hydraulic. J Geophys Res 102:493–514

Proshutinsky A, Bourke RH, Mclaughlin FA (2002) The role of the Beaufort Gyre in Arctic climate variability: seasonal to decadal climate scales. Geophys Res Lett 29:2100. https://doi.org/10.1029/2002GL015847

Proshutinsky A, Krishfield R, Timmermans M et al (2009) Beaufort Gyre freshwater reservoir: state and variability from observations. J Geophys Res 114:C00A10. https://doi.org/10.1029/2008JC005104

Proshutinsky A, Dukhovskoy D, Timmermans M et al (2015) Arctic circulation regimes. Phil Trans R Soc A 373:20140160

Rabe B, Karcher M, Schauer U et al (2011) An assessment of Arctic Ocean freshwater content changes from the 1990s to the 2006–2008 period. Deep Res I 58:173–185. https://doi.org/10.1016/j.dsr.2010.12.002

Rabe B, Karcher M, Kauker F et al (2014) Arctic Ocean basin liquid freshwater storage trend 1992–2012. Geophys Res Lett 41:961–968. https://doi.org/10.1002/2013GL058121

Rahmstorf S (2007) Projecting future sea-level rise. Science 315:368–371

Rennermalm AK, Wood EF, Weaver AJ et al (2007) Relative sensitivity of the Atlantic meridional overturning circulation to river discharge into Hudson Bay and the Arctic Ocean. J Geophys Res 112:G04S48. https://doi.org/10.1029/2006JG000330

Rhein M, Kieke D, Hüttl-Kabus S et al (2011) Deep water formation, the subpolar gyre, and the meridional overturning circulation in the subpolar North Atlantic. Deep Res II 58:1819–1832. https://doi.org/10.1016/j.dsr2.2010.10.061

Riser SC, Freeland HJ, Roemmich D et al (2016) Fifteen years of ocean observations with the global Argo array. Nat Clim Chang 6:145–153. https://doi.org/10.1038/nclimate2872

Rothrock DA, Percival DB, Wensnahan M (2008) The decline in Arctic Sea-ice thickness: separating the spatial, annual, and interannual variability in a quarter century of submarine data. J Geophys Res 113:C05003. https://doi.org/10.1029/2007JC004252

Rudels B (2009) Arctic Ocean circulation. In: Steele JH, Turekian KK, Thorpe SA (eds) Encyclopedia of ocean sciences, 2nd edn. Academic, San Diego, pp 211–225

Rudels B, Schauer U, Björk G et al (2013) Observations of water masses and circulation with focus on the Eurasian Basin of the Arctic Ocean from the 1990s to the late 2000s. Ocean Sci 9:147–169. https://doi.org/10.5194/os-9-147-2013

Schaffer J, von Appen W-J, Dodd PA et al (2017) Warm water pathways toward Nioghalvfjerdsfjorden Glacier, Northeast Greenland. J Geophys Res Ocean 122:4004–4020. https://doi.org/10.1002/2016JC012264.Received

Schauer U, Beszczynska-Möller A (2009) Problems with estimation and interpretation of oceanic heat transport – conceptual remarks for the case of Fram Strait in the Arctic Ocean. Ocean Sci 5:487–494. https://doi.org/10.5194/os-5-487-2009

Schauer U, Loeng H, Rudels B et al (2002) Atlantic water flow through the Barents and Kara Seas. Deep Sea Res I 49:2281–2298

Schauer U, Beszczynska-Möller A, Walczowski W et al (2008) Variation of measured heat flow through the Fram Strait between 1997 and 2006. In: Dickson B, Meincke J, Rhines P (eds) Arctic-subarctic ocean fluxes. Springer, Dordrecht, pp 65–85

Semmler T, Stulic L, Jung T et al (2016) Seasonal atmospheric responses to reduced Arctic Sea ice in an ensemble of coupled model simulations. J Clim 29:5893–5913. https://doi.org/10.1175/JCLI-D-15-0586.1

Serreze MC, Barry RG (2011) Processes and impacts of Arctic amplification: a research synthesis. Glob Planet Chang 77:85–96. https://doi.org/10.1016/j.gloplacha.2011.03.004

Serreze MC, Barry RG (2014) The Arctic climate system, 2nd edn. Cambridge University Press, Cambridge

Serreze MC, Francis JA (2006) The arctic amplification debate. Clim Chang 76:241–264. https://doi.org/10.1007/s10584-005-9017-y

Serreze MC, Barrett AP, Slater AG et al (2006) The large-scale freshwater cycle of the Arctic. J Geophys Res 111:C11010. https://doi.org/10.1029/2005JC003424

Serreze MC, Barrett AP, Slater AG et al (2007) The large-scale energy budget of the Arctic. J Geophys Res 112:D11122. https://doi.org/10.1029/2006JD008230

Serreze MC, Barrett AP, Stroeve JC et al (2009) The emergence of surface-based Arctic amplification. Cryosphere 3:11–19

Slagstad D, Wassmann PFJ, Ellingsen I (2015) Physical constrains and productivity in the future Arctic Ocean. Front Mar Sci 2:85. https://doi.org/10.3389/fmars.2015.00085

Smedsrud LH, Ingvaldsen R, Environmental N et al (2010) Heat in the Barents Sea: transport, storage, and surface fluxes. Ocean Sci 6:219–234

Smedsrud LH, Esau I, Ingvaldsen RB et al (2013) The role of the Barents Sea in the Arctic climate system. Rev Geophys 51:415–449. https://doi.org/10.1002/rog.20017.1.INTRODUCTION

Smedsrud LH, Halvorsen MH, Stroeve JC et al (2017) Fram Strait sea ice export variability and September Arctic Sea ice extent over the last 80 years. Cryosphere 11:65–79. https://doi.org/10.5194/tc-11-65-2017

Stroeve J, Holland MM, Meier W et al (2007) Arctic Sea ice decline: faster than forecast. Geophys Res Lett 34:L09501. https://doi.org/10.1029/2007GL029703

Stroeve JC, Serreze MC, Holland MM et al (2012) The Arctic's rapidly shrinking sea ice cover: a research synthesis. Clim Chang 110:1005–1027. https://doi.org/10.1007/s10584-011-0101-1

Suo L, Gao Y, Guo D et al (2017) Sea-ice free Arctic contributes to the projected warming minimum in the North Atlantic. Environ Res Lett 12:074004. https://doi.org/10.1088/1748-9326/aa6a5e

Swingedouw D, Mignot J, Braconnot P et al (2009) Impact of freshwater release in the North Atlantic under different climate conditions in an OAGCM. J Clim 22:6377–6403. https://doi.org/10.1175/2009JCLI3028.1

Talley LD, Pickard GL, Emery WJ et al (2011) Descriptive physical oceanography: an introduction, 6th edn. Elsevier, Amsterdam

Taylor FW (2009) Elementary climate physics. Oxford University Press, Oxford

Taylor PC, Cai M, Hu A et al (2013) A decomposition of feedback contributions to polar warming amplification. J Clim 26:7023–7043. https://doi.org/10.1175/JCLI-D-12-00696.1

Thomas DN, Dieckmann GS (2010) Sea ice, 2nd edn. Wiley-Blackwell, Oxford

Thompson DWJ, Wallace JM (1998) The Arctic oscillation signature in the wintertime geopotential height and temperature fields. Geophys Res Lett 25:1297–1300

Thyng BKM, Greene CA, Hetland RD et al (2016) True colors of oceanography: guidelines for effective and accurate colormap selection. Oceanography 29:9–13

Timmermans M-L (2015) The impact of stored solar heat on Arctic Sea ice growth. Geophys Res Lett 42:6399–6406. https://doi.org/10.1002/2015GL064541.Abstract

Tomczak M, Godfrey JS (1994) Regional oceanography: an introduction, 1st edn. Elsevier, Amsterdam

Tsubouchi T, Bacon S, Garabato ACN et al (2012) The Arctic Ocean in summer: a quasi-synoptic inverse estimate of boundary fluxes and water mass transformation. J Geophys Res 117:C01024. https://doi.org/10.1029/2011JC007174

Turner J, Marshall GJ (2011) Climate change in the polar regions. Cambridge University Press, Cambridge

van den Broeke MR, Bamber JL, Ettema J et al (2009) Partitioning recent greenland mass loss. Science 326:984–986

Velicogna I (2009) Increasing rates of ice mass loss from the Greenland and Antarctic ice sheets revealed by GRACE. Geophys Res Lett 36:L19503. https://doi.org/10.1029/2009GL040222

Vihma T (2014) Effects of Arctic Sea ice decline on weather and climate: a review. Surv Geophys 35:1175–1214. https://doi.org/10.1007/s10712-014-9284-0

Wang M, Overland JE (2009) A sea ice free summer Arctic within 30 years? Geophys Res Lett 36:L07502. https://doi.org/10.1029/2009GL037820

Wekerle C, Wang Q, Danilov S et al (2013) The Canadian Arctic Archipelago throughflow in a multiresolution global model: model assessment and the driving mechanism of interannual variability. J Geophys Res 118:4525–4541. https://doi.org/10.1002/jgrc.20330

Winton M (2008) Sea Ice – Albedo feedback and nonlinear Arctic climate change. In: ET DW, Bitz CM, Tremblay L-B (eds) Arctic

Sea ice decline: observations, projections, mechanisms, and implications. American Geophysical Union, Washington, DC, pp 111–132

Woodgate RA, Weingartner TJ, Lindsay R (2012) Observed increases in Bering Strait oceanic fluxes from the Pacific to the Arctic from 2001 to 2011 and their impacts on the Arctic Ocean water column. Geophys Res Lett 39:L24603. https://doi.org/10.1029/2012GL054092

Yang Q, Dixon TH, Myers PG et al (2016) Recent increases in Arctic freshwater flux affects Labrador Sea convection and Atlantic overturning circulation. Nat Commun 7:10525. https://doi.org/10.1038/ncomms10525

Zhang R (2015) Mechanisms for low-frequency variability of summer Arctic Sea ice extent. Proc Natl Acad Sci U S A 112:4570–4575. https://doi.org/10.1073/pnas.1422296112

Marine Optics and Ocean Color Remote Sensing

Veloisa Mascarenhas and Therese Keck

Abstract

Light plays an important role in aquatic ecosystems, both marine and freshwater. Penetration of light underwater influences various biogeochemical processes and also influences activities and behavioral patterns of marine organisms. In addition, dissolved and particulate water constituents present in the water column absorb and scatter light, giving water its characteristic color. The concentration or abundance of these constituents, referred to as optically active constituents (OACs) also determine light availability underwater. Thus color being an indicator of water column content, serves as a water quality parameter. Monitoring of the ocean color variables, such as the OAC concentrations and their optical properties, therefore, allows assessment of the health of an ecosystem. Advances in optical methodologies have improved the understanding of our ecosystems through multispectral and hyperspectral in situ measurements and observations. However, the ocean environment is vast and dynamic and so limitations of spatial and temporal coverage have been overcome with satellite remote sensing that provides oceanographers with repeated synoptic coverage. Being recognized as an essential climate variable (ECV) ocean color is monitored as part of the climate change initiative (CCI) of the European Space Agency (ESA). This chapter aims to provide the reader with an overview of the science of ocean color, introducing involved common terminologies and concepts and its global coverage using satellite remote sensing.

V Mascarenhas (✉)
Institut für Chemie und Biologie des Meeres, Universität Oldenburg, Wilhelmshaven, Germany
e-mail: veloisa.john.mascarenhas@uni-oldenburg.de

T. Keck
Institute for Space Sciences, Freie Universität Berlin, Berlin, Germany
e-mail: therese.keck@wew.fu-berlin.de

Introduction to Ocean Color, Fundamental Concepts, and Optical Tools

Veloisa Mascarenhas

Role of Light in Water

Sunlight plays a key role in the ecology of aquatic ecosystems. Its interaction with water, dissolved and particulate suspended materials is an important physical phenomenon and influences several biogeochemical processes in the global ocean. Penetration of sunlight below water surface facilitates associated biological processes like primary production or plankton distribution in the water column (Kirk 1994). In addition to facilitating photosynthetic processes which form the base of ecological food chain, sunlight also influences the behavioral patterns and activities of marine organisms which are affected by the ambient light field that undergoes vertical changes within the water column (Frank et al. 2012). Mesopelagic fish and zooplankton abundances in different coastal locations are known to correlate with light availability (Aksnes et al. 2004) and changes in light availability conditions are in turn known to have implications for mesopelagic regime shifts (Aksnes et al. 2009).

Fate of Light in Water: Optically Active Constituents and Optical Properties

Fresh and marine waters are a witch's brew of dissolved and particulate matter, both organic and inorganic (Mobley 1994). The dissolved and particulate materials in addition to water molecules interact with light and are therefore known as optically active constituents (OACs). Phytoplankton, colored dissolved organic matter (CDOM), and suspended particulate matter (SPM) being variable in time and space are the three OACs extensively studied across fresh and marine water eco-

S. Jungblut et al. (eds.), *YOUMARES 8 – Oceans Across Boundaries: Learning from each other*,
https://doi.org/10.1007/978-3-319-93284-2_4

systems (Binding et al. 2008; Garaba et al. 2014; Holinde and Zielinski 2016; Mascarenhas et al. 2017). The OACs in the medium interact with the ambient light via processes of absorption and scattering, which gives water its characteristic color. The processes of absorption and scattering are referred to as inherent optical properties (IOPs) of water and depend solely on the OACs present in water. Spatial and temporal variability in the type and abundance of these OACs subsequently induces variability in the IOPs of water. In addition to the IOPs, water bodies are also characterized in terms of their apparent optical properties (AOPs). The AOPs depend both on the OACs and the incident light field.

Phytoplankton are drifting microscopic algae that photosynthesize and form the base of food webs in aquatic (marine and freshwater) ecosystems. Chlorophyll, a green pigment in the phytoplankton absorbs preferentially the blue and red wavelengths of the visible light spectra and reflects green. Therefore, oceans with high concentrations of phytoplankton appear in shades of blue-green depending on the type and density of the phytoplankton population (e.g., North Sea water during algal blooms in Fig.1). Although small in size, these organisms cause large scale impacts. For example, it has been proposed that phytoplankton can steer Pacific tropical cyclones (Gnanadesikan et al. 2010). CDOM, the optically active component of the dissolved organic matter pool, absorbs UV light in the surface waters which is harmful for phytoplankton (Kirk 1994). However, phytoplankton also compete with CDOM for light in the shorter visible wavelength spectra. Also known as yellow substances, gilvin, or *gelbstoff*, CDOM occurs naturally in aquatic environments primarily as a result of tannin-stained waters released from decaying detritus (Coble 2007). Waters comprising of high concentrations of CDOM range from yellow-green to brown (e.g., lake water with dead organic material in Fig. 1). Inorganic suspended matter (ISM), the inorganic component of the SPM, strongly scatters longer (red) wavelengths, thereby giving waters with high sediment concentrations a reddish-brown color (e.g., Wadden Sea in Fig. 1). Pure water, however, absorbs longer wavelength red light. Therefore, open ocean waters with very low concentrations of OACs appear blue (e.g., Atlantic Ocean and North Sea water in Fig. 1). Hence, the OACs influence light availability underwater and determine the color of the oceans (Fig. 1).

Sunlight at the ocean surface is partly reflected (governed by Snell's law and Fresnel equations), while the rest is transmitted through the water column. Underwater light is then either absorbed and/or scattered by water molecules and the OACs present in the water column. The backwards-scattered light then gives water its characteristic color and carries information of ocean constituents, which is captured by satellite sensors hundreds of kilometers above the earth's surface (see section "Space-borne remote sensing"). Detailed understanding of light interactions with the OACs of a medium and its propagation in the medium is fundamental to radiative transfer studies in aquatic ecosystems. Therefore, optical oceanography, i.e., the study of light interactions in the oceans, is vital in understanding the underwater light field, bio-optical relations, and related ecosystem dynamics.

Fig. 1 Various colors observed in fresh and marine waters influenced by the presence of varying optically active constituents. (Reproduced with permission from Marcel Wernard, NIOZ)

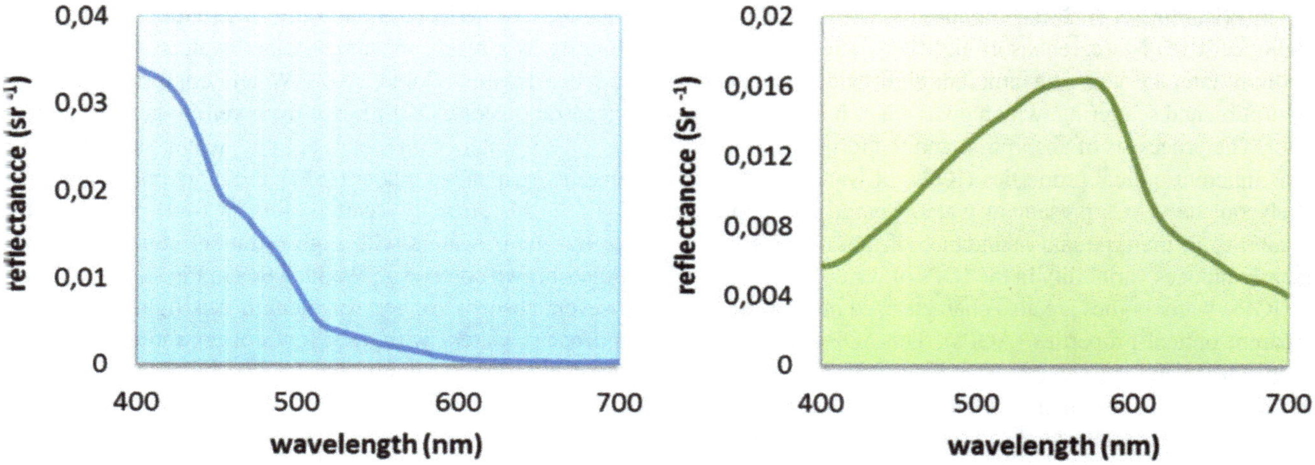

Fig. 2 Spectral reflectance in case-1 (Pacific Ocean) and case-2 waters (Norwegian Fjord). Case-1 waters consisting of very low optically active constituents (OACs), reflect light in the blue region. Case-2 waters with high concentrations of OACs (here chlorophyll *a* in phytoplankton), reflect strongly the green wavelengths. (Data: Daniela Voss, ICBM, University of Oldenburg)

In waters (mostly open ocean) consisting of very low phytoplankton abundances, most visible light is scattered by the water molecules. Water selectively scatters and absorbs certain wavelengths of visible light (Pope and Fry 1997). Longer wavelengths are quickly absorbed from water while shorter wavelengths penetrate deeper, which gives the deep open oceans their characteristic blue color (Fig. 2, blue spectra). In coastal waters (influenced by terrestrial runoff) with higher proportions of dissolved and particulate matter, both absorption and scattering increases, making them appear green (Fig. 2, green spectra) or brown depending on its constituents (Morel and Prieur 1977). Detailed and accurate understanding of the water constituents and their interaction with light is essential in studies of radiative transfer (Chang and Dickey 2004).

Light Penetration and Euphotic Depth

Only the surface layer of the ocean receives sufficient light to allow phytoplankton growth through primary production. Sunlight entering the ocean may travel up to 1000 m deep but there is barely any significant light beyond 200 m. Based on light availability, water columns are divided into 3 different zones. The upper 150–200 m layer of the ocean is called the 'sunlit' or the 'euphotic' zone. The extent of this layer is determined by the depth at which the Photosynthetically Active Radiation (PAR) reduces to 1% of its surface value. In bio-optical literature, PAR values are given in units of mol photons s^{-1} m^{-2} or einst s^{-1} m^{-2}, where one einstein is one mole of photons (6.023×10^{23} photons). PAR is a broadband quantity, often estimated using only the visible wavelengths, 400–700 nm (Mobley 1994). Beyond approximately 200 m depth, the intensity of light decreases rapidly with increasing depth and is insufficient to support any photosynthetic activity. From about 200–1000 m the zone is referred to as 'twilight' or 'dysphotic'. Below 1000 m the zone is known as 'aphotic' or 'midnight' zone and is entirely dark.

The depth of the euphotic zone (Zeu) depends highly on the turbidity of the water column caused by varying concentrations of organic and inorganic optically active constituents (OACs) present either in dissolved form or in suspension. Phytoplankton populations, dead organic matter, CDOM, and inorganic sediments diminish the amount of light available for photosynthetic activity causing the depth of light penetration to differ dramatically between oceanic and coastal waters (Fig. 3a). In open ocean waters with relatively low phytoplankton, the blue-green wavelengths penetrate deeper in the water column. In contrast, high concentrations of both suspended particulate (phytoplankton and sediments) and dissolved matter strongly absorb the blue-green wavelengths in coastal waters thereby restricting penetration in deeper waters. The longer red wavelengths, however, are quickly absorbed by water molecules in near surface waters irrespective of the water optical type (Fig. 3b). In estuarine and fjordal ecosystems, with different fresh and saltwater mixing zones, the euphotic depth reduces gradually with increase in turbidity from the outer (downstream) to inner region (upstream) (Mascarenhas et al. 2017). It is in the euphotic zone, that the majority of primary production takes place.

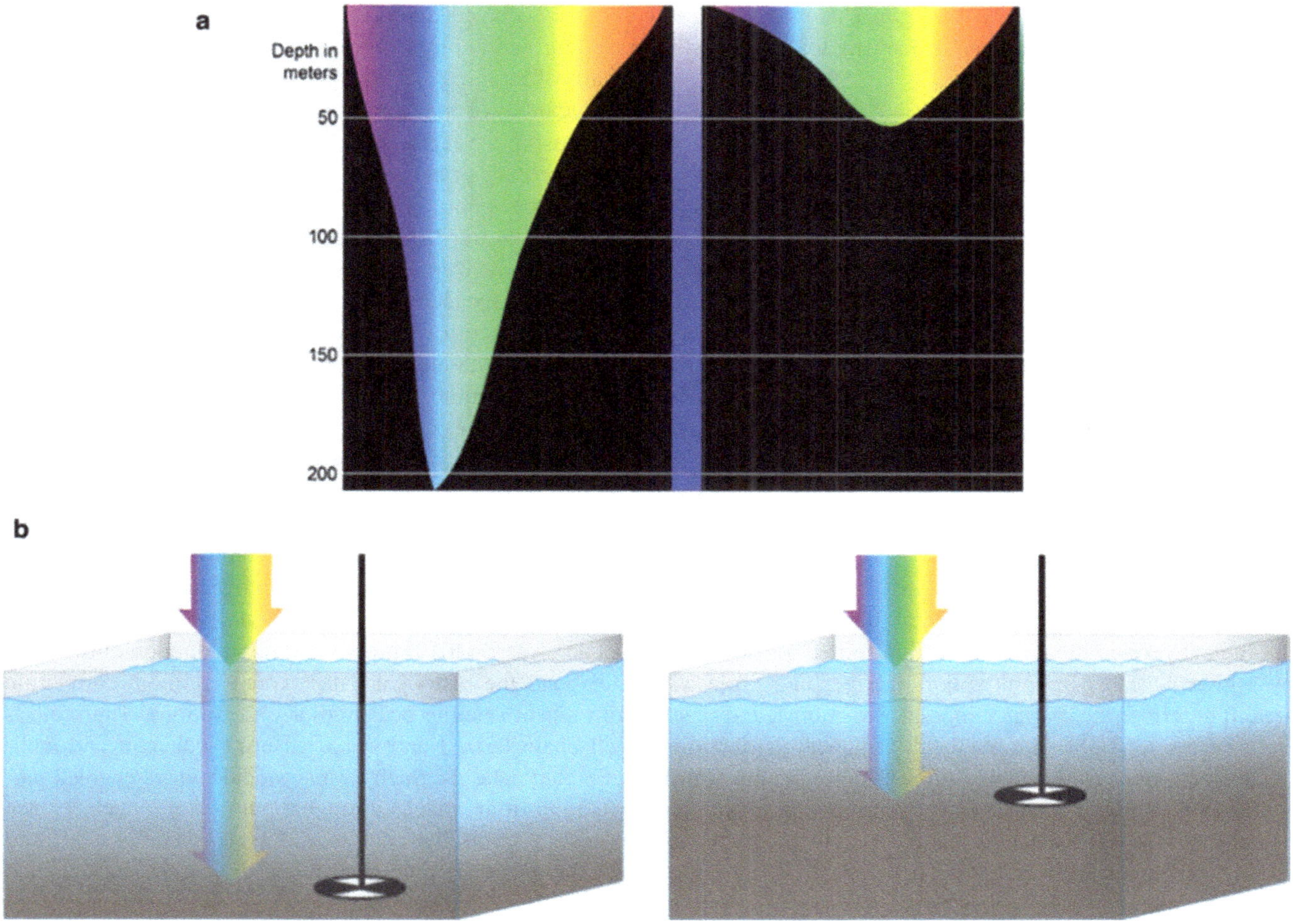

Fig. 3 Penetration of light underwater. (**a**) Spectral attenuation of visible light wavelengths (400–700 nm) in open and coastal waters. (*Image courtesy of Kyle Carothers, NOAA-OE*) (http://oceanexplorer.noaa.gov/explorations/04deepscope/background/deeplight/media/diagram3.html). (**b**) Light attenuation in clear (open) and turbid (coastal) waters. Low concentration of OACs allows deeper light penetration in open ocean waters, while higher concentrations limit light penetration in turbid coastal waters. Measured with a Secchi disk (the black and white disk), the extent of light penetration, referred to as the Secchi disk depth, is an index of water clarity. (Courtesy of the Integration and Application Network, University of Maryland Center for Environmental Science (ian.umces.edu/symbols/))

Optical Classification of Water Types

Natural waters vary highly in their composition and thus also in the extent to which they transmit light underwater. Thus, oceanographers introduced ocean classification schemes based on the optical properties of water. The classification provided a broad indication of the water optical character. Jerlov first attempted to classify open ocean waters into 5 (I, IA, IB, II and III) and coastal waters into 9 different optical water types (numbered from 1 to 9) based on spectral light transmission curves (Jerlov 1976). Morel and Prieur (1977) subsequently introduced the Case 1 (optical properties dominated by phytoplankton and covarying material) and Case 2 (optical properties dominated by suspended sediments and CDOM that vary independently of phytoplankton) classification schemes (based on the shape of reflectance spectra, Fig. 2).

Nature of Light and Light from the Sun

Light consists of numerous localized packets of electromagnetic energy, called photons moving with a velocity of 2.998×10^{8} ms^{-1} in vacuum. Each photon carries a linear and an angular momentum. In addition, they also have an associated wavelength and frequency. Thus, photons exhibit both particle and wave nature and both aspects are fundamental for a proper understanding of light. Energy of a photon is inversely proportional to its wavelength. That means shorter wavelengths possess more energy than the longer (Watson and Zielinski 2013).

How Is Radiation Measured?

In ocean (and also in freshwater ecosystems) optics, radiant energy is measured using two classes of light detectors: thermal and quantum. Thermal detectors (thermometers, thermocouples, bolometers, and pyranometers) absorb radiant energy and convert it into heat energy, wherein the detector responds to consequent changes in temperature of the absorbing medium. Quantum detectors (photographic films, photovoltaic, photoconductive, and photoemissive) react directly to the number of incident photons and not on the cumulative energy of the photons (Mobley 1994; Cunningham and McKee 2013).

Spectral radiance (unit: W sr^{-1} m^{-2} nm^{-1}) is the fundamental radiometric quantity of interest in aquatic optics. It is the radiant flux emitted, reflected, transmitted, or absorbed by a given surface, per unit solid angle per unit projected area. It describes the spatial, temporal, directional, and spectral structure of light. However, full radiance distributions are difficult to measure and assimilate. Therefore, quantities such as total scalar irradiance (W m^{-2} nm^{-1}), downward and upward (planar and scalar) irradiances are obtained by integrating radiances over defined intervals of solid angle. Profiling radiometer assemblies enable precise descriptions of radiative transfer in natural waters (Moore et al. 2009).

Reflectance (Fig. 2), an important AOP fundamental to remote sensing of the oceans, is computed from the above mentioned radiance and irradiance measurements. Earlier, ocean color remote sensing scientists used irradiance reflectance (the ratio of upwelling irradiance to downwelling irradiance) to develop algorithms for IOPs and other ocean parameter retrievals (Morel and Prieur 1977). However, recently, remote sensing reflectance (ratio of upwelling radiance to downwelling irradiance, measured just above the water surface) is more preferred by optical oceanographers (O'Reilly et al. 1998), as it is less sensitive to conditions such as sun angle and sky conditions. Radiative transfer studies relate water AOPs to IOPs.

Optical Tools

Optical oceanography relies strongly on field observations. Although the use of optics in the study of oceans dates back to ancient times, advances in optical technology have played a crucial role in improving our understanding and exploration of the aquatic environments via means of imaging, vision, and sensing. Some of the earliest ocean color measurements were those of Secchi disc depth using a Secchi disk (Fig. 3b) named after the nineteenth century priest and astronomer Pietro Angelo Secchi aboard the papal yacht L'Immacolata Concezione to determine water transparency (Wernand 2010). These measurements were made using white discs of 0.4–3.75 m diameter to measure ocean clarity. Observations of light penetration depth were also made during Britain's 1872–1876 HMS Challenger expedition (Wernand 2013). The depth is determined by lowering the disc in water until it disappears from view.

In 1887, Francois Alphonse Forel introduced his ocean color comparator scale ranging from blue to green for identification of ocean color, later extended by Willie Ule from green to brown. Referred to as the Forel-Ule scale, it is well known and most commonly used in oceanography and limnology to determine color of natural waters. Wernand and van der Woerd (2010) proposed a reintroduction of the scale to expand the historical datasets and facilitate correlation with recent satellite ocean color observations. The scale is well characterized and stable ensuring coherent and well-calibrated datasets. Such simple methods have enabled participation from citizens through a number of citizen science projects such as the citclops (http://www.citclops.eu/) and eye on water (http://eyeonwater.org/) across Europe and beyond (Busch et al. 2016).

Optical sensors measure interaction of light (via absorption and scattering) with water constituents and thereby enable an assessment of the variability in water optical properties in relation to the observed OAC concentrations (Zielinski et al. 2009; Busch et al. 2013). Such observations are fundamental in the establishments of bio-optical models that relate OACs to their optical properties. Via methods of bio-optical inversion, these models enable determination of bio-geo-chemical parameters form remotely sensed signals (see section "Why do we use satellite measurements?"). Commonly used measurements of ocean color parameters include those of light transmission, absorption, scattering, fluorescence, and radiance distribution via methods of spectrophotometry, fluorometry, and radiometry respectively (Dickey et al. 2011). Sensors with selective membranes have enabled additional in situ monitoring of parameters like nutrients, dissolved oxygen, and carbon dioxide (Moore et al. 2009). However, field observations are limited in space and time and thereby lack regular or repeated global coverage. Therefore, satellite missions, which began monitoring the Earth in the 1960s, play an essential role by remotely monitoring the global oceans and providing oceanographers with repeated synoptic coverage.

The next section of the chapter introduces the topic of satellite remote sensing of ocean color. It discusses briefly the developments in ocean color remote sensing over the last few decades, the challenges and processes involved and its applications.

Space-Borne Remote Sensing

Therese Keck

Why Do We Use Satellite Measurements?

Remote sensing is a technique describing properties of an object without having physical contact. Human eyes are sensible to the solar electro-magnetic spectrum from 400 to 700 nm ranging from violet to red (visible spectrum, VIS). Similar to the cones in our eyes, which detect different "colors", water color measurement instruments are designed mostly within the optical spectrum in the visible and near-infra-red from 380 to 800 nm. Beyond these borders, water is strongly absorbing and the instruments receive no signal anymore. A monochromatic measurement may contain information about specific properties and a combination of certain bands can result, for instance, in an RGB image. Most of the instruments measure in a passive way by receiving reflected and back-scattered light from the water.

Generally, one of the most common questions in satellite remote sensing is "Why do we spend so much effort in converting electro-magnetic signals sensed with expensive and complex instruments which are far away in space"? Indeed, in situ and field measurements directly offer properties of the observed matter (e.g., algae content, temperature). Similar results from remote sensing require planning and operation of expensive sensors and their platforms as well as sophisticated algorithms to retrieve physical "products" (e.g., chlorophyll *a* concentration, water vapor content, temperature) from the satellite sensor signals. Nevertheless, the advantage is a relatively high and continuous spatial and temporal coverage of the entire globe.

For example, in Lake Erie (Fig. 4) at the border of Canada and the United States, large algae blooms appear every summer that can vary quickly in spatial and temporal dimension (Rowe et al. 2016). Harmful algae blooms (HABs) have a strong impact on the environment and are toxic to animals and humans. Satellite remote sensing enables us to investigate such events without being at the location or taking in situ samples. Therefore, measurements from even hardly or seldom reachable areas such as the open ocean or at high latitudes can be provided. Analyzing satellite sensor images,

Fig. 4 The western Lake Erie at the border of Canada and U.S. is known for extreme algae blooms. The OLCI RGB image shows a large bloom from 15 September 2017. (OLCI data provided from Copernicus/Eumetsat, RGB image produced with the freely available software SNAP (http://step.esa.int/main/toolboxes/snap/))

information about the spatial extent, location, and chlorophyll concentration are retrieved alongside other parameters. These data can be used to create climatologies and warnings.

Additionally, it is possible to detect a pattern's temporal and spatial variability because satellites revisit the same geographic area every few days (e.g., the polar-orbiting satellites Terra and Aqua have a revisiting time of 1–2 days) or scan the area every few minutes (e.g., the geostationary Meteosat Second Generation MSG-10; EUMETSAT 2017). We are able to observe the atmosphere, the Earth's surface, and the waters with space-borne remote sensing since more than 50 years on a daily to weekly base in a reasonable spatial resolution ranging from a few meters to several kilometers covering the entire earth. However, there is also remote sensing on Earth conducted in the field (e.g., on ships or at the Aeronautic Robotic Network (AERONET) stations) or in the air with instruments mounted on planes.

The knowledge of short- and long-term variability in the oceans and their constituents measured by remote sensing techniques serves as an important resource in oceanographic science. Since the 1960s, space-borne remote sensing supports human needs. "Satellite product users" (e.g., governmental administrations, environmental agencies, or scientific institutions) use "satellite products" to monitor freshwater pools and to warn against pollution. Both play an important role for health and the environment.

The knowledge of the water constituents allows the prediction of fishing grounds and thereby providing economic benefits and sustainable exploitation of the oceans. In the case of natural disasters, satellite imagery provides a quick analysis of the extent and the impact finding quick ways for evacuation and first aid.

For example, the people in Cape Town, South Africa, suffer from severe drought since 2017 ongoing until now and the fresh water supply is strongly restricted since the beginning of 2018 due to a decrease of the largest reservoir, the Theewaterskloof Dam, to around 13% of its average capacity (A. Voiland, 2018-01-30, https://earthobservatory.nasa.gov/IOTD/view.php?id=91649, accessed 09 February 2018).

Environmental changes are observed by the variation in the constituents, the water extent, or the water level. Satellite images show erosion changes along coastlines or the growth of islands. Tracking phytoplankton supports fisheries, the transportation industry, and tourism industry identifying regions of high fish content (Moreira and Pires 2016), which they can either systematically avoid or locate. In order to warn against the toxicity of the HABs, governmental institutes are interested in identifying and tracking phytoplankton using remote sensing (Schaeffer et al. 2015).

The water availability and the water cycle play an important role in climate change. Climate models benefit from an improved understanding of changes and mechanisms analyzed from data retrieved from satellite remote sensing. Additionally, performing photosynthesis, phytoplankton is part of the carbon cycle and consume carbon dioxide, which is a major contributing agent in the frame of global warming. Field et al. (1998) reported that around 47% of the total net primary production is performed in marine ecosystems. Thus, it is of high importance to retrieve and understand the variability of the water constituents on all available temporal and spatial scales which already can almost be covered by remote sensing.

Overview of Technical Details

Platforms

The sensors that measure signals from the Earth's atmosphere, waters and land surfaces are mounted on satellite platforms. Each satellite flies in a specific orbit around the Earth and is loaded with power supplies, navigation tools, and support systems for the instruments. Generally, the most common satellite orbits are geostationary or polar-orbiting, which leads to differences in spatio-temporal resolutions. Geostationary satellites continuously monitor specified geographical locations above the Earth's surface in height of approximately 36,000 km. Therefore, they cannot cover the complete globe. For example, the Geostationary Ocean Color Imager GOCI onboard the Communication Ocean and Meteorological Satellite 1 (COMS) captures images over Korean waters eight times a day (Ryu et al. 2012). Usually, television and communication satellites operate in this orbit due to the stable location.

Polar-orbiting satellites circle around the globe in approximately 100 min at a height of about 700–800 km. Their sensors are capable to cover the entire surface of the earth. The time to receive a full coverage depends on the sensor's swath (the scanning line or area on the ground) and can last from 2 to several days. The sensor Moderate Resolution Imaging Spectrometer (MODIS) onboard the platforms Aqua and Terra has a revisiting time of less than 3 days due to its large swath of 2330 km (Xiong et al. 2005). Polar-orbiting satellites are usually sun-synchronous: They cross the equator at the same local time (LT). Aqua passes the equator from South to North (ascending node) at 1:30 p.m. LT and Terra has an equator-crossing time of 10:30 a.m. LT in a descending node (Xiong et al. 2005). Figure 5a illustrates the product chlorophyll *a* calculated from MODIS/Terra for all available orbits for 2017-07-28.

Instruments

The measurement sensors or instruments are installed onboard the platform. There are two main measurement techniques. MODIS is a whiskbroom scanner, which oscillates across the satellite flight direction. Subsequently, it scans a part of the swath area from one side to the other and backwards while the satellite continues moving (Xiong et al. 2005). A sensor with a pushbroom measuring technique does not rotate: The whole swath width is scanned at once and pushed forward with the satellite flight direction and movement. The Medium Resolution Imaging Spectrometer MERIS onboard the Environmental Satellite (Envisat) is a prominent example (ESA 2006).

Most of the remote sensing instruments in space have multiple measuring bands or channels to detect a certain spectral range of light and its intensity. Mainly, a channel is defined by its central wavelength and the band width described by an individual response function. The function determines the ability of a band to detect a specific part of the electro-magnetic spectrum. For instance, the MODIS band 1 ranges from 620–670 nm detecting all photons within this wavelength range (Xiong et al. 2005). The response function defines how much of an infinitesimal wavelength interval contributes to the finally measured signal at this band. Chlorophyll *a* exhibits interesting features with an increase in absorption towards 670 nm in this spectral range (Bricaud et al. 1998). Using measuring bands with a large band width leads to a loss of specific spectral features which particularly reduces the information quality in water bodies. Therefore, it is important to carefully specify the spectral settings of a channel depending on the sensor's objective.

There are spectrometers with a higher spectral resolution than MERIS or MODIS. For example, the latest space-borne

Fig. 5 Global coverage of the MODIS (Terra) Level-3 product Chlorophyll *a* concentration (OCl-Algorithm) on a daily base (28 July 2017) and monthly mean for June 2017. (Pictures provided by Ocean Biology Processing Group/NASA, downloaded from https://oceancolor.gsfc.nasa.gov/cgi/l3 (29 July 2017))

sensors HICO (Hyperspectral Imager for the Coastal Ocean) or the future HSI (Hyperspectral Imager) onboard EnMAP (The Environmental Mapping and Analysis Program) are hyperspectral sensors exhibiting measurement bands every 5–10 nm, which enables us to investigate even narrow spectral features (e.g., the phytoplankton peak near 683 nm).

Measurement

The satellites operate at a height where the atmosphere is already extremely thin and, therefore, negligible. Hence, we distinguish between at-sensor-measurements at top of the atmosphere (TOA) and bottom of atmosphere (BOA) measurements which are back-calculated from the TOA measurements. The TOA measurement signal contains information about the atmosphere and the underlying water body or land surface. Depending on the atmospheric composition (gas mixtures, aerosols, clouds), the intensity reduces by absorption and scattering in the incident direction. It can also increase if diffuse sky light scatters in the direction of the incoming solar radiation.

At the water surface, the incident radiation is partly reflected in the atmosphere and the other part penetrates the water. Depending on the water conditions defined by the water properties and the constituents, most of the radiation is absorbed or scattered. A little part is scattered backwards and leaves the water again. The amount of water-leaving photons depends on the atmospheric and water conditions and the radiation has to pass the atmosphere again to approach at the sensors at top at the atmosphere. Mathematically, the measured signal is a function of all conditions and constituents of the atmosphere and the water. We can access these parameters by inversion (exemplary, see inversion techniques described in Rodgers 2000). Generally, the retrieved TOA signal contains approximately 90% atmospheric and 10% oceanic information.

Additionally, the TOA measurement changes with the spectrum and the viewing geometry. As outlined beforehand, in the near infrared we expect a very low or no signal above (clear) water bodies due to the strong absorptive properties of waters beyond 750 nm but a high response in the blue visible range. Most of the optical instruments have a nadir viewing geometry where the measurement sensor looks directly downwards. An off-nadir measurement with a viewing zenith angle (VZA) greater than 0° from the normal axis between satellite and surface increases the path between the location of the water-leaving radiation and the sensor. The signal can increase by diffuse scattering in the atmosphere and/or attenuates due to more opportunities for absorption and scattering by molecules and particles.

The main natural obstacles (along others) in optical remote sensing are clouds, sunglint, and the interfering atmosphere. Clouds appear thick and white to the human eyes and also to optical sensors. In different wavelengths regimes, for example for microwave measurements, clouds are transparent and the sensors can measure the underlying surface. Usually, microwave instruments are used for the detection of sea surface temperature, surface height or land applications. For optical measurements, certain algorithms ("cloud masks") exclude pixels with expected cloud coverage. Sunglint occurs at smooth and highly reflective surfaces such as water or fresh snow if the solar light is directly reflected into the sensor. The bright reflection usually oversaturates a sensor's measurement capability and also contains very low or no information about the water body. However, a change in the viewing geometry reduces or even avoids the measurement of sunglint.

The measured TOA radiation has to pass the atmosphere, which highly changes the received signal that leaves the water. The measurement can be "back-calculated" to a BOA-measurement, which is ideally equal to the water-leaving signal. Therefore, it is necessary to estimate the influence of the atmosphere on the TOA-signal by proxies and additional measurements. Using the estimation, the signal can be corrected for the atmosphere ("Atmospheric Correction").

Selected Sensors for Water Remote Sensing

Historically, scientific Earth observation started in the late 1950s to support weather forecasts and to analyze weather phenomena. In the following, we present some selected sensors that have or had the main mission to observe water bodies. Therefore, each sensor's bands were carefully chosen for water applications. However, they are also used above land and most of the introduce sensors also have land and atmospheric missions.

The Coastal Zone Color Scanner (CZCS) onboard the US-platform NIMBUS-7, operational from 1978 to 1987, was one of the first satellite sensors mainly designed to observe the oceans. The CZCS measurements were a first step towards global mapping of chlorophyll *a* concentration and the impact of the oceans on the carbon cycle. In 1996, the Sea-viewing Wide Field-of-view Sensor (SeaWiFS) onboard Seastar began sensing the ocean in eight channels within 400 nm to 900 nm. SeaWiFS operated until 2010 and was slightly tilted to avoid sun glint. MODIS, introduced in section "Instruments", is mounted on board the satellites Aqua and Terra operating from 1999 and 2002, respectively, until present time. The Medium Resolution Imaging Spectrometer MERIS was one of 11 instruments onboard the Environmental Satellite Envisat that operated from 2002 until a technical platform failure in 2012. MERIS supported the chlorophyll *a* fluorescence investigation with a band at 681 nm nearly to 683 nm where the fluorescence peaks and some bands usable for chlorophyll *a* algorithms (ESA 2006).

The Ocean and Land Color Imager (OLCI) on board Sentinel-3 continues the heritage of MERIS with 6 additional bands (ESA 2013). Sentinel-3A was launched in 2016 and Sentinel-3B is planned for 2018 (https://earth.esa.int/web/guest/missions/esa-eo-missions/sentinel-3, 29 July 2017). Hyperspectral imagers usable for water measurements are the Hyperspectral Imager for the Coastal Ocean (HICO) installed on the International Space Station (ISS) and the Hyperspectral Imager (HSI) onboard EnMAP. HICO operated from 2008 to 2014 (http://hico.coas.oregonstate.edu/, 29 July 2017-07-29) and EnMAP is planned for launch in 2019 (http://www.enmap.org/).

Using Remote Sensing Measurements

Preprocessing

Before the space-borne measurements are available for the user, they are usually preprocessed. The state of processing is defined by its level. The processing is mostly done by the operating space agency and, hence, the expressions may sometimes vary slightly and the agencies may not provide all levels for all sensors. Referring to Martin (2014) the levels (L) are briefly introduced:

Level 0 data sets contain the raw measurements without any correction besides measurement or transfer artifacts.

Level 1 data sets contain temporal and spatial information. Level 1B data provide measurements converted to a radiometric unit (e.g., radiance).

Level 2 data sets contain physical parameters calculated from L1B data (e.g., sea surface temperature). L2 data require the application of multiple channels, land-sea-masks and cloud masks, and usually an atmospheric correction that accounts for the influence of the atmosphere on the signal.

Sometimes, Levels 3 products are available for specific locations or a specific gridding. They may also contain temporally merged products, e.g., monthly means, to reduce data gaps due to clouds or other obstacles. Figure 5a illustrates daily Chlorophyll *a* product from 28 July 2017 and a monthly mean for the month of June 2017 is given in Fig. 5b. Level 4 data may incorporate match-up data of in situ and field measurements. In oceanic applications, the atmospheric correction for the Level 2 data is mainly applied to convert the TOA measurements into values that would have been measured if the atmosphere were absent. There is a wide range of applications and algorithms that can conduct atmospheric corrections.

Applications

All levels are usually provided in a scientific binary data format and are mostly available for free (e.g., MODIS data) or on request for scientific purposes. Oceanic remote sensing data can be downloaded via the following selected webpages:

- https://oceancolor.gsfc.nasa.gov/ for SeaWiFS, MODIS, MERIS, CZCS, and others. They provide Chlorophyll-concentration, sea surface temperature, and a quasi-RGB image per orbit or on a temporal averaged base.
- https://scihub.copernicus.eu/s3/#/home for OLCI data as Level 1 or Level 2

EumetView provides a quick access to OLCI data with orbital RGB images (access via http://eumetview.eumetsat.int/mapviewer/). The RGB image can be downloaded. The WorldView page for the MODIS measurements (https://worldview.earthdata.nasa.gov/) supports RGB images, reflectance, and several layers (e.g., sea ice and chlorophyll concentration) for an easy overview of the entire globe. Both views provide images in near-real time.

The free software SNAP (download via http://step.esa.int/main/toolboxes/snap/) by ESA is a useful tool for statistical analyses of satellite data and display results per band or as RGB image (see Figs. 4 and 6). For some sensors, it provides atmospheric corrections, conversion to a higher level and generation of products like chlorophyll concentration.

For example, the two band ratio "blue-green-ratio" gives a first estimation of the amount of algae and chlorophyll, respectively, in the ocean in arbitrary units (Martin 2014). In clear waters, the maximum reflectance is located in the blue part of the visible spectrum. The peak shifts to the green regime in the presence of phytoplankton. The blue-green-ratio BG = "blue channel"/"green channel" = R(440 nm)/R(555 nm) compares these two extremes. A higher blue-green-ratio indicates a "more blue" water and we expect low or no algal content. Unfortunately, this ratio is easily disturbed by additional substances (CDOM, sediments) that change the shape of a reflectance spectrum. Therefore, this ratio is only applicable for a first guess in clear waters with phytoplankton dominance.

Bio-optical Models

Bio-optical models link optical measurements of reflectance or radiance and biological parameters like chlorophyll *a* concentration, water quality, euphotic depth, and others. These biogeochemical variables are a main interest of the end-users who want to decide or analyze specific issues. Depending on the observed water, the complexity of a bio-optical model can vary from a ratio to an extensive non-linear function. Morel and Prieur (1977) introduced optically simple waters, which only contain phytoplankton (case-1) and optically complex waters (case-2). Exemplarily in case-1 waters, several chlorophyll *a* concentration algorithms are based on blue-green ratios (BGs). BG is the relation between a "green" and "blue" measurement band in the VIS providing a qualitative estimate of the relative presence of phytoplankton.

According to Martin (2014), the MODIS Ocean-Color-3-band-Algorithm OC3M is an empirical relation between the maximal ratio of some blue-green ratios from measured reflectance and chlorophyll *a* [mg m^{-3}] measurements:

$$R_L = \log_{10}\left(\max\left[R_{RS}(443\text{nm},488\text{nm})/R_{RS}(551\text{nm})\right]\right)$$
$$\log_{10}(\text{Chla}) = 0.2424 - 2.742 * R_L + 1.802 * R_L^2 + 0.002 R_L^3 - 1.228 R_L^4$$

Where R_{RS} refers to remote sensing reflectance, R_L to the maximal ratio and Chla to chlorophyll a concentration.

There is a wide range of "OC" algorithms depending on the instrument and the degree of the polynomial. Bio-optical models also can describe the spectral shape of IOPs based on only a few measurements, for example,

$$a_{\text{CDOM}} = a_{\text{CDOM}}(440) * \exp\left(-S * (\lambda - 440)\right)$$

where a_{CDOM} refers to absorption by CDOM, and S to the slope of the absorption spectra. CDOM absorption (m^{-1}) exponentially decreases with longer wavelengths and a bio-optical model for spectral CDOM is often based on a measurement at approximately 440 nm. The equations for a_{cdom} and phytoplankton absorption (a_{ph}) are commonly used models to describe CDOM in case-2 waters and phytoplankton absorption in case-1 waters (Gilerson et al. 2008; Brewin et al. 2011; McKee et al. 2014). Usually, the shape factor S

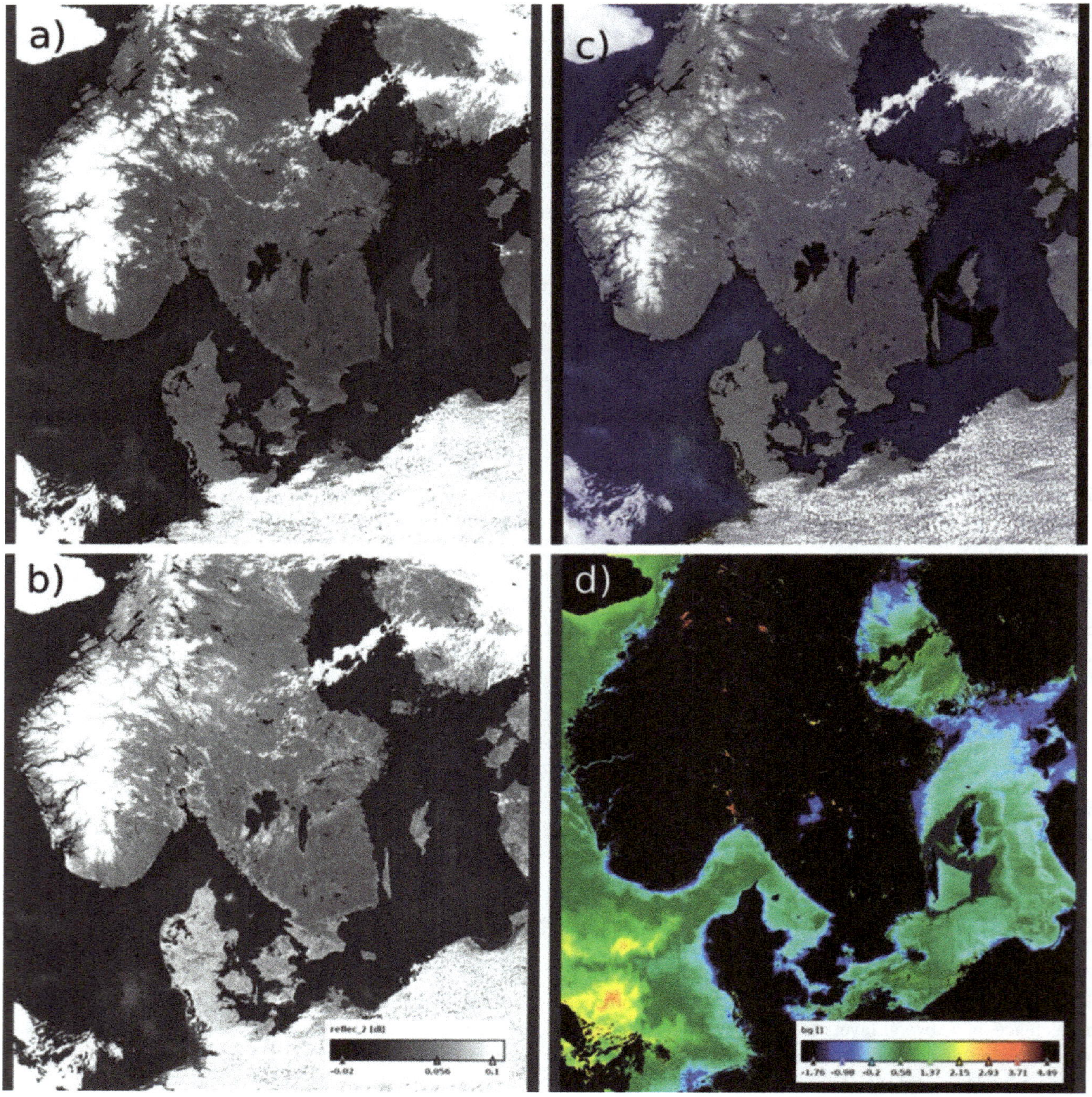

Fig. 6 MERIS captured the North Sea and Baltic Sea on 06 June 2008. (**a**) and (**b**) Level-2 reflectance for the channels 442 nm and 560 nm, respectively, share the color bar from panel. (**c**) RGB image calculated with several bands. (**d**) blue-green-ratio between the 442 nm and the 560 nm band. (Produced with the freely available software SNAP (http://step.esa.int/main/toolboxes/snap/) and the MERIS data are provided from the Ocean Biology Processing Group/NASA)

of the exponential decrease ranges from 0.005 nm^{-1} to 0.031 nm^{-1} (Brewin et al. 2015; Bricaud et al. 2012; Chen et al. 2017) and the reference is usually set to an available blue wavelength (440 nm).

In case-2 waters, Stedmon et al. (2000) and Kowalczuk et al. (2006) analyzed CDOM in Danish coastal waters and in the Baltic Sea and found a dependency on the origination of CDOM and seasonal stratification, respectively. This shows the complexity of case-2 waters. Bricaud et al. (1995) and Bricaud et al. (1998) found an empirical equation for chlorophyll concentration and absorption with spectral dependent coefficients A [m^3 mg^{-1}] and B [dl] in open oceans. Therefore, the model is mainly valid for case-1 waters:

$$a_{ph}(\lambda) = A(\lambda) * \text{chl}a^{1-B(\lambda)}$$

where a_{ph} refers to phytoplankton absorption, chl *a* is the chlorophyll a concentrations, A and B wavelength dependent parameters. Retrieving results from this bio-optical model, the chlorophyll concentration is needed. Therefore, the model is highly dependent on accurate measurement results of chlorophyll, which can be complicated in case-2 waters due to the possible presence of additional water constituents such as CDOM.

Alongside the relations between optical properties and biogeochemical constituents and the spectral expression of IOPs, bio-optical models can also relate different IOPs to each other. For instance, the degradation products of phytoplankton have similar optical properties to CDOM and the amount increases with increasing chlorophyll *a* (e.g., Bricaud et al. 2012). Thus, there is a natural correlation of both absorption coefficients. Based on data of the North Sea merged from Nechad et al. (2015), a bio-optical model relating CDOM and a_{ph}440 yields for 440 nm.

$$\text{cdom}(440\text{nm}) = 0.24 * a_{ph}(440\text{nm})^{0.43}$$

There are also bio-optical models related to scattering of non-algal particles of phytoplankton. However, estimating or measuring scattering coefficients is more difficult than absorption coefficients due to a high dependence on the viewing angle and the anisotropic behavior of the scattering phase function constraining the direction of the scattered light (Petzold 1972).

Pure water absorption and scattering coefficients have been measured and analyzed in laboratory experiments and, for instance, are provided by Pope and Fry (1997). Bio-optical models are a highly important part in modeling of water bodies and the simulations and prediction can differ significantly due to the choice of the models. There are a wide range of bio-optical models that can be found in literature due to the difficulty in measurement (e.g., measurement technique and site selection) and various empirical and statistical relations and concepts.

Conclusions and Outlook

Ocean color is an index of ecosystem health. Changes in ocean color indicate changes in its optical constituents that contribute to ocean color. Regular monitoring of these changes is important as it allows the health of ecosystems to be kept in check. Advances in optical methodologies have greatly improved our understanding of the oceanic environment. However, the ever increasing effects of anthropogenic influence and climate change, repeated spatial and temporal coverage is of utmost relevance. Satellite remote sensing, therefore, plays an important role by providing opportunities of global monitoring of the vast and dynamic oceanic ecosystem. Being recognized as an essential climate variable (ECV) ocean color is monitored as part of the climate change initiative (CCI) project of the European Space Agency (ESA) in the global climate observing system (GCOS).

However, accurate interpretation of the remotely sensed signal is challenging and requires good estimation of atmospheric corrections. Furthermore, the complexities are amplified in the complex case-2 waters owing to the contribution of non-varying optical components like CDOM and inorganic suspended sediments. Development of region specific algorithms therefore becomes necessary. Hence, in situ observations still continue to play an important role in bio-optical algorithm development and validation purposes. Moreover, satellite observations of the surface ocean in combination with bio-optical algorithms (derived from in situ autonomous profiling systems, e.g., buoys, floats) are being incorporated into the development of 3D bio-optical ocean models with potential applications in physics and biogeochemistry of the dynamic environment at a number of relevant scales.

For further reading we recommend 'Ocean optics web book' (http://www.oceanopticsbook.info/) and the 'IOCCG Report Series' (http://ioccg.org/what-we-do/ioccg-publications/ioccg-reports/).

Acknowledgments Authors are extremely thankful to the reviewer for providing helpful comments, which greatly improved the chapter. The chapter contains modified Copernicus Sentinel data (2017) and Ocean Biology Processing Group (OBPG) data products and images.

Appendix

This article is related to the YOUMARES 8 conference session no. 7: "Ocean Optics and Ocean Color Remote Sensing". The original Call for Abstracts and the abstracts of the presentations within this session can be found in the appendix "Conference Sessions and Abstracts", chapter "3 Ocean Optics and Ocean Color Remote Sensing", of this book.

References

Aksnes DL, Nejstgaard J, Sædberg E et al (2004) Optical control of fish and zooplankton populations. Limnol Oceanogr 49:233–238. https://doi.org/10.4319/lo.2004.49.1.0233

Aksnes DL, Dupont N, Staby A et al (2009) Coastal water darkening and implications for mesopelagic regime shifts in Norwegian fjords. Mar Ecol Prog Ser 384:39–49

Binding CE, Jerome JH, Bukata RP et al (2008) Spectral absorption properties of dissolved and particulate matter in Lake Erie. Remote Sens Environ 112(4):1702–1711

Brewin RJ, Hardman-Mountford NJ, Lavender SJ et al (2011) An intercomparison of bio-optical techniques for detecting dominant phytoplankton size class from satellite remote sensing. Remote Sens Environ 115(2):325–339

Brewin RJ, Raitsos DE, Dall'Olmo G et al (2015) Regional ocean-colour chlorophyll algorithms for the Red Sea. Remote Sens Environ 165:64–85

Bricaud A, Babin M, Morel A et al (1995) Variability in the chlorophyll-specific absorption coefficients of natural phytoplankton: analysis and parameterization. J Geophys Res 100(C7):13321–13332

Bricaud A, Morel A, Babin M et al (1998) Variations of light absorption by suspended particles with chlorophyll *a* concentration in oceanic (case 1) waters: analysis and implications for bio-optical models. J Geophys Res 103(C13):31033–310444

Bricaud A, Ciotti A, Gentili B (2012) Spatial-temporal variations in phytoplankton size and colored detrital matter absorption at global and regional scales, as derived from twelve years of SeaWiFS data (1998–2009). Global Biogeochem Cycles 26:GB1010

Busch JA, Zielinski O, Cembella AD (2013) 8 – optical assessment of harmful algal blooms (HABs). In: Watson J, Zielinski O (eds) Subsea optics and imaging. Woodhead Publishing, Oxford, pp 171–214e. https://doi.org/10.1533/9780857093523.2.171

Busch JA, Price I, Jeansou E, Zielinski O, van der Woerd HJ (2016) Citizens and satellites: assessment of phytoplankton dynamics in a NW Mediterranean aquaculture zone. Int J Appl Earth Obs Geoinf 47:40–49

Chang GC, Dickey TD (2004) Coastal ocean optical influences on solar transmission and radiant heating rate. J Geophys Res 109:C01020

Chen J, He X, Zhou B, Pan D (2017) Deriving colored dissolved organic matter absorption coefficient from ocean color with a neural quasi-analytical algorithm. J Geophys Res 122:8543–8556

Coble PG (2007) Marine optical biogeochemistry: the chemistry of ocean color. Chem Rev 107(2):402–418

Cunningham A, McKee D (2013) 4 – Measurement of hyperspectral underwater light fields. In: Watson J, Zielinski O (eds) Subsea optics and imaging. Woodhead Publishing, Oxford, pp 83–97. https://doi.org/10.1533/9780857093523.2.83

Dickey TD, Kattawar GW, Voss KJ (2011) Shedding new light on light in the ocean. Phys Today 64(4):44–49

ESA (2006) Meris product handbook. URL http://envisat.esa.int/pub/ESA_DOC/ENVISAT/MERIS/

ESA (2013) Sentinel-3 user handbook, Tech Rep. 1. URL https://sentinel.esa.int

EUMETSAT (2017) MSG level 1.5 image data format description, Tech. Rep. 105-v8, Darmstadt. URL https://www.eumetsat.int/website/home/Data/TechnicalDocuments/index.html

Field CB, Behrenfeld MJ, Randerson JT et al (1998) Primary production of the biosphere: integrating terrestrial and oceanic components. Science 281(5374):237–240

Frank TM, Johnsen S, Cronin TW (2012) Light and vision in the deep-sea benthos: II. Vision in deep-sea crustaceans. J Exp Biol 215(19):3344–3353. https://doi.org/10.1242/jeb.072033

Garaba SP, Voß D, Zielinski O (2014) Physical, bio-optical state and correlations in north–western European shelf seas. Remote Sens 6(6):5042–5066

Gilerson A, Zhou J, Hlaing S et al (2008) Fluorescence component in the reflectance spectra from coastal waters. II. Performance of retrieval algorithms. Opt Express 16(4):2446–2460

Gnanadesikan A, Emanuel K, Vecchi GA et al (2010) How ocean color can steer Pacific tropical cyclones. Geophys Res Lett 37:L18802

Holinde L, Zielinski O (2016) Bio-optical characterization and light availability parameterization in Uummannaq Fjord and Vaigat–Disko Bay (West Greenland). Ocean Sci 12:117–128

Jerlov NG (1976) Marine optics, Elsevier oceanography series 14. Elsevier Scientific Publishing Company, Amsterdam

Kirk JT (1994) Light and photosynthesis in aquatic ecosystems. Cambridge University Press, Cambridge

Kowalczuk P, Stedmon CA, Markager S (2006) Modeling absorption by CDOM in the Baltic Sea from season, salinity and chlorophyll. Mar Chem 101(1–2):1–11

Martin S (2014) An introduction to ocean remote sensing. Cambridge University Press, Cambridge

Mascarenhas V, Voß D, Wollschlaeger J et al (2017) Fjord light regime: bio-optical variability, absorption budget, and hyperspectral light availability in Sognefjord and Trondheimsfjord, Norway. J Geophys Res 122:3828–3847

McKee D, Röttgers R, Neukermans G et al (2014) Impact of measurement uncertainties on determination of chlorophyll-specific absorption coefficient for marine phytoplankton. J Geophys Res 119:9013–9025

Mobley CD (1994) Light and water: radiative transfer in natural waters. Academic, Dan Diego

Moore C, Barnard A, Fietzek P et al (2009) Optical tools for ocean monitoring and research. Ocean Sci 5:661–684

Moreira D, Pires JC (2016) Atmospheric CO_2 capture by algae: negative carbon dioxide emission path. Bioresour Technol 215:371–379

Morel A, Prieur L (1977) Analysis of variations in ocean color. Limnol Oceanogr 22(4):709–722

Nechad B, Ruddick K, Schroeder T et al (2015) CoastColour Round Robin data sets: a database to evaluate the performance of algorithms for the retrieval of water quality parameters in coastal waters. Earth Syst Sci Data 7(2):319–348

O'Reilly JE, Maritorena S, Mitchell BG et al (1998) Ocean color chlorophyll algorithms for SeaWiFS. J Geophys Res 103:24937–24953

Petzold TJ (1972) Volume scattering functions for selected ocean waters. Scripps Institution of Oceanography, San Diego

Pope RM, Fry ES (1997) Absorption spectrum (380–700 nm) of pure water. II. Integrating cavity measurements. Appl Opt 36(33):8710–8723

Rodgers CD (2000) Inverse methods for atmospheric sounding: theory and practice. World scientific, Singapore

Rowe M, Anderson E, Wynne TT et al (2016) Vertical distribution of buoyant *Microcystis* blooms in a Lagrangian particle tracking model for short-term forecasts in Lake Erie. J Geophys Res 121:5296–5314

Ryu J-H, Han H-J, Cho S et al (2012) Overview of geostationary ocean color imager (GOCI) and GOCI data processing system (GDPS). Ocean Sci 47(3):223–233

Schaeffer BA, Loftin KA, Stumpf RP et al (2015) Agencies collaborate, develop a cyanobacteria assessment network. EOS, Earth Space Sci News 96. https://doi.org/10.1029/2015EO038809

Stedmon C, Markager S, Kaas H (2000) Optical properties and signatures of chromophoric dissolved organic matter (CDOM) in Danish coastal waters. Estuar Coast Shelf Sci 51(2):267–278

Watson J, Zielinski O (2013) Subsea optics and imaging. Woodhead Publishing, Oxford

Wernand M (2010) On the history of the Secchi disc. J Eur Opt Soc Rap Publ 5:10013s

Wernand MR (2013) 3 – The history of subsea optics. In: Watson J, Zielinski O (eds) Subsea optics and imaging. Woodhead Publishing, Oxford, pp 35–79. https://doi.org/10.1533/9780857093523.1.35

Wernand M, van der Woerd HJ (2010) Spectral analysis of the Forel-Ule Ocean colour comparator scale. J Eur Opt Soc Rap Publ 5:10014s

Xiong J, Toller G, Chiang V et al (2005) MODIS level 1B algorithm theoretical basis document, NASA MODIS characterization support team, Washington, DC

Zielinski O, Busch JA, Cembella AD et al (2009) Detecting marine hazardous substances and organisms: sensors for pollutants, toxins, and pathogens. Ocean Sci 5:329–349

Phytoplankton Responses to Marine Climate Change – An Introduction

Laura Käse and Jana K. Geuer

Abstract

Phytoplankton are one of the key players in the ocean and contribute approximately 50% to global primary production. They serve as the basis for marine food webs, drive chemical composition of the global atmosphere and thereby climate. Seasonal environmental changes and nutrient availability naturally influence phytoplankton species composition. Since the industrial era, anthropogenic climatic influences have increased noticeably – also within the ocean. Our changing climate, however, affects the composition of phytoplankton species composition on a long-term basis and requires the organisms to adapt to this changing environment, influencing micronutrient bioavailability and other biogeochemical parameters. At the same time, phytoplankton themselves can influence the climate with their responses to environmental changes. Due to its key role, phytoplankton has been of interest in marine sciences for quite some time and there are several methodical approaches implemented in oceanographic sciences. There are ongoing attempts to improve predictions and to close gaps in the understanding of this sensitive ecological system and its responses.

L. Käse (✉)
Alfred Wegener Institute (AWI), Helmholtz Centre for Polar and Marine Research, Biologische Anstalt Helgoland, Helgoland, Germany
e-mail: laura.kaese@awi.de

J. K. Geuer (✉)
Alfred Wegener Institute (AWI), Helmholtz Centre for Polar and Marine Research, Bremerhaven, Germany
e-mail: jana.geuer@awi.de

Introduction

Phytoplankton are some of the smallest marine organisms. Still, they are one of the most important players in the marine environment. They are the basis of many marine food webs and, at the same time, sequester as much carbon dioxide as all terrestrial plants together. As such, they are important players when it comes to ocean climate change.

In this chapter, the nature of phytoplankton will be investigated. Their different taxa will be explored and their ecological roles in food webs, carbon cycles, and nutrient uptake will be examined. A short introduction on the range of methodology available for phytoplankton studies is presented. Furthermore, the concept of ocean-related climate change is introduced. Examples of seasonal plankton variability are given, followed by an introduction to time series, an important tool to obtain long-term data. Finally, some predictions of phytoplankton community shifts related to climate change will be presented.

This review aims to give an introduction of phytoplankton, climate models and the interaction of phytoplankton with the environment. We want to point out small scale changes caused by seasonality as well as examples of whole ecosystem changes.

What Is Phytoplankton?

Plankton play a key role in the ocean as they provide the foundation of marine food webs. In general, the term plankton ("*planktos*" = wandering or drifting) indicates that these organisms dwell in water as they are not able to move against the currents (Hensen 1887). Nekton, on the contrary, can move freely and include mostly organisms bigger than around 2 cm. The broad range of planktonic organisms divides into several trophic levels and size classes as proposed by Sieburth et al. (1978). They belong to all different types of taxonomic groups such as viruses, archaea, bacteria,

S. Jungblut et al. (eds.), *YOUMARES 8 – Oceans Across Boundaries: Learning from each other*,
https://doi.org/10.1007/978-3-319-93284-2_5

fungi, algae, protozoa, and animals. Viruses and bacteria (virio- and bacterioplankton) as well as archaea belong to femto- and picoplankton, which range from 0.02 to 0.2 μm and 0.2 to 2.0 μm in size, respectively. Mycoplankton (fungi) can mostly be found within nanoplankton (2.0–20 μm). Phytoplankton spans from picoplankton up to microplankton (2–200 μm), whereas zooplankton, in rare cases, can reach up to 200 cm (megaplankton).

The high diversity of phytoplankton extends from prokaryotes (cyanobacteria) to several groups of eukaryotes. Classification of phytoplankton groups constantly changes due to the increasing amount of molecular phylogenetic studies and is under constant flux of opinion (e.g., Parfrey et al. 2006). Cyanobacteria have been traditionally classified using morphological features. However, due to the different scientific communities, the bacterial classification is not easily comparable with the phycological taxonomy. In the last decades, several new concepts have been introduced (see e.g., Hoffmann et al. 2005; Komárek 2010; Komárek et al. 2014). With new approaches that are based on molecular techniques and the arising problems to integrate this new information into the classification, there have been several approaches for reaching a consensus in both communities (e.g., Komárek 2006; Palinska and Surosz 2014). So far, all major cyanobacterial groups, even cyanobacteria that have been categorized as freshwater species, can be found in the marine environment (Burja et al. 2001; Paerl 2012).

Adl et al. (2005) revised the classification of protozoa from Levine et al. (1980) and expanded it to other protists in the name of the International Society of Protistologists. They compared modern morphological approaches, biochemical pathways and molecular phylogenetics data to create a new classification. Only 7 years later Adl et al. (2012) revised this classification. This new revision proposes a division into six super-groups: Archaeplastida, Amoebozoa, Opisthokonta, Excavata, and SAR (Stramenopila, Alveolata, and Rhizaria). Throughout the last years, the concept of different super-groups has been applied for the eukaryotic phytoplankton. Changes and uncertainties are still present in the super-groups that are named here. Additionally, several groups of organisms exist, which do not belong to any of the super-groups, for example some groups of flagellates.

Phytoplankton Taxonomy and Morphology

Depending on area, season, and size class, different groups can act as dominating organisms in the food web and, therefore, regulate the seasonality of the predators as well. The most frequent dominating eukaryotic phytoplankton belong to diatoms (Stramenopila), dinoflagellates (Alveolata) or haptophytes (also called prymnesiophytes, no super-group) (Fig. 1). Other groups include Chlorophyta (Archaeplastida), Cryptophyta, Centrohelida and Telonemia, with the last three not belonging to any of the super-groups (e.g., Paerl 1988; Arrigo et al. 1999; Adl et al. 2012).

Diatoms (Bacillariophyta) possess a so-called frustule of silica that consists of two overlapping valves (hypotheca and epitheca) and a girdle (cingulum). Reproduction is mostly asexual. The old cell divides and each daughter cell builds up a new smaller theca inside the parent wall. If the theca gets too small for further reproduction the cell dies. Prior to death, the cell releases auxospores, which grow into new cells. Another characteristic feature is the symmetry of diatoms. They are either centric or pennate symmetric. They occur as single cells or more often in colonies (Gross 1937). Diatoms are mainly autotrophs, with several heterotrophic strategies to survive during darkness (e.g., Tuchman et al. 2006; McMinn and Martin 2013).

Dinoflagellates consist of thecate and athecate groups. Thecate dinoflagellates possess a cover of cellulose plates in contrast to athecate dinoflagellates, which are more variable in shape. Both groups possess two characteristic parts: episome and hyposome. The cells also feature two grooves. A

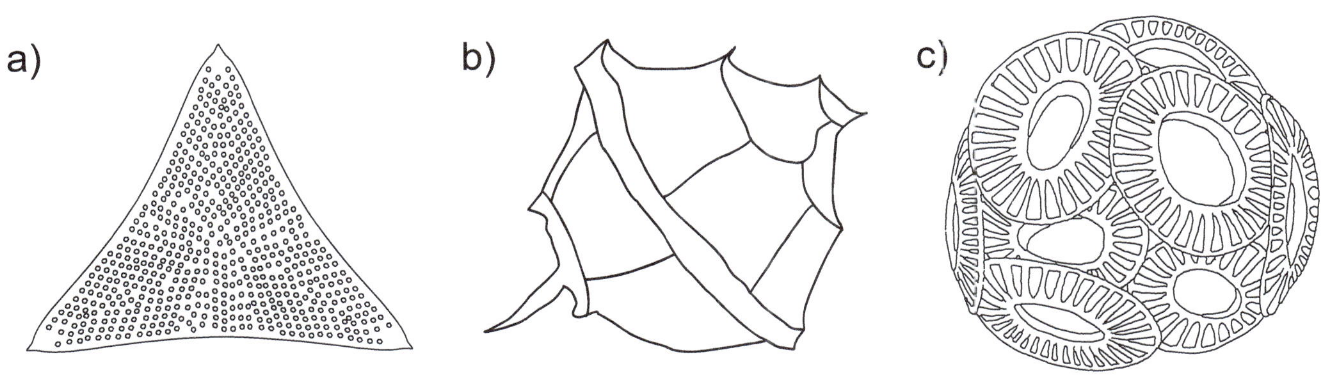

Fig. 1 Exemplary schematic drawings of three important phytoplankton groups. (**a**) Triangular diatom *Trigonium* sp., (**b**) dinoflagellate *Pyrodinium bahamense* and (**c**) coccolithophorid *Emiliania huxleyi* (prymnesiophytes). (Adapted from the open source Plankton*Net Data Provider at the Alfred Wegener Insitute for Polar and Marine Research (**a**) and (**c**), and from Landsberg et al. (2006) (**b**))

cingulum divides the cells into two parts and houses one transverse flagellum and the sulcus houses a second longitudinal flagellum. Dinoflagellates can be auto-, mixo-, and heterotrophs (e.g., Carvalho et al. 2008; McMinn and Martin 2013). Several species can cause so-called "red-tides" and harmful algae blooms (Loeblich 1976; Taylor et al. 2008).

Haptophytes belong to flagellates and consist of different groups and genera. They include, for example, coccolithophorids and the potentially toxic algae genus *Prymnesium*, which includes some cytotoxin, ichthyotoxin, neurotoxin, and haemolytic toxin producing species (Seoane et al. 2017). Motile haptophytes possess two flagella and a haptonema. The haptonema is a characteristic cell organelle and resembles a third flagellum. In contrast to the other two flagella, it is not used for swimming but to capture particles and to attach to surfaces (Hibberd 1976; Kawachi et al. 1991; Jordan and Chamberlain 1997; Andersen 2004).

Primary Production and Essential Elements

Due to its broad distribution and abundance in the ocean, phytoplankton is the fundamental primary producer and serves as a food source at the base of oceanic food webs. It is part of the microbial loop due to its interaction with bacteria and its decomposition by viral lysis and bacteria.

In general, phytoplankton is dependent on the availability of nutrients, light, and other prevalent conditions such as regional and seasonal changes both physically (temperature, salinity, currents, mixing of water layers, precipitation) as well as biologically (e.g., parasites, grazing of potential predators) (Falkowski and Oliver 2007; Racault et al. 2012, further reading: Mackas et al. 1985; Fenchel 1988; Reid et al. 1990).

Phytoplankton uses photosynthesis as energy source and, doing so, contributes with 48% noticeably to global carbon fixation by taking up and incorporating carbon from carbon dioxide. Another important environmental function of phytoplankton is the production of oxygen during photosynthesis (Field et al. 1998). Since photosynthesis requires light, active phytoplankton can only be found in the euphotic zone of the ocean (Fig. 2). Depth of the euphotic zone may differ enormously depending on the presence of biological and non-biological substances absorbing and scattering light within the water column. However, phytoplankton itself often narrows the euphotic zone (Lorenzen 1972).

Phytoplankton as primary producers are part of the biological carbon pump, since they take up carbon dioxide (CO_2) from the atmosphere and bind the carbon in their cells, which are then taken up by higher trophic levels or become part of sinking particles and remineralisation. Time scales for the carbon to re-enter the cycle and to be reused can vary from days, over weeks and years up to several millennia, especially for carbon reaching the sediment surface (Emerson and Hedges 1988; Shen and Benner 2018). Sinking particles that originate from fragmentation, aggregation or egestion after consumption by higher trophic levels such as zooplankton can either be consumed again or be decomposed by microbial processes. At the same time, active vertical migration by the organisms distributes the carbon further within different water layers and therefore has a significant impact on the oceanic carbon cycle and productivity (Azam 1998; Buesseler et al. 2007). As consequence, phytoplankton are subject to high fluctuations and show seasonality as well as a spatial heterogeneity.

To produce biomass, phytoplankton need certain nutrients, the most important being carbon (C), nitrogen (N) and phosphorous (P). For marine primary production, Redfield (1958) calculated the ratio in which these essential nutrients are required as C:N:P = 106:16:1.

Important sources of nitrogen are nitrate and ammonium. Ammonium can be taken up effectively by phytoplankton and provides up to 35% of nitrogen assimilated depending on species and location (Eppley et al. 1971, 1979). Nitrate uptake as nitrogen source requires a higher amount of energy. Thus, ammonium uptake is generally preferred (Thompson et al. 1989). Furthermore, nitrate uptake is relatively slow. Phytoplankton show a great metabolic diversity. For example some phytoplankton species are incapable of nitrate uptake, whereas other species even prefer the uptake of nitrate to ammonium. Ammonium can, in high concentrations, even suppress growth (Glibert et al. 2016; Van Oostende et al. 2017). Nitrogen can be taken up faster by amino acids and fastest via ammonium (Dortch 1982), though only some phytoplankton species are able to take up amino acids (Wheeler et al. 1974).

In competitive environments, however, organic nitrogen such as urea can serve as valuable source to phytoplankton (Bradley et al. 2010). The availability of nitrogen in different forms can also have an influence on the respective species composition (Glibert et al. 2016; Van Oostende et al. 2017).

Phosphorus is also essential for phytoplankton and is usually taken up via phosphate, which frequently acts as limiting nutrient (Perry 1976). Both nitrogen and phosphorus can act as limiting nutrients for primary production (Smith 2006). Some phytoplankton species are capable of reducing their phosphorus demand by producing substitute lipids instead of phospholipids (Van Mooy et al. 2009). Marine diatoms, which can make up large fractions of phytoplankton communities, are furthermore dependent on silicate to form their characteristic external shell (Harvey 1939; Paasche 1973a, b; Treguer et al. 1995; Turner et al. 1998).

Apart from these crucial elements, a range of trace metals is required for phytoplankton growth. Morel and Price (2003) made a first attempt to calculate a stoichiometry for essential trace metals including iron, manganese, zinc, copper, cobalt,

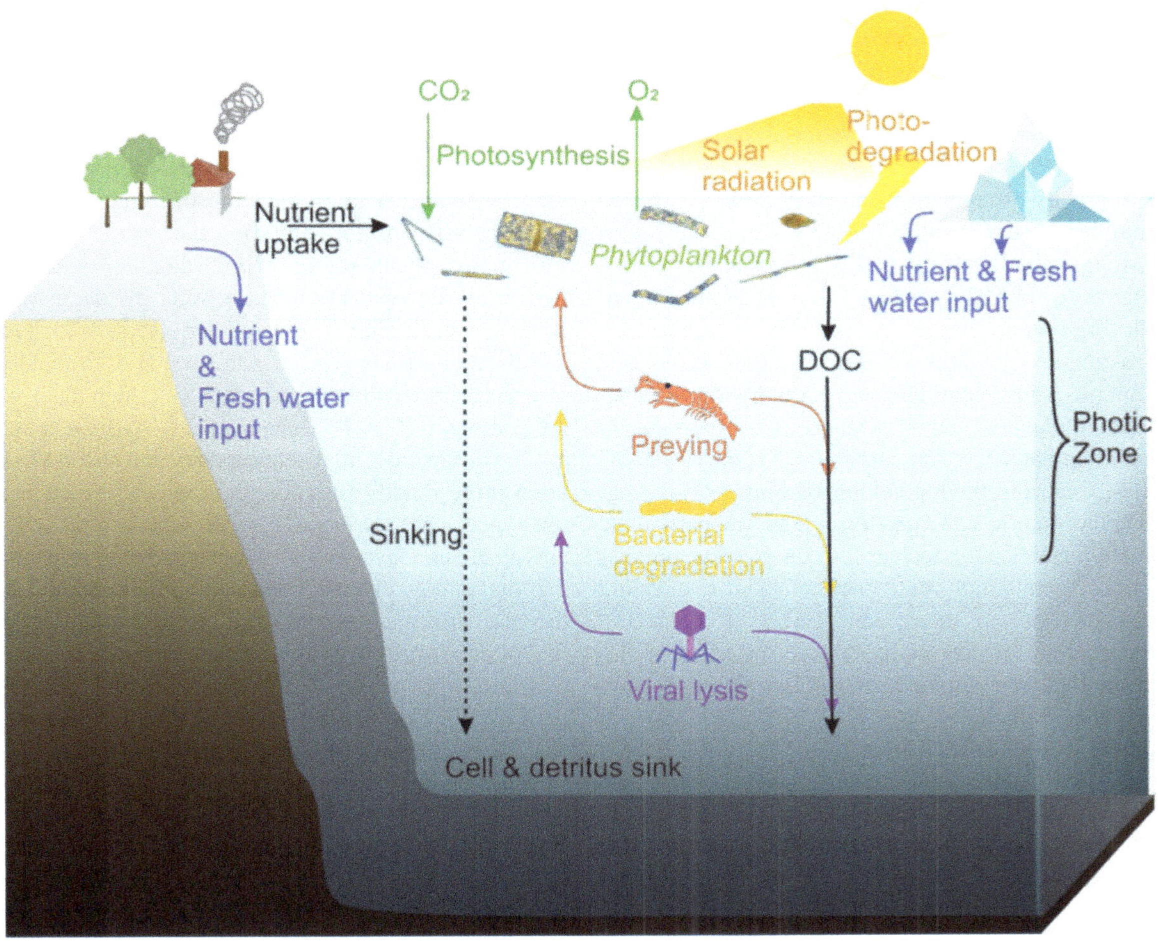

Fig. 2 Cycling of marine phytoplankton. Phytoplankton live in the photic zone of the ocean, where photosynthesis is possible. During photosynthesis, they assimilate carbon dioxide and release oxygen. If solar radiation is too high, phytoplankton may fall victim to photodegradation. For growth, phytoplankton cells depend on nutrients, which enter the ocean by rivers, continental weathering, and glacial ice meltwater on the poles. Phytoplankton release dissolved organic carbon (DOC) into the ocean. Since phytoplankton are the basis of marine food webs, they serve as prey for zooplankton, fish larvae and other heterotrophic organisms. They can also be degraded by bacteria or by viral lysis. Although some phytoplankton cells, such as dinoflagellates, are able to migrate vertically, they are still incapable of actively moving against currents, so they slowly sink and ultimately fertilize the seafloor with dead cells and detritus

and cadmium. Particularly iron is a crucial trace metal that is strongly affecting the productivity of phytoplankton in vast areas of the ocean (Martin and Gordon 1988; Morel et al. 1991). To facilitate trace metal uptake, phytoplankton can make use of ligands, which are organic molecules that are able to complex metals and help to keep them in solution. Especially ligands complexing iron, so called siderophores, are beneficial for phytoplankton (Hassler et al. 2011; Boiteau et al. 2016).

Due to the strong effect iron has on the productivity of phytoplankton, its role was assessed in large scale experiments. After the first successful iron fertilization experiments, which tested the importance of iron in situ on a large scale (e.g., Martin et al. 1994; Coale et al. 1996), the possibility to reduce inorganic carbon with iron fertilization was defined, yielding in sequestering of carbon dioxide during blooms (Bakker et al. 2001, 2005; Boyd et al. 2007). While Buesseler et al. (2004) showed that the "Southern Ocean Iron Experiment" caused a small increase in carbon flux in the region, the "Kerguelen Ocean and Plateau compared Study" could prove an even higher carbon sequestration efficiency (Blain et al. 2007).

Other, more complex molecules are even more important for phytoplankton growth. Some species require exogenous vitamins to grow. Especially vitamin-B depletion can negatively influence phytoplankton productivity (Gobler et al. 2007).

Oceanic dissolved organic carbon (DOC) is one of the largest marine carbon reservoirs. Kirchman et al. (1991) calculated turnover rates of DOC using its bacterial uptake. DOC and dissolved organic nitrogen (DON) cycle differently from each other. During phytoplankton blooms, more

DOC than DON is produced, presumably by phytoplankton (Kirchman et al. 1991). The amount of DOC bacteria can assimilate depends on the phytoplankton species releasing it (Malinsky-Rushansky and Legrand 1996). Phytoplankton release of DOC alone cannot meet bacterial needs and thus allochthonous DOC sources as well as sloppy feeding, viral lysis, hydrolysis by exoenzymes, and zooplankton excretion play a role in releasing additional DOC into the ocean (Fig. 2) (Mopper and Lindroth 1982; Baines and Pace 1991; Xiao and Azam 2011). DOC produced by phytoplankton contains both high and low molecular weight substances. Bacteria assimilate these low molecular weight substances, such as amino acids, peptides, and carbohydrates rather quickly. High molecular weight substances are only slowly or not at all assimilated and can contribute to refractory DOC (Sundh 1992). During phytoplankton blooms, polysaccharide particle formation can transform DOC to particulate organic matter. Such polysaccharides can provide binding sites for trace metals and could participate in controlling their residence time in the ocean (Engel et al. 2004). Therefore, a variety of potentially relevant bioactive molecules exists within the complex DOC pool produced by phytoplankton that influences the ecological interplay of phytoplankton with its environment.

Methods for Studying Phytoplankton Species Composition

Several comprehensive reviews providing good overviews over a variety of methods are available for plankton research. Techniques to assess phytoplankton diversity were collected by Johnson and Martiny (2015). Applications of flow cytometry have been reviewed by Dubelaar and Jonker (2000). A revision of case studies for molecular methods to estimate diversity is available from Medlin and Kooistra (2010). Reviews for nutrient quantification, pigment analysis and remote sensing are also available (Cloern 1996; Jeffrey et al. 1999; Roy et al. 2011; Blondeau-Patissier et al. 2014).

Methods that yield useful approaches to help understanding phytoplankton species composition and its interconnection to environmental conditions are summarized in Fig. 3.

Climate Influences on Phytoplankton

Since the beginning of the industrial era, anthropogenic influences on the climate have steadily increased. Covering more than two thirds of the Earth's surface, the area for exchange between the atmosphere and sea surface is large. Apart from that, the ocean is subject to several effects triggered by climate change.

Climate Change in the Ocean

The two most prominent changes to the ocean triggered by climate change are ocean warming and acidification. Both aspects affect the ocean globally. Increasing anthropogenic carbon dioxide emissions have increased partial pressure of carbon dioxide, both, in the atmosphere and the ocean. The ocean acts as sink for anthropogenic carbon dioxide and is, by increasingly taking up carbon dioxide, gradually acidified. It is estimated that surface water pH decreased by 0.1 since the beginning of the industrial era. With increasing acidification, ocean surface water becomes gradually corrosive to calcium carbonate minerals, of which many seashells are composed (Fig. 4) (Ciais et al. 2013; Rhein et al. 2013).

The ocean has a high heat capacity and absorbs solar radiation more readily than ice. It is virtually certain that the upper ocean has warmed. This warming dominates the global energy change inventory and accounts for more than 90% of the total energy change inventory, while melting ice, warming of continents, and the warming of the atmosphere play only a minor role. Warming of the upper ocean is an important factor that has led to an average sea level rise of 0.19 m between 1901 and 2010 and it is likely that the sea level rise will accelerate (Fig. 4) (Rhein et al. 2013).

Furthermore, there are plenty of regional changes connected to climate change such as patterns of salinity trends. The IPCC report defines a region as a territory characterized by specific geographical and climatological features, whose climate is affected by scale features (e.g., topography, land use characteristics, and lakes) and remote influences from other regions (IPCC 2013). Local changes in salinity are expected (Fig. 4). In general, a higher contrast between fresh and salty regions is expected with salty regions becoming saltier and vice versa. Sea level rise in combination with wind stress is expected to result in high waves in some regions. Intermediate and deep water changes are yet difficult to assess, since long-term data are lacking. Generally, changes in salinity, density, and temperature appear to occur regionally. Anthropogenic influences on coastal runoff and atmospheric deposition of nutrients are another important regional factor. Changing nutrients, such as the input of nitrogen fertilizers, can influence the biological carbon pump and ultimately lead to an increasing eutrophication of waters (Fig. 4) (Ciais et al. 2013; Rhein et al. 2013).

Seasonality and Future Changes in Phytoplankton Communities

Phytoplankton communities undergo seasonal changes. Depending on regional properties like climatic or biogeographic conditions, the changes can differ greatly. While

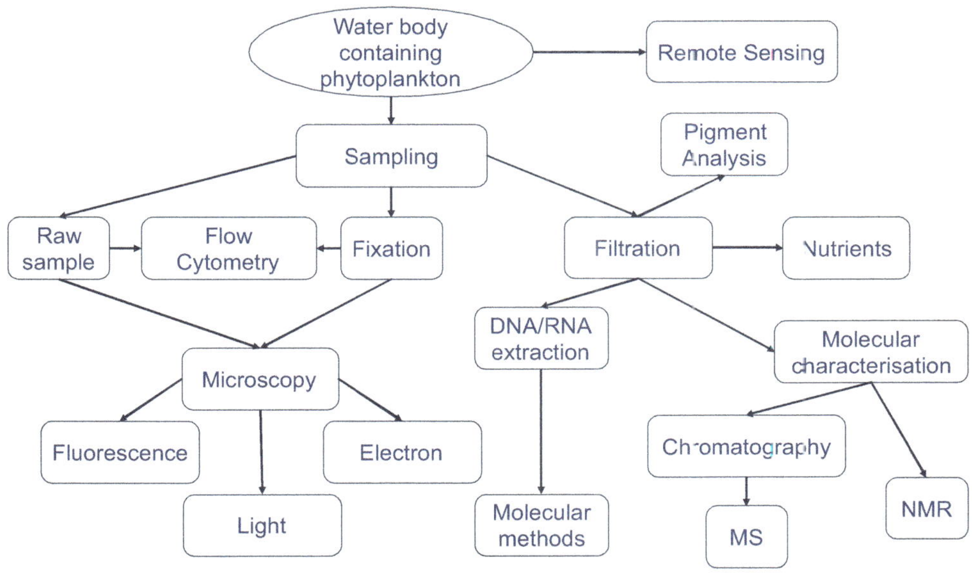

Fig. 3 Schematic overview of the methods used for phytoplankton studies. Three different possibilities to process the sample are using raw samples, fixation or preservation, and filtration. For microscopy and flow cytometry raw samples either are measured immediately or have to be fixed for later measurements. Since molecular methods, pigment analysis and detection of molecular tracers usually require concentrated cells, filter residues serve for phytoplankton measurements. Molecular characterization and quantification of trace molecules is performed using chromatography, mass spectrometry (MS), and nuclear magnetic resonance (NMR) spectroscopy

regions near the equator undergo relatively small changes in temperature during the year, the poles are influenced by large changes caused by severe differences in sunshine intensity and daylight duration. As the environmental factors are already highly influenced by these changes, phytoplankton communities need to adapt to these different conditions as well. Specifically useful study sites are the poles, since seasonal climate variability is very distinct. Other sites that are under continuous and alternating changes are shelf and coastal systems, which are for example influenced by freshwater inflow from the mainland as well as tides and wave actions. The following examples of different regional seasonal changes over the globe and the corresponding phytoplankton community successions shall give a small overview about the vast influence of climate conditions on phytoplankton communities.

In the Arctic summer, glacial ice melt water adds iron and other nutrients into the Labrador Sea (Fig. 2). Apart from the coastal summer blooms resulting from that input, glacial meltwater nutrients travel distances of up to 300 km on the ocean's surface (Arrigo et al. 2017). In the western Arctic, even at closely located sites, different stages of seasonal development could be observed for local phytoplankton communities. The considerable variability in quantitative abundances and biomass values of local phytoplankton species is highly dependent on the irregularity of seasonal processes in the physical environment, ice melting, heating, and the dynamics of stratification (Sukhanova et al. 2009). Furthermore, massive and widespread phytoplankton blooms could occur under the Arctic sea ice, given regional nitrogen concentrations higher than 10 μmol L^{-1} in 50% of the ice covered continental shelf. Those under-ice blooms are also an important factor to be taken into account when estimating changes in the arctic environment (Arrigo et al. 2012).

In the Antarctic, species abundance and composition are largely influenced by seasons in the distinct regions subjected to differences in environmental factors and processes (Deppeler and Davidson 2017). Tréguer and Jacques (1992) divided the Southern Ocean into four different zones, without considering the Permanent Ice Zone, with regard to their different nutrient regimes, physical parameters, and extents of primary production. While diatom-dominated blooms and severe nutrient decreases can be observed in the Coastal and Continental Shelf Zone, the Seasonal Ice Zone is characterized by a very variable hydrological system depending on the ice cover retreat and growth. The Permanently Open Ocean Zone is a nutrient rich region, while the northern part is characterized by a silicate limitation and the Polar Front Zone can harbor high amounts of phytoplankton. There are, however, vast regions, which are suffering under iron limitations. In general, nanoplankton dominates within sea ice and open water unless diatom blooms occur, which happens in May, November and December at the bottom of the ice as well as in January and February in open water (e.g., Perrin et al. 1987; Swart et al. 2015).

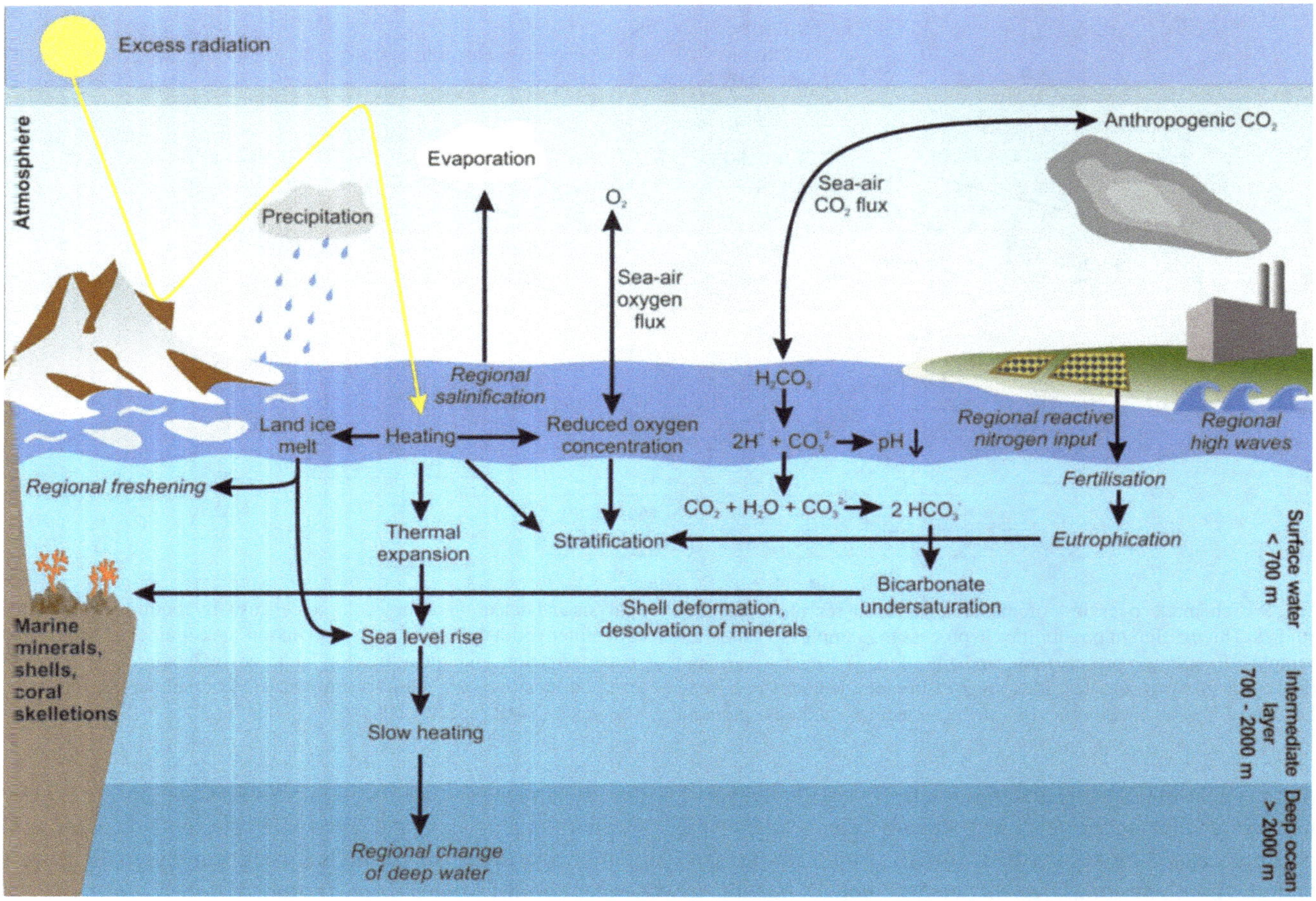

Fig. 4 Overview about climatic changes and their effects on the ocean after Ciais et al. (2013) and Rhein et al. (2013). Regional effects are displayed in italics. Excess solar radiation enters the atmosphere. Ice reflects this radiation, but it is taken up by the surface ocean, leading to its warming. Ocean warming results in land ice melt and thermal expansion, which both result in a sea level rise. Heating of vast areas of the surface ocean also slowly heats up the intermediate water layer which, among others, can ultimately lead to regional changes of deep water. Regional freshening occurs on sites with melting land ice. Regional salinification on the contrary happens in areas of vast evaporation. Surface ocean warming also decreases the solubility of gases, leading to a reduced oxygen concentration and thus changes in the sea-oxygen flux. Excess anthropogenic carbon dioxide enhances its uptake by the ocean and leads to a gradual acidification of the ocean. A decreasing pH results in bicarbonate undersaturation, which causes dissolving of shells and other minerals. Regional input of reactive nitrogen can lead to fertilization and eutrophication. Another regional effect is the occurrence of high waves. Heating, reduced oxygen concentrations and eutrophication lead to higher stratification of water masses

In the Cooperation Sea, which is located in the Seasonal Ice Zone as defined by Tréguer and Jacques (1992), nutrient concentrations increase throughout the year until December, where nitrate and silicate drop notably followed by a phosphate decrease. Here, the largest species diversity occurs during summer, while plenty of dead diatoms are observable in winter (Perrin et al. 1987).

On the Weddell Sea ice edge, huge differences between spring and autumn can be observed. During spring, long chains of vegetative cells form and diatoms as well as haptophytes build up gelatinous colonies in the open water, while only short chains and few single cells occur under the ice. The same conditions hold true for autumn, where short chains and single cells dominate. Furthermore, diatom spores with storage products and enlarged cells are produced then, as results of their sexual reproduction cycle. Thus, the ice edge serves as boundary of different life stages (Fryxell 1989).

One of the most productive regions in the Southern Ocean is the Western Antarctic Peninsula, where phytoplankton blooms occur around November to December, after the sea ice retreats in October. In 2012, the bloom was dominated by diatoms and *Phaeocystis* sp., with diatoms being the most dominant group at the peak of the bloom. Mixing events can cause a crash in the phytoplankton population, which happened in mid-December in this region. Afterwards, the population consisted of large groups of cryptophytes and *Phaeocystis* sp. In March, a second, smaller bloom, dominated by diatoms and *Phaeocystis* sp., could be observed (Goldman et al. 2015).

Within the Eastern English Channel, diatoms, Chrysophyceae, Raphidophyceae and Prymnesiophyceae contribute most to carbon biomass. 40 species of diatoms, and two species for Chrysophyceae and Raphidophyceae can be found, respectively. A yearly occurring *Phaeocystis* sp. spring bloom represents the group Prymnesiophyceae. During summer, mostly large diatoms (>100 μm) dominated the community, whereas during the rest of the year mostly small cells could be found. Furthermore, Cryptophyceae (seven genera) could be found in early spring and autumn, Dinophyceae (26 genera or species) were found with the highest abundance in summer as well as Chlorophyceae and Prasinophyceae (Breton et al. 2000).

Not et al. (2004) found that in the eukaryotic picoplankton the Prasinophyceae *Micromonas pusilla* was the dominating species in the Western English Channel. In contrast to bigger size classes, picoplankton shows a high abundance throughout the year. The microphytoplankton bloom was dominated by a few diatom species like *Guinardia delicatula, Chaetoceros socialis, Pseudo-nitzschia* spp. and *Thalassiosira* spp. during late spring and had maximum abundances during late summer (Ward et al. 2011).

Long-term data from Helgoland in the German Bight suggested interactions of different environmental conditions with phytoplankton seasonality. Increase in sunshine hours correlates with increasing Secchi depths (measure of water transparency) and water temperature. Less turbulence in the water body leads to increasing Secchi depths. Higher temperatures improve growth rates of phytoplankton, but cause lower abundances in early spring. Increased river discharge causes a decrease in salinity in spring, which negatively correlates with Secchi depth. Increasing Secchi depth and thus a bigger euphotic zone benefits the growth of phytoplankton. Concentrations of nutrients such as nitrate, phosphate, and silicate decline rapidly during spring, when the phytoplankton bloom starts and stay at low levels until autumn, when another phytoplankton bloom occurs. Depletion of nutrients causes inhibition of phytoplankton growth. In autumn and winter, new nutrients are released, causing concentrations to increase again. High zooplankton abundances cause belated phytoplankton blooms during spring. Higher grazing pressure during winter decreases phytoplankton abundances, which then need a longer recovery time (Wiltshire et al. 2015). The phytoplankton community is dominated by diatoms in spring and early summer according to daily counts (Wiltshire et al. 2008). Dinoflagellate abundance rose from spring and reached maximum values during summer, where *Noctiluca scintillans*, *Gyrodinium* spp., and *Protoperidinium* spp. dominated. Mixotrophic dinoflagellates occurred in lower abundances than heterotrophs, which correlate with phytoplankton availability. However, during summer 2007, a bloom could be observed, in which several dinoflagellates such as *Lepidodinium chlorophorum*, *Scrippsiella/Pentapharsodinium* spp., and *Akashiwo sanguinea* occurred (Löder et al. 2012). Cryptophytes could be found throughout the year with decline during diatom dominated times (Metfies et al. 2010).

As a sub-tropical region, the estuaries of the Gulf of Mexico are representing a warmer temperate region with long periods of warm temperatures as well as tropical storms (Georgiou et al. 2005; D'sa et al. 2011; Turner et al. 2017). In a study in the Pensacola Bay (Florida, USA) from 1999 to 2001, Murrell and Lores (2004) investigated the role of cyanobacteria on the seasonal dynamics. The three most abundant taxa were belonging to diatoms (*Thalassiosira* sp., Pennales, and *Cyclotella* sp.), and diatoms represented over 50% of total abundance of phytoplankton counts. During December and January, dinoflagellates had high abundances (*Prorocentrum minimum*, *Gymnodinium* sp.), whereas high abundances of chlorophytes and cryptophytes were found during the spring and summer months. Cyanobacteria showed a strong correlation with high water temperatures and had highest abundances in summer. Further characterization indicated that the cyanobacteria belonged to the *Synechococcus* genus. In total, cyanobacteria made up of a large percentage of total chlorophyll (on average 43%) and dominated the chlorophyll biomass during their summer peak (Murrell and Lores 2004).

Another study showing similar results was conducted by Dorado et al. (2015) in Galveston Bay (Texas, USA) from February 2008 to December 2009. North of Galveston Bay high phytoplankton biomass could be observed, with diatoms being the dominating phytoplankton group, followed by dinoflagellates, cryptophytes, and green algae. In comparison, the phytoplankton biomass was lower in the southern part of the bay and dominated by cyanobacteria. Cyanobacteria and green algae correlated inter alia to temperature and chlorophyll a. Results of a multivariate analysis also showed that dinoflagellates and cyanobacteria are more abundant in areas where vertical mixing is limited (mid- and lower region of the bay). Seasonal patterns showed that diatoms, dinoflagellates, and cryptophyte abundances were highest during winter and spring, whereas cyanobacteria were most abundant in summer. It was found that high freshwater discharge correlated with diatom growth, indicating that a decrease of freshwater is accompanied with lower nutrient concentrations. These conditions coupled with the temperature changes are then more favorable for cyanobacteria growth.

Time Series Monitoring of Phytoplankton Diversity

Due to the importance of phytoplankton for the environment and their large seasonal variability, long-term studies dealing with phytoplankton diversity are a very important feature to monitor changes and to yield predictions for the future (Zingone et al. 2015).

Examples for European time series are Plymouth Station L4 (Harris 2010) in the western English Channel, Helgoland Roads in the south-eastern North Sea (Wiltshire et al. 2010) or the HELCOM surveys in the Baltic Sea (Wasmund et al. 2011). One example for an automated system is the Continuous Plankton Recorder survey, which collects information about plankton communities in the North Atlantic basin (Reid et al. 2003; McQuatters-Gollop et al. 2015).

In general, time series use different time scales and sampling intervals, depending on the methods chosen, the amount and variety of parameters and sampling area. Therefore, sampling can range from daily sampling (e.g., Helgoland Roads) over monthly sampling to sampling during certain periods like phytoplankton spring blooms. A distinction can be made between manual sampling and automatic systems like ferry boxes, floats, gliders, and moorings for measurements in the open ocean or other places that are difficult to access. The latter are being implemented more and more, especially during the last decades (Wiltshire et al. 2010; Church et al. 2013).

The responses monitored depend on the focus of the respective phytoplankton studies. Short-term responses caused by nutrient changes can be tested in lab experiments as well as in situ during short time cruises. The observation of responses to habitat changes, regime shifts, climate change and other permanent adaptions require studies that cover one or more stations over a longer time period, which for climate change related studies is at least 30 years (e.g., Walther et al. 2002). Therefore, plankton time series are an important component in the study of long-term changes in marine biodiversity and the obtained data serve as a first indicator for changes in the ecosystem. They can help understanding changes in species distributions and, if explicit enough, provide working hypotheses, which can be tested in the laboratory. Applied benefits in using time series are a better understanding and prediction of the occurrence of possible toxic as well as invasive organisms. For these reasons, time series serve as important tool in marine ecological research (Boero et al. 2015).

A good example for using time series for predictions is the Continuous Plankton Recorder (CPR) survey of 50 years of monitoring dinoflagellate and diatom compositions in the northeast Atlantic Ocean and the North Sea, which could help to predict the following compositional changes. In this area, the ratio is shifting towards a larger diatom proportion. Increasing winds and resulting turbulences yielding better conditions for diatoms compared to dinoflagellates reinforce this assumption (Hinder et al. 2012). These composition shifts of the past and trends in combination with modelling approaches can therefore be used as a forecasting system.

However, the complexity of phytoplankton communities and a high analogy in their morphology make it difficult to identify in particular small sized nano- and picoplankton and to distinguish potentially toxic from non-toxic species. Most conventional time series still use traditional microscopy techniques. Due to their size, small protists are usually underreported or cannot be resolved to species level in these time series. During the last years scientists tried to implement new methods into these long-term studies to include yet underreported organisms. Whereas pigment analyses using HPLC or chlorophyll analyses are already part of many long-term studies (e.g., Karl et al. 2001; Harding Jr. et al. 2015), molecular methods such as DNA microarrays and next-generation sequencing have only been implemented in short-termed studies so far (e.g., Gescher et al. 2008; Medinger et al. 2010; Charvet et al. 2012).

Predictions of Phytoplankton Community Changes in Response to Climate Change

Phytoplankton can serve as indicator for climate or environmental change-induced shifts in the plankton community. Early studies showed that climate change does have an observable influence on the ocean (Madden and Ramanathan 1980; Manabe and Wetherald 1980; Cess and Goldenberg 1981; Hansen et al. 1981; Ramanathan 1981; Etkins and Epstein 1982). Enhanced carbon dioxide levels result in a climatic change all over the globe, influencing precipitation and temperature. Higher global temperatures ultimately lead to higher ocean temperatures and thus a reduction of sea ice in both coverage and thickness (Manabe and Stouffer 1980; Rhein et al. 2013). This results in a local desalination of the ocean, to which phytoplankton cells have to respond. Higher carbon dioxide saturation in the atmosphere will furthermore lead to a shift in equilibrium between air and water and result in elevated carbon dioxide concentrations in the ocean. As a result, the marine environment will become more acidic, potentially influencing sensitive molecular interactions (Kuma et al. 1996).

Physical and biological changes concerning the oceanic carbon sink have been predicted by Sarmiento et al. (1998). They predicted a possible reduction of carbon downward flux in the Southern Ocean due to increasing rainfall and stratification. Their simulations hinted at already occurring physical and biological changes due to climate change and atmosphere-ocean interactions. More recent studies and models show that already small changes in the Southern Ocean can induce feedbacks in the climate system due to extensive changes in the net atmosphere-ocean balance of carbon dioxide (Gruber et al. 2009). The authors also noted the possibly important role of other oceanic regions that could be large contributors to feedbacks in the climate system.

Primary production in the ocean has declined in the last decades and corresponds with increasing sea surface temperature and decreasing iron input. Since especially in high latitudes the ocean acts as important carbon sink, a climate change related further decline in primary production suggests major implications for the carbon cycle (Gregg et al. 2003). The same trend was predicted for many regions using a global model due to increasing stratification and nutrient limiting conditions in the ocean, with exception of the poles (Henson et al. 2018). Reduced sea ice and longer bloom periods in the Arctic have already lead to an increase in net primary production (Arrigo and van Dijken 2015). In contrast, net primary production decreases were also predicted with simulations from nine Earth system models within the framework of the fifth phase of the Coupled Model Intercomparison Project (CMIP5) (Fu et al. 2016).

Useful tools are one-dimensional biogeochemical models such as MEDUSA (Model of Ecosystem Dynamics, nutrient Utilisation, Sequestration and Acidification) that can globally simulate multi-decadal plankton ecosystem scenarios (Yool et al. 2011). In a global approach, the model was used to investigate spring bloom timing related to climate change in a high resolution. The change in bloom initiation timing was substantial, which could lead to food shortages for predators. Additionally, increasing ocean stratification and nutrient limiting conditions will likely result in less total primary production (Henson et al. 2018). Detailed future predictions using this one-dimensional biogeochemical model exist for the Ross Sea in the Antarctic. Primary production for the twenty-first century was estimated and presumably increases 5% in the early and 14% in the late twenty-first century. Melting ice, increased radiance, and decreasing mixed layer depths influence primary production during the first half, diatom mass likely stays constant while *Phaeocystis antarctica* multiplies, which then switches for the second half. Shallower mixed layer depths will change phytoplankton composition and carbon export (Kaufman et al. 2017).

On the Patagonian coast, average primary production will likely increase and phytoplankton communities sequester significant carbon amounts important for secondary production. However, these predictions cannot be made for open ocean areas without restrictions (Villafañe et al. 2015). Furthermore, changes will vastly differ regionally, showing increasing primary production in some areas and decreasing primary production in others. Another critical value influencing phytoplankton variability and competition is the increase of stratified conditions within the water column (Yoshiyama et al. 2009).

Many studies have been conducted to gather more information about phytoplankton community changes and their effects on the food web (e.g., Edwards and Richardson 2004; Schlüter et al. 2012; Harding Jr. et al. 2015). Fu et al. (2016) used a model to simulate climate change impacts on net primary production and export production. Using an intense warming scenario, the net primary production was critically dependent on the phytoplankton community structure. This model gives a good insight in the importance of community-based studies in order to monitor changes in this sensitive system. Changes in phytoplankton communities have, for example, already been observed under changing environmental conditions in the Arctic regions. Shifts in certain protist abundances indicate an enhanced presence of potentially toxic *Alexandrium* dinoflagellate species (Elferink et al. 2017).

Changes in the phytoplankton composition also cause the whole food web to change since predators might have to adapt to new food sources. Alternating environmental factors can facilitate the invasion of new species, which can migrate naturally inside the water masses or might be introduced via ballast water. These atypical range expansions cause structural changes in the food web, especially if the invasive species can adapt well or even better than indigenous species and may even become dominating (Walther et al. 2002; Olenina et al. 2010).

Other effects include the shifts of bloom events, mainly due to temporal and long-term climatic changes, or the timing of phyto- and zooplankton growth. These changes in timing could result in drastic consequences of ecosystem functionality. Existing studies on trophic mismatching in the plankton community are mostly focused on interactions between spring blooms (e.g., Edwards and Richardson 2004; Wiltshire et al. 2008). Therefore, not much is known about the ecological impacts and the functioning of the marine ecosystem (Thackeray 2012).

The floral composition of Chesapeake Bay at the US coast of the Atlantic Ocean revealed a shift in phytoplankton community. With nitrogen being the limiting nutrient but diatoms requiring relatively large amounts, the local community will likely shift to a smaller diatom proportion. Anthropogenic nutrient input might trigger changes as well as climate-related shifts in phytoplankton composition (Harding Jr. et al. 2015). Seasonal variability studies can provide useful insights into future climate changes, as they can give an impression about the mechanisms leading to changes. Studies at the coast of Patagonia in Argentina exposed seasonally different phytoplankton communities to possible future conditions, like enhanced temperatures and solar radiance, nutrient enrichment, and ocean acidification. Increasing ocean temperature has little effect on pre-bloom communities. However, ultraviolet radiance during blooms leads to photochemical inhibition of phytoplankton. Increasing temperatures might lead to a decreasing mixed layer depth, which would expose the community to higher radiations (Villafañe et al. 2013). Shallower mixed layers combined with stronger solar radiance as future condition might also result in cellular stress. Chlorophyll a can decline in phytoplankton cells as response to light stress. The cells can contract and move their chloroplasts, which leads to a temporary photoinhibition of photosynthesis (Kiefer 1973). Diatoms are more prone to ozone-related negative solar UV-B radiation, which can affect aquatic systems, thus generally likely dominating future communities (Häder et al. 2007).

Ocean acidification might lead to a shift in nutrient requirements and C:N:P stoichiometry, thus influencing biogeochemical cycles (Bellerby et al. 2008). In addition, decreasing pH influences micronutrient bioavailability such as leading to decreased concentrations of iron bioavailability and increase phytoplankton iron stress (Shi et al. 2010).

Besides phytoplankton being influenced by nutrients and other environmental factors, they themselves influence the climate and environment they are living in. One example is the role of phytoplankton in the formation of former ice ages. Iron-rich dust was transported to the Southern Ocean, where water masses were rich of nutrients such as nitrate and phosphate but lacked iron. This natural iron fertilization of phytoplankton in the Subantarctic could partly explain atmospheric carbon dioxide changes over the last 1.1 million years. Measurements of foraminifera-bound nitrogen isotopes from sediment cores taken in the Subantarctic Atlantic indicated dust flux, productivity and the degree of nitrate consumption as characterizing factors for peak glacial times and millennial cold events. Triggering blooms and changes in the Southern Ocean's food web and biological pump can be seen as the cause of the full emergence of ice age conditions. However, the main drivers for the initial carbon dioxide decrease were most likely physical processes, such as surface water stratification, wind changes and changes in sea ice extent (Martínez-Garcia et al. 2009, 2014; Jaccard et al. 2013).

Another example for phytoplankton impacts on the climate is dimethylsulfide (DMS). DMS is the degradation product of dimethylsulfoniopropionate (DMSP), which is produced by phytoplankton as an osmoprotectant and degraded by marine bacteria (Yoch 2002). The main DMSP producing phytoplankton belong to the groups of dinoflagellates and prymnesiophytes, but also include some diatoms and Chrysophyceae species (Keller et al. 1989). Important phytoplankton include *Phaeocystis* sp., *Emiliania huxleyi*, *Prorocentrum* sp. and *Gymnodinium* sp. (Yoch 2002). Since atmospheric DMS is an important sulfur source for the global environment and its oxidation causes reflection of solar radiation, it can have a cooling effect on the Earth's temperature (for further reading, see Yoch 2002; Stefels et al. 2007; Lana et al. 2012).

Effects like these make phytoplankton blooms interesting candidates to actively help reversing the effects of climate change, for example by trying to trigger carbon sequestering blooms (Bakker et al. 2005). However, large scale blooms may have unforeseen ecological effects, such as becoming toxic (Silver et al. 2010).

Harmful Algal Blooms

Harmful algal blooms (HABs) refer to blooms of diatoms, dinoflagellates, raphidophytes, haptophytes, cyanobacteria, and certain macroalgae perceived as harmful due a negative impact on the environment or public health. Some have the capability to express toxins under certain circumstances. Other blooms are harmful not due to toxins but because the build-up of high biomass leads to disruption of food webs and development of anoxic zones (Kudela et al. 2017).

Apart from ecological effects, HABs can affect human health upon exposure to poisoned seawater, food or marine aerosols and can have severe socio-economic impacts (Pierce et al. 2003; Fleming et al. 2007). Most frequent HAB related illness worldwide is Ciguatera Fish Poisoning (CFP), which occurs manly in the tropics and subtropics. A variety of dinoflagellate species can produce toxins, such as the *Gambierdiscus toxicus* species complex, which can produce the toxins maitotoxin and ciguatoxin (Murata et al. 1992).

Other illnesses are Paralytic Shellfish Poisoning (PSP) and Diarrhetic Shellfish Poisoning (DSP), which can occur worldwide (Berdalet et al. 2016). *Prorocentrum lima* can produce a variety of toxins, i.a. DSP causative okadaic acid (Murakami et al. 1982). HABs can have extensive ecological effects, such as mass mortality of whales suffering from PSP by feeding on mackerels poisoned with saxitoxins from dinoflagellates or enhanced fish kills (Geraci et al. 1989; Glibert et al. 2001; Nash et al. 2017).

Furthermore, some diatoms can also express toxins. Several species of the genus *Pseudo-nitzschia* are, for example, capable of producing domoic acid (Rao et al. 1988).

HABs can have widespread occurrences. They can occur at coastlines all over the world and have been reported throughout history from Canada, Japan, Scotland, Australia, and many other places (e.g., White 1977; Murakami et al. 1982; Bruno et al. 1989; Nash et al. 2017). Although toxic blooms are a natural phenomenon, they can also be a reaction to environmental shifts and the production of toxins can be connected to environmental conditions (Etheridge and Roesler 2005). Experiments with toxin producing *Alexandrium* sp. showed that increasing radiance and temperature significantly enhanced toxin production (Lim et al. 2006). Nutrient changes in particular can have distinct effects in triggering toxin production. Iron fertilization can lead to formation of a toxic *Pseudo-nitzschia* spp. bloom and natural iron fertilization might have the same effect (Silver et al.

2010). Also low ammonium concentrations and low salinities that can be found in estuaries can lead to enhanced toxin production in *Alexandrium* sp. (Hamasaki et al. 2001).

Because of the damages HABs may cause, their detection and prediction is an on-going scientific challenge, which is approached with different techniques, such as molecular methods, chromatographic pigment analysis, optical spectroscopy, and remote sensing (Millie et al. 1997; John et al. 2005; Trainer et al. 2009). Automated monitoring showed promising predictions (Campbell et al. 2010). Besides establishing a monitoring network, Wells et al. (2015) suggested parameters for routine measurements, including physical parameters, nutrient concentrations, phytoplankton identification, and toxin concentrations.

Linking the effects of climate change with changes in global HAB occurrence and developing monitoring strategies has been the subject of many studies up to date (e.g., Edwards et al. 2006; Moore et al. 2008; Hallegraeff 2010; Hinder et al. 2012; Kudela et al. 2017).

Conclusions

Phytoplankton are a very diverse and important player in the ocean due to their many roles in different marine cycles. Phytoplankton are highly dependent on a diversity of nutrients and influenced by physical and chemical properties in the ocean. Anthropogenic influences on the climate will change these conditions. Some of these effects are global, some remain regional. As diverse as these effects can be, changes to phytoplankton communities will occur as well. One of these examples are harmful algae blooms, which are a hot topic regarding ecological impacts. Another example are possible shifts of ecological niches, which influence the whole marine food web. When predicting such changes, a solid data base is crucial. A wide range of methods targeting different parameters are just as crucial as obtaining data over a long time period.

Seasonal variations in community shifts and changes of the cell morphology show, that phytoplankton adapt to changing environmental conditions regularly. Some of these seasonally observed changes can be extrapolated to future scenarios.

Climate change related conditions in the ocean will change phytoplankton composition and adaption, as they will have to deal with differing nutrient and trace metal bioavailability, physical conditions or temperatures. However, blooms triggered by such conditions can have an opposite effect by influencing the climate themselves.

Apart from species composition, cell physiology is another important aspect that can be changed by climate. Chemicals produced by phytoplankton, such as toxins, can have vast ecological impacts and are one of the most pressing topics when predicting phytoplankton changes.

In conclusion, phytoplankton are an important connecting element within the sensitive marine system. Therefore, accurate predictions are difficult to make, but the existing methods and models are a good way to improve the local understanding. In addition, new models and different approaches looking at factor interactions shall give new and better insights.

Appendix

This article is related to the YOUMARES 8 conference session no. 10: "Phytoplankton in a Changing Environment – Adaptation Mechanisms and Ecological Surveys". The original Call for Abstracts and the abstracts of the presentations within this session can be found in the appendix "Conference Sessions and Abstracts", chapter "4 Phytoplankton in a Changing Environment – Adaptation Mechanisms and Ecological Surveys", of this book.

References

Adl SM, Simpson AGB, Farmer MA et al (2005) The new higher level classification of eukaryotes with emphasis on the taxonomy of protists. J Eukaryot Microbiol 52:399–451. https://doi.org/10.1111/j.1550-7408.2005.00053.x

Adl SM, Simpson AG, Lane CE et al (2012) The revised classification of eukaryotes. J Eukaryot Microbiol 59:429–493. https://doi.org/10.1111/j.1550-7408.2012.00644.x

Andersen RA (2004) Biology and systematics of Heterokont and haptophyte algae. Am J Bot 91:1508–1522. https://doi.org/10.3732/ajb.91.10.1508

Arrigo KR, van Dijken GL (2015) Continued increases in Arctic Ocean primary production. Prog Oceanogr 136:60–70. https://doi.org/10.1016/j.pocean.2015.05.002

Arrigo KR, Robinson DH, Worthen DL et al (1999) Phytoplankton community structure and the drawdown of nutrients and CO_2 in the Southern Ocean. Science 283:365–367. https://doi.org/10.1126/science.283.5400.365

Arrigo KR, Perovich DK, Pickart RS et al (2012) Massive phytoplankton blooms under Arctic Sea ice. Science 336:1408. https://doi.org/10.1126/science.1215065

Arrigo KR, van Dijken GL, Castelao RM et al (2017) Melting glaciers stimulate large summer phytoplankton blooms in Southwest Greenland waters. Geophys Res Lett 44:6278–6285. https://doi.org/10.1002/2017GL073583

Azam F (1998) Microbial control of oceanic carbon flux: the plot thickens. Science 280:694–696. https://doi.org/10.1126/science.280.5364.694

Baines SB, Pace ML (1991) The production of dissolved organic matter by phytoplankton and its importance to bacteria: patterns across marine and freshwater systems. Limnol Oceanogr 36:1078–1090. https://doi.org/10.4319/lo.1991.36.6.1078

Bakker DCE, Watson AJ, Law CS (2001) Southern Ocean iron enrichment promotes inorganic carbon drawdown. Deep Res Pt II 48:2483–2507. https://doi.org/10.1016/S0967-0645(01)00005-4

Bakker DCE, Bozec Y, Nightingale PD et al (2005) Iron and mixing affect biological carbon uptake in SOIREE and EisenEx, two Southern Ocean iron fertilisation experiments. Deep Res Pt I 52:1001–1019. https://doi.org/10.1016/j.dsr.2004.11.015

Bellerby RGJ, Schulz KG, Riebesell U et al (2008) Marine ecosystem community carbon and nutrient uptake stoichiometry under varying ocean acidification during the PeECE III experiment. Biogeosciences 5:1517–1527. https://doi.org/10.5194/bg-5-1517-2008

Berdalet E, Fleming LE, Gowen R et al (2016) Marine harmful algal blooms, human health and wellbeing: challenges and opportunities in the 21st century. J Mar Biol Assoc UK 96:61–91. https://doi.org/10.1017/S0025315415001733

Blain S, Quéguiner B, Armand L et al (2007) Effect of natural iron fertilization on carbon sequestration in the Southern Ocean. Nature 446:1070–1074. https://doi.org/10.1038/nature05700

Blondeau-Patissier D, Gower JFR, Dekker AG et al (2014) A review of ocean color remote sensing methods and statistical techniques for the detection, mapping and analysis of phytoplankton blooms in coastal and open oceans. Prog Oceanogr 123:123–144. https://doi.org/10.1016/j.pocean.2013.12.008

Boero F, Kraberg AC, Krause G et al (2015) Time is an affliction: why ecology cannot be as predictive as physics and why it needs time series. J Sea Res 101:12–18. https://doi.org/10.1016/j.seares.2014.07.008

Boiteau RM, Mende DR, Hawco NJ et al (2016) Siderophore-based microbial adaptations to iron scarcity across the eastern Pacific Ocean. Proc Natl Acad Sci U S A 113:14237–14242. https://doi.org/10.1073/pnas.1608594113

Boyd PW, Jickells T, Law CS et al (2007) Mesoscale iron enrichment experiments 1993–2005: synthesis and future directions. Science 315:612–617. https://doi.org/10.1126/science.1131669

Bradley PB, Sanderson MP, Frischer ME et al (2010) Inorganic and organic nitrogen uptake by phytoplankton and heterotrophic bacteria in the stratified Mid-Atlantic Bight. Estuar Coast Shelf Sci 88:429–441. https://doi.org/10.1016/j.ecss.2010.02.001

Breton E, Brunet C, Sautour B et al (2000) Annual variations of phytoplankton biomass in the eastern English Channel: comparison by pigment signatures and microscopic counts. J Plankton Res 22:1423–1440. https://doi.org/10.1093/plankt/22.8.1423

Bruno DW, Dear G, Seaton DD (1989) Mortality associated with phytoplankton blooms among farmed Atlantic salmon, *Salmo salar* L., in Scotland. Aquaculture 78:217–222. https://doi.org/10.1016/0044-8486(89)90099-9

Buesseler KO, Andrews JE, Pike SM et al (2004) The effects of iron fertilization. Science 304:414–417. https://doi.org/10.1126/science.1086895

Buesseler KO, Antia AN, Chen M et al (2007) An assessment of the use of sediment traps for estimating upper ocean particle fluxes. J Mar Res 65:345–416. https://doi.org/10.1357/002224007781567621

Burja AM, Banaigs B, Abou-Mansour E et al (2001) Marine cyanobacteria – a profilic source of natural products. Tetrahedron 57:9347–9377

Campbell L, Olson RJ, Sosik HM et al (2010) First harmful dinophysis (Dinophyceae, Dinophysiales) bloom in the U.S. is revealed by automated imaging flow cytometry. J Phycol 75:66–75. https://doi.org/10.1111/j.1529-8817.2009.00791.x

Carvalho WF, Minnhagen S, Granéli E (2008) *Dinophysis norvegica* (Dinophyceae), more a predator than a producer? Harmful Algae 7:174–183. https://doi.org/10.1016/j.hal.2007.07.002

Cess RD, Goldenberg SD (1981) The effect of ocean heat capacity upon global warming due to increasing atmospheric carbon dioxide. J Geophys Res 86:498–502. https://doi.org/10.1029/JC086iC01p00498

Charvet S, Vincent WF, Lovejoy C (2012) Chrysophytes and other protists in high Arctic lakes: molecular gene surveys, pigment signatures and microscopy. Polar Biol 35:733–748. https://doi.org/10.1007/s00300-011-1118-7

Church MJ, Lomas MW, Muller-Karger F (2013) Sea change: charting the course for biogeochemical ocean time-series research in a new millennium. Deep Res Pt II 93:2–15. https://doi.org/10.1016/j.dsr2.2013.01.035

Ciais P, Sabine C, Bala G et al (2013) Carbon and other biogeochemical cycles. In: Stocker TF, Qin D, Plattner G-K et al (eds) Climate change 2013: the physical science basis. Contribution of Working Group I to the Fifth Assessment Report of the Intergovernmental Panel on Climate Change. Cambridge University Press, Cambridge/New York

Cloern JE (1996) Phytoplankton bloom dynamics in coastal ecosystems: a review with some general lessons from sustained investigation of San Francisco Bay, California. Rev Geophys 34:127–168. https://doi.org/10.1029/96RG00986

Coale KH, Johnson KS, Fitzwater SE et al (1996) A massive phytoplankton bloom induced by an ecosystem-scale iron fertilization experiment in the equatorial Pacific Ocean. Nature 383:495–501

D'sa EJ, Korobkin M, Ko DS (2011) Effects of Hurricane Ike on the Louisiana-Texas coast from satellite and model data. Remote Sens Lett 2:11–19. https://doi.org/10.1080/01411161.2010.489057

Deppeler SL, Davidson AT (2017) Southern Ocean phytoplankton in a changing climate. Front Mar Sci 4:40. https://doi.org/10.3389/fmars.2017.00040

Dorado S, Booe T, Steichen J et al (2015) Towards an understanding of the interactions between freshwater inflows and phytoplankton communities in a subtropical estuary in the Gulf of Mexico. PLoS One 10:e0130931. https://doi.org/10.1371/journal.pone.0130931

Dortch Q (1982) Effect of growth conditions on accumulation of internal nitrate, ammonium, amino acids, and protein in three marine diatoms. J Exp Mar Bio Ecol 61:243–264. https://doi.org/10.1016/0022-0981(82)90072-7

Dubelaar GBJ, Jonker RR (2000) Flow cytometry as a tool for the study of phytoplankton. Sci Mar 64:135–156. https://doi.org/10.3989/scimar.2000.64n2135

Edwards M, Richardson AJ (2004) Impact of climate change on marine pelagic phenology and trophic mismatch. Nature 430:881–884. https://doi.org/10.1038/nature02808

Edwards M, Johns DG, Leterme SC et al (2006) Regional climate change and harmful algal blooms in the Northeast Atlantic. Limnol Oceanogr 51:820–829

Elferink S, Neuhaus S, Wohlrab S et al (2017) Molecular diversity patterns among various phytoplankton size-fractions in West Greenland in late summer. Deep Res Pt I 121:54–69. https://doi.org/10.1016/j.dsr.2016.11.002

Emerson S, Hedges JI (1988) Processes controlling the organic carbon content of open ocean sediments. Paleoceanography 3:621–634. https://doi.org/10.1029/PA003i005p00621

Engel A, Thoms S, Riebesell U et al (2004) Polysaccharide aggregation as a potential sink of marine dissolved organic carbon. Nature 428:929–932. https://doi.org/10.1038/nature02453

Eppley RW, Carlucci AF, Holm-Hansen O et al (1971) Phytoplankton growth and composition in shipboard cultures supplied with nitrate, ammonium, or urea as the nitrogen source. Limnol Oceanogr 16:741–751. https://doi.org/10.4319/lo.1971.16.5.0741

Eppley RW, Renger EH, Hurrison WG et al (1979) Ammonium distribution in southern California coastal waters and its role in the growth of phytoplankton. Limnol Oceanogr 24:495–509. https://doi.org/10.4319/lo.1979.24.3.0495

Etheridge SM, Roesler CS (2005) Effects of temperature, irradiance, and salinity on photosynthesis, growth rates, total toxicity, and toxin composition for *Alexandrium fundyense* isolates from the Gulf of Maine and Bay of Fundy. Deep Res Pt II 52:2491–2500. https://doi.org/10.1016/j.dsr2.2005.06.026

Etkins R, Epstein ES (1982) The rise of global mean sea level as an indication of climate change. Science 215:287–289

Falkowski PG, Oliver MJ (2007) Mix and match: how climate selects phytoplankton. Nat Rev Microbiol 5:813–819. https://doi.org/10.1038/nrmicro1751

Fenchel T (1988) Marine plankton food chains. Annu Rev Ecol Syst 19:19–38. https://doi.org/10.1146/annurev.es.19.110188.000315

Field CB, Behrenfeld MJ, Randerson JT et al (1998) Primary production of the biosphere: integrating terrestrial and oceanic components. Science 281:237–240. https://doi.org/10.1126/science.281.5374.237

Fleming LE, Kirkpatrick B, Pierce R et al (2007) Aerosolized red-tide toxins (Brevetoxins) and asthma. Chest 131:187–194. https://doi.org/10.1378/chest.06-1830

Fryxell GA (1989) Marine phytoplankton at the Weddell Sea ice edge: seasonal changes at the specific level. Polar Biol 10:1–18. https://doi.org/10.1007/BF00238285

Fu W, Randerson JT, Moore JK (2016) Climate change impacts on net primary production (PP) and export production (EP) regulated by increasing stratification and phytoplankton community structure in the CMIP5 models. Biogeosciences 13:5151–5170. https://doi.org/10.5194/bg-13-5151-2016

Georgiou IY, FitzGerald DM, Stone GW (2005) The impact of physical processes along the Louisiana coast. J Coast Res SI 44:72–89. https://doi.org/10.2307/25737050

Geraci JR, Anderson DM, Timperi RJ et al (1989) Humpback Whales (*Megaptera novaeangliae*) fatally poisoned by dinoflagellate toxin. Can J Fish Aquat Sci 46:1895–1898. https://doi.org/10.1139/f89-238

Gescher C, Metfies K, Frickenhaus S et al (2008) Feasibility of assessing the community composition of prasinophytes at the Helgoland roads sampling site with a DNA microarray. Appl Environ Microbiol 74:5305–5316. https://doi.org/10.1128/AEM.01271-08

Glibert PM, Magnien R, Lomas MW et al (2001) Harmful algal blooms in the Chesapeake and coastal bays of Maryland, USA: comparison of 1997, 1998, and 1999 events. Estuaries 24:875–883. https://doi.org/10.2307/1353178

Glibert PM, Wilkerson FP, Dugdale RC et al (2016) Pluses and minuses of ammonium and nitrate uptake and assimilation by phytoplankton and implications for productivity and community composition, with emphasis on nitrogen-enriched conditions. Limnol Oceanogr 61:165–197. https://doi.org/10.1002/lno.10203

Gobler CJ, Norman C, Panzeca C et al (2007) Effect of B-vitamins (B1, B12) and inorganic nutrients on algal bloom dynamics in a coastal ecosystem. Aquat Microb Ecol 49:181–194. https://doi.org/10.3354/ame01132

Goldman JAL, Kranz SA, Young JN et al (2015) Gross and net production during the spring bloom along the western Antarctic peninsula. New Phytol 205:182–191. https://doi.org/10.1111/nph.13125

Gregg WW, Conkright ME, Ginoux P et al (2003) Ocean primary production and climate: global decadal changes. Geophys Res Lett 30:10–13. https://doi.org/10.1029/2003GL016889

Gross F (1937) The life history of some marine planktonic diatoms. Philos Trans R Soc Lond B 228:1–47. https://doi.org/10.1098/rstb.1937.0008

Gruber N, Gloor M, Mikaloff Fletcher SE et al (2009) Oceanic sources, sinks, and transport of atmospheric CO_2. Global Biogeochem Cycles 23:GB1005. https://doi.org/10.1029/2008GB003349

Häder D-P, Kumar HD, Smith RC et al (2007) Effects of solar UV radiation on aquatic ecosystems and interactions with climate change. Photochem Photobiol Sci 6:267–285. https://doi.org/10.1039/B700020K

Hallegraeff GM (2010) Ocean climate change, phytoplankton community responses, and harmful algal blooms: a formidable predictive challenge. J Phycol 46:220–235. https://doi.org/10.1111/j.1529-8817.2010.00815.x

Hamasaki K, Horie M, Tokimitsu S et al (2001) Variability in toxicity of the dinoflagellate *Alexandrium tamarense* isolated from Hiroshima Bay, western Japan, as a reflecion of changing environmental conditions. J Plankton Res 23:271–278. https://doi.org/10.1093/plankt/23.3.271

Hansen J, Johnson D, Lacis A et al (1981) Climate impact of increasing atmospheric carbon dioxide. Science 213:957–966. https://doi.org/10.1126/science.213.4511.957

Harding LW Jr, Adolf JE, Mallonee ME et al (2015) Climate effects on phytoplankton floral composition in Chesapeake Bay. Estuar Coast Shelf Sci 162:53–68. https://doi.org/10.1016/j.ecss.2014.12.030

Harris R (2010) The L4 time-series: the first 20 years. J Plankton Res 32:577–583. https://doi.org/10.1093/plankt/fbq021

Harvey HW (1939) Substances controlling the growth of a diatom. J Mar Biol Assoc UK 23:499–520. https://doi.org/10.1017/S0025315400014041

Hassler CSC, Schoemann V, Nichols CM et al (2011) Saccharides enhance iron bioavailability to Southern Ocean phytoplankton. Proc Natl Acad Sci U S A 108:1076–1081. https://doi.org/10.1073/pnas.1010963108

Hensen V (1887) Über die Bestimmung des Planktons oder des im Meere treibenden Materials an Pflanzen und Thieren. Ber Komm Wiss Unters dt Meere 5:1–108

Henson SA, Cole HS, Hopkins J et al (2018) Detection of climate change-driven trends in phytoplankton phenology. Glob Chang Biol 24:e101–e111. https://doi.org/10.1111/gcb.13886

Hibberd DJ (1976) The ultrastructure and taxonomy of the Chrysophyceae and Prymnesiophyceae (Haptophyceae): a survey with some new observations on the ultrastructure of the Chrysophyceae. Bot J Linn Soc 72:55–80. https://doi.org/10.1111/j.1095-8339.1976.tb01352.x

Hinder SL, Hays GC, Edwards M et al (2012) Changes in marine dinoflagellate and diatom abundance under climate change. Nat Clim Chang 2:271–275. https://doi.org/10.1038/nclimate1388

Hoffmann L, Komárek J, Kaštovský J (2005) System of cyanoprokaryotes (cyanobacteria) – state in 2004. Arch Hydrobiol Suppl Algol Stud 117:95–115. https://doi.org/10.1127/1864-1318/2005/0117-0095

IPCC (2013) Annex III: glossary [Planton, S. (ed.)]. In: Stocker TF, Qin D, Plattner G-K et al (eds) Climate change 2013: the physical science basis. Contribution of Working Group I to the Fifth Assessment Report of the Intergovernmental Panel on Climate Change. Cambridge University Press, Cambridge/New York

Jaccard SL, Hayes CT, Hodell DA et al (2013) Two modes of change in Southern Ocean productivity over the past millions years. Science 339:1419–1423

Jeffrey SW, Wright SW, Zapata M (1999) Recent advances in HPLC pigment analysis of phytoplankton. Mar Freshw Res 50:879–896. https://doi.org/10.1071/MF99109

Jiao N, Azam F (2011) Microbial carbon pump and its significance for carbon sequestration in the ocean. In: Jiao N, Azam F, Sanders S (eds) Microbial carbon pump in the ocean. Science/AAAS Business Office, Washington, DC, pp 43–45

John U, Medlin LK, Groben R (2005) Development of specific rRNA probes to distinguish between geographic clades of the *Alexandrium tamarense* species complex. J Plankton Res 27:199–204. https://doi.org/10.1093/plankt/fbh160

Johnson ZI, Martiny AC (2015) Techniques for quantifying phytoplankton biodiversity. Annu Rev Mar Sci 7:299–324. https://doi.org/10.1146/annurev-marine-010814-015902

Jordan RWR, Chamberlain AHL (1997) Biodiversity among haptophyte algae. Biodivers Conserv 6:131–152. https://doi.org/10.1023/A:1018383817777

Karl DM, Bidigare RR, Letelier RM (2001) Long-term changes in plankton community structure and productivity in the North Pacific subtropical gyre: the domain shift hypothesis. Deep Res Pt II 48:1449–1470. https://doi.org/10.1016/S0967-0645(00)00149-1

Kaufman DE, Friedrichs MAM, Smith WO et al (2017) Climate change impacts on southern Ross Sea phytoplankton composition, productivity, and export. J Geophys Res Ocean 122:2339–2359. https://doi.org/10.1002/2016JC012514

Kawachi M, Inouye I, Maeda O et al (1991) The haptonema as a food-capturing device: observations on *Chrysochromulina hirta* (Prymnesiophyceae). Phycologia 30:563–573. https://doi.org/10.2216/i0031-8884-30-6-563.1

Keller MD, Bellows WK, Gulliard RL (1989) Dimethyl sulfide production in marine phytoplankton. Am Chem Soc 81:168–182

Kiefer DA (1973) Chlorophyll α fluorescence in marine centric diatoms: responses of chloroplasts to light and nutrient stress. Mar Biol 23:39–46. https://doi.org/10.1007/BF00394110

Kirchman DL, Suzuki Y, Garside C et al (1991) High turnover rates of dissolved organic carbon during a spring phytoplankton bloom. Nature 352:612–614. https://doi.org/10.1038/352612a0

Komárek J (2006) Cyanobacterial taxonomy: current problems and prospects for the integration of traditional and molecular approaches. Algae 21:349–375. https://doi.org/10.4490/ALGAE.2006.21.4.349

Komárek J (2010) Recent changes (2008) in cyanobacteria taxonomy based on a combination of molecular background with phenotype and ecological consequences (genus and species concept). Hydrobiologia 639:245–259. https://doi.org/10.1007/s10750-009-0031-3

Komárek J, Kaštovský J, Mareš J et al (2014) Taxonomic classification of cyanoprokaryotes (cyanobacterial genera) 2014, using a polyphasic approach. Preslia:295–335

Kudela RM, Berdalet E, Enevoldsen H et al (2017) GEOHAB – the global ecology and oceanography of harmful algal blooms program: motivation, goals, and legacy. Oceanography 30:12–21. https://doi.org/10.5670/oceanog.2017.106

Kuma K, Nishioka J, Matsunaga K (1996) Controls on iron(III) hydroxide solubility in seawater: the influence of pH and natural organic chelators. Limnol Oceanogr 41:396–407. https://doi.org/10.4319/lo.1996.41.3.0396

Lana A, Simó R, Vallina SM et al (2012) Potential for a biogenic influence on cloud microphysics over the ocean: a correlation study with satellite-derived data. Atmos Chem Phys 12:7977–7993. https://doi.org/10.5194/acp-12-7977-2012

Landsberg JH, Hall S, Johannessen JN et al (2006) Saxitoxin puffer fish poisoning in the United States, with the first report of *Pyrodinium bahamense* as the putative toxin source. Environ Health Perspect 114:1502–1507. https://doi.org/10.1289/ehp.8998

Levine NDD, Corliss JOO, Coc FEG et al (1980) A newly revised classification of the protozoa. J Protozool 27:37–58. https://doi.org/10.1111/j.1550-7408.1980.tb04228.x

Lim PT, Leaw CP, Usup G et al (2006) Effects of light and temperature on growth, nitrate uptake, and toxin production of two tropical dinoflagellates: *Alexandrium tamiyavanichii* and *Alexandrium minutum* (Dinophyceae). J Phycol 42:786–799. https://doi.org/10.1111/j.1529-8817.2006.00249.x

Löder MGJ, Kraberg AC, Aberle N et al (2012) Dinoflagellates and ciliates at Helgoland roads, North Sea. Helgol Mar Res 66:11–23. https://doi.org/10.1007/s10152-010-0242-z

Loeblich AR (1976) Dinoflagellate evolution: speculation and evidence. J Protozool 23:13–28. https://doi.org/10.1111/j.1550-7408.1976.tb05241.x

Lorenzen CJ (1972) Extinction of light in the ocean by phytoplankton. ICES J Mar Sci 34:262–267. https://doi.org/10.1093/icesjms/34.2.262

Mackas DL, Denman KL, Abbott MR (1985) Plankton patchiness: biology in the physical vernacular. Bull Mar Sci 37:652–674

Madden RA, Ramanathan V (1980) Detecting climate change due to increasing carbon dioxide. Science 209:763–768. https://doi.org/10.1126/science.209.4458.763

Malinsky-Rushansky NZ, Legrand C (1996) Excretion of dissolved organic carbon by phytoplankton of different sizes and subsequent bacterial uptake. Mar Ecol Prog Ser 132:249–255. https://doi.org/10.3354/meps132249

Manabe S, Stouffer RJ (1980) Sensitivity of a global climate model to an increase of CO_2 concentration in the atmosphere. J Geophys Res 85:5529–5554. https://doi.org/10.1029/JC085iC10p05529

Manabe S, Wetherald RT (1980) On the distribution of climate change resulting from an increase in CO_2 content of the atmosphere. J Atmos Sci 37:99–118. https://doi.org/10.1175/1520-0469(1980)037<0099:OTDOCC>2.0.CO;2

Martin JH, Gordon RM (1988) Northeast Pacific iron distributions in relation to phytoplankton productivity. Deep Sea Res Pt A 35:177–196. https://doi.org/10.1016/0198-0149(88)90035-0

Martin JH, Coale KH, Johnson KS et al (1994) Testing the iron hypothesis in ecosystems of the equatorial Pacific Ocean. Nature 371:123–129. https://doi.org/10.1038/371123a0

Martínez-García A, Rosell-Melé A, Geibert W et al (2009) Links between iron supply, marine productivity, sea surface temperature, and CO_2 over the last 11 Ma. Paleoceanography 24:1–14. https://doi.org/10.1029/2008PA001657

Martínez-García A, Sigman DM, Ren H et al (2014) Iron fertilization of the subantarctic ocean during the last ice age. Science 343:1347–1350. https://doi.org/10.1126/science.1246848

McMinn A, Martin A (2013) Dark survival in a warming world. Proc R Soc B 280:20122909. https://doi.org/10.1098/rspb.2012.2909

McQuatters-Gollop A, Edwards M, Helaouët P et al (2015) The continuous plankton recorder survey: how can long-term phytoplankton datasets contribute to the assessment of good environmental status? Estuar Coast Shelf Sci 162:88–97. https://doi.org/10.1016/j.ecss.2015.05.010

Medinger R, Nolte V, Pandey RV et al (2010) Diversity in a hidden world: potential and limitation of next-generation sequencing for surveys of molecular diversity of eukaryotic microorganisms. Mol Ecol 19:32–40. https://doi.org/10.1111/j.1365-294X.2009.04478.x

Medlin LK, Kooistra WHCF (2010) Methods to estimate the diversity in the marine photosynthetic protist community with illustrations from case studies: a review. Diversity 2:973–1014. https://doi.org/10.3390/d2070973

Metfies K, Gescher C, Frickenhaus S et al (2010) Contribution of the class cryptophyceae to phytoplankton structure in the German bight. J Phycol 46:1152–1160. https://doi.org/10.1111/j.1529-8817.2010.00902.x

Millie DF, Schofield OM, Kirkpatrick GJ et al (1997) Detection of harmful algal blooms using photopigments and absorption signatures: a case study of the Florida red tide dinoflagellate, *Gymnodinium breve*. Limnol Oceanogr 42:1240–1251. https://doi.org/10.4319/lo.1997.42.5_part_2.1240

Moore SK, Trainer VL, Mantua NJ et al (2008) Impacts of climate variability and future climate change on harmful algal blooms and human health. Environ Health 7:S4. https://doi.org/10.1186/1476-069X-7-S2-S4

Mopper K, Lindroth P (1982) Diel and depth variations in dissolved free amino acids and ammonium in the Baltic Sea determined by shipboard HPLC analysis. Limnol Oceanogr 27:336–347. https://doi.org/10.4319/lo.1982.27.2.0336

Morel FMM, Price NM (2003) The biogeochemical cycles of trace metals. Science 300:944–947. https://doi.org/10.1126/science.1083545

Morel FMM, Hudson RJM, Price NM (1991) Limitation of productivity by trace metals in the sea. Limnol Oceanogr 36:1742–1755. https://doi.org/10.4319/lo.1991.36.8.1742

Murakami Y, Oshima Y, Yasumoto T (1982) Identification of okadaic acid as a toxic component of a marine dinoflagellate *Prorocentrum lima*. Bull Japanese Soc Sci Fish 48:69–72. https://doi.org/10.2331/suisan.48.69

Murata M, Iwashita T, Yokoyama A et al (1992) Partial structures of Maitotoxin, the most potent marine toxin from dinoflagellate *Gambierdiscus toxicus*. J Am Chem Soc 114:6594–6596. https://doi.org/10.1021/ja00042a070

Murrell MC, Lores EM (2004) Phytoplankton and zooplankton seasonal dynamics in a subtropical estuary: importance of cyanobacteria. J Plankton Res 26:371–382. https://doi.org/10.1093/plankt/fbh038

Nash SMB, Baddock MC, Takahashi E et al (2017) Domoic acid poisoning as a possible cause of seasonal cetacean mass stranding events in Tasmania, Australia. Bull Environ Contam Toxicol 98:8–13. https://doi.org/10.1007/s00128-016-1906-4

Not F, Latasa M, Marie D et al (2004) A single species, *Micromonas pusilla* (Prasinophyceae), dominates the eukaryotic picoplankton in the Western English Channel. Appl Environ Microbiol 70:4064–4072. https://doi.org/10.1128/AEM.70.7.4064

Olenina I, Wasmund N, Hajdu S et al (2010) Assessing impacts of invasive phytoplankton: the Baltic Sea case. Mar Pollut Bull 60:1691–1700. https://doi.org/10.1016/j.marpolbul.2010.06.046

Paasche E (1973a) Silicon and the ecology of marine plankton diatoms. I. *Thalassiosira pseudonana* (*Cyclotella nana*) grown in a chemostat with silicate as limiting nutrient. Mar Biol 19:117–126. https://doi.org/10.1007/BF00353582

Paasche E (1973b) Silicon and the ecology of marine plankton diatoms. II. Silicate-uptake kinetics in five diatom species. Mar Biol 19:262–269. https://doi.org/10.1007/BF02097147

Paerl HW (1988) Nuisance phytoplankton blooms in coastal, estuarine, and inland waters. Limnol Oceanogr 33:823–843. https://doi.org/10.4319/lo.1988.33.4_part_2.0823

Paerl HW (2012) Marine plankton. In: Whitton BA (ed) Ecology of cyanobacteria ii: their diversity in space and time. Springer Science+Business Media B.V, Dordrecht, pp 127–153

Palinska KA, Surosz W (2014) Taxonomy of cyanobacteria: a contribution to consensus approach. Hydrobiologia 740:1–11. https://doi.org/10.1007/s10750-014-1971-9

Parfrey LW, Barbero E, Lasser E et al (2006) Evaluating support for the current classification of eukaryotic diversity. PLoS Genet 2:2062–2073. https://doi.org/10.1371/journal.pgen.0020220

Perrin RA, Lu P, Marchant HJ (1987) Seasonal variation in marine phytoplankton and ice algae at a shallow antarctic coastal site. Hydrobiologia 146:33–46. https://doi.org/10.1007/BF00007575

Perry MJ (1976) Phosphate utilization by an oceanic diatom in phosphorus-limited chemo-stat culture and in the oligotrophic waters of the central North Pacific. Limnol Oceanogr 21:88–107. https://doi.org/10.4319/lo.1976.21.1.0088

Pierce RH, Henry MS, Blum PC et al (2003) Brevetoxin concentrations in marine aerosol: human exposure levels during a Karenia brevis harmful algal bloom. Bull Environ Contam Toxicol 70:161–165. https://doi.org/10.1007/s00128-002-0170-y.Brevetoxin

Plankton*Net Data Provider at the Alfred Wegener Insitute for Polar and Marine Research Plankton*Net. hdl:10013/de.awi.planktonnet

Racault MF, Le Quéré C, Buitenhuis E et al (2012) Phytoplankton phenology in the global ocean. Ecol Indic 14:152–163. https://doi.org/10.1016/j.ecolind.2011.07.010

Ramanathan V (1981) The role of ocean-atmosphere interactions in the CO_2 climate problem. J Atmos Sci 38:918–930

Rao DVS, Quilliam MA, Pocklington R (1988) Domoic acid – a neurotoxic amino acid produced by the marine diatom *Nitzschia pungens* in culture. Can J Fish Aquat Sci 45:2076–2079. https://doi.org/10.1139/f88-241

Redfield AC (1958) The biological control of chemical factors in the environment. Am Sci 46:205–221. https://doi.org/10.5194/bg-11-1599-2014

Reid PC, Lancelot C, Gieskes WWC et al (1990) Phytoplankton of the North Sea and its dynamics: a review. Neth J Sea Res 26:295–331. https://doi.org/10.1016/0077-7579(90)90094-W

Reid PC, Colebrook JM, Matthews JBL et al (2003) The continuous plankton recorder: concepts and history, from plankton Indicator to undulating recorders. Prog Oceanogr 58:117–173. https://doi.org/10.1016/j.pocean.2003.08.002

Rhein M, Rintoul SR, Aoki S et al (2013) Observations: ocean. In: Stocker TF, Qin D, Plattner G-K et al (eds) Climate change 2013: the physical science basis. Contribution of Working Group I to the Fifth Assessment Report of the Intergovernmental Panel on Climate Change. Cambridge University Press, Cambridge/New York

Roy S, Llewellyn CA, Egeland ES et al (2011) Phytoplankton pigments: characterization, chemotaxonomy and applications in oceanography. Cambridge University Press, Cambridge

Sarmiento JL, Hughes TMC, Stouffer RJ et al (1998) Simulated response of the ocean carbon cycle to anthropogenic climate warming. Nature 393:245–249. https://doi.org/10.1038/30455

Schlüter MH, Kraberg A, Wiltshire KH (2012) Long-term changes in the seasonality of selected diatoms related to grazers and environmental conditions. J Sea Res 67:91–97. https://doi.org/10.1016/j.seares.2011.11.001

Seoane S, Riobó P, Franco J (2017) Haemolytic activity in different species of the genus *Prymnesium* (Haptophyta). J Mar Biol Assoc UK 97:491–496. https://doi.org/10.1017/S0025315416001077

Shen Y, Benner R (2018) Mixing it up in the ocean carbon cycle and the removal of refractory dissolved organic carbon. Sci Rep 8:2542. https://doi.org/10.1038/s41598-018-20857-5

Shi D, Xu Y, Hopkinson BM et al (2010) Effect of ocean acidification on iron availability to marine phytoplankton. Science 327:676–679. https://doi.org/10.1126/science.1183517

Sieburth JM, Smetacek V, Lenz J (1978) Pelagic ecosystem structure: heterotrophic compartments of the plankton and their relationship to plankton size fractions. Limnol Oceanogr 23:1256–1263. https://doi.org/10.4319/lo.1978.23.6.1256

Silver MW, Bargu S, Coale SL et al (2010) Toxic diatoms and domoic acid in natural and iron enriched waters of the oceanic Pacific. Proc Natl Acad Sci U S A 107:20762–20767. https://doi.org/10.1073/pnas.1006968107

Smith VH (2006) Responses of estuarine and coastal marine phytoplankton to nitrogen and phosphorus enrichment. Limnol Oceanogr 51:377–384. https://doi.org/10.4319/lo.2006.51.1_part_2.0377

Stefels J, Steinke M, Turner S et al (2007) Environmental constraints on the production and removal of the climatically active gas dimethylsulphide (DMS) and implications for ecosystem modelling. Biogeochemistry 83:245–275. https://doi.org/10.1007/978-1-4020-6214-8_18

Sukhanova IN, Flint MV, Pautova LA et al (2009) Phytoplankton of the western Arctic in the spring and summer of 2002: structure and seasonal changes. Deep Res Pt II 56:1223–1236. https://doi.org/10.1016/j.dsr2.2008.12.030

Sundh I (1992) Biochemical composition of dissolved organic carbon derived from phytoplankton and used by heterotrophic bacteria. Appl Environ Microbiol 58:2938–2947

Swart S, Thomalla SJ, Monteiro PMS (2015) The seasonal cycle of mixed layer dynamics and phytoplankton biomass in the Sub-Antarctic Zone: a high-resolution glider experiment. J Mar Syst 147:103–115. https://doi.org/10.1016/j.jmarsys.2014.06.002

Taylor FJR, Hoppenrath M, Saldarriaga JF (2008) Dinoflagellate diversity and distribution. Biodivers Conserv 17:407–418. https://doi.org/10.1007/s10531-007-9258-3

Thackeray SJ (2012) Mismatch revisited: what is trophic mismatching from the perspective of the plankton? J Plankton Res 34:1001–1010. https://doi.org/10.1093/plankt/fbs066

Thompson PA, Levasseur ME, Harrison PJ (1989) Light-limited growth on ammonium vs. nitrate: what is the advantage for marine phytoplankton? Limnol Oceanogr 34:1014–1024. https://doi.org/10.4319/lo.1989.34.6.1014

Trainer VL, Hickey BM, Lessard EJ et al (2009) Variability of *Pseudo-nitzschia* and domoic acid in the Juan de Fuca eddy region and its adjacent shelves. Limnol Oceanogr 54:289–308. https://doi.org/10.4319/lo.2009.54.1.0289

Tréguer P, Jacques G (1992) Dynamics of nutrients and phytoplankton, and fluxes of carbon, nitrogen and silicon in the Antarctic Ocean. Polar Biol 12:149–162. https://doi.org/10.1007/BF00238255

Tréguer P, Nelson DM, Vanbennekom AJ et al (1995) The silica balance in the world ocean – a reestimate. Science 268:375–379. https://doi.org/10.1126/science.268.5209.375

Tuchman NC, Schollett MA, Rier ST et al (2006) Differential heterotrophic utilization of organic compounds by diatoms and bacteria under light and dark conditions. Hydrobiologia 561:167–177. https://doi.org/10.1007/s10750-005-1612-4

Turner RE, Qureshi N, Rabalais NN et al (1998) Fluctuating silicate:nitrate ratios and coastal plankton food webs. Proc Natl Acad Sci U S A 95:13048–13051. https://doi.org/10.1073/pnas.95.22.13048

Turner RE, Rabalais NN, Justić D (2017) Trends in summer bottom-water temperatures on the northern Gulf of Mexico continental shelf from 1985 to 2015. PLoS One 12:e0184350. https://doi.org/10.1371/journal.pone.0184350

Van Mooy BAS, Fredricks HF, Pedler BE et al (2009) Phytoplankton in the ocean use non-phosphorus lipids in response to phosphorus scarcity. Nature 458:69–72. https://doi.org/10.1038/nature07659

Van Oostende N, Fawcett SE, Marconi D et al (2017) Variation of summer phytoplankton community composition and its relationship to nitrate and regenerated nitrogen assimilation across the North Atlantic Ocean. Deep Res Pt I 121:79–94. https://doi.org/10.1016/j.dsr.2016.12.012

Villafañe VE, Banaszak AT, Guendulain-García SD et al (2013) Influence of seasonal variables associated with climate change on photochemical diurnal cycles of marine phytoplankton from Patagonia (Argentina). Limnol Oceanogr 58:203–214. https://doi.org/10.4319/lo.2013.58.1.0203

Villafañe VE, Valiñas MS, Cabrerizo MJ et al (2015) Physio-ecological responses of Patagonian coastal marine phytoplankton in a scenario of global change: role of acidification, nutrients and solar UVR. Mar Chem 177:411–420. https://doi.org/10.1016/j.marchem.2015.02.012

Walther GR, Post E, Convey P et al (2002) Ecological responses to recent climate change. Nature 416:389–395. https://doi.org/10.1038/416389a

Ward BB, Rees AP, Somerfield PJ et al (2011) Linking phytoplankton community composition to seasonal changes in f-ratio. ISME J 5:1759–1770. https://doi.org/10.1038/ismej.2011.50

Wasmund N, Tuimala J, Suikkanen S et al (2011) Long-term trends in phytoplankton composition in the western and Central Baltic Sea. J Mar Syst 87:145–159. https://doi.org/10.1016/j.jmarsys.2011.03.010

Wells ML, Trainer VL, Smayda TJ et al (2015) Harmful algal blooms and climate change: learning from the past and present to forecast the future. Harmful Algae 49:68–93. https://doi.org/10.1016/j.hal.2015.07.009

Wheeler PA, North BB, Stephens GC (1974) Amino acid uptake by marine phytoplankton. Limnol Oceanogr 19:249–259. https://doi.org/10.4319/lo.1974.19.2.0249

White AW (1977) Dinoflagellate toxins as probable cause of an Atlantic herring (*Clupea harengus harengus*) kill, and pteropods as apparent vector. J Fish Res Board Can 34:2421–2424. https://doi.org/10.1139/f77-328

Wiltshire KH, Malzahn AM, Wirtz K et al (2008) Resilience of North Sea phytoplankton spring bloom dynamics: an analysis of long-term data at Helgoland Roads. Limnol Oceanogr 53:1294–1302. https://doi.org/10.4319/lo.2008.53.4.1294

Wiltshire KH, Kraberg A, Bartsch I et al (2010) Helgoland roads, north sea: 45 years of change. Estuar Coasts 33:295–310. https://doi.org/10.1007/s12237-009-9228-y

Wiltshire KH, Boersma M, Carstens K et al (2015) Control of phytoplankton in a shelf sea: determination of the main drivers based on the Helgoland Roads Time Series. J Sea Res 105:42–52. https://doi.org/10.1016/j.seares.2015.06.022

Yoch DC (2002) Dimethylsulfoniopropionate: its sources, role in the marine food web, and biological degradation to dimethylsulfide. Appl Environ Microbiol 68:5804–5815. https://doi.org/10.1128/AEM.68.12.5804

Yool A, Popova EE, Anderson TR (2011) MEDUSA-1.0: a new intermediate complexity plankton ecosystem model for the global domain. Geosci Model Dev 4:381–417. https://doi.org/10.5194/gmd-4-381-2011

Yoshiyama K, Mellard JP, Litchman E et al (2009) Phytoplankton competition for nutrients and light in a stratified water column. Am Nat 174:190–203. https://doi.org/10.1086/600113

Zingone A, Harrison PJ, Kraberg A et al (2015) Increasing the quality, comparability and accessibility of phytoplankton species composition time-series data. Estuar Coast Shelf Sci 162:151–160. https://doi.org/10.1016/j.ecss.2015.05.024

Reading the Book of Life – Omics as a Universal Tool Across Disciplines

Jan David Brüwer and Hagen Buck-Wiese

Abstract

In the last centuries, new high-throughput technologies, including sequencing and mass-spectrometry, have emerged and are constantly refurbished in order to decipher the molecular code of life. In this review, we summarize the physiological background from genes via transcriptome to proteins and metabolites and discuss the variety of dimensions in which a biological entity may be studied. Herein, we emphasize regulatory processes which underlie the plasticity of molecular profiles on different ome layers. We discuss the four major fields of omic research, namely genomics, transcriptomics, proteomics, and metabolomics, by providing specific examples and case studies for (i) the assessment of functionality on molecular, organism, and community level; (ii) the possibility to use omic research for categorization and systematic efforts; and (iii) the evaluation of responses to environmental cues with a special focus on anthropogenic influences. Thereby, we exemplify the knowledge gains attributable to the integration of information from different omes and the enhanced precision in predicting the phenotype. Lastly, we highlight the advantages of combining multiple omics layers in assessing the complexity of natural systems as meta-communities and -organisms.

J. D. Brüwer (✉)
Red Sea Research Center, Division of Biological and Environmental Science and Engineering (BESE), King Abdullah University of Science and Technology (KAUST), Thuwal, Saudi Arabia

Faculty of Biology and Chemistry, University of Bremen, Bremen, Germany

Max Planck Institute for Marine Microbiology, Bremen, Germany
e-mail: bruewer_j@gmx.de

H. Buck-Wiese (✉)
Faculty of Biology and Chemistry, University of Bremen, Bremen, Germany

Max Planck Institute for Marine Microbiology, Bremen, Germany
e-mail: h.buckwiese@googlemail.com

Introduction and Historical Background

The discovery of nucleic acids in 1896 by Friedrich Miescher and the suggestion of deoxyribonucleic acid (DNA) as the genetic material by Avery, MacLeod, and McCarty in 1943 revolutionized the life sciences (Avery et al. 1943; Dahm 2005). Genomics, from the suggested word "genome" for haploid chromosome sets by Hans Winkler (Noguera-Solano et al. 2013), arose with the aim to decipher the molecular language. It took another 10 years before Franklin, Wilkins, Watson, and Crick unraveled the double-helical structure of DNA in 1953 (Dahm 2005). The conversion from nucleotide sequence into amino acid was first recognized, when Heinrich Matthaei and Marshall Nirenberg discovered that the RNA sequence of three Uracil bases codes for the amino acid phenylalanine with their so-called Poly-U experiment (Nirenberg 2004; Dahm 2005). Five years later, in 1966, the translation of all base combinations into the 20 protein-forming amino acids had been resolved (Nirenberg 2004). For nucleotide sequence analysis, Frederick Sanger and colleagues developed the first widely applied method, the Sanger sequencing, in 1977 and, thus, established the foundation for modern genomic and transcriptomic research (Box 1) (Sanger et al. 1977). In more recent years, high-throughput molecular technologies, e.g., next-generation sequencing (NGS) (Box 1) and mass spectrometry (Box 2) have developed, enabling genome-scale deciphering of the molecular signatures, which encode life on earth.

These technologies provide the opportunity for a wide range of studies which can be divided into four major fields according to the targeted molecules: genomics, transcriptomics, proteomics, and metabolomics. In definition, genomics describes the analysis of any genetic material (DNA) isolated from an organism or the environment. It includes, for example, whole genome sequencing and detection methods such as environmental DNA (eDNA). Transcriptomics is the study of any form of RNA, including messenger RNA (mRNA), transfer RNA (tRNA), ribosomal RNA (rRNA),

S. Jungblut et al. (eds.), *YOUMARES 8 – Oceans Across Boundaries: Learning from each other*,
https://doi.org/10.1007/978-3-319-93284-2_6

Box 1: Nucleic Acid Sequence Analysis

Background

The nucleic acids contain information in the shape of a code constituting of two purines, Adenine A and Guanine G, and two pyrimidine bases, Cytosine C and either Thymine T in DNA or Uracil U in RNA. Selective pairing of A with T and G with C gives rise to the stable double strand structure of DNA and confers a mechanism to pass on the information in the coded sequence via polymerization, i.e. DNA replication and RNA transcription (in this case, substituting U for T) (Alberts et al. 2008; Klug et al. 2012). The widely applied DNA/RNA sequencing methods to read the nucleotide code are based on this selective binding.

The first sequencing developed by Sanger in the 1970s required four separate polymerizations, each with a fraction of dideoxynucleotides (ddNTPs) which would terminate the elongation – hence the name 'chain-termination method' (Lu et al. 2016). Parallel size separation (using gel-electrophoresis) of the synthesized strands, each with a specific dd-nucleotide at the end, and subsequent radioactive detection allowed to infer the order of the different bases in the template's sequence. Modern techniques for Sanger sequencing are based on fluorescently labeled ddNTPs, emitting differentiable signals, which can be detected by a laser and evaluated electronically (Schuster 2008). Recently developed second-generation sequencing (such as Illumina) use dNTPs which emit a base-specific fluorescent signal when the phosphordiester bond is formed and the DNA elongated. Different to the traditional Sanger sequencing, the process does not require termination and every elongation process yields a signal per nucleic acid. The advantages of these sequencing methods lie in high-throughput through the simultaneous sequencing of multiple DNA/RNA fragments (e.g., from environmental samples) from a variety of organisms with usually reliable high-quality results (Schuster 2008). The drawbacks belay in comparatively short sequence strands (about 100–300 bp), demanding assemblies to solve the 'puzzle' of different short fragments. However, third-generation sequencing (such as offered by PacBio with the SMRT cell) make use of double-stranded DNA with two hairpin structures at the end, the so-called SMRTbell. This way, fragments of several thousand base pairs may be sequenced, which may subsequently be complemented by shorter fragments to maintain the quality standard via high coverage (Rhoads and Au 2015).

The emerging fourth generation sequencing technique, the nanopore sequencing (such as the MinIION by Oxford Nanopore Techniques), does not require previous amplification but aims at directly sequencing single molecules and promises to sequence tens of kilobases (kb). A membrane is equipped with nanopores that is selectively permeable for DNA and RNA. An electric force is driving the electrophoresis of the negatively charged fragments towards the anode and, thus, into the membrane. A motor protein is ratcheting the fragment through the membrane. This causes different perturbations of the membrane current depending on the nucleotide, which may be computationally translated into base sequences (Cherf et al. 2012; Feng et al. 2015). Different from previous sequencing methods, the fourth generation nanopore sequencing may even be used to analyze proteins, polymers, and other single-strand macromolecules (Feng et al. 2015).

Strategies

To target a particular portion of the queried nucleotide sequence, e.g., targeting the 16S rRNA/rDNA of microorganisms for phylogenetic assessment, specific primer sequences can be used.

A variety of techniques grouped under the description of restriction site-associated DNA sequencing (RADseq) is currently in scope for assessing genotypic differences of a range of organisms, including those with largely unknown genomes. These techniques are based on digestion of isolated DNA with one or few restriction enzymes and subsequent sequencing of resulting fragments. As most restriction sites prevail among specimen and closely related species, predominantly similar sets of loci are sequenced, at which different alleles can be identified (Andrews et al. 2016).

In case of whole genome sequencing using NGS, short fragments of DNA of few hundred base pair length are inserted into vectors, called library. To aid in later assembly, libraries with shotgun mate pair fragments of specified greater lengths complement the short vector sequences, which consist of a high fragment coverage. After standard quality controls of the reads (including adapter and primer removal), the assembly of the genome from the multitude of small sequences relies on overlapping regions and mate pairs (e.g., Baumgarten et al. 2015).

Prior to RNA sequencing, the RNA-template has to be transcribed into a cDNA, using a reverse transcriptase. A quantitative interpretation of transcriptome and

(continued)

Box. 1: (continued)
metagenome data analyses has to be treated with caution due to exponential amplification steps. However, normalization steps to account for differential amplification within samples, as well as differential sequencing depth across samples, may be used to gain better estimates of quantities as well as maintaining data comparable. This may be achieved by calculation of "Fragments Per Kilobase of exon model per million Mapped reads" (FPKM). Further biostatistic normalization to eliminate sequencing biases, e.g., using nCounter (Geiss et al. 2008), may be helpful in evaluation of the data (Liu et al. 2016).

Box 2: Mass-Per-Charge of Peptides and Metabolites

Protein and targeted metabolite analyses, including antibody, ionization, and spectroscopy approaches, date back more than a century. Technical advances in the field of mass spectrometry (MS) are, however, revolutionizing the possibilities in these fields, now supporting proteome-wide peptide sequence identification and untargeted metabolome characterization and comparison.

Protein studies have traditionally been relying on the usage of antibodies on a small scale but as a precisely localizing method. Nevertheless, limited availability of antibodies for different protein structures, comparatively low throughput, high costs for antibody production, and low quantitative comparability due to lacking standards have hampered proteome-scale assessments. Deep high-throughput MS has emerged as an opportunity to read-out relative and absolute concentrations of proteins genome-wide. Label-free quantification via tandem mass spectrometry (MS/MS) allows the recognition of individual peptide spectra. These are compared to entries in databases, optimally containing all peptide sequences expected to be present, but few irrelevant ones. Current developmental and research efforts, though, target the *de novo* determination only from the peptide's spectrum (Liu et al. 2016; Ruggles et al. 2017).

Current-standard for untargeted metabolome analysis is a liquid chromatography coupled with mass spectrometry. Since theoretically every type of small molecule possesses a unique retention time and a unique mass-per-charge ratio, this procedure separates and characterizes each metabolite. Adjustments in liquid phases regarding hydrophilic and hydrophobic components and their directions can improve the resolution achieved by retention. The experimental approach requires a comparison of the metabolic profile yielded by the mass spectrometer either to a standard or between two or more samples. A bioinformatic overlay of the produced profiles provides information on significant differences in abundance and thereby delineates molecules of interest. Their mass-per-charge ratios now serve to find reference molecules in databases. However, due to the novelty of metabolome-wide studies, there is a considerable number of molecules, which remains to be identified and entered into the repositories. If there is a mass-to-charge hit and standards are available for the molecules of interest, the identity can be confirmed via retention times and MS/MS profiles (Patti et al. 2012).

and micro RNA (miRNA). The study of the protein content of an organism and its respective functions is comprised by proteomics. Metabolomics deals with any small molecules that are produced or ingested by an organism (Handelsman 2004; Patti et al. 2012; Pascault et al. 2015; Beale et al. 2016; Liu et al. 2016).

In this review, we will delineate the physiological background of omics research and will exemplify the wide spectrum of applicability under aspects of functionality, systematics, and response to environmental cues. Finally, we aim to highlight the significance of multi-omics for an in-depth understanding of complex systems.

Physiological Background

The genome depicts the inherited foundation within a cell and is – apart from epigenetic changes – consistent in almost every healthy somatic cell of a multicellular organism. It encodes for the high variety of proteins, as well as non-protein coding sequences, such as **r**ibosomal RNA (rRNA), **t**ransfer RNA (tRNA), and **mi**cro RNA (miRNA) (Alberts et al. 2008).

Gene expression begins with the transcription of a DNA sequence into a pre-mRNA. The newly synthesized nucleotide sequence constitutes a reverse complement of the coding strand with ribose phosphates instead of deoxyribose phosphates forming the backbone, and Uracil pairing with Adenine instead of Thymine (Alberts et al. 2008).

Promoter sequences upstream of open reading frames, the DNA region to be transcribed, contribute significantly to expression by recruiting the RNA polymerase. However, expression profiles remain a complex puzzle due to influences of *cis*- and *trans*-regulatory motifs and binding of transcription factors. Further, epigenetic modifications as cytosine methylation, histone acetylation, and changes in chromatin structure may lead to a subsequently altered transcriptome (Alberts et al. 2008).

Due to the translation of mRNA into amino acids via the triplet code, proteins are in a qualitative sense direct product of genes with mRNA transcripts as intermediates. This allows functional predictions of genes via comparison of sequence similarities to annotated genes in a highly curated database, such as NCBI RefSeq (O'Leary et al. 2016).

In eukaryotes, the RNA sequence is, nevertheless, subject to possible modifications, which may impede the recognition of a gene-protein pair. Variable intron removal from maturing mRNAs by splicing may lead to multiple isoforms from a single pre-mRNA (Alberts et al. 2008). Further, RNA editing (see example in section "Response to environmental cues") may introduce sequence alterations as a co- or post-transcriptional modification, not to be confused with decapping, splicing, and poly(A)-removal (see e.g., Klug et al. 2012; Liew et al. 2017).

Sequence Alterations Influence Protein Functioning

Non-synonymous sequence alterations, i.e. single nucleotide exchanges, deletions, or insertions, may significantly influence or disrupt protein functioning. Firstly, a protein's physiological role is sensitive to secondary and tertiary structure formation and stability (e.g., α-helix and cysteine double bounds, respectively), which may be significantly altered due to aforementioned non-synonymous sequence alterations. Secondly, the phosphorylation of serine, threonine, and tyrosine, as well as acetylation and ubiquitylation of lysine are major post-translational modifications, which are involved in triggering activation and degradation (reviewed in Klug et al. 2012; Ruggles et al. 2017). Thus, sequence alterations which lead to the exchange of one of these four amino acids are likely to affect the protein's performance. Lastly, guiding and localization sequences are essential to position proteins in cellular compartments or membranes. For example, the nuclear membrane of most eukaryotic cells is freely permeable to molecules up to 9 nm. Macromolecules of greater sizes depend on a specific nuclear localization sequence (NLS), which mediates the transport. Alteration of a single amino acid may result in a dysfunction of the NLS and the decreased transport efficiency of the macromolecule into the nucleus (Zanta et al. 1999).

Consequently, complex reactions such as protein-protein interactions, transcription cascades, signaling networks, and metabolic pathways may be altered by single nucleotide exchanges (Kim et al. 2016).

Quantitative Regulation of the Proteome

The physiological roles of RNA reach far beyond the gene to protein transmission, where (pre)-mRNA, rRNA, and tRNA are allocated. For instance, the translation-regulatory roles of miRNAs have been discovered in 1993 (Almeida et al. 2011; see section "Functionality"). In humans, for example, at least 70% of the genome is transcribed into RNA, but only about 2% are effectively translated into protein (Pheasant and Mattick 2007). Consequently, immense proportions of the genome are suggested to encode for quantitative regulation, which can be detected with current omics approaches (Klug et al. 2012). The current state of knowledge considers the abundance of mRNA transcripts to explain up to 84% of the respective protein concentration. This value may vary depending on the respective mRNA, mainly attributable to sequence- or splice isoform-dependent translation rates (Liu et al. 2016). Additionally, induced changes in gene expression, e.g., due to environmental cues, may only be detectable in the proteome after a lag phase (e.g., 6–7 h in mammals; see also section "Response to environmental cues").

The number of copies per gene does not generally define respective transcript nor protein abundances. Genetic diseases or tumors may induce gene copy number alterations (CNAs). In such cases, transcriptome and proteome do mostly not exhibit the same fold changes as could be expected from the CNAs in the genome. Negative feedback loops, called buffering, may occur on the transcriptional and translational level. There are, however, plenty of sequence-specific exceptions to this general pattern, which are, therefore, possibly involved in the symptomatic (Liu et al. 2016 and references therein).

Metabolomics

The entirety of small molecules within an organism, the metabolome, constitutes a biochemical representation. It is substance to continuous turn-over, alteration, and relocation by the physiological machinery of RNAs and, most of all, proteins (e.g., Patti et al. 2012; Beale et al. 2016). While targeted metabolomics assesses only a fraction of particular interest, newly emerged technologies enable untargeted detection and quantification of almost the entire metabolome (Patti et al. 2012).

Untargeted metabolomics combined with genomic and/or transcriptomic data may allow the inference of gene and protein function, as well as metabolic cascades and pathways. It becomes possible to detect physiological attributes such as the use of substrates, secondary metabolite secretion, or possible inter-individual signaling, and connect these to the presence or expression of genes (Freilich et al. 2011;

Llewellyn et al. 2015; Kim et al. 2016). In combination with information on intrinsic or even environmental ontology, it may provide insights into the plastic phenotypic range and might suggest possible adaptation or acclimatization responses (Dick 2017).

Functionality

A genome-wide survey on potential open reading frames and prediction of gene function can help to characterize an organism or study its ecological background. An example from marine plant genetics is the recently published genome of the true seaweed *Zostera marina* (commonly referred to as eelgrass). It contains 20,450 genes, of which a majority (86.6%) were validated using a transcriptomic approach (Olsen et al. 2016). Functional annotation revealed gene losses and gains that could be attributed to the marine habitat. Those included losses of stomatal differentiation, airborne communication, and immune-system-related genes, to name only three examples (Olsen et al. 2016).

Using next-generation sequencing or quantitative PCR (qPCR) approaches, transcript abundances may be assessed (Liu et al. 2016; see also Box 1). As such, this provides a good possibility to estimate biological activity rather than the mere presence and abundance. In microbial ecology, for example, the *nifH* gene is a common biomarker for nitrogen-fixing bacteria, i.e. diazotrophs (Gaby and Buckley 2012). Pogoreutz et al. (2017) queried gene and transcript abundance of *nifH* in order to investigate nitrogen fixation in the coral holobiont (see Box 3 for details on the metaorganism/holobiont concept). They detected autotrophic corals to exhibit a higher *nifH* gene abundance, correlated with increased expression rates. Consequently, the authors suggested that low nitrogen-uptake via heterotrophy may be compensated by the microbial component of the holobiont.

Transcriptomes are interesting in another regard, as some RNA species have regulatory functions, e.g. miRNAs which are short (about 22 base pairs) single-stranded RNA molecules. They have the potential to align with mRNA via sequence identity and thereby either inhibit the translation or induce degradation (Gottlieb 2017). A single miRNA may bind to several different mRNAs and *vice versa* (Selbach et al. 2008). In humans, the Chromosome 19 miRNA cluster (C19mc) is almost exclusively expressed in the extra-embryonic tissue of the placenta (Luo et al. 2009) and seems to be an important component of the immune system during viral infections (Delorme-Axford et al. 2013). C19mc has been suggested to be a key component of embryonic-maternal communication, as well as essential to suppress a maternal immune response (Gottlieb 2017).

Box 3: Metaorganisms

Evidence supports the notion that all multicellular organisms live in synergistic interdependence with a variety of microorganisms, including bacteria, archaea, and viruses. Together, they form a complex entity, termed holobiont or metaorganism (McFall-Ngai et al. 2013; Bosch and Miller 2016). The microbial community of a metaorganism constantly influences the performance of a metaorganism. The human gastrointestinal tract (GI) microbiota, for example, is of great importance for the digestion and ingestion of nutrients and metabolites. In addition, the human microbiota have been suggested to have great impacts on the behavior and even neurological functions (Turnbaugh et al. 2007; Biagi et al. 2012; Sampson and Mazmanian 2015).

Various studies can show that a metaorganism's microbiome changes and that at least part of the metaorganism can compensate for environmental stressors (Bosch and Miller 2016; Buck-Wiese et al. 2016; Hernandez-Agreda et al. 2016; Ziegler et al. 2017). In order to study the biology of a multicellular organism, the whole metaorganism should therefore be respected.

In a metagenomics and -proteomics study, Leary et al. (2014) assessed the microbial community of biofilms on two different navy ships. The metagenomics data revealed prokaryotic signature to be most abundant on both ships. However, the meta-proteome on the first ship hull was dominated by eukaryotic cytoskeleton proteins, while diatom carbon fixation and photosystem related proteins were most abundant on the second hull. The authors argue that observed differences between metagenomics and -proteomic results may be due to retention of prokaryotic DNA in the biofilm, especially of inactivated or dead bacteria. Further, the eukaryotic proteome is usually larger in size and may exhibit a higher dynamic range of gene expression. In this case, a single omics approach may have provided misleading results (Leary et al. 2014; Beale et al. 2016).

The relatively novel field of untargeted metabolomics can provide sufficiently broad information to infer previously unknown functions. For example, stony corals are in constant association with a variety of microorganisms (Rohwer et al. 2002). About a decade ago, a study by Ritchie (2006) could show that bacteria isolated from the coral mucus microbiome inhibit the growth of several gram-positive and -negative bacteria, and suggested an antimicrobial activity. In addition, Shnit-Orland et al. (2012) could show that *Pseudoalteromonas* spp., a frequent coral mucus symbiont, secrets antimicrobial agents against gram-positive strains (see also Brüwer et al. within the abstracts related to the chapter "Tropical Aquatic Ecosystems across Time, Space, and Disciplines"). Untargeted metabolomics could be applied to investigate these inferred substances as putative components for medicinal use.

In fact, omic and multi-omic studies on functionality elucidate the significance of the molecular code on the phenotype level. They thereby contribute to the body of knowledge by which we can extrapolate information from molecular reads. Ultimately, they enable us to "write" in the book of life.

Systematics

Scientists often aim to explain complex natural phenomena with comprehensive models, which are constantly adapted and expanded. The species model, for example, is – if not updated – at least constantly discussed in the scientific literature, especially regarding prokaryotes (e.g., Stackebrandt et al. 2002; Wilmes et al. 2009; Amann and Rosselló-Móra 2016).

Mutations are an essential source of genetic variability, which are estimated to occur at (region-specific) constant rates per replication for closely related species (Gillooly et al. 2005) and are used to resolve phylogenetic relationships. Polymorphisms (mostly single nucleotide polymorphism, SNP) in the genome create different alleles that may be targeted by specific restriction enzymes. Some techniques and methods, such as restriction site-associated DNA sequencing (RADseq), enable the detection of many SNPs on a population level in order to study the genetic background of populations and migratory dynamics (Andrews et al. 2016; see also Box 1).

As some parts of the genome are more prone to mutations which could lead to lethal dysfunctions of the encoded molecule, they depict highly preserved regions with significantly lower mutation rates compared to other parts of the genome. Non-lethal mutations that do happen within these regions usually remain in the genome and may be queried by amplification and/or sequencing for phylogenetic assignment.

The DNA regions coding for the small subunit of ribosomal RNA in prokaryotes, the 16S rRNA gene, and the mitochondrially encoded cytochrome c oxidase I in eukaryotes, the COI gene, constitute such highly conserved areas of genetic information, which are commonly used for taxonomic characterization by barcoding approaches (Pimm et al. 2014; see also McCarthy et al. within the abstracts related to this chapter for an ancient DNA example).

Traditional microbial characterization approaches require a cultivation prior to phenotypic classification. However, most marine microbes are very challenging to or not at all cultivable (Pedrós-Alió 2012; Epstein 2013; Amann and Rosselló-Móra 2016). The advances of next-generation sequencing methods provide the possibility to detect and phylogenetically classify a vast amount of microbes simultaneously, including those that are not cultivable (see also Weinheimer et al. within the abstracts related to this chapter). In a recently published microbial ecology study, Röthig et al. (2016) aimed to characterize the microbiome of the model metaorganism *Aiptasia*. Besides a metagenomics approach by DNA isolation of *Aiptasia* tissue and subsequent high throughput 16S rRNA gene sequencing, the authors applied a culture dependent approach. They detected 295 different taxa in the metagenomic data, while they were only able to culture 14 of those (with a 100% match in the 16S rRNA) (Röthig et al. 2016; see also Slaby et al. within the abstracts related to this chapter for a marine sponge microbiome).

Similarly, the gastrointestinal (GI) microbial community is an important component of a vertebrate metaorganism (see Box 3). Dewar et al. (2013) assessed the residual GI microbiota in feces of the king, gentoo, macaroni, and little penguins by 16S rRNA gene sequencing. The authors detected a diverse microbial community (>5 k operational taxonomic units (OTUs) identified) with significant differences in relative abundances of microbial taxa across penguin species. They further identified known human pathogens (including *Helicobacter, Veillonella, Mycoplasma*, etc.), although their virulence in penguins or other sea-birds remains questionable (Dewar et al. 2013, 2014).

Samples of environmental DNA (eDNA) may be subject to similar analysis of highly conserved genome regions. Such studies would aim to detect DNA traces in the environment to extrapolate on the presence and potentially even abundance of the corresponding organisms (Taberlet et al. 2012; Valentini et al. 2016; see also Mauvisseau et al. within the abstracts related to this chapter). A recent comparison of the efficiency of traditional surveys and eDNA approaches aimed to detect amphibians and fishes in natural aquatic environments by designing group (i.e. amphibian and fish) specific primers of mitochondrial DNA (mtDNA) (Valentini et al. 2016). Two amphibian species (*Triturus marmoratus* and *Pelophylax* sp.) were observed with conventional methods but not detected via barcoding, whilst a total of 64 species could exclusively be recorded in the eDNA samples. The fish survey prompted a similar result. Thus, ecological surveys using eDNA appear to be more thorough and accurate. In addition, they are less destructive, are more (cost) efficient, and do not fully rely on the taxonomic knowledge and species identification of experts, compared to commonly applied surveys. Furthermore, it will be less challenging to establish standardized protocols, thus, making survey studies more comparable, especially if they are conducted across various scientific laboratories (Valentini et al. 2016).

In virology, characterization efforts demand the verification of newly identified viruses by visual evidence (e.g.,

electron microscopy), as well as molecular methods (e.g., sequencing). However, a panel of leading virologists recently proposed that viruses only known from metagenomic samples need to be incorporated into the official classification scheme by the International Committee on Taxonomy of Viruses (ICTV) (Fauquet and Mayo 2001; Simmonds et al. 2017). Paez-Espino et al. (2016), for example, analyzed around 5 terabyte metagenomics data of diverse samples, which led to the discovery of 125,000 new predicted viral genomes as well as a massive increase in the number of putatively identified viral genes (Paez-Espino et al. 2016; Simmonds et al. 2017).

To conclude, metagenomic and metatranscriptomic data may be used to detect, characterize, and taxonomically rank all 'lifeforms', including previously unknown and uncultured organisms and viruses.

Metabolomic analyses, for example of lipids, are commonly performed in marine plankton research to identify the dietary preferences of plankton species. These analyses are based on the fact that some fatty acids (biomarkers) are characteristic of specific groups of phyto- or zooplankton (e.g., 16:1(n-7) for diatoms, 18:4(n-3) for dinoflagellates and 18:1(n-9) for metazoans) and are incorporated into the consumers body tissue largely unmodified, thus retaining a dietary signature (Graeve et al. 1994; Dalsgaard et al. 2003). For example, the fatty acid pattern of the Arctic copepod *Calanus finmarchicus* typically reflects a dinoflagellate nutrition. However, in *in vitro* feeding experiments with the diatom *Thalassiosira antarctica* over several weeks, the fatty acid composition depicted a change towards a diatom-like signature/profile (Graeve et al. 1994) showing the unchanged incorporation of dietary fatty acids. Thus, the fatty acid biomarker concept might allow differentiating the source of phytoplankton (diatom vs. dinoflagellate) and if a species mostly feeds on phytoplankton (herbivory) or other zooplankton (carnivory) giving hints on its trophic position and role in the food web. In this regard, studying the proteome or metabolome can provide clues on the functional role of an organism and might even provide a glimpse on the phylogeny (Jones et al. 2014; Llewellyn et al. 2015).

Functional groups of proteins are not necessarily linked to phylogeny. The Gene Ontology (GO) provides a hierarchical structure based on functional grouping of the gene products. Thus, it helps to identify the physiological role based on molecular function, cellular component, and biological process (The Gene Ontology Consortium 2004, 2017). The Kyoto Encyclopedia of Genes and Genomes (KEGG) has been established to interrelate genes based on their high-level function. Identifying a queried gene in the KEGG PATHWAY database integrates it in a corresponding pathway and may show its connectivity to other genes, thus allows to extrapolate on its physiological impact (Kanehisa et al. 2016, 2017). In summary, GO-terms and the KEGG database may help to identify involved physiological processes, as well as potentially easing the comparison amongst different organisms or species.

Response to Environmental Cues

The recent geological era is called the Anthropocene, because human activities have severely impacted geology and ecosystems, including the alteration of carbon fluxes. Long-term temperature rise and increased short-term fluctuations (IPCC 2007, 2014), as well as pollution accidents, such as oil spills (e.g., McNutt et al. 2012), challenge the adaptive capacities of species. However, technological advances allow us to utilize those capacities to increase the yield of biological products (e.g., Park et al. 2015). To subsequently suggest strategies for conservation and bioengineering, the different omics fields can be a useful tool to identify genetic, transcriptomic, proteomic, or metabolic responses to environmental cues.

A consortium of marine geneticists proposed to investigate the adaptability and resilience to environmental stress in a three-step approach (Voolstra et al. 2015). Firstly, species along a natural gradient should be queried for genetic, epigenetic, or transcriptional differences, to secondly experimentally test the resilience of specimen from each extremum to the corresponding opposite poles. Thirdly, possible (epi-) genetic or transcriptional differences should be investigated and their impacts regarding the environmental parameter evaluated (Voolstra et al. 2015). Following this approach, Ziegler et al. (2017) assessed the plasticity of the microbial community within a metaorganism in response to temperature regimes (Box 3). They were able to show that the microbiome significantly contributed to thermal-stress resilience (Ziegler et al. 2017). This underlines the role of a metaorganism's microbial community when reacting to and coping with environmental change (Bosch and Miller 2016; Buck-Wiese et al. 2016).

Liew et al. (2017) exposed a facultative endosymbiotic dinoflagellate to acute light and temperature stress and sequenced the transcriptome, to investigate stress-induced RNA editing (see section "Physiological background"). They observed base exchanges from RNA editing most prominently responsive to a heat stress treatment (Liew et al. 2017; see also Olschowsky et al. within the abstracts related to this chapter). RNA editing may induce non-synonymous substitutions which could alter protein functioning as well as stability and thereby contribute to acclimatization (see section "Physiological background").

The exact same dataset was used by Brüwer et al. (2017), who *in silico* detected a diverse viral community associated with the dinoflagellate and observed differential viral gene expression upon heat treatment. As in this example, next-generation sequencing often contains "by-catch" of host-associated microorganisms and viruses, which may be assessed separately (see also Levin et al. 2017; Brüwer and Voolstra 2018; Brüwer and Voolstra within the abstracts related to this chapter).

Human-induced catastrophes, such as oil spills, result in immediate long-lasting changes of an ecosystem. In April 2010, the off-shore drilling platform Deepwater Horizon sank in the Gulf of Mexico, which resulted in 650 million liters of oil and gas being released into the deep sea (McNutt et al. 2012). All of the emissions combined could cover the Vatican state with an about 147 m thick layer. Kimes et al. (2013) assessed the oil spill affected deep sea-sediments, as commonly observed aerobic oil degradation may be hindered in anaerobic environments. Using a metabolomics approach, they detected an increase of benzyl succinates, a typical product of anaerobic oil biodegradation. An additional metagenomics analysis revealed an aggregation of anaerobic bacteria, in particular *Deltaproteobacteria*, in the respective sediments. This points towards an anoxic catabolism of hydrocarbons, thus suggesting a breakdown of oil (Kimes et al. 2013).

Metabolomics are frequently applied for process optimization and yield increase in bioengineering. In the field of algae-based biofuel production, nitrogen starvation of the algae *Chlamydomonas reinhardtii* has been shown to increase carbon assimilation, nitrogen metabolism, and triacylglycerol production (Park et al. 2015), triacylglycerol being the targeted metabolite.

In conclusion, the omics-toolbox provides the possibility to evaluate adaptation and acclimatization on a metaorganism scale. Due to its broad approach and the high throughput methods, rather unexpected pathways may be revealed.

Complex Systems and Multi-meta-omics

The individual omic techniques can only display a fraction of the biological complexity, as outlined above, since the measurable appearance of life (i.e., an ome) varies greatly depending on the layer accessed (e.g., genome, proteome, etc.). This constitutes, for example, in the turn-over and dynamic changes of the many intracellular components in response to the environment. Single-cell studies aim to tackle this complexity within individual cells by integrating multi-omic data (Bock et al. 2016; see Kalita et al. within the abstracts related to this chapter). On a larger scale, meta-data from many co-existing organisms are valuable to study the composition and interactions of communities.

Metagenomic and metatranscriptomic data may delineate community compositions. This information combined with a metabolic profile can reveal the ecological interactions *exempli gratia* in a microbial consortium (Freilich et al. 2011), or depict the metabolites in an environment as a functional trait of the respective community (Llewellyn et al. 2015). From such insights, the contributions of species to an ecosystem's functioning and productivity can be deduced. Teeling et al. (2012) collected samples of the North Sea twice a week over the course of a year for multi-omics analysis. The succession of different algal substrate degrading bacterial communities responded to occurring phytoplankton blooms. They concluded that high bacterioplankton diversity in a relatively homogeneous habitat as 'the ocean' may result from temporally distributed niches (Teeling et al. 2012).

Comparisons of species in a community, which are very different for example in size, trophic level, or lifestyle, often lack precision. Although two taxa may share an ecological feature, their relative abundances may be very different in the size of their impact on the environment (e.g., sea-weed grazing of gastropoda and dugongs). And even though two taxa may perform a comparable task, their ecological function can be entirely different (e.g., free-living and endosymbiotic dinoflagellates). Multi-omics can overcome these discrepancies, as it can provide simultaneous information on multiple layers (see Leary et al. 2014; Moran 2015; Thiele et al. 2017 for marine microbiome).

Conclusion

The rapid development of new techniques, software, and decreasing costs for high throughput methods allow unprecedented experimental designs. The rising field of omic studies provides a toolbox for a better understanding of the complexity of life on earth. It is now possible to characterize organisms on a genome-wide scale. Omics have revealed a diversity of previously undetected species and simplify quantitative and functional analyses. However, the transition between different ome-layers is highly variable and requires the integration of multiple omic approaches. This can improve our understanding of the link between the genotype and phenotype. With multi-omics, we could find out how complex biological networks function.

Appendix

This article is related to the YOUMARES 8 conference session no. 2: "Reading the Book of Life – Omics as a Universal Tool Across Disciplines". The original Call for Abstracts and the abstracts of the presentations within this session can be found in the appendix "Conference Sessions and Abstracts", chapter "5 Reading the Book of Life – Omics as a Universal Tool Across Disciplines", of this book.

References

Alberts B, Johnson A, Lewis J et al (2008) Molecular biology of the cell. Garland Science, New York

Almeida MI, Reis RM, Calin GA (2011) MicroRNA history: discovery, recent applications, and next frontiers. Mutat Res – Fundam Mol Mech Mutagen 717:1–8

Amann R, Rosselló-Móra R (2016) After all, only millions? MBio 7(4):e00999–16

Andrews KR, Good JM, Miller MR et al (2016) Harnessing the power of RADseq for ecological and evolutionary genomics. Nat Rev Genet 17:81–92

Avery OT, Macleod CM, McCarty M (1943) Studies on the chemical nature of the substance inducing transformation of pneumococcal types: induction of transformation by a Desoxyribonucleic acid fraction isolated from Pneumococcus type III. J Exp Med 79:137–158

Baumgarten S, Simakov O, Esherick LY et al (2015) The genome of *Aiptasia*, a sea anemone model for coral symbiosis. Proc Natl Acad Sci USA 112:11893–11898

Beale DJ, Karpe AV, Ahmed W (2016) Beyond metabolomics: a review of multi-omics-based approaches. In: Beale DJ, Kouremenos KA, Palombo A (eds) Microbial metabolomics. Springer, Cham, pp 289–312

Biagi E, Candela M, Fairweather-Tait S et al (2012) Ageing of the human metaorganism: the microbial counterpart. Age 34:247–267

Bock C, Farlik M, Sheffield NC (2016) Multi-omics of single cells: strategies and applications. Trends Biotechnol 34:605–608

Bosch TCG, Miller DJ (2016) The Holobiont imperative. Springer, Vienna

Brüwer JD, Voolstra CR (2018) First insight into the viral community of the cnidarian model metaorganism Aiptasia using RNA-Seq data. PeerJ 6:e4449

Brüwer JD, Agrawal S, Liew YJ et al (2017) Association of coral algal symbionts with a diverse viral community responsive to heat shock. BMC Microbiol 17:174

Buck-Wiese H, Voolstra CR, Brüwer JD (2016) The metaorganism frontier – incorporating microbes into the organism's response to environmental change. In: Bode M, Jessen C, Golz V (eds) YOUMARES 7 conference proceedings, pp 94–102

Cherf GM, Lieberman KR, Rashid H et al (2012) Automated forward and reverse ratcheting of DNA in a nanopore at 5-A precision. Nat Biotechnol 30:344–348

Dahm R (2005) Friedrich Miescher and the discovery of DNA. Dev Biol 278:274–288

Dalsgaard J, John MS, Kattner G et al (2003) Fatty acid trophic markers in the pelagic marine environment. Adv Mar Biol 46:225–340

Delorme-Axford E, Donker RB, Mouillet J-F et al (2013) Human placental trophoblasts confer viral resistance to recipient cells. Proc Natl Acad Sci U S A 110:12048–12053

Dewar ML, Arnould JPY, Dann P et al (2013) Interspecific variations in the gastrointestinal microbiota in penguins. MicrobiologyOpen 2:195–204

Dewar ML, Arnould JPY, Krause L et al (2014) Influence of fasting during moult on the faecal microbiota of penguins. PLoS One 9:e99996

Dick GJ (2017) Embracing the mantra of modellers and synthesizing omics, experiments and models. Environ Microbiol Rep 9:18–20

Epstein SS (2013) The phenomenon of microbial uncultivability. Curr Opin Microbiol 16:636–642

Fauquet CM, Mayo MA (2001) The 7th ICTV report. Arch Virol 146:189–194

Feng Y, Zhang Y, Ying C et al (2015) Nanopore-based fourth-generation DNA sequencing technology. Genomics Proteomics Bioinf 13:4–16

Freilich S, Zarecki R, Eilam O et al (2011) Competitive and cooperative metabolic interactions in bacterial communities. Nat Commun 2:589

Gaby JC, Buckley DH (2012) A comprehensive evaluation of PCR primers to amplify the *nifH* gene of nitrogenase. PLoS One 7:e42149

Geiss GK, Bumgarner RE, Birditt B et al (2008) Direct multiplexed measurement of gene expression with color-coded probe pairs. Nat Biotechnol 26:317–325

Gillooly JF, Allen AP, West GB et al (2005) The rate of DNA evolution: effects of body size and temperature on the molecular clock. Proc Natl Acad Sci U S A 102:140–145

Gottlieb A (2017) Untersuchungen zur Expression ausgewählter Stammzellgene an embryonalen Geweben und Tumoren. Dissertation, Universität Bremen

Graeve M, Kattner G, Hagen W (1994) Diet-induced changes in the fatty acid composition of Arctic herbivorous copepods: experimental evidence of trophic markers. J Exp Mar Biol Ecol 182:97–110

Handelsman J (2004) Metagenomics: application of genomics to uncultured microorganisms. Microbiol Mol Biol Rev 68:669–685

Hernandez-Agreda A, Leggat W, Bongaerts P, Ainsworth D (2016) The microbial signature provides insight into the mechanistic basis of coral success across reef habitats. MBio 7:e00560–e00516

IPCC (2007) Climate change 2007: The physical science basis. Summary for policymakers. Contribution of working group I to the fourth assessment report of the intergovernmental panel on climate change. Cambridge University Press, Cambridge

IPCC (2014) Climate change 2014: Synthesis report. Contribution of working groups I, II and III to the fifth assessment report of the intergovernmental panel on climate change. IPCC, Geneva

Jones OAH, Sdepanian S, Lofts S et al (2014) Metabolomic analysis of soil communities can be used for pollution assessment. Environ Toxicol Chem 33:61–64

Kanehisa M, Sato Y, Kawashima M et al (2016) KEGG as a reference resource for gene and protein annotation. Nucleic Acids Res 44:D457–D462

Kanehisa M, Furumichi M, Tanabe M et al (2017) KEGG: new perspectives on genomes, pathways, diseases and drugs. Nucleic Acids Res 45:D353–D361

Kim BM, Kim J, Choi IY et al (2016) Omics of the marine medaka (*Oryzias melastigma*) and its relevance to marine environmental research. Mar Environ Res 113:141–152

Kimes NE, Callaghan AV, Aktas DF et al (2013) Metagenomic analysis and metabolite profiling of deep-sea sediments from the Gulf of Mexico following the Deepwater Horizon oil spill. Front Microbiol 4:50

Klug WS, Spencer CA, Palladino MA (2012) Concepts of genetics, 10th edn. Pearson Education, San Francisco

Leary DH, Li RW, Hamdan LJ et al (2014) Integrated metagenomic and metaproteomic analyses of marine biofilm communities. Biofouling 30:1211–1223

Levin RA, Voolstra CR, Weynberg KD et al (2017) Evidence for a role of viruses in the thermal sensitivity of coral photosymbionts. ISME J 11:808–812

Liew YJ, Li Y, Baumgarten S et al (2017) Condition-specific RNA editing in the coral symbiont *Symbiodinium microadriaticum*. PLoS Genet 13:1–22

Liu Y, Beyer A, Aebersold R (2016) On the dependency of cellular protein levels on mRNA abundance. Cell 165:535–550

Llewellyn CA, Sommer U, Dupont CL et al (2015) Using community metabolomics as a new approach to discriminate marine microbial particulate organic matter in the western English Channel. Prog Oceanogr 137:421–433

Lu H, Giordano F, Ning Z (2016) Oxford Nanopore MinION sequencing and genome assembly. Genomics Proteomics Bioinf 14:265–279

Luo S-S, Ishibashi O, Ishikawa G et al (2009) Human villous trophoblasts express and secrete placenta-specific microRNAs into maternal circulation via exosomes. Biol Reprod 81:717–729

McFall-Ngai M, Hadfield MG, Bosch TCG et al (2013) Animals in a bacterial world, a new imperative for the life sciences. Proc Natl Acad Sci U S A 110:3229–3236

McNutt MK, Camilli R, Crone TJ et al (2012) Review of flow rate estimates of the Deepwater Horizon oil spill. Proc Natl Acad Sci U S A 109:20260–20267

Moran MA (2015) The global ocean microbiome. Science 347:aac8455

Nirenberg M (2004) Historical review: deciphering the genetic code – a personal account. Trends Biochem Sci 29:46–54

Noguera-Solano R, Ruiz-Gutierrez R, Rodriguez-Caso JM (2013) Genome: twisting stories with DNA. Endeavour 37:213–219

O'Leary NA, Wright MW, Brister JR et al (2016) Reference sequence (RefSeq) database at NCBI: current status, taxonomic expansion, and functional annotation. Nucleic Acids Res 44:D733–D745

Olsen JL, Rouzé P, Verhelst B et al (2016) The genome of the seagrass *Zostera marina* reveals angiosperm adaptation to the sea. Nature 530:331–335

Paez-Espino D, Eloe-Fadrosh EA, Pavlopoulos GA et al (2016) Uncovering Earth's virome. Nature 536:425–430

Park JJ, Wang H, Gargouri M et al (2015) The response of *Chlamydomonas reinhardtii* to nitrogen deprivation: a systems biology analysis. Plant J 81:611–624

Pascault N, Loux V, Derozier S et al (2015) Technical challenges in metatranscriptomic studies applied to the bacterial communities of freshwater ecosystems. Genetica 143:157–167

Patti GJ, Yanes O, Siuzdak G (2012) Metabolomics: the apogee of the omics trilogy. Int J Pharm Pharm Sci 13:263–269

Pedrós-Alió C (2012) The rare bacterial biosphere. Annu Rev Mar Sci 4:449–466

Pheasant M, Mattick JS (2007) Raising the estimate of functional human sequences. Genome Res 17:1245–1253

Pimm SL, Jenkins CN, Abell R et al (2014) The biodiversity of species and their rates of extinction, distribution, and protection. Science 344:1246752

Pogoreutz C, Rädecker N, Cárdenas A et al (2017) Nitrogen fixation aligns with *nifH* abundance and expression in two coral trophic functional groups. Front Microbiol 8:1187

Rhoads A, Au KF (2015) PacBio sequencing and its applications. Genomics Proteomics Bioinformatics 13:278–289

Ritchie KB (2006) Regulation of microbial populations by coral surface mucus and mucus associated bacteria. Mar Ecol Prog Ser 322:1–14

Rohwer F, Seguritan V, Azam F et al (2002) Diversity and distribution of coral-associated bacteria. Mar Ecol Prog Ser 243:1–10

Röthig T, Costa RM, Simona F et al (2016) Distinct bacterial communities associated with the coral model *Aiptasia* in Aposymbiotic and symbiotic states with *Symbiodinium*. Front Mar Sci 3:234

Ruggles KV, Krug K, Wang X et al (2017) Methods, tools and current perspectives in proteogenomics. Mol Cell Proteomics 16:959–981

Sampson TR, Mazmanian SK (2015) Control of brain development, function, and behavior by the microbiome. Cell Host Microbe 17:565–576

Sanger F, Nicklen S, Coulson AR (1977) DNA sequencing with chain-terminating inhibitors. Proc Natl Acad Sci U S A 74:5463–5467

Schuster SC (2008) Next-generation sequencing transforms today's biology. Nat Methods 5:16–18

Selbach M, Schwanhäusser B, Thierfelder N et al (2008) Widespread changes in protein synthesis induced by microRNAs. Nature 455:58–63

Shnit-Orland M, Sivan A, Kushmaro A (2012) Antibacterial activity of *Pseudoalteromonas* in the coral Holobiont. Microb Ecol 64:851–859

Simmonds P, Adams MJ, Benkö M et al (2017) Consensus statement: virus taxonomy in the age of metagenomics. Nat Rev Microbiol 15:161–168

Stackebrandt E, Frederiksen W, Garrity GM et al (2002) Report of the *ad hoc* comittee for the re-evaluation of the species definition in bacteriology. Int J Syst Evol Microbiol 52:1043–1047

Taberlet P, Coissac E, Hajibabaei M et al (2012) Environmental DNA. Mol Ecol 21:1789–1793

Teeling H, Fuchs BM, Becher D et al (2012) Substrate-controlled succesion of marine Bacterioplankton populations induced by a phytoplankton bloom. Science 336:608–611

The Gene Ontology Consortium (2004) The Gene Ontology (GO) database and informatics resource. Nucleic Acids Res 32:D258–D261

The Gene Ontology Consortium (2017) Expansion of the gene ontology knowledgebase and resources: The gene ontology consortium. Nucleic Acids Res 45:D331–D338

Thiele S, Richter M, Balestra C et al (2017) Taxonomic and functional diversity of a coastal planktonic bacterial community in a river-influenced marine area. Mar Genomics 32:61–69

Turnbaugh PJ, Ley RE, Hamady M et al (2007) The human microbiome project. Nature 449:804–810

Valentini A, Taberlet P, Miaud C et al (2016) Next-generation monitoring of aquatic biodiversity using environmental DNA metabarcoding. Mol Ecol 25:929–942

Voolstra CR, Miller DJ, Ragan MA et al (2015) The ReFuGe 2020 consortium—using "omics" approaches to explore the adaptability and resilience of coral holobionts to environmental change. Front Mar Sci 2:68

Wilmes P, Simmons SL, Denef VJ et al (2009) The dynamic genetic repertoire of microbial communities. FEMS Microbiol Rev 33:109–132

Zanta MA, Belguise-Valladier P, Behr JP (1999) Gene delivery: a single nuclear localization signal peptide is sufficient to carry DNA to the cell nucleus. Proc Natl Acad Sci U S A 96:91–96

Ziegler M, Seneca FO, Yum LK et al (2017) Bacterial community dynamics are linked to patterns of coral heat tolerance. Nat Commun 8:14213

Bio-telemetry as an Essential Tool in Movement Ecology and Marine Conservation

Brigitte C. Heylen and Dominik A. Nachtsheim

Abstract

Marine top predators represent an essential part of marine ecosystems. They are generally regarded as "sentinels of the sea" since their presence reflects high biological productivity. However, many populations are experiencing dramatic declines attributed to various human-induced threats (e.g., pollution, climate change, overfishing), highlighting the need for effective conservation. In this review, we show that bio-telemetry can be an essential tool, not only to improve knowledge about the animals' ecology, but also for conservation purposes. As such, we will first discuss the most important state-of-the-art devices (e.g., time-depth recorders, accelerometers, satellite tags) and illustrate how they can improve our understanding of movement ecology. We will then examine the challenges and ethical issues related to bio-telemetry, and lastly, demonstrate its enormous value in resolving present and future conservation issues.

Introduction

Marine top predators are widely regarded as potential qualitative indicators of the health and status of marine ecosystems (Burger et al. 2004; Piatt et al. 2007; Boersma 2008; Campbell et al. 2012; Wikelski and Tertitski 2016). Their responses to changes in the environment can be measured by examining different aspects of their ecology using corresponding methods, for example, foraging behavior (e.g., satellite-linked logging devices and time-depth recorders), energy expenditure (e.g., double-labelled water), stress levels (e.g., corticosteroid hormone concentrations), and diet trends (e.g., stable isotopes and fatty acid analyses) (Votier et al. 2010). The decline of top predators can result in trophic downgrading, which has far reaching consequences on the structure and dynamics of marine ecosystems (Heithaus et al. 2008; Estes et al. 2011; Boaden and Kingsford 2015). At the moment, marine top predators face unprecedented challenges and their future existence is threatened, due to the effects of rapid environmental changes, overfishing, pollution, and many other anthropogenic disturbances (Robinson et al. 2005). If we consider seabirds, for instance, their conservation status has deteriorated faster than any other bird group over recent decades (Croxall et al. 2012; Paleczny et al. 2015). There are also countless examples available for other marine top predators, such as cetaceans (Rosenbaum et al. 2014; McKenna et al. 2015), pinnipeds (Antonelis et al. 2006; Costa et al. 2010a), elasmobranchs (Baum et al. 2003; Graham et al. 2012), and large teleosts (Block et al. 2001; Boyce et al. 2008).

In this review, we focus on the knowledge that can be derived from bio-telemetry and its efficacy in movement ecology and conservation studies. Bio-telemetry can be defined as the remote recording of behavioral, physiological, and environmental data by means of electronic tags, attached to animals (Hays et al. 2016). Here, we use the term bio-telemetry synonymously with bio-logging; the latter includes loggers that must be recovered to download the data, which were stored on the device. We aim to give a general overview of existing devices, assess specifically how bio-telemetry can improve our understanding of movement ecology, while taking the negative impacts on the animals into account, and discuss how bio-telemetry can help in recommending conservation measures.

B. C. Heylen (✉)
Behavioural Ecology and Ecophysiology, University of Antwerp, Antwerp, Belgium

Terrestrial Ecology Unit, Ghent University, Ghent, Belgium
e-mail: brigitte.heylen@UGent.be

D. A. Nachtsheim
Institute for Terrestrial and Aquatic Wildlife Research, University of Veterinary Medicine Hannover, Büsum, Germany

BreMarE – Bremen Marine Ecology, Marine Zoology, University of Bremen, Bremen, Germany
e-mail: Dominik.Nachtsheim@tiho-hannover.de

S. Jungblut et al. (eds.), *YOUMARES 8 – Oceans Across Boundaries: Learning from each other*,
https://doi.org/10.1007/978-3-319-93284-2_7

Existing Bio-telemetry Devices

We are undoubtedly living in the golden age of bio-telemetry studies (McIntyre 2014; Hussey et al. 2015; Hays et al. 2016). The following section provides an overview of the most commonly used animal-borne bio-telemetry devices and their potential in the context of studying animal behaviors. We start with archival loggers, which can be defined as

Box 1: Quick Guide to Bio-telemetry Terminology

Acoustic telemetry: tagged animals are detected and recorded by acoustic receivers at fixed moorings that are recovered periodically; this enables tracking of individual animals

Archival logger: animal-borne instrument that records and stores data on-board; must be relocated for data download

Argos satellite tag: animal-borne device that communicates with polar-orbiting Argos satellites to determine its location; typical location errors range between 500 m and 10 km

Bio-telemetry: method of remote recording of behavioral, physiological and environmental data by electronic tags attached to animals

Fastloc® GPS: takes a snapshot of relevant satellite information in a fraction of a second, when a diving animal surfaces; the calculation of a GPS position is performed on-board the tag, even when not in view of satellites

GPS tag: device that determines an animal's position via the Global Positioning System (GPS); typical location errors range between 20 m and 50 m

GSM: Global System for Mobile Communications; a cellular network that is the global standard for mobile communication, however, also frequently used to relay data from animal-borne instruments (e.g., GPS tags)

Hydrophone: an underwater microphone

Jaw movement sensor: device consisting of Hall sensor and magnet, which are attached to upper and lower mandible, respectively; detects mouth openings and hence feeding events

Pop-up archival transmitting tag (PAT): satellite-linked data logger that is commonly deployed on sharks; logger records and stores temperature, depth and ambient light levels over pre-programmed period until it pops up to the surface and delivers data via the Argos satellite system

Satellite-linked data logger: combination of a satellite tag (Argos and/or GPS) with an archival logger; records an animal's position as well as information on different behaviors and ambient conditions

Stomach temperature logger: device that monitors internal body temperature; a sharp drop in temperature can be attributed to the ingestion of relatively cold prey, enabling the detection of feeding events

Time-depth recorder (TDR): device that records a time-series of dive depths, resulting in time-depth profiles

Tri-axial accelerometer: measures acceleration caused by earth's gravitational field as well as acceleration by the animal in all three space dimensions

Tri-axial magnetometer: measures the orientation towards the earth's magnetic field in all three space dimensions instruments that record and store data on-board. This holds true not only for standard time-depth recorders (TDRs), but also for cameras, accelerometers and magnetometers. As they only store the data on-board, one is usually required to recover the instrument and download the data. In the next section we introduce location-only satellite tags. Collecting animal locations is straightforward for Argos satellite tag users, as once they are deployed, the data is relayed and provided online to the user, which enables near real-time tracking. However, this fundamentally differs from global positioning system (GPS) devices, since the position determination works differently. With Argos, the satellite system determines the device's position, while with GPS, the device determines its own position. The latter has the disadvantage that one usually needs to recover the GPS tags, or couple them with satellite or mobile phone networks. Finally, there are tags, which consist of both an archival tag (e.g., TDR, accelerometer, magnetometer) and a satellite tag (e.g., Argos, GPS). These satellite-linked data loggers are the most sophisticated ones and also the ones most commonly used. A 'Quick guide to bio-telemetry terminology' can be found in Box 1.

Archival Loggers

Archival loggers are deployed on animals to record, in general, movements, specific behaviors, physiological processes or environmental conditions. It is necessary to retrieve these instruments to download data, which can be challenging in highly mobile marine predators. This disadvantage is offset by deploying archival loggers on species with high site fidelity that haul out or breed on land (e.g., elephant seals, seabirds during breeding season), or by combining them with a radio transmitter (Very High Frequency – VHF, or Ultra High Frequency – UHF) to relocate the tag when it falls off (Wilson et al. 2002; Dragon et al. 2012; Villegas-Amtmann et al. 2013). Since data transmission is not a constraint for archival loggers, these devices are able to collect data in high-resolution, which are otherwise impossible to obtain and extremely valuable to study fine-scale processes.

Per Scholander is generally regarded as the pioneer of bio-logging; he deployed the first archival loggers on marine animals in the 1940s (Ropert-Coudert and Wilson 2005; Kooyman 2007; McIntyre 2014). In his famous monograph on diving physiology, he recorded the maximum dive depths of whales, dolphins, and seals by using a capillary tube depth gauge attached to the animal (Scholander 1940; Ponganis 2013). Later, maximum dive depth recorders were used on Weddell seals *Leptonychotes weddellii* in Antarctica, to study their maximum diving and breath-holding capacities (DeVries and Wohlschlag 1964). A major step in the history of bio-logging was the development of the first TDR by Gerald Kooyman, which was able to record full time-depth profiles of a dive (Kooyman 1965, 1966). This invention led to fascinating opportunities to study the diving behavior of marine top predators and the related physiological adaptations (Kooyman 1973). At the time, these devices were still quite heavy and, for instance, too large for most bird species. However, due to rapid technological advances, their dimensions have decreased quickly, which means that they can presently be used on a variety of animals (Ropert-Coudert and Wilson 2005; McIntyre 2014; Hussey et al. 2015).

At the same time, improved and additional sensors were included in the classic TDR deployment, enabling diving behavior to be studied in even greater detail. For instance, magnetometers were incorporated, which measure the orientation towards the earth's magnetic field in three axes (heading/yaw, pitch, and roll) and are especially sensitive to record angular rotations (Fig. 1b). Tri-axial magnetometry enables researchers to track three-dimensional movements of diving animals via dead-reckoning, and thus to reconstruct three-dimensional profiles of a dive (Davis et al. 1999; Mitani et al. 2003; Wilson et al. 2007; Williams et al. 2017). This was an important improvement, as marine animals inhabit a three-dimensional space and respond to environmental cues in all three dimensions. For diving predators both the horizontal and vertical distribution of prey patches is important. Foraging chinstrap penguins *Pygoscelis antarctica*, for instance, choose to pass by shallow and dense prey aggregations and reach for deeper and more homogenously distributed prey fields with higher encounter probabilities (Zamon et al. 1996)—a finding which would be overlooked in a conventional, two-dimensional analysis of predator-prey distribution. Therefore, taking all three dimensions into consideration is essential to understand fine-scale habitat use or foraging behavior.

At present, it is also common to incorporate accelerometers into archival loggers. These devices measure acceleration, which is caused by earth's gravity (static component) and a change in the animal's speed (dynamic component). It usually records accelerations in three dimensions, the x-, y-, and z-axes or surge, sway, and heave (Fig. 1a). When positioned on the head and/or jaw of a marine predator, accelerometers can provide information about rapid head movements, indicating prey capture attempts (Naito et al. 2010; Kokubun et al. 2011; Gallon et al. 2013). However, accelerometers are often deployed close to the animal's center of gravity, i.e., in the center of the trunk, which is more useful to record overall movement patterns (e.g., swimming, resting, and flying). As such, tri-axial accelerometry can be used to identify and quantify different behaviors and activity patterns, and subsequently, put them in relation to energy

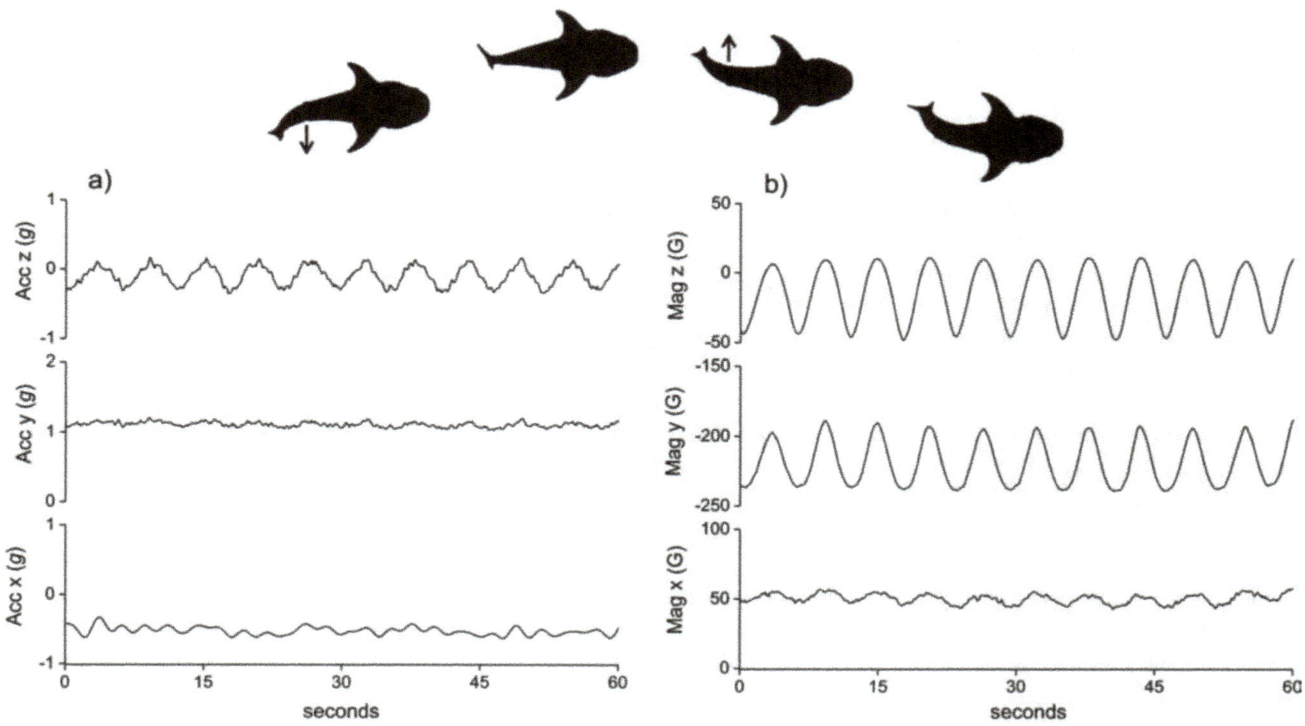

Fig. 1 Swimming behavior of a whale shark *Rhincodon typus* as indicated by tri-axial (**a**) accelerometry, and (**b**) magnetometry. One oscillation corresponds to one tail stroke. Note the weak signal and high degree of noise in the accelerometer data (due to the low stroke frequency). The magnetometer is less susceptible to this noise and is, therefore, better to resolve the angular rotation of the tail strokes. (Reproduced from Williams et al. (2017) (CC-BY 4.0))

expenditure (Wilson et al. 2006, 2008; Shepard et al. 2008; Sakamoto et al. 2009a). For instance, different at-sea activities (i.e., diving, transiting, resting, and surfacing) during foraging trips of lactating northern fur seals *Callorhinus ursinus* and Antarctic fur seals *Arctocephalus gazella* were classified based on accelerometer and dive data. Using these classified behaviors, time-activity budgets were determined and activity-specific energy expenditures were accurately calculated from accelerometer data (Jeanniard-du-Dot et al. 2017). These various applications make accelerometers a powerful and promising tool for future employments.

Video and still-picture cameras are another example of archival loggers, commonly deployed on marine top predators (Fig. 2a) (Moll et al. 2007). These devices not only take photos at regular intervals or record video sequences, but additional incorporated sensors are also able to gather data on environmental conditions (e.g., dive depths and ambient temperature) (Ponganis et al. 2000; Moll et al. 2007; Naito et al. 2010). Over the years, the quality of video footage and photographs has increased substantially, with high-definition cameras being the current status quo (Chapple et al. 2015; Krause et al. 2015; Machovsky-Capuska et al. 2016). High quality recordings require enormous memory capacities; thus the recording time ranges between hours and a few days, when a duty cycle is activated. Nevertheless, camera loggers can be extremely valuable tools to obtain direct observations of difficult to observe, and therefore rarely documented, animal behaviors (Takahashi et al. 2004; Sakamoto et al. 2009b; Handley and Pistorius 2016). For example, it could be demonstrated that black-browed albatrosses *Thalassarche melanophris* actively follow killer whales *Orcinus orca* and possibly feed on prey remains that were left over by them (Sakamoto et al. 2009b). Such observations are crucial to understand how far-ranging animals locate prey patches in the vast ocean. Camera loggers are also frequently deployed to investigate a predator's foraging behavior in greater detail. Animal-borne imaging can reveal foraging strategies and hunting behavior (Davis et al. 1999; Watanuki et al. 2008; Goldbogen et al. 2012; Krause et al. 2015), quantify prey intake (e.g., Ponganis et al. 2000; Watanabe et al. 2003), or validate prey capture events derived from accelerometers or dive characteristics (Watanabe and Takahashi 2013; Volpov et al. 2015, 2016). Furthermore, cameras attached to diving predators can serve as remote sensors to monitor the surrounding environment. They can, for instance, provide information on the behavior and occurrence of prey species (Fuiman et al. 2002), or detect hitherto unknown faunal communities (Watanabe et al. 2006).

Fig. 2 (**a**) Adult Weddell seal *Leptonychotes weddellii* equipped with an infrared camera logger. (**b**) Adélie penguin *Pygoscelis adeliae* with GPS and dive logger. (**c**) Young grey seal *Halichoerus grypus* with a GSM-relayed Fastloc® GPS data logger, tagged on Helgoland, Germany. (**d**) Male southern elephant seal *Mirounga leonina* instrumented with a CTD-Satellite Relay Data Logger on King George Island/Isla 25 de Mayo. (**e**) Lesser black-backed gull *Larus fuscus* tagged with a solar-powered GPS logger and tri-axial accelerometer. (**f**) Harbor porpoise *Phocoena phocoena* with a Digital Sound Recording-Tag (DTAG), attached by suction cups. Photos used with permission from Dominik Nachtsheim (a), Nina Dehnhard (b), Abbo van Neer (c), Alfred-Wegener-Institut/Horst Bornemann (CC-BY 4.0) (d), Brigitte Heylen (e), and Jonas Teilmann (f)

Besides these commonly used archival loggers, there are a variety of alternative devices, which record very specific behaviors or physiological processes. Jaw movement sensors are able to detect mouth-opening-events, and can therefore provide a proxy for prey capture attempts (Wilson et al. 2002; Ropert-Coudert et al. 2004; Liebsch et al. 2007). Another measurement used to detect prey capture can be the drop in internal body temperature when relatively cold prey is ingested by marine endotherms. As such, esophagus or stomach temperature loggers have been regularly deployed in seabirds and marine mammals to record feeding activities (Wilson et al. 1992; Ropert-Coudert et al. 2001; Austin et al. 2006a; Ropert-Coudert and Kato 2006). Other sensors are able to record heart rate (Woakes et al. 1995; Hindell and Lea 1998; Froget et al. 2004; Chaise et al. 2017) or flipper strokes (Sato et al. 2003; Insley et al. 2008; Jeanniard-du-Dot et al. 2017) to study field metabolic rates and energy expenditures. There are certainly many more archival loggers available for various applications and research questions, however, a discussion of those is beyond the scope of this review.

Argos Satellite Tags

The Argos satellite system was initiated in the late 1970s and represents a cooperative project between Centre National d'Études Spatiales (CNES) in France, and the National Oceanic and Atmospheric Administration (NOAA) and National Aeronautics and Space Administration (NASA) in the USA. Argos is operated and managed by Collecte Localisation Satellites (CLS) in Toulouse, France. Soon after its initiation, the high importance of Argos satellites for wildlife telemetry studies was recognized. Satellite telemetry was developed as a means of overcoming the logistical difficulties and high costs of conventional VHF radio telemetry (Fancy et al. 1988). Animal locations are determined through Doppler shift via communication between an animal-borne satellite transmitter and polar-orbiting satellites. The estimated locations are provided online by CLS and allow quasi-live tracking of tagged individuals (Fancy et al. 1988; Costa et al. 2012).

Argos satellite tags represent the first reliable system to track horizontal movements of marine animals. This information is essential to analyze habitat use or migration patterns. The first successful deployment was conducted on a basking shark *Cetorhinus maximus* and only provided locations over the course of two weeks (Priede 1984). Nevertheless, this was sufficient to conclude that the shark was probably feeding on zooplankton along a frontal system (Priede 1984; Priede and Miller 2009). Since then, enormous improvements in both the Argos satellite system and satellite tags have been achieved (e.g., more polar-orbiting satellites, better sensitivity of satellite sensors, longer transmitter battery life, more streamlined tag shapes), ultimately leading to a larger quantity of collected data (Hays et al. 2007). Argos satellite tags have been deployed on a diverse assemblage of marine top predator species, including seabirds (Jouventin and Weimerskirch 1990; Spencer et al. 2014; Pistorius et al. 2017), sharks (Priede 1984; Eckert and Stewart 2001; Weng et al. 2005), pinnipeds (Costa et al. 2010a; Dietz et al. 2013; Arcalís-Planas et al. 2015) and cetaceans (Andrews et al. 2008; Edrén et al. 2010; Hauser et al. 2010; Reisinger et al. 2015).

GPS Tags

Despite the many advantages of Argos satellite tags, one of the major drawbacks is the relatively low location accuracy, with errors generally ranging between 500 m and 10 km (Costa et al. 2010b). Qualitatively poor Argos locations are especially prevalent in studies involving diving top predators, since the time spent at the surface to enable successful uplinks to the satellite is limited (Vincent et al. 2002; Costa et al. 2010b; Patterson et al. 2010). GPS tags provide a much better accuracy, usually with errors less than 50 m (Costa et al. 2010b; Dujon et al. 2014). Despite this higher location accuracy, most researchers have, for quite some time, refrained from using GPS tags, mainly due to the length of time (10–30 min) and high energy demand required to fix a GPS position (Tomkiewicz et al. 2010; Costa et al. 2012). This meant that they were well suited for seabirds (Fig. 2e) (Ryan et al. 2004; Pinaud and Weimerskirch 2007; Votier et al. 2010), but less so for diving animals. This problem has more recently been overcome by the development of a Fastloc® GPS, for which GPS positions can be obtained within milliseconds, which enables a successful location fix even within a short surfacing event (Costa et al. 2010b). Thus, GPS tags are now also increasingly used on marine mammals (Heide-Jørgensen et al. 2013; Villegas-Amtmann et al. 2013; McKenna et al. 2015). The GPS positions are either stored on-board the device and must be downloaded from a recovered tag, or can be transmitted via the Argos satellite system (Costa et al. 2010b; Patterson et al. 2010). GPS locations can also be relayed through communication with the Global Systems for Mobile Communication (GSM)—the mobile phone network (McConnell et al. 2004; Cronin and McConnell 2008). The locations are stored internally and transmitted as a text message, together with ancillary information, when the animal is within the coverage of the GSM network (Cronin and McConnell 2008). These devices represent a promising tool for the relatively inexpensive and accurate tracking of coastal top predator species (Fig. 2c) (Jessopp et al. 2013; Wilson et al. 2017), especially in the light of the rapidly expanding GSM network around the globe.

Satellite-linked Data Loggers

For many applications, simultaneous information on both horizontal movements and specific behaviors, for example, diving behavior, is required to better understand how marine animals respond to their environment and use their habitat. This is achieved by combining a satellite tag (Argos and/or GPS) with an archival logger (e.g., time-depth recorders, accelerometers), i.e., a satellite-linked data logger. These devices not only record an animal's position, but also log information on different behaviors and ambient conditions. Some instruments are able to transmit these data via Argos satellites, while others need to be recovered for data download. Remotely collected data via satellite only provide compressed and reduced information due to bandwidth limitations, whereas retrievable instruments offer data in high resolution. However, it is expected that the impact of this constraint will continue to lessen with the on-going rapid technological advance and further developments in the field of bio-telemetry (see, for example, Cox et al. 2017).

The first satellite-linked data logger was a combination of a satellite transmitter and a TDR (Merrick et al. 1994). This provided the opportunity to combine location data with concurrent behavioral data, and thus enabled the analysis of horizontal and vertical movements (Merrick et al. 1994; Ryan et al. 2004; Burns et al. 2008; Bestley et al. 2015; Heerah et al. 2016). The first satellite-linked dive recorders were effective to study habitat use in relation to diving behavior, but had restricted applicability, due to the limited information available about each dive (Merrick et al. 1994; Burns 1999; Davis et al. 2007; Nachtsheim et al. 2017). The development of the satellite relay data logger (SRDL) revolutionized the study of top predator movements, providing locations and compressed time-depth profiles for each dive via satellite communication (Fedak et al. 2002). These devices can also incorporate high precision environmental sensors, which are able to record valuable CTD (Conductivity, Temperature, Depth) data. The collected temperature and salinity profiles have a relatively good quality and accuracy, compared to traditional oceanographic measurements, such as floats and moorings (Boehme et al. 2009). This development meant that studying foraging behavior in relation to actual environmental conditions, as experienced by the animals, became possible (Fig. 2d) (Biuw et al. 2007; McIntyre et al. 2011; Lowther et al. 2013; Blanchet et al. 2015; Labrousse et al. 2015).

Accelerometers and magnetometers can also be coupled with Argos or GPS devices, providing even more powerful tools to study animal behavior (Fig. 2e) (Wilson et al. 2008; Bouten et al. 2013; Cox et al. 2017). These instruments are able to give extremely detailed information about three-dimensional movements and behaviors. More sophisticated devices have also incorporated a hydrophone for sound recordings of diving predators, such as cetaceans, which enables researchers to relate their movements to acoustic behavior as well as to the surrounding soundscape (Fig. 2f) (Nowacek et al. 2001; Johnson and Tyack 2003; Aguilar Soto et al. 2008; Wisniewska et al. 2016). Such instruments usually have to be recovered for data retrieval, but recently other systems have been developed, which allow the remote downloading of concurrent GPS and accelerometer data at ground base stations (Bouten et al. 2013). Another exciting and promising approach is the recent launch of the ICARUS initiative (Icarus Initiative 2018). Specifically-designed ICARUS satellite tags will be able to record GPS positions as well as tri-axial accelerometer data and transmit the data to the ICARUS antenna on-board the International Space Station. The data will then be made available to the users in near real time via the database Movebank (see Box 2 for an overview on databases for animal movement data). These remotely-operating systems enable the study of migration patterns and habitat use of migratory species over large spatial and temporal scales in great detail (Wikelski et al. 2007; Bouten et al. 2013; Stienen et al. 2016; Wikelski and Tertitski 2016).

> **Box 2: Links to Online Databases Hosting Marine Animal Movement Data**
>
> Biodiversity.aq—Antarctic biodiversity data base: http://www.biodiversity.aq
>
> Lifewatch.be—A virtual laboratory for biodiversity research: http://www.lifewatch.be/
>
> OBIS-SEAMAP—Ocean Biogeographic Information System Spatial Ecological Analysis of Megavertebrate Populations: http://seamap.env.duke.edu/
>
> Ocean Tracking network: https://members.oceantrack.org/projects
>
> OCEARCH's Global Shark Tracker: http://www.ocearch.org
>
> MEOP—Marine Mammals Exploring the Oceans from Pole to Pole: http://www.meop.net
>
> MMT—Marine Mammal Tracking: https://www.pangaea.de/?q=project%3Alabel%3AMMT
>
> Movebank: A Database for Animal Tracking Data: https://www.movebank.org

For animals that do not regularly come to the surface, such as fish, pop-up archival transmitting tags (PATs) can be used (Carlson et al. 2010; Jorgensen et al. 2010; Campana et al. 2011; Hammerschlag et al. 2011). PATs usually constitute an archival logger and an Argos satellite transmitter, and, in the case of sharks and other elasmobranchs, are anchored in the dorsal fin or dorsal musculature (Hammerschlag et al. 2011). The logger records and stores temperature, depth, and ambient light levels over several months to years. After a preprogrammed period, the tag detaches from the animal and pops up to the ocean surface. While floating, it transmits the recorded data to Argos satellites (Hammerschlag et al. 2011). The animal's horizontal movements are reconstructed from the *in situ* measured sea surface temperature and light levels, resulting in a daily estimate of latitude and longitude. Since the location estimate is relatively imprecise, with mean errors ranging between 60 and 180 km, only large-scale movements and migrations patterns can be investigated (Block et al. 2011; Campana et al. 2011; Hammerschlag et al. 2011). A comparable technology is used to track large-scale and often multi-year movements of pinnipeds and flying seabirds, however, this system is not linked to satellites. Similar to PATs, light level geolocators record light levels from which locations can be derived. They are relatively inexpensive and the spatial resolution (ca. 100–200 km) is often sufficient for wide-ranging species, such as petrels, albatrosses, terns, or elephant seals (Afanasyev 2004; Bradshaw et al. 2004; Phillips et al. 2004; Egevang et al. 2010; Weimerskirch et al. 2014).

Acoustic Telemetry

Tracking animals by means of acoustic telemetry was specifically developed for use in marine and freshwater ecosystems. It is based on the idea that tagged animals, most often fish, are registered by submerged receiving stations (Donaldson et al. 2014; Hussey et al. 2015). The tag is usually a transmitter, which is either surgically implanted or attached externally to the animal and emits signals with a given pulse rate. The transmitters can be miniscule (<0.5 g in some cases), and can even be used to tag small or juvenile fish (McMichael et al. 2010). The presence of the animal is recorded when the transmissions are detected by a hydrophone in the receivers. However, the detection efficiency, i.e., the attenuation of sound, varies between different environments and water conditions, which represent a major limitation of acoustic telemetry (Donaldson et al. 2014). Often, multiple acoustic receivers are organized in a comprehensive array or network to cover the pre-defined study area, enabling the tracking of each individual's movements. The data are stored within the receiver station and can either be downloaded by recovering the device or via wireless technology (Dagorn et al. 2007; Donaldson et al. 2014; Hussey et al. 2015). Such acoustic receivers can also be installed on mobile platforms, such as predators, to detect encounters with tagged individuals. This means that interactions between a predator and its prey, as well as spatiotemporal patterns of predator and prey distribution can be studied (Lidgard et al. 2014).

Movement Ecology

Movement is a fundamental characteristic of many species, and as such, plays an important role in the survival and reproduction of individuals. This, in turn, affects the structure and dynamics of populations and ecosystems. Therefore, to manage marine ecosystems properly, it is imperative to understand the causes, patterns, mechanisms, and consequences of individual movement. Data derived from biotelemetry can provide insight into animals' movement ecology and show interactions within the ecosystem they inhabit (Cagnacci et al. 2010). This enables a mechanistic understanding of movement ecology, including foraging behavior and seasonal migration. However, interpreting biotelemetry data remains a challenge in certain applications, (e.g., habitat modelling), due to the inherent features of telemetry data. These features include spatial and temporal auto-correlation, uneven sampling intervals, uneven sampling effort across individuals and uneven detectability across different habitats (Aarts et al. 2008). Choosing an appropriate analytical approach for the respective research questions is, therefore, crucial. An overview of the currently available methods is described in Carter et al. (2016).

Foraging Behavior

Optimal foraging theory predicts that every individual strives to minimize its foraging effort, while maximizing its foraging success to assure its survival and reproduction—the drivers for natural selection (MacArthur and Pianka 1966; Schoener 1971; Pyke 1984). For instance, Antarctic fur seals have distinct foraging strategies, with associated trade-offs related to habitat availability, travel costs, prey accessibility and prey quality (Arthur et al. 2016). We can differentiate between top predators that are central place foragers (CPFs), which regularly return to a specific location between foraging trips to feed young, store food, or rest, and thus face spatial constraints on foraging (Orians and Pearson 1979), and roaming foragers, a term we use for all non-CPFs. CPFs prefer foraging habitat near their central place, because, when travel time between the central site and the prey resource increases, their net energy gain decreases (Andersson 1978). For both CPFs and roaming foragers, efficiency is determined by the trade-offs between energy expenditures and gains (Shoji et al. 2016). Individuals may further optimize their foraging efficiency by concentrating on a specific prey type or the exploitation of a specific habitat. In stable environments, specialization in foraging strategies can be highly advantageous as individuals decrease search and handling costs, and reduce their niche overlap with other individuals, thus minimizing competition (Bolnick et al. 2003). A potential cost of specialization is that individuals may lack the flexibility to respond to environmental change (Bolnick et al. 2003; McIntyre et al. 2017). Within the light of rapid environmental change as a result of anthropogenic activities, behavioral plasticity in foraging behavior becomes particularly important. Climate change may affect top predators' geographic ranges and energy balance by altering the distribution and abundance of prey populations (MacLeod 2009; Hazen et al. 2013). Additionally, food resources may change at the local scale due to other anthropogenic influences. Local depletion and competition with fisheries can be regarded as general examples of the latter (Lidgard et al. 2014; Cronin et al. 2016). A topical issue is the future European Union ban on fisheries discards, currently providing a major resource for a wide range of seabird species (Garthe et al. 1996; Furness 2003; Votier et al. 2010, 2013; Bicknell et al. 2013; Krüger et al. 2017).

Specialization may require specific foraging behaviors (Patrick and Weimerskirch 2014). Single individuals may apply multiple foraging strategies or only a single one, or they may temporarily switch between strategies. When individuals use different foraging strategies, their choice may depend on a range of intrinsic and extrinsic factors (Patrick et al. 2013; Camphuijsen et al. 2015). These factors are not mutually exclusive, as the underlying processes influence each other notably. As such, it is difficult to identify the causal effects of each of these factors separately. Therefore, a multifactorial approach is required to investigate their influence on foraging ecology. Integrating multiple types of data at different scales enhances our understanding of a predator's foraging behavior as well as the circumstances that lead to foraging success, as these can differ not only between and within species, but even within individuals depending on the conditions (Austin et al. 2006b; Watanabe and Takahashi 2013; McIntyre et al. 2017).

Intrinsic factors can strongly influence foraging efficiency. One of these factors is age or experience. It is generally

assumed that older individuals have learned to improve their energy gain and lower their effort per feeding attempt, as illustrated in the Caspian Gull *Larus cachinnans* (Skórka and Wójcik 2008). Another, more frequently observed defining intrinsic factor of foraging efficiency is sex, which can be observed in many taxa as a possible mechanism to reduce intra-specific competition. Sexually distinct foraging strategies were observed, amongst others, in lesser black-backed gulls *Larus fuscus* and harbor seals *Phoca vitulina* (Camphuijsen et al. 2015; Wilson et al. 2015b). A healthy body condition is also vital for efficient foraging. African penguins in low body condition after an energy-demanding breeding season may have difficulties in gaining enough fat reserves to molt, and consequently forage less efficiently as feather quality deteriorates, eventually leading to starvation (Crawford et al. 2011). Lastly, brood demand is also an important factor. In general, when offspring is present, net energy demand is higher, which will affect parental feeding strategies (Pinaud et al. 2005; O'Dwyer et al. 2007; Rishworth et al. 2014; Shoji et al. 2015). Additionally, the increasing energetic demands of developing progeny may make it even harder to deliver sufficient food as the breeding season progresses.

Movement, hence foraging ecology, is also significantly determined by extrinsic factors, such as prey distribution and availability, environmental features, intra- and interspecific competition, and the presence of anthropogenic food sources (Lewis et al. 2001; Grémillet et al. 2004; Biuw et al. 2007; Dragon et al. 2010; Labrousse et al. 2015). For the latter, fisheries discards are one of the most important drivers of seabird distributions (Garthe et al. 1996; Bartumeus et al. 2010; Patrick et al. 2015; Krüger et al. 2017), while recreational fisheries, offshore wind farms, and terrestrial refuse tips can play a role too (Griffiths et al. 2004). Environmental factors, such as sea surface temperature (SST), oceanographic features, sea ice conditions, and atmospheric conditions, are also highly important, since they are the main drivers for prey distribution and availability (Tremblay et al. 2009; Labrousse et al. 2015; Cox et al. 2016). For instance, a study using satellite-linked dive recorders revealed that the distance to the continental shelf break and sea ice concentration were the most important drivers of crabeater seal's *Lobodon carcinophaga* distribution in the Weddell Sea in the summer of 1998 (Fig. 3) (Nachtsheim et al. 2017). Both their distribution and foraging behavior aligned well with the life

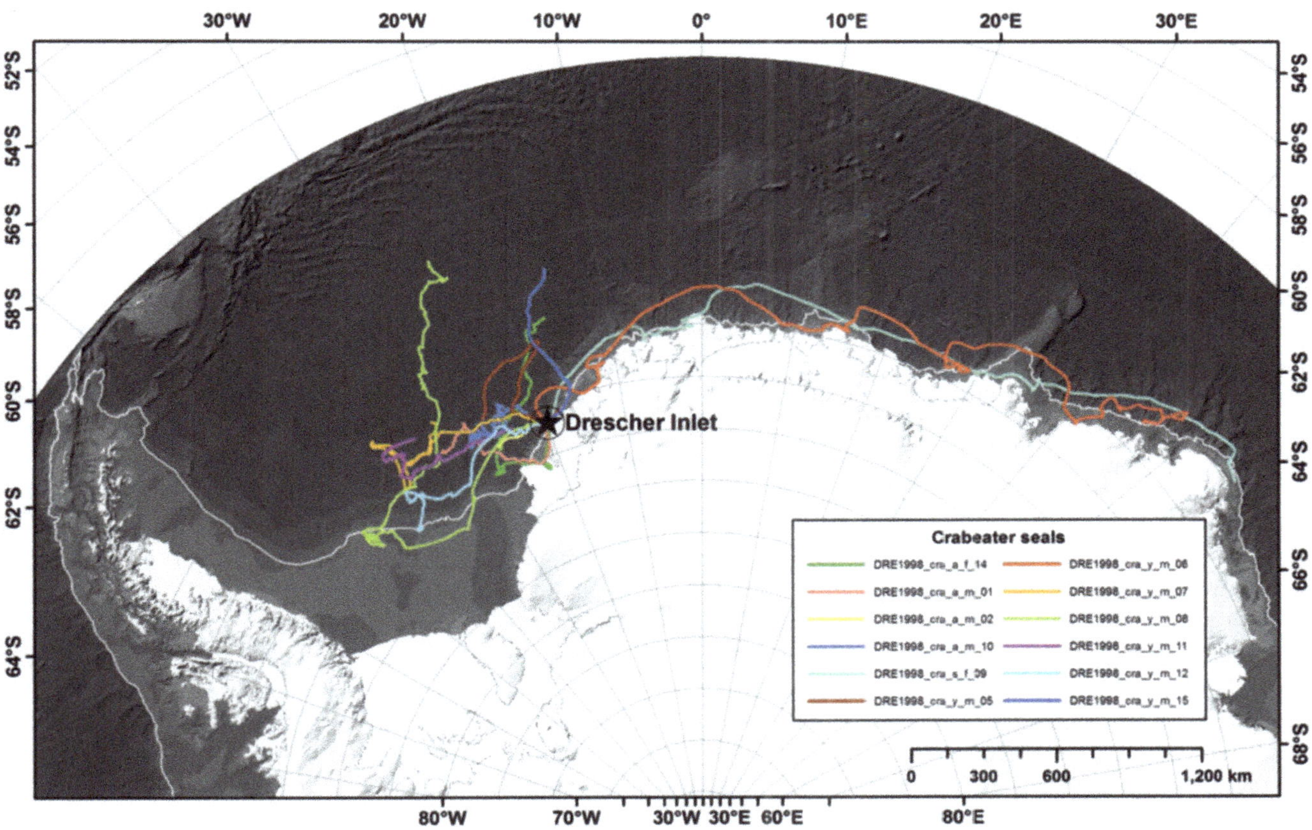

Fig. 3 Tracks of 12 crabeater seals *Lobodon carcinophaga* in the Weddell Sea dispersing from the tagging location in the Drescher Inlet (star). Each colored line represents an individual track. Bathymetry is indicated by various shades of grey (light = shallow, dark = deep). The white line shows the 1000 m isobath defined as continental shelf break. Ten seals explored the eastern and central Weddell Sea, while two animals moved far eastwards up to 45°E along the coast. (Reproduced from Nachtsheim et al. (2017) (CC-BY 4.0))

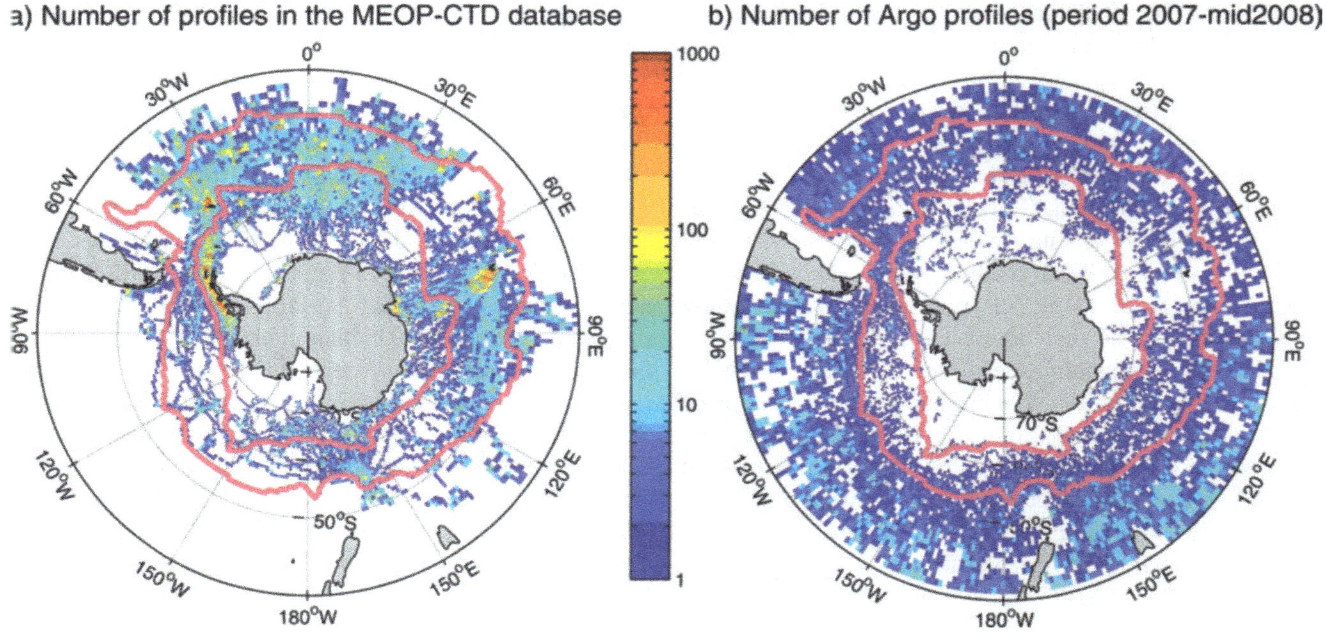

Fig. 4 Number of temperature-salinity profiles of (**a**) seal-derived data from the Marine mammal Exploring the Oceans Pole to Pole (MEOP) program, and (**b**) Argo float profiles (see Gould et al. (2004) for more details). Superimposed in pink are the Antarctic Circumpolar Current borders. This figure shows clearly how animal-borne devices can complement traditional observations, particularly at places where access is limited (e.g., due to sea ice). (Reproduced from Roquet et al. (2013) (CC-BY 4.0))

history of Antarctic krill *Euphausia superba*, their preferred prey (Nachtsheim et al. 2017).

Oceanographic and environmental data are readily available online, with the NOAA and the Copernicus program being the largest providers (Copernicus 2017; NOAA 2017), or from the animal-borne devices themselves. The latter sometimes provide physical oceanographic data from areas that cannot be sampled using other conventional approaches (Fig. 4) (Årthun et al. 2012; Roquet et al. 2013). Following this further, detailed oceanographic data could provide information about how marine top predators will probably respond to climatic change (Costa et al. 2010a; McIntyre et al. 2011). For instance, CTD satellite tags deployed on southern elephant seals *Mirounga leonina*, crabeater seals, and Weddell seals in the Western Antarctic Peninsula have shown that these three species occupy very different habitat types, hence trophic niches, within this region, and will therefore be affected differently by climate change (Costa et al. 2010a). Additionally, many marine top predators feed at depth and several studies have demonstrated the association between oceanographic features of the water column and predator's diving behavior (Fig. 5) (Biuw et al. 2010; Heerah et al. 2013; Guinet et al. 2014). Quantifying foraging effort at depth, based on the detection of changes in diving behavior, can relate the actual behavior of the predator in three dimensions to the heterogeneous environment they respond to (Heerah et al. 2016). For southern elephant seals, the switch from transit to hunting mode was associated with colder water temperatures, relatively short dive bottom time and rapid descent rates (Bestley et al. 2013). As mentioned before, foraging efficiency should be investigated by looking at both horizontal and vertical movements, as this can reveal interesting inter-specific behavioral differences, for example, resource partitioning through different dive behavior in closely related predator species (Wilson 2010; Villegas-Amtmann et al. 2013; Bestley et al. 2015). Information on foraging effort, which is, among others and depending on the species, determined by the number of dives, dive duration, vertical and horizontal travel distance, and the time a foraging trip takes, can be readily derived from conventional tracking data (Boyd et al. 2014). However, the study of movement ecology is lifted to another level by using tri-axial accelerometers combined with GPS loggers, which are able to provide high-resolution behavioral data on the level of decision-making (Wilson et al. 2008; Watanabe and Takahashi 2013; Bidder et al. 2014). For example, by linking GPS and accelerometer data of lesser black-backed gulls, a recent study was able to provide insight into how a flight generalist (i.e., a bird who has the ability to radically alter its flight mode in response to external conditions) can reduce the energetic cost of movement (Shamoun-Baranes et al. 2016).

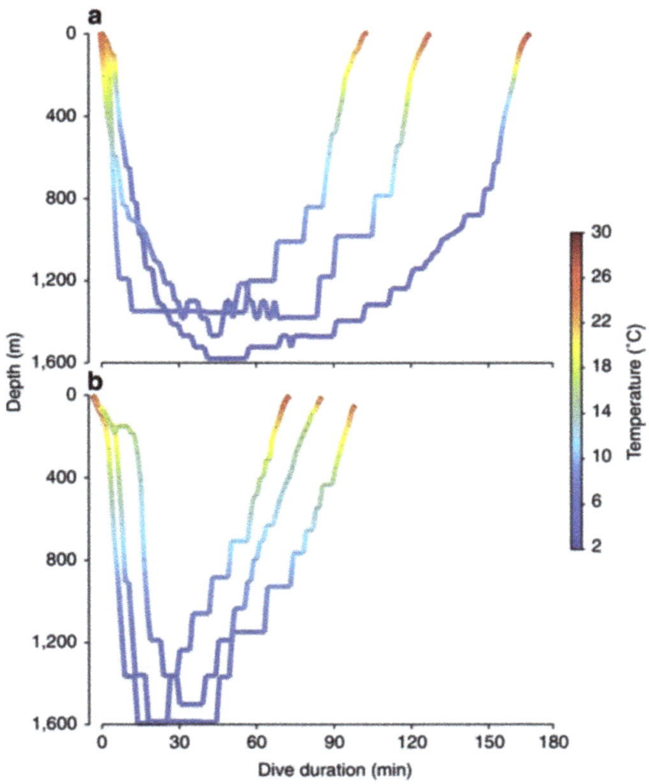

Fig. 5 Depth and temperature profiles from Chilean devil rays *Mobula tarapacana*, tagged with pop-up satellite archival transmitters during (**a**) daylight hours (6 a.m. – 6 p.m.), and (**b**) night-time hours (6 p.m. – 6 a.m.). These profiles show that devil rays are among the deepest-diving animals, especially at night, albeit shorter in time. (Reproduced from Thorrold et al. (2014) (CC-BY 4.0))

Hydroacoustic surveys can reveal the link between foraging habitat and the distribution of prey, and provide more insight into different aspects of foraging behavior. Although one might assume that predictable prey abundance determines foraging habitats, accessibility is an important factor too, especially for surface foragers, as is the case for Peruvian booby *Sula variegata* and guanay cormorant *Phalacrocorax bougainvilliorum* (Boyd et al. 2015). A specific foraging strategy may be influenced by several factors. Foraging range may be determined by the overall distribution of prey, the predators' ability to detect prey may be influenced by the distance between prey patches, and prey capture efficiency may be affected by individual patch characteristics (Carroll et al. 2017). Little penguins *Eudyptula minor* caught more prey where aggregations were relatively dense, compact and shallow (Carroll et al. 2017). Another study revealed that masked boobies *Sula dactylatra* from Phillip Island, Australia, showed a trade-off between strong foraging site fidelity around their colony where less prey is available, and more distant foraging trips with less predictable but larger prey patches (Sommerfeld et al. 2015). Temporal differences in foraging behavior can also be observed. For instance, dive depths of most northern elephant seal *Mirounga angustirostris* showed a clear diel pattern, consistent with targeting vertically migrating prey species (Fig. 5) (Robinson et al. 2012). Pursuing this further by deploying the hydroacoustic devices directly on large marine mammals, a recently developed sonar tag is able to record acoustic backscatter in front of a diving predator, and as such, quantify their prey field (Lawson et al. 2015).

By combining bio-telemetry data with other fine-scale measurements, such as vessel monitoring systems (VMS), we further enhance our multifactorial approach. These VMS are used globally, and since 2005, all fishing vessels in the European Union longer than 15 m are required to transmit their position through VMS. By looking at the fine-scale overlap between seabirds and fisheries, the ecological effect of foraging in association with fishing vessels can be determined, including carry-over effects, as not all species respond in the same way. For instance, foraging royal albatrosses *Diomedea sanfordi* showed low rates of overlap with fisheries and it provided them with no ecological advantage (Sugishita et al. 2015). Conversely, fisheries discards are an important part of lesser black-backed gulls' diet in the North Sea (Garthe et al. 1996; Sommerfeld et al. 2016). This is also the case for southern giant petrels *Macronectes giganteus*, whose non-breeding distribution largely overlap with zones of high fishing intensity off the coast of South America (Krüger et al. 2017). Northern gannets *Morus bassanus*, on the other hand, showed clear individual differences in discard consumption and foraging behavior (Votier et al. 2010).

To date, as a result of this multifactorial approach, marine scientists encounter new challenges in coordinating and analyzing high resolution datasets gathered across large spatial scales, (e.g., ocean basins). These challenges can only be overcome by means of multidisciplinary collaborations between biologists, oceanographers, statisticians and engineers (Hussey et al. 2015; Hays et al. 2016). Such collaborations can foster the development of new, innovative and cost-effective bio-telemetry approaches and promote cutting-edge analytical techniques.

Migration

Migration is defined as long-distance movement of individuals, with a temporal recurrence. Some marine top predators move across vast expanses of the marine environment to acquire spatiotemporal variable resources several times a year, annually or across multiple years. One of the most important uses of bio-telemetry is to identify migration routes and their overlap with anthropogenic features. This application is extremely useful for conservation purposes, and will, therefore, be discussed in more detail in section "Conservation".

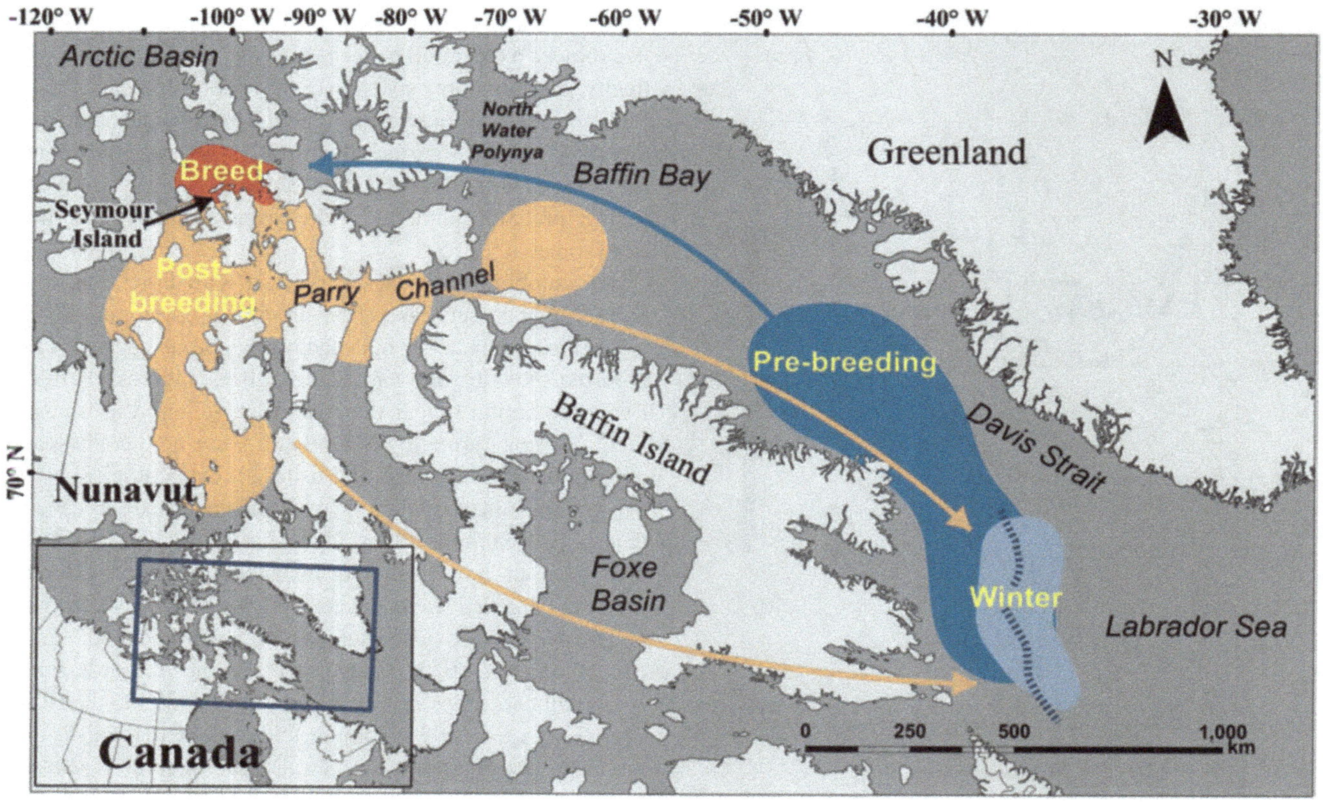

Fig. 6 Annual distribution and migration routes of the ivory gull *Pagophila eburnea* in the Canadian Arctic. The 50% kernels represent the general distribution during breeding (red), post-breeding (orange), winter (light blue) and pre-breeding seasons (dark blue). General direction of post-breeding migration is indicated by the orange arrows and direction of pre-breeding migration is indicated by the blue arrow. The dashed line through the winter kernel represents a composite of the typical edge of the pack ice, 2010–2013 (December through April). (Reproduced from Spencer et al. (2014) (CC-BY 4.0))

Various spatiotemporal scales of movement are tied to different life-history functions (Bestley et al. 2009; Block et al. 2011; Putman et al. 2014). Many top predators have evolved life histories that involve travelling large distances between predator-free breeding colonies and areas with large prey abundance (Corkeron and Connor 1999; Costa et al. 2012; Weimerskirch et al. 2012). For example, chinook salmon *Oncorhynchus tshawytscha* hatch in freshwater, then migrate to the sea where they spend most of their adult life, to ultimately migrate back to freshwater where they spawn and die (Quinn 2005). Long-term observations of these predators' movements provide not only information on the spatial extent of their populations and potential rates of exchange among them, but also expose detailed characteristics of the habitats they use and clues to their navigation abilities (Block et al. 2011; Costa et al. 2012). Bio-telemetry represents a crucial approach to gain this valuable knowledge on migration routes and patterns.

Environmental features undoubtedly influence migration. Analysis of migratory behavior of ivory gulls *Pagophila eburnea* revealed considerable individual variation of post-breeding migratory route selection, and suggested that the timing of formation/recession and extent of sea ice could play an important role in this (Fig. 6) (Spencer et al. 2014). González-Solís et al. (2009) showed that winds are a major determinant of the migratory routes of three shearwater species, the Manx *Puffinus puffinus*, the Cory's *Calonectris borealis*, and the Cape Verde *Calonectris edwardsii*.

For some species, migration appears to evolve through social learning, for instance in humpback whales *Megaptera novaeangliae* (Weinrich 1998). However, for many groups, the processes that shape migration routes remain enigmatic, despite the fact that satellite tracking can detail many migratory facets (Block et al. 2011). A particular challenge lies in explaining how juvenile animals, with no prior migratory experience, are able to locate specific oceanic feeding habitats (Lohmann et al. 2008; Gould and Gould 2012). One study showed that juvenile chinook salmon respond to magnetic fields at the latitudinal extremes of their ocean range, which lead towards their marine feeding grounds (Putman et al. 2014). The authors concluded that fish may use a combination of magnetic intensity and inclination angle to assess their geographic location (Putman et al. 2014). Whether this is the case for all migratory species remains to be determined. Studies on navigational ability in migratory seabirds showed that there can be many potential cues, for instance, olfactory, hydrodynamic, sky polarization, sun and star positions (Muheim et al. 2006; Lohmann et al. 2008; Gagliardo et al. 2013). However, detecting the geomagnetic field was, for example, in black-browed albatross *Diomedea melanophris* not one of them (Bonadonna et al. 2003). As such, we need to remind ourselves that experimental studies are never able to look at the full spectrum of natural conditions that occur during migration and, therefore, cannot eliminate particular cues completely.

Bio-telemetry already enables scientists to study the migratory routes of marine top predators in detail. However, at present, many questions remain, especially concerning the drivers and the processes that shape migration. This opens up interesting opportunities for future research and invites scientists to think outside the box to tackle current practical problems. Again, inter- and multidisciplinary collaborations could provide solutions.

Challenges and Ethics of Bio-telemetry

Attaching a device to an animal necessarily raises ethical questions (Wilson and McMahon 2006), which still remain largely understudied (Vandenabeele et al. 2011). Is it appropriate to equip an animal with an instrument that may impact its natural behavior or may even affect its breeding success and survival? What is regarded as acceptable practice? Does the acquisition of knowledge outweigh the potential adverse effects of a tag? In this section we will first highlight, separately for each top predator group, the difficulties of attaching a tag to an animal, and then list the known tag effects for each group. Finally, we mention approaches for improvements in the field of bio-telemetry and give recommendations from the perspective of animal welfare.

First of all, it is far from easy to equip an animal with a tag. It usually involves capturing and physically restraining the animal for a certain period of time. Sometimes even chemical immobilization is necessary, particularly in large pinnipeds (Gales and Mattlin 1998; Bornemann et al. 2013). The capture itself causes short-term stress for the animal, but does not usually have long-term consequences (Baker and Johanos 2002; McMahon et al. 2005; Blanchet et al. 2014). The tag attachment techniques vary considerably between, and even within, the different taxa, but can generally be divided into invasive and non-invasive attachments.

Most bio-telemetry devices for pinnipeds are non-invasive and commonly glued to the fur by using epoxy resin (Fedak et al. 1983; Lowther et al. 2013; Nachtsheim et al. 2017), or more recently, superglue (Cronin et al. 2016). The devices are either actively recovered or remain on the animal for up to one year until they fall off during the next annual molt. The attachment procedure itself, i.e., gluing the device to the fur, can cause small superficial abrasions or even lesions in a few cases, however, these injuries heal completely soon after the tag has been shed (Field et al. 2012). There is evidence that the placement of particularly large and bulky tags increases the hydrodynamic drag, potentially causing elevated metabolic costs and leading to behavioral changes in the short term (Walker and Boveng 1995; Hazekamp et al. 2010; Blanchet et al. 2014; Maresh et al. 2015; Rosen et al. 2017). However, long-term studies on the mass gain and survival of seals could not find adverse effects of bio-telemetry devices (Baker and Johanos 2002; McMahon et al. 2008; Mazzaro and Dunn 2010).

The attachment of non-invasive telemetry devices on cetaceans has proven to be challenging. The smooth and rapidly regenerating skin of whales, dolphins and porpoises hampers the attachment of external tags. A commonly used non-invasive method is deploying tags, such as DTAGs, with suction cups (Fig. 2f). These devices usually stay on the animal for a few hours or days (Aguilar Soto et al. 2008; Wisniewska et al. 2016). This method is obviously not sufficient to study large-scale migration patterns and hence, invasive tags are frequently used on cetaceans. The tag is either anchored in the blubber or muscle tissue, especially on large whales and dolphins (Weller 2008; Hauser et al. 2010; Reisinger et al. 2014; Gendron et al. 2015), or pinned through the dorsal fin of small dolphins and porpoises (Irvine et al. 1982; Teilmann et al. 2007; Balmer et al. 2014). While the former can be remotely deployed (e.g., using a crossbow), the latter requires physically capturing the animals. The wound-healing and long-term effects of invasive tagging have rarely been studied, particularly due to the elusive nature of most cetaceans. Although swellings, cavities, and scars are frequently observed at the tag site in large whales, even over a period of multiple years, these injuries, however, had no impact on the body condition, overall health or reproductive success of the animals (Best and Mate 2007; Mate et al. 2007; Weller 2008; Gendron et al. 2015; Norman et al. 2017). In small cetaceans, dorsal fin-mounted tags can fall off due to corrosion of the pins. In some cases, mild inflammatory responses have been reported around the pin holes, but they usually heal without complication, resulting in the formation of scar tissue (Sonne et al. 2012; Heide-Jørgensen et al. 2017). More severe is the migration of dorsal fin tags through the tissue over time due to the constant hydrodynamic drag, ultimately leading to tag loss and fin damage (Irvine et al. 1982; Martin et al. 2006; Balmer et al. 2014). This raises the concern of whether or not such tags negatively affect the natural behavior of small cetaceans (van der Hoop et al. 2014). However, on the long term, body condition, survival rates and reproductive success of tagged small cetaceans were not affected (Martin et al. 2006; Heide-Jørgensen et al. 2017)

Fig. 7 Different steps of deploying a video camera logger on a Magellanic penguin *Spheniscus magellanicus*. (**a**) First, a template is placed on the back of the penguin and Tesa® tape is placed in layers between the feathers. (**b**) The logger is placed on the back and the tape is wrapped tightly around the device before releasing the penguin back to the nest. (**c**) After the foraging trip, the tape is removed. (Photos used with permission from William Kay)

Similar problems occur when tagging sharks or other elasmobranchs. Due to their placoid scale epidermis, a non-invasive attachment procedure is almost impossible (Hammerschlag et al. 2011). Most PATs are forcefully implanted into the well-developed dorsal musculature, which reduces the probability of premature release (Carlson et al. 2010; Campana et al. 2011; Hammerschlag et al. 2011). As with dolphins, satellite transmitters are often attached to the shark's dorsal fin using bolts and pins, which can be a source of infection or lead to fin damage (Meyer et al. 2010; Hammerschlag et al. 2011; Jewell et al. 2011). Therefore, new non-invasive attachment methods are currently being tested and further developed. For example, a clamp system was successfully used to attach a combined accelerometer and magnetometer to the second dorsal fin of a whale shark (Fig. 1) (Gleiss et al. 2009; Williams et al. 2017).

Most deployments on seabirds involve the attachment of non-invasive bio-telemetry devices. Most commonly the tags are taped or glued to the back or tail feathers (Fig. 7), carried in a harness or attached to the leg in the form of a band ring (Wilson 1997; Phillips et al. 2003; Shaffer et al. 2005). Depending on the target species, certain attachment methods and locations are discouraged due to known adverse effects. For instance, the use of harnesses may have detrimental effects on petrels and albatrosses, which is why taping to the back feathers is the preferred method (Phillips et al. 2003; Weimerskirch et al. 2007; Mallory and Gilbert 2008). In contrast to other top predators, the effect of tagging has been relatively well studied in seabirds—possibly due to their small size and alarmingly large tags in the past (Wilson et al. 1986). Adult mass, breeding productivity and success, as well as foraging behavior are seemingly not altered and do not differ between tagged and untagged individuals (Phillips et al. 2003; Votier et al. 2004; Chivers et al. 2015; Thaxter et al. 2016). However, for the great skua *Stercorarius skua*, the over-winter return rate as a measure of survival were substantially lower for tagged than for untagged birds, whereas for the sympatric lesser black-backed gull neither short nor long term effects were reported (Thaxter et al. 2016). Thus, even morphologically similar species can react quite differently to the same animal-borne device. Most of the problems which occur can be attributed to the size of the device, which is traditionally aimed to be below 5% of the body weight of the bird (Cochran 1980; Phillips et al. 2003), but also to the shape, position and attachment method (Bannasch et al. 1994; Vandenabeele et al. 2014). This extra weight means an increase in energy expenditure during flight of approximately 5%, without taking into account the increase in drag (Vandenabeele et al. 2012). Thus, large tags can have an influence on the activity budget of seabirds, leading to a reduction of time spent flying (Chivers et al. 2015). Even though humans' movement ecology is somewhat different from that of other animals, we can use the following example to illustrate the possible influence of a tags' weight. If an average sized human with a body weight of 80 kg was equipped with a device that makes up 5% of his body weight, he would carry an extra 4 kg—this is equal the weight of two six packs of canned drinks. If the device was only 1% of the body weight, the extra weight would constitute 800 g, which is comparable to a small laptop. Since most studies define 3–5% of the individual's body weight as acceptable practice, but do not take into account the substantial extra energetic costs, we recommend reducing the tag weight load to well below these values, as suggested by Phillips et al. (2003) and Vandenabeele et al. (2012), optimally below 1% of the body weight.

The tagging of charismatic marine top predators still remains a controversial topic, also within the scientific community (Hazekamp et al. 2010; van der Hoop et al. 2014). However, no adverse long-term effects of bio-telemetry devices were reported for the vast majority of study species. Both the quality and quantity of bio-telemetry data are otherwise impossible to obtain, and tagging studies can immensely enhance our understanding of movements and behavior of marine predators, leading to informed management decisions. These facts advocate the right of bio-telemetry existence (McMahon et al. 2012). Nevertheless, there are various valid points in the light of animal welfare that are of concern and must be addressed appropriately. In addition to the increase of drag caused by the attachment of an external device, the extra weight load still represents a major issue for many marine predators.

We believe that, as a further improvement, the scientific community should foster the development of new technologies and the on-going miniaturization of tags. To reduce hydrodynamic drag, Computational Fluid Dynamics (CFD) has proven to be a crucial tool to assess and improve the design of bio-telemetry devices (Pavlov et al. 2007; Balmer et al. 2014). In addition, CFD minimizes the risk of injury and other potentially adverse effects. Furthermore, potential capture and attachment methods for different animal groups should be reviewed and discussed among researchers. It is evident that tag effects must be more thoroughly investigated, and in particular the focus should be directed towards metabolic costs and fitness consequences (McIntyre 2015). It must not be forgotten that we, as scientists, have the responsibility to ensure the wellbeing of the animal. This is not only relevant from an animal welfare viewpoint, but also from a scientific position, as we aim to record and understand the natural behavior of animals in their environment.

Conservation

The marine environment is predicted to face marked changes by the year 2040 (Brierley and Kingsford 2009), so there is no time to lose. The current challenges require multidisciplinary collaboration to develop effective strategies that rely on an improved understanding of the threat that climate change imposes on species, and the way that it interacts with their natural coping mechanisms (Dawson et al. 2011). Telemetry-derived data have a tremendous potential to inform resource management and conservation, but unfortunately, relatively few examples of their application exist (Wilson et al. 2015a). The financial cost of collecting telemetry data is relatively high, which makes it essential to critically evaluate the conservation benefit of the currently used strategies (McGowan et al. 2017).

One of these strategies, is to combine high resolution bio-telemetry data with environmental data and data on anthropogenic activities to build individual-based models, with the aim of predicting habitat use in the (near) future (Grimm et al. 1999; Stillman et al. 2003; Bestley et al. 2013; Stillman et al. 2015; van der Vaart et al. 2016). In these models, different scenarios of anticipated anthropogenically-driven environmental change can be simulated. Once predictive models have been implemented, the outcome can be used for conservation purposes. One extensive study found that modelled physical changes under the Intergovernmental Panel on Climate Change A2 scenario predict increased species overlap and a potential for niche compression across the North Pacific (Hazen et al. 2013). For species that are already under threat, such changes could exacerbate population declines or inhibit recovery. These models can undoubtedly be used, among other applications, to design marine protected areas (MPAs), which can be a powerful tool for attenuating anthropogenic threats. Several studies have already conducted extensive tracking programs on seabirds to identify foraging hotspots, and used those spatial patterns to inform MPA design (Le Corre et al. 2012; Thaxter et al. 2012; Lascelles et al. 2016).

However, to ensure that these MPAs are highly efficient, adaptive and dynamic management is necessary (Agardy et al. 2003; Maxwell et al. 2015). At present, the majority of marine management approaches (e.g., quota setting, total allowable catches, and MPAs) are relatively static in contrast to the ocean itself and the majority of ocean uses (Agardy 1994; Hyrenbach et al. 2000; Crowder and Norse 2008). Some marine predators are highly mobile and travel great distances over different temporal scales, as is the case, for example, for Arctic terns *Sterna paradisaea* (Egevang et al. 2010), and gray whales *Eschrichtius robustus* (Mate et al. 2015). Others, like Atlantic bluefin tuna *Thunnus thynnus* show complex spatial dynamics, (e.g., homing) and population structure (e.g., several subpopulations of different sizes) (Fromentin and Lopuszanski 2014). To effectively manage this highly dynamic marine system, conservation measures must become more flexible in space and time in the same way as both the environment and the resource users have (Hyrenbach et al. 2000).

A robust understanding of the spatiotemporal distribution and ecology of migratory species is necessary for successful conservation. Conditions experienced during the non-breeding period may have carry-over effects on breeding performance, which in turn may affect population dynamics (Harrison et al. 2011). Consequently, the anthropogenic impact outside the breeding season also needs to be taken into consideration when making management decisions. Conservation efforts can be targeted at areas where the majority of the population congregate, as is the case for little auks *Alle alle* at two key areas located in the Greenland Sea and off Newfoundland (Fort et al. 2013).

Fast-acquisition satellite telemetry can provide evidence-based information on individual animal movements to delineate interspecific relationships, and can be used to increase the efficacy of conservation planning (Gredzens et al. 2014). Complimentary co-management, customized for each location, would be advisable when different species use similar habitats (Gredzens et al. 2014). By including data from different species, we obtain more information about commensal foraging associations or multispecies feeding flocks and intra- and interspecific interactions (Barlow et al. 2002; Elliott et al. 2010; Lidgard et al. 2012). Areas of predator overlap often occur in regions with high ecological importance, and are therefore useful in designing MPA boundaries. Raymond et al. (2015) used combined tracking data of six species (Adélie *Pygoscelis adeliae* and emperor *Aptenodytes forsteri* penguins, light-mantled albatross *Phoebetria palpebrata*, Antarctic fur seals, southern elephant seals, and Weddell seals) to identify areas of particular ecological significance. These areas were characterized by their proximity to breeding colonies and by sea-ice dynamics, and were therefore considered in the MPA proposal for East Antarctica (Raymond et al. 2015). A similar study used tracking data from 14 different predator species to identify their foraging habitats around the Prince Edward Islands (Reisinger et al. 2018). Amongst others, the results of this study can support the conservation and management of Subantarctic ecosystems and the marine predators they sustain (Reisinger et al. 2018).

Another effective aspect of protective measures can be to establish corridors that link the breeding and foraging grounds of migratory species. One study used a geospatial approach to design a corridor along the north-west coast of Australia that incorporates 11 existing MPAs and overlaps with humpback whale migratory tracks (Pendoley et al. 2014). The study concluded that the proposed network would be beneficial for at least 20 other marine vertebrates, although not all at the same temporal scale (Pendoley et al. 2014).

Establishing pelagic MPAs for oceanic species is another possible measure in adaptive and dynamic management. Bio-telemetry can be a useful tool to acknowledge their necessity and to determine their flexible boundaries (Game et al. 2009). For instance, tiger sharks *Galeocerdo cuvier* were considered to be a reef-associated coastal species, but bio-telemetry actually showed that they display directional movements across ocean basins (Holland et al. 1999; Lowe et al. 2006; Heithaus et al. 2007). Another study determined the primary migratory corridor of gray whales, and concluded that they face a wider range of industrial activities and developments than previously thought (Ford et al. 2012). Protecting far-ranging species can present a major challenge for spatial management, but they are not always equally vulnerable over their entire range (Game et al. 2009). These species often exhibit increased vulnerability in a small number of demographically critical areas, as is the case in wandering albatrosses *Diomedea exulans* (Weimerskirch et al. 2006). Other species potentially overlap with different human activities during each stage of their migration (e.g., humpback whales) (Rosenbaum et al. 2014). This example highlights the need for adaptive and dynamic management once more, for it would be economically advisable to move the pelagic MPA in space and time synchronously with the whales' migration.

Tracking data can also identify high-use areas and coordinate policy actions that mitigate anthropogenic risks such as those associated with ship strikes, offshore wind farms, oil spills, or bycatch. For instance, manta ray *Manta birostris* aggregations coincide with some of the busiest shipping lanes (Halpern et al. 2008). This, together with the expansion of megafauna tourism industry, could have an impact on their population numbers (Berman-Kowalewski et al. 2010). However, despite the fact that manta rays forage over large spatial scales (~100 km) far offshore, they also show high site fidelity and associate with frontal zones (Graham et al. 2012). This knowledge could be used to establish new dynamically protected areas overlaying the frontal region (Graham et al. 2012). Bio-telemetry can also be used to estimate seabirds' vulnerability to offshore wind farms by determining activity-specific and spatially explicit flight heights and collision risks (Cleasby et al. 2015). Another example revealed that 25% of the North American northern gannet populations migrate annually to the Gulf of Mexico and suffered from severe oiling in the aftermath of the Deepwater Horizon explosion (Montevecchi et al. 2012a, b). These findings contrasted discernibly with available mark-recapture data, and showed that tracking research can be extremely useful when little information on animal distribution in pollution zones is available (Montevecchi et al. 2012a). Tracking data combined with survival and reproduction measurements from their colonies can reveal many possible repercussions of marine pollution, and inform management about conservation concerns (Montevecchi et al. 2012a). Another widespread anthropogenic problem is air-breathing megafauna bycatch. This bycatch intensity varies substantially within and between catch gear and regions (Lewison et al. 2014). Tracking data of marine predators can be overlapped with fisheries data for a dynamic management approach, which could minimize bycatch. For example, an improved understanding of the horizontal and vertical spatiotemporal distribution of North Pacific albatrosses in relation to pelagic fisheries could improve management protocols (e.g., time-area closures and gear mitigation), to reduce the bycatch of these endangered and threatened species (Costa et al. 2012). Lastly, one of the most interesting applications to date is the automated, near-real-time density prediction tool for Eastern North Pacific blue whales *Balaenoptera musculus*, which enables a more accurate examination of the year-round spatiotemporal overlap of the whales with potentially harmful human activities, such as shipping (Hazen et al. 2016). This study identified high interannual variability in occurrence, emphasizing again the benefit of a dynamic approach (Hazen et al. 2016). Tools like this allow a finer-scale management, which is more economically feasible and socially acceptable (Hazen et al. 2016).

Undoubtedly, there is a necessity for innovative and interdisciplinary approaches, monitoring programs and research initiatives to inform decision makers (Cooke 2008), but it is not only those people who need to be informed. In light of current rapid environmental changes, it is also imperative to engage the general public, and to gain their support for conservation. One of the biggest advantages of telemetry is its usefulness in outreach, as this data can be presented in a highly attractive way. However, it is not only their support that is vital. Web-based citizen science projects are a prime example of how volunteers can help process the enormous amount of telemetry-derived data (e.g., through the Zooniverse portal with links to Seabirdwatch, Weddell Seal Count, Snapshots at Sea, Penguin Watch, etc.) (Zooniverse 2017). In general, results from bio-telemetry studies on marine top predators can inform local, regional, and international conservation management, and the general public, which in turn, could trigger actions to mitigate potential anthropogenic threats. To be effective, these management measures should be highly flexible according to the very dynamic nature of marine ecosystems.

Conclusion

It is evident that tagging marine top predators with bio-telemetry devices has an enormous value – not only for studying their movement ecology but also for resolving marine conservation issues. Data from these devices have already answered numerous fundamental questions, and will continue to do so in the future (Hays et al. 2016). The current rapid technological advances will certainly lead to an ever growing variety of new devices and methodological approaches. Nevertheless, animal welfare may not be overlooked. Although no negative long-term effects of bio-telemetry devices were reported for most species, there is still potential for improvements. Tag effects must be thoroughly assessed and the miniaturization of bio-telemetry devices must continue to ensure the wellbeing of tagged animals.

In conclusion, adaptive and dynamic management is necessary to guarantee healthy marine ecosystems and this can benefit substantially from monitoring top predators with bio-telemetry devices (Maxwell et al. 2015). However, regardless of the management approach, it is clear that multidisciplinary collaboration is key and sharing data is imperative, if conservation is to be successful (Cvitanovic et al. 2015). Additionally, scientists need new tools and frameworks to link animal telemetry-derived data to conservation and management, while maximizing the outcome of the global investment in bio-telemetry devices (McGowan et al. 2017). This will enable us to address the relevant key questions in top predator movement ecology—in conjunction with governments and society—and ultimately lead to effective conservation and management of marine ecosystems.

Appendix

This article is related to the YOUMARES 8 conference session no. 1: “Sentinels of the Sea: Ecology and Conservation of Marine Top Predators”. The original Call for Abstracts and the abstracts of the presentations within this session can be found in the appendix “Conference sessions and Abstracts”, chapter “6 Sentinels of the Sea: Ecology and Conservation of Marine Top Predators”, of this book.

References

Aarts G, MacKenzie M, McConnell B et al (2008) Estimating space-use and habitat preference from wildlife telemetry data. Ecography 31:140–160. https://doi.org/10.1111/j.2007.0906-7590.05236.x

Afanasyev V (2004) A miniature daylight level and activity data recorder for tracking animals over long periods. Mem Natl Inst Polar Res Spec Issue 58:227–233

Agardy MT (1994) Advances in marine conservation: the role of marine protected areas. Trends Ecol Evol 9:267–270. https://doi.org/10.1016/0169-5347(94)90297-6

Agardy MT, Bridgewater P, Crosby MP et al (2003) Dangerous targets? unresolved issues and ideological clashes around marine protected areas. Aquat Conserv Mar Freshwat Ecosyst 13:353–367. https://doi.org/10.1002/aqc.583

Aguilar Soto N, Johnson MP, Madsen PT et al (2008) Cheetahs of the deep sea: deep foraging sprints in short-finned pilot whales off Tenerife (Canary Islands). J Anim Ecol 77:936–947. https://doi.org/10.1111/j.1365-2656.2008.01393.x

Andersson M (1978) Optimal foraging area: size and allocation of search effort. Theor Popul Biol 13:397–409

Andrews RD, Pitman RL, Ballance LT (2008) Satellite tracking reveals distinct movement patterns for Type B and Type C killer whales in the southern Ross Sea, Antarctica. Polar Biol 31:1461–1468. https://doi.org/10.1007/s00300-008-0487-z

Antonelis GA, Baker JD, Johanos TC et al (2006) Hawaiian monk seal (*Monachus schauinslandi*): status and conservation. Atoll Res Bull 543:75–101

Arcalís-Planas A, Sveegaard S, Karlsson O et al (2015) Limited use of sea ice by the Ross seal (*Ommatophoca rossii*), in Amundsen Sea, Antarctica, using telemetry and remote sensing data. Polar Biol 38:445–461. https://doi.org/10.1007/s00300-014-1602-y

Årthun M, Nicholls KW, Makinson K et al (2012) Seasonal inflow of warm water onto the southern Weddell Sea continental shelf, Antarctica. Geophys Res Lett 39:2–7. https://doi.org/10.1029/2012GL052856

Arthur B, Hindell M, Bester MN et al (2016) South for the winter? within-dive foraging effort reveals the trade-offs between divergent foraging strategies in a free-ranging predator. Funct Ecol 30:1623–1637. https://doi.org/10.1111/1365-2435.12636

Austin D, Bowen WD, McMillan JI et al (2006a) Stomach temperature telemetry reveals temporal patterns of foraging success in a free-ranging marine mammal. J Anim Ecol 75:408–420. https://doi.org/10.1111/j.1365-2656.2006.01057.x

Austin D, Bowen WD, McMillan JI et al (2006b) Linking movement, diving, and habitat to foraging success in a large marine predator. Ecology 87:3095–3108

Baker JD, Johanos TC (2002) Effects of research handling on the endangered Hawaiian monk seal. Mar Mamm Sci 18:500–512. https://doi.org/10.1111/j.1748-7692.2002.tb01051.x

Balmer BC, Wells RS, Howle LE et al (2014) Advances in cetacean telemetry: a review of single-pin transmitter attachment techniques on small cetaceans and development of a new satellite-linked transmitter design. Mar Mamm Sci 30:656–673. https://doi.org/10.1111/mms.12072

Bannasch R, Wilson RP, Culik B (1994) Hydrodynamic aspects of design and attachment of a back-mounted device in penguins. J Exp Biol 194:83–96

Barlow KE, Boyd IL, Croxall JP et al (2002) Are penguins and seals in competition for Antarctic krill at South Georgia? Mar Biol 140:205–213. https://doi.org/10.1007/s00227-001-0691-7

Bartumeus F, Giuggioli L, Louzao M et al (2010) Fishery discards impact on seabird movement patterns at regional scales. Curr Biol 20:215–222. https://doi.org/10.1016/j.cub.2009.11.073

Baum JK, Myers RA, Kehler DG et al (2003) Collapse and conservation of shark populations in the Northwest Atlantic. Science 299:389–392. https://doi.org/10.1126/science.1079777

Berman-Kowalewski M, Gulland FMD, Wilkin S et al (2010) Association between blue whale (*Balaenoptera musculus*) mortality and ship strikes along the California coast. Aquat Mamm 36:59–66. https://doi.org/10.1578/AM.36.1.2010.59

Best PB, Mate B (2007) Sighting history and observations of southern right whales following satellite tagging off South Africa. J Cetacean Res Manag 9:111–114

Bestley S, Gunn JS, Hindell MA (2009) Plasticity in vertical behaviour of migrating juvenile southern bluefin tuna (*Thunnus maccoyii*) in relation to oceanography of the south Indian Ocean. Fish Oceanogr 18:237–254. https://doi.org/10.1111/j.1365-2419.2009.00509.x

Bestley S, Jonsen ID, Hindell MA et al (2013) Integrative modelling of animal movement: incorporating in situ habitat and behavioural information for a migratory marine predator. Proc R Soc B Biol Sci 280:2012–2262. https://doi.org/10.1098/rspb.2012.2262

Bestley S, Jonsen ID, Hindell MA et al (2015) Taking animal tracking to new depths: synthesizing horizontal – vertical movement relationships for four marine predators. Ecology 96:417–427. https://doi.org/10.1890/14-0469.1

Bicknell AWJ, Oro D, Camphuijsen CJ et al (2013) Potential consequences of discard reform for seabird communities. J Appl Ecol 50:649–658. https://doi.org/10.1111/1365-2664.12072

Bidder OR, Campbell HA, Gómez-Laich A et al (2014) Love thy neighbour: automatic animal behavioural classification of acceleration data using the k-nearest neighbour algorithm. PLoS One 9:e88609. https://doi.org/10.1371/journal.pone.0088609

Biuw M, Boehme L, Guinet C et al (2007) Variations in behavior and condition of a Southern Ocean top predator in relation to in situ oceanographic conditions. Proc Natl Acad Sci U S A 104:13705–13710

Biuw M, Nøst OA, Stien A et al (2010) Effects of hydrographic variability on the spatial, seasonal and diel diving patterns of southern elephant seals in the Eastern Weddell Sea. PLoS One 5:e13816. https://doi.org/10.1371/journal.pone.0013816

Blanchet M-A, Lydersen C, Biuw M et al (2014) Instrumentation and handling effects on Antarctic fur seals (*Arctocephalus gazella*). Polar Res 33:21630. https://doi.org/10.3402/polar.v33.21630

Blanchet M-A, Lydersen C, Ims RA et al (2015) Seasonal, oceanographic and atmospheric drivers of diving behaviour in a temperate seal species living in the high Arctic. PLoS One 10:e0132686. https://doi.org/10.1371/journal.pone.0132686

Block BA, Dewar H, Blackwell SB et al (2001) Migratory movements, depth preferences, and thermal biology of Atlantic bluefin tuna. Science 293:1310–1314. https://doi.org/10.1126/science.1061197

Block BA, Jonsen ID, Jorgensen SJ et al (2011) Tracking apex marine predator movements in a dynamic ocean. Nature 475:86–90. https://doi.org/10.1038/nature10082

Boaden AE, Kingsford MJ (2015) Predators drive community structure in coral reef fish assemblages. Ecosphere 6:1–33

Boehme L, Lovell P, Biuw M et al (2009) Technical note: animal-borne CTD-satellite relay data loggers for real-time oceanographic data collection. Ocean Sci 5:685–695. https://doi.org/10.5194/osd-6-1261-2009

Boersma PD (2008) Penguins as Marine Sentinels. Bioscience 58:597–607. https://doi.org/10.1641/B580707

Bolnick DI, Svanbäck R, Fordyce JA et al (2003) The ecology of individuals: incidence and implications of individual specialization. Am Nat 161:1–28. https://doi.org/10.1086/343878

Bonadonna F, Chamaillé-Jammes S, Pinaud D et al (2003) Magnetic cues: are they important in Black-browed Albatross *Diomedea melanophris* orientation? Ibis 145:152–155. https://doi.org/10.1046/j.1474-919X.2003.00117.x

Bornemann H, de Bruyn PJN, Reisinger RR et al (2013) Tiletamine/zolazepam immobilization of adult post-moult southern elephant seal males. Polar Biol 36:1687–1692. https://doi.org/10.1007/s00300-013-1378-5

Bouten W, Baaij EW, Shamoun-Baranes J et al (2013) A flexible GPS tracking system for studying bird behaviour at multiple scales. J Ornithol 154:571–580. https://doi.org/10.1007/s10336-012-0908-1

Boyce DG, Tittensor DP, Worm B (2008) Effects of temperature on global patterns of tuna and billfish richness. Mar Ecol Prog Ser 355:267–276. https://doi.org/10.3354/meps07237

Boyd C, Punt AE, Weimerskirch H et al (2014) Movement models provide insights into variation in the foraging effort of central place foragers. Ecol Modell 286:13–25. https://doi.org/10.1016/j.ecolmodel.2014.03.015

Boyd C, Castillo R, Hunt GL et al (2015) Predictive modelling of habitat selection by marine predators with respect to the abundance and depth distribution of pelagic prey. J Anim Ecol 84:1575–1588. https://doi.org/10.1111/1365-2656.12409

Bradshaw CJA, Hindell MA, Sumner MD et al (2004) Loyalty pays: potential life history consequences of fidelity to marine foraging regions by southern elephant seals. Anim Behav 68:1349–1360. https://doi.org/10.1016/j.anbehav.2003.12.013

Brierley AS, Kingsford MJ (2009) Impacts of climate change on marine organisms and ecosystems. Curr Biol 19:R602–R614. https://doi.org/10.1016/j.cub.2009.05.046

Burger J, Gochfeld M, Serra-Sogas N et al (2004) Marine birds as sentinels of environmental pollution. Ecohealth 1:263–274

Burns JM (1999) The development of diving behavior in juvenile Weddell seals: pushing physiological limits in order to survive. Can J Zool 77:737–747

Burns JM, Hindell MA, Bradshaw CJA et al (2008) Fine-scale habitat selection of crabeater seals as determined by diving behavior. Deep Res Part II 55:500–514. https://doi.org/10.1016/j.dsr2.2007.11.012

Cagnacci F, Boitani L, Powell RA et al (2010) Animal ecology meets GPS-based radiotelemetry: a perfect storm of opportunities and challenges. Philos Trans R Soc Lond B Biol Sci 365:2157–2162. https://doi.org/10.1098/rstb.2010.0107

Campana SE, Dorey A, Fowler M et al (2011) Migration pathways, behavioural thermoregulation and overwintering grounds of blue sharks in the Northwest Atlantic. PLoS One 6:e16854. https://doi.org/10.1371/journal.pone.0016854

Campbell SJ, Hoey AS, Maynard J et al (2012) Weak compliance undermines the success of no-take zones in a large government-controlled marine protected area. PLoS One 7:1–12. https://doi.org/10.1371/journal.pone.0050074

Camphuijsen CJ, Shamoun-Baranes J, Van Loon EE et al (2015) Sexually distinct foraging strategies in an omnivorous seabird. Mar Biol 162:1417–1428. https://doi.org/10.1007/s00227-015-2678-9

Carlson JK, Ribera MM, Conrath CL et al (2010) Habitat use and movement patterns of bull sharks *Carcharhinus leucas* determined using pop-up satellite archival tags. J Fish Biol 77:661–675. https://doi.org/10.1111/j.1095-8649.2010.02707.x

Carroll G, Jonsen I, Cox M et al (2017) Hierarchical influences of prey distribution on patterns of prey capture by a marine predator. Funct Ecol. https://doi.org/10.1111/1365-2435.12873

Carter MID, Bennett KA, Embling CB et al (2016) Navigating uncertain waters: a critical review of inferring foraging behaviour from location and dive data in pinnipeds. Mov Ecol 4:25. https://doi.org/10.1186/s40462-016-0090-9

Chaise LL, Paterson W, Laske TG et al (2017) Implantation of subcutaneous heart rate data loggers in southern elephant seals (*Mirounga leonina*). Polar Biol. https://doi.org/10.1007/s00300-017-2144-x

Chapple TK, Gleiss AC, Jewell OJD et al (2015) Tracking sharks without teeth: a non-invasive rigid tag attachment for large predatory sharks. Anim Biotelem 3:14. https://doi.org/10.1186/s40317-015-0044-9

Chivers LS, Hatch SA, Elliott KH (2015) Accelerometry reveals an impact of short-term tagging on seabird activity budgets. Condor Ornithol Appl 118:159–168. https://doi.org/10.1650/CONDOR-15-66.1

Cleasby IR, Wakefield ED, Bearhop S et al (2015) Three-dimensional tracking of a wide-ranging marine predator: flight heights and vulnerability to offshore wind farms. J Appl Ecol 52:1474–1482. https://doi.org/10.1111/1365-2664.12529

Cochran WW (1980) Wildlife telemetry. In: Schemnitz SD (ed) Wildlife management techniques manual. Wildlife Society, Washington, DC, pp 507–520

Cooke SJ (2008) Biotelemetry and biologging in endangered species research and animal conservation: relevance to regional, national, and IUCN Red List threat assessments. Endanger Species Res 4:165–185. https://doi.org/10.3354/esr00063

Copernicus (2017) Satellite Earth observation and in situ (non-space) data. http://www.copernicus.eu/. Accessed 11 Nov 2017

Corkeron PJ, Connor RC (1999) Why do baleen whales migrate. Mar Mamm Sci 15:1228–1245. https://doi.org/10.1111/j.1748-7692.1999.tb00887.x

Costa DP, Huckstadt LA, Crocker DE et al (2010a) Approaches to studying climatic change and its role on the habitat selection of Antarctic pinnipeds. Integr Comp Biol 50:1018–1030. https://doi.org/10.1093/icb/icq054

Costa DP, Robinson PW, Arnould JPY et al (2010b) Accuracy of ARGOS locations of pinnipeds at-sea estimated using Fastloc GPS. PLoS One 5:e8677. https://doi.org/10.1371/journal.pone.0008677

Costa DP, Breed GA, Robinson PW (2012) New insights into pelagic migrations: implications for ecology and conservation. Annu Rev Ecol Evol Syst 43:73–96. https://doi.org/10.1146/annurev-ecolsys-102710-145045

Cox SL, Miller PI, Embling CB et al (2016) Seabird diving behaviour reveals the functional significance of shelf-sea fronts as foraging hotspots. R Soc Open Sci 3:160317

Cox SL, Orgeret F, Gesta M et al (2017) Processing of acceleration and dive data on-board satellite relay tags to investigate diving and foraging behaviour in free-ranging marine predators. Methods Ecol Evol. https://doi.org/10.1111/2041-210X.12845

Crawford R, Altwegg R, Barham B et al (2011) Collapse of South Africa's penguins in the early 21st century. African J Mar Sci 33:139–156. https://doi.org/10.2989/1814232X.2011.572377

Cronin MA, McConnell BJ (2008) SMS seal: a new technique to measure haul-out behaviour in marine vertebrates. J Exp Mar Bio Ecol 362:43–48. https://doi.org/10.1016/j.jembe.2008.05.010

Cronin MA, Gerritsen H, Reid D et al (2016) Spatial overlap of grey seals and fisheries in Irish waters, some new insights using telemetry technology and VMS. PLoS One 11:e0160564. https://doi.org/10.1371/journal.pone.0160564

Crowder L, Norse E (2008) Essential ecological insights for marine ecosystem-based management and marine spatial planning. Mar Policy 32:772–778. https://doi.org/10.1016/j.marpol.2008.03.012

Croxall JP, Butchart SHM, Lascelles B et al (2012) Seabird conservation status, threats and priority actions: a global assessment. Bird Conserv Int 22:1–34. https://doi.org/10.1017/S0959270912000020

Cvitanovic C, Hobday AJ, van Kerkhoff L et al (2015) Improving knowledge exchange among scientists and decision-makers to facilitate the adaptive governance of marine resources: a review of knowledge and research needs. Ocean Coast Manag 112:25–35. https://doi.org/10.1016/j.ocecoaman.2015.05.002

Dagorn L, Pincock D, Girard C et al (2007) Satellite-linked acoustic receivers to observe behavior of fish in remote areas. Aquat Living Resour 20:307–312. https://doi.org/10.1051/alr:2008001

Davis RW, Fuiman LA, Williams TM et al (1999) Hunting behavior of a marine mammal beneath the Antarctic fast ice. Science 283:993–996. https://doi.org/10.1126/science.283.5404.993

Davis RW, Jaquet N, Gendron D et al (2007) Diving behavior of sperm whales in relation to behavior of a major prey species, the jumbo squid, in the Gulf of California, Mexico. Mar Ecol Prog Ser 333:291–302. https://doi.org/10.3354/meps333291

Dawson TP, Jackson ST, House JI et al (2011) Beyond predictions: biodiversity conservation in a changing climate. Science 332:53–58. https://doi.org/10.1126/science.1200303

DeVries AL, Wohlschlag DE (1964) Diving depths of the Weddell seal. Science 145:292. https://doi.org/10.1126/science.145.3629.292

Dietz R, Teilmann J, Andersen SM et al (2013) Movements and site fidelity of harbour seals (*Phoca vitulina*) in Kattegat, Denmark, with implications for the epidemiology of the phocine distemper virus. ICES J Mar Sci 70:186–195

Donaldson MR, Hinch SG, Suski CD et al (2014) Making connections in aquatic ecosystems with acoustic telemetry monitoring. Front Ecol Environ 12:565–573. https://doi.org/10.1890/130283

Dragon AC, Monestiez P, Bar-Hen A et al (2010) Linking foraging behaviour to physical oceanographic structures: Southern elephant seals and mesoscale eddies east of Kerguelen Islands. Prog Oceanogr 87:61–71. https://doi.org/10.1016/j.pocean.2010.09.025

Dragon AC, Bar-Hen A, Monestiez P et al (2012) Horizontal and vertical movements as predictors of foraging success in a marine predator. Mar Ecol Prog Ser 447:243–257. https://doi.org/10.3354/meps09498

Dujon AM, Lindstrom RT, Hays GC (2014) The accuracy of Fastloc-GPS locations and implications for animal tracking. Methods Ecol Evol 5:1162–1169. https://doi.org/10.1111/2041-210X.12286

Eckert SA, Stewart BS (2001) Telemetry and satellite tracking of whale sharks, *Rhincodon typus*, in the Sea of Cortez, Mexico, and the north Pacific Ocean. Environ Biol Fishes 60:299–308

Edrén SMC, Wisz MS, Teilmann J et al (2010) Modelling spatial patterns in harbour porpoise satellite telemetry data using maximum entropy. Ecography 33:698–708. https://doi.org/10.1111/j.1600-0587.2009.05901.x

Egevang C, Stenhouse IJ, Phillips RA et al (2010) Tracking of Arctic terns *Sterna paradisaea* reveals longest animal migration. Proc Natl Acad Sci 107:2078–2081. https://doi.org/10.1073/pnas.0909493107

Elliott KH, Gaston AJ, Crump D (2010) Sex-specific behavior by a monomorphic seabird represents risk partitioning. Behav Ecol 21:1024–1032. https://doi.org/10.1093/beheco/arq076

Estes JA, Terborgh J, Brashares JS et al (2011) Trophic downgrading of planet Earth. Science 333:301–306. https://doi.org/10.1126/science.1205106

Fancy SG, Pank LF, Douglas DC et al (1988) Satellite telemetry: a new tool for wildlife research and management. Fish Wildl Serv 172:1–54

Fedak MA, Anderson SS, Curry MG (1983) Attachment of a radio tag to the fur of seals. J Zool 200:298–300. https://doi.org/10.1111/j.1469-7998.1983.tb05794.x

Fedak MA, Lovell P, McConnell B et al (2002) Overcoming the constraints of long range radio telemetry from animals: getting more useful data from smaller packages. Integr Comp Biol 42:3–10. https://doi.org/10.1093/icb/42.1.3

Field IC, Harcourt RG, Boehme L et al (2012) Refining instrument attachment on phocid seals. Mar Mammal Sci 28:1–9. https://doi.org/10.1111/j.1748-7692.2011.00519.x

Ford JKB, Durban JW, Ellis GM et al (2012) New insights into the northward migration route of gray whales between Vancouver Island, British Columbia, and southeastern Alaska. Mar Mamm Sci 29:325–337. https://doi.org/10.1111/j.1748-7692.2012.00572.x

Fort J, Moe B, Strøm H et al (2013) Multicolony tracking reveals potential threats to little auks wintering in the North Atlantic from marine pollution and shrinking sea ice cover. Divers Distrib 19:1322–1332. https://doi.org/10.1111/ddi.12105

Froget G, Butler PJ, Woakes AJ et al (2004) Heart rate and energetics of free-ranging king penguins (*Aptenodytes patagonicus*). J Exp Biol 207:3917–3926. https://doi.org/10.1242/jeb.01232

Fromentin J-M, Lopuszanski D (2014) Migration, residency, and homing of bluefin tuna in the western Mediterranean Sea. ICES J Mar Sci 71:510–518. https://doi.org/10.1093/icesjms/fst157

Fuiman LA, Davis RW, Williams TM (2002) Behavior of midwater fishes under the Antarctic ice: observations by a predator. Mar Biol 140:815–822. https://doi.org/10.1007/s00227-001-0752-y

Furness RW (2003) Impacts of fisheries on seabird communities. Sci Mar 67:33–45. https://doi.org/10.3989/scimar.2003.67s233

Gagliardo A, Bried J, Lambardi P et al (2013) Oceanic navigation in Cory's shearwaters: evidence for a crucial role of olfactory cues for homing after displacement. J Exp Biol 216:2798–2805. https://doi.org/10.1242/jeb.085738

Gales NJ, Mattlin RH (1998) Fast, safe, field-portable gas anesthesia for otariids. Mar Mamm Sci 14:355–361. https://doi.org/10.1111/j.1748-7692.1998.tb00727.x

Gallon S, Bailleul F, Charrassin JB et al (2013) Identifying foraging events in deep diving southern elephant seals, *Mirounga leonina*, using acceleration data loggers. Deep Res Part II Top Stud Oceanogr 88–89:14–22. https://doi.org/10.1016/j.dsr2.2012.09.002

Game ET, Grantham HS, Hobday AJ et al (2009) Pelagic protected areas: the missing dimension in ocean conservation. Trends Ecol Evol 24:360–369. https://doi.org/10.1016/j.tree.2009.01.011

Garthe S, Camphuijsen CJ, Furness RW (1996) Amounts of discards by commercial fisheries and their significance as food for seabirds in the North Sea. Mar Ecol Prog Ser 136:1–11. https://doi.org/10.3354/meps136001

Gendron D, Martinez Serrano I, Ugalde de la Cruz A et al (2015) Long-term individual sighting history database: an effective tool to monitor satellite tag effects on cetaceans. Endanger Species Res 26:235–241

Gleiss AC, Norman B, Liebsch N et al (2009) A new prospect for tagging large free-swimming sharks with motion-sensitive data-loggers. Fish Res 97:11–16. https://doi.org/10.1016/j.fishres.2008.12.012

Goldbogen JA, Calambokidis J, Friedlaender AS et al (2012) Underwater acrobatics by the world's largest predator: 360° rolling manoeuvres by lunge-feeding blue whales. Biol Lett 9:20120986. https://doi.org/10.1098/rsbl.2012.0986

González-Solís J, Felicísimo A, Fox JW et al (2009) Influence of sea surface winds on shearwater migration detours. Mar Ecol Prog Ser 391:221–230. https://doi.org/10.3354/meps08128

Gould J, Roemmich D, Wijffels S, Freeland H, Ignaszewsky M, Jianping X, Takeuchi K et al (2004) Argo profiling floats bring new era of in situ ocean observations. Eos Trans AGU 85(19):185–191

Gould JL, Gould CG (2012) Nature's compass. The mystery of animal navigation. Princeton University Press, Princeton

Graham RT, Witt MJ, Castellanos DW et al (2012) Satellite tracking of manta rays highlights challenges to their conservation. PLoS One 7:e36834. https://doi.org/10.1371/journal.pone.0036834

Gredzens C, Marsh H, Fuentes MMPB et al (2014) Satellite tracking of sympatric marine megafauna can inform the biological basis for species co-management. PLoS One 9:e98944. https://doi.org/10.1371/journal.pone.0098944

Grémillet D, Dell'Omo G, Ryan PG et al (2004) Offshore diplomacy or how seabirds mitigate intra-specific competition: a case study based on GPS tracking of Cape Gannets from neighbouring breeding colonies. Mar Ecol Prog Ser 268:265–279. https://doi.org/10.3354/meps268265

Griffiths CL, Van Sittert L, Best PB et al (2004) Impacts of human activities on marine animal life in the Benguela: a historical overview. Oceanogr Mar Biol Annu Rev 42:303–392

Grimm V, Wyszomirski T, Aikman D et al (1999) Individual-based modelling and ecological theory: synthesis of a workshop. Ecol Modell 115:275–282. https://doi.org/10.1016/S0304-3800(98)00186-0

Guinet C, Vacquié-Garcia J, Picard B et al (2014) Southern elephant seal foraging success in relation to temperature and light conditions: insight into prey distribution. Mar Ecol Prog Ser 499:285–301. https://doi.org/10.3354/meps10660

Halpern BS, Walbridge S, Selkoe KA et al (2008) A global map of human impact on marine ecosystems. Science 319:948–952. https://doi.org/10.1126/science.1149345

Hammerschlag N, Gallagher AJ, Lazarre DM (2011) A review of shark satellite tagging studies. J Exp Mar Bio Ecol 398:1–8. https://doi.org/10.1016/j.jembe.2010.12.012

Handley JM, Pistorius P (2016) Kleptoparasitism in foraging gentoo penguins *Pygoscelis papua*. Polar Biol 39:391–395. https://doi.org/10.1007/s00300-015-1772-2

Harrison XA, Blount JD, Inger R et al (2011) Carry-over effects as drivers of fitness differences in animals. J Anim Ecol 80:4–18. https://doi.org/10.1111/j.1365-2656.2010.01740.x

Hauser N, Zerbini AN, Geyer Y et al (2010) Movements of satellite-monitored humpback whales, *Megaptera novaeangliae*, from the Cook Islands. Mar Mamm Sci 26:679–685. https://doi.org/10.1111/j.1748-7692.2009.00363.x

Hays GC, Bradshaw CJA, James MC et al (2007) Why do Argos satellite tags deployed on marine animals stop transmitting? J Exp Mar Bio Ecol 349:52–60. https://doi.org/10.1016/j.jembe.2007.04.016

Hays GC, Ferreira LC, Sequeira AMM et al (2016) Key questions in marine megafauna movement ecology. Trends Ecol Evol 31:463–475. https://doi.org/10.1016/j.tree.2016.02.015

Hazekamp AAH, Mayer R, Osinga N (2010) Flow simulation along a seal: the impact of an external device. Eur J Wildl Res 56:131–140. https://doi.org/10.1007/s10344-009-0293-0

Hazen EL, Jorgensen S, Rykaczewski RR et al (2013) Predicted habitat shifts of Pacific top predators in a changing climate. Nat Clim Chang 3:234–238. https://doi.org/10.1038/nclimate1686

Hazen EL, Palacios DM, Forney KA et al (2016) WhaleWatch: a dynamic management tool for predicting blue whale density in the California Current. J Appl Ecol 54:1415–1428. https://doi.org/10.1111/1365-2664.12820

Heerah K, Andrews-Goff V, Williams G et al (2013) Ecology of Weddell seals during winter: influence of environmental parameters on their foraging behaviour. Deep Res Part II 88–89:23–33. https://doi.org/10.1016/j.dsr2.2012.08.025

Heerah K, Hindell MA, Andrew-Goff V et al (2016) Contrasting behaviour between two populations of an ice-obligate predator in East Antarctica. Ecol Evol 7:606–618. https://doi.org/10.1002/ece3.2652

Heide-Jørgensen MP, Laidre KL, Nielsen NH et al (2013) Winter and spring diving behavior of bowhead whales relative to prey. Anim Biotelem 1:15. https://doi.org/10.1186/2050-3385-1-15

Heide-Jørgensen MP, Nielsen NH, Teilmann J et al (2017) Long-term tag retention on two species of small cetaceans. Mar Mamm Sci 33:713–725. https://doi.org/10.1111/mms.12394

Heithaus MR, Wirsing AJ, Dill LM et al (2007) Long-term movements of tiger sharks satellite-tagged in Shark Bay, Western Australia. Mar Biol 151:1455–1461. https://doi.org/10.1007/s00227-006-0583-y

Heithaus MR, Frid A, Wirsing AJ et al (2008) Predicting ecological consequences of marine top predator declines. Trends Ecol Evol 23:202–210. https://doi.org/10.1016/j.tree.2008.01.003

Hindell MA, Lea M-A (1998) Heart rate, swimming speed, and estimated oxygen consumption of a free-ranging Southern elephant seal. Physiol Zool 71:74–84. https://doi.org/10.1086/515890

Hindell MA, Harcourt R, Waas JR et al (2002) Fine-scale three-dimensional spatial use by diving, lactating female Weddell seals *Leptonychotes weddellii*. Mar Ecol Prog Ser 242:275–284. https://doi.org/10.3354/Meps242275

Holland KN, Wetherbee BM, Lowe CG et al (1999) Movements of tiger sharks (*Galeocerdo cuvier*) in coastal Hawaiian waters. Mar Biol 134:665–673. https://doi.org/10.1007/s002270050582

Hussey NE, Kessel ST, Aarestrup K et al (2015) Aquatic animal telemetry: a panoramic window into the underwater world. Science 348:1255642. https://doi.org/10.1126/science.1255642

Hyrenbach KD, Forney KA, Dayton PK (2000) Marine protected areas and ocean basin management. Aquat Conserv Mar Freshwat Ecosyst 10:437–458. https://doi.org/10.1017/CBO9781107415324.004

Icarus Initiative (2018) International cooperation for animal research using space. https://icarusinitiative.org/. Accessed 20 Nov 2017

Insley SJ, Robson BW, Yack T et al (2008) Acoustic determination of activity and flipper stroke rate in foraging northern fur seal females. Endanger Species Res 4:147–155. https://doi.org/10.3354/esr00050

Irvine AB, Wells RS, Scott MD (1982) An evaluation of techniques for tagging small Odontocete Cetaceans. Fish Bull 80:135–143

Jeanniard-du-Dot T, Guinet C, Arnould JPY et al (2017) Accelerometers can measure total and activity-specific energy expenditures in free-ranging marine mammals only if linked to time-activity budgets. Funct Ecol 31:377–386. https://doi.org/10.1111/1365-2435.12729

Jessopp M, Cronin M, Hart T (2013) Habitat-mediated dive behavior in free-ranging grey seals. PLoS One 8:e63720. https://doi.org/10.1371/journal.pone.0063720

Jewell OJD, Wcisel MA, Gennari E et al (2011) Effects of smart position only (SPOT) tag deployment on white sharks *Carcharodon carcharias* in South Africa. PLoS One 6:e27242. https://doi.org/10.1371/journal.pone.0027242

Johnson MP, Tyack PL (2003) A digital acoustic recording tag for measuring the response of wild marine mammals to sound. IEEE J Ocean Eng 28:3–12. https://doi.org/10.1109/JOE.2002.808212

Jorgensen SJ, Reeb CA, Chapple TK et al (2010) Philopatry and migration of Pacific white sharks. Proc Biol Sci 277:679–688. https://doi.org/10.1098/rspb.2009.1155

Jouventin P, Weimerskirch H (1990) Satellite tracking of Wandering albatrosses. Nature 343:746–748

Kokubun N, Kim J, Shin H et al (2011) Penguin head movement detected using small accelerometers: a proxy of prey encounter rate. J Exp Biol 214:3760–3767. https://doi.org/10.1242/jeb.058263

Kooyman GL (1965) Techniques used in measuring diving capacities of Weddell seals. Polar Rec 12:391–394

Kooyman GL (1966) Maximum diving capacities of the Weddell seal, *Leptonychotes weddelli*. Science 151:1553–1554. https://doi.org/10.1126/Science.151.3717.1553

Kooyman GL (1973) Respiratory adaptations in marine mammals. Am Zool 13:457–468

Kooyman GL (2007) Animal-borne instrumentation systems and the animals that bear them: then (1939) and now (2007). Mar Technol Soc J 41:6–8. https://doi.org/10.4031/002533207787441935

Krause DJ, Goebel ME, Marshall GJ et al (2015) Novel foraging strategies observed in a growing leopard seal (*Hydrurga leptonyx*) population at Livingston Island, Antarctic Peninsula. Anim Biotelem 3:24. https://doi.org/10.1186/s40317-015-0059-2

Krüger L, Paiva VH, Petry MV et al (2017) Seabird breeding population size on the Antarctic Peninsula related to fisheries activities in non-breeding ranges off South America. Antarct Sci. https://doi.org/10.1017/S0954102017000207

Labrousse S, Vacquié-Garcia J, Heerah K et al (2015) Winter use of sea ice and ocean water mass habitat by southern elephant seals: the length and breadth of the mystery. Prog Oceanogr 137:52–68. https://doi.org/10.1016/j.pocean.2015.05.023

Lascelles BG, Taylor PR, Miller MGR et al (2016) Applying global criteria to tracking data to define important areas for marine conservation. Divers Distrib 22:422–431

Lawson GL, Hückstädt LA, Lavery AC et al (2015) Development of an animal-borne "sonar tag" for quantifying prey availability: test deployments on northern elephant seals. Anim Biotelem 3:22. https://doi.org/10.1186/s40317-015-0054-7

Le Corre M, Jaeger A, Pinet P et al (2012) Tracking seabirds to identify potential Marine Protected Areas in the tropical western Indian Ocean. Biol Conserv 156:83–93. https://doi.org/10.1016/j.biocon.2011.11.015

Lewis S, Sherratt TN, Hamer KC et al (2001) Evidence of intra-specific competion for food in a pelagic seabird. Nature 412:816–818

Lewison RL, Crowder LB, Wallace BP et al (2014) Global patterns of marine mammal, seabird, and sea turtle bycatch reveal taxa-specific and cumulative megafauna hotspots. Proc Natl Acad Sci USA 111:5271–5276. https://doi.org/10.1073/pnas.1318960111

Lidgard DC, Bowen WD, Jonsen ID et al (2012) Animal-borne acoustic transceivers reveal patterns of at-sea associations in an upper-trophic level predator. PLoS One 7:e48962. https://doi.org/10.1371/journal.pone.0048962

Lidgard DC, Bowen WD, Jonsen ID et al (2014) Predator-borne acoustic transceivers and GPS tracking reveal spatiotemporal patterns of encounters with acoustically tagged fish in the open ocean. Mar Ecol Prog Ser 501:157–168. https://doi.org/10.3354/meps10670

Liebsch N, Wilson RP, Bornemann H et al (2007) Mouthing off about fish capture: jaw movement in pinnipeds reveals the real secrets of ingestion. Deep Res Part II 54:256–269. https://doi.org/10.1016/j.dsr2.2006.11.014

Lohmann KJ, Lohmann CMF, Endres CS (2008) The sensory ecology of ocean navigation. J Exp Biol 211:1719–1728. https://doi.org/10.1242/jeb.015792

Lowe CG, Wetherbee BM, Meyer CG (2006) Using acoustic telemetry monitoring techniques to quantify movement patterns and site fidelity of sharks and giant trevally around French Frigate Shoals and Midway Atoll. Atol Res Bull 543:281–303

Lowther AD, Harcourt RG, Page B et al (2013) Steady as he goes: at-sea movement of adult male Australian sea lions in a dynamic marine environment. PLoS One 8:e74348. https://doi.org/10.1371/journal.pone.0074348

MacArthur RH, Pianka ER (1966) On optimal use of a patchy environment. Am Nat 100:603–609

Machovsky-Capuska GE, Priddel D, Leong PHW et al (2016) Coupling bio-logging with nutritional geometry to reveal novel insights into the foraging behaviour of a plunge-diving marine predator. New Zeal J Mar Freshwat Res 50:418–432. https://doi.org/10.1080/00288330.2016.1152981

MacLeod CD (2009) Global climate change, range changes and potential implications for the conservation of marine cetaceans: a review and synthesis. Endanger Species Res 7:125–136. https://doi.org/10.3354/esr00197

Mallory ML, Gilbert CD (2008) Leg-loop harness design for attaching external transmitters to seabirds. Mar Ornithol 36:183–188

Maresh JL, Adachi T, Takahashi A et al (2015) Summing the strokes: energy economy in northern elephant seals during large-scale foraging migrations. Mov Ecol 3:22. https://doi.org/10.1186/s40462-015-0049-2

Martin AR, Da Silva VMF, Rothery PR (2006) Does radio tagging affect the survival or reproduction of small cetaceans? a test. Mar Mamm Sci 22:17–24. https://doi.org/10.1111/j.1748-7692.2006.00002.x

Mate BR, Mesecar R, Lagerquist B (2007) The evolution of satellite-monitored radio tags for large whales: one laboratory's experience. Deep Res Part II Top Stud Oceanogr 54:224–247. https://doi.org/10.1016/j.dsr2.2006.11.021

Mate BR, Ilyashenko VY, Bradford AL et al (2015) Critically endangered western gray whales migrate to the eastern North Pacific. Biol Lett 11:20150071. https://doi.org/10.1098/rsbl.2015.0071

Maxwell SM, Hazen EL, Lewison RL et al (2015) Dynamic ocean management: defining and conceptualizing real-time management of the ocean. Mar Policy 58:42–50. https://doi.org/10.1016/j.marpol.2015.03.014

Mazzaro LM, Dunn JL (2010) Descriptive account of long-term health and behavior of two satellite-tagged captive harbor seals *Phoca vitulina*. Endanger Species Res 10:159–163. https://doi.org/10.3354/esr00190

McConnell B, Beaton R, Bryant E et al (2004) Phoning home-a new GSM mobile phone telemetry system to collect mark-recapture data. Mar Mamm Sci 20:274–283. https://doi.org/10.1111/j.1748-7692.2004.tb01156.x

McGowan J, Beger M, Lewison RL et al (2017) Integrating research using animal-borne telemetry with the needs of conservation management. J Appl Ecol 54:423–429. https://doi.org/10.1111/1365-2664.12755

McIntyre T (2014) Trends in tagging of marine mammals: a review of marine mammal biologging studies. African J Mar Sci 36:409–422. https://doi.org/10.2989/1814232X.2014.976655

McIntyre T (2015) Animal telemetry: tagging effects. Science 349:596–597

McIntyre T, Ansorge IJ, Bornemann H et al (2011) Elephant seal dive behaviour is influenced by ocean temperature: implications for climate change impacts on an ocean predator. Mar Ecol Prog Ser 441:257–272. https://doi.org/10.3354/meps09383

McIntyre T, Bester M, Bornemann H et al (2017) Slow to change? individual fidelity to three-dimensional foraging habitats in southern elephant seals *Mirounga leonina*. Anim Behav 127:91–99

McKenna MF, Calambokidis J, Oleson EM et al (2015) Simultaneous tracking of blue whales and large ships demonstrates limited behavioral responses for avoiding collision. Endanger Species Res 27:219–232. https://doi.org/10.3354/esr00666

McMahon CR, van den Hoff J, Burton HR (2005) Repeated handling and invasive research methods in wildlife research: impacts at the population level. Ambio 34:426–429

McMahon CR, Field IC, Bradshaw CJA et al (2008) Tracking and data-logging devices attached to elephant seals do not affect individual mass gain or survival. J Exp Mar Bio Ecol 360:71–77. https://doi.org/10.1016/j.jembe.2008.03.012

McMahon CR, Harcourt R, Bateson P et al (2012) Animal welfare and decision making in wildlife research. Biol Conserv 153:254–256. https://doi.org/10.1016/j.biocon.2012.05.004

McMichael GA, Eppard MB, Carlson TJ et al (2010) The juvenile salmon acoustic telemetry system: a new tool. Fisheries 35:9–22

Merrick RL, Loughlin TR, Antonelis GA et al (1994) Use of satellite-linked telemetry to study Steller sea lion and northern fur seal foraging. Polar Res 13:105–114

Meyer CG, Papastamatiou YP, Holland KN (2010) A multiple instrument approach to quantifying the movement patterns and habitat use of tiger (*Galeocerdo cuvier*) and Galapagos sharks (*Carcharhinus galapagensis*) at French Frigate Shoals, Hawaii. Mar Biol 157:1857–1868. https://doi.org/10.1007/s00227-010-1457-x

Mitani, Y., Sato, K., Ito, S., Cameron, M. F., Siniff, D. B., & Naito, Y. (2003). A method for reconstructing three-dimensional dive profiles of marine mammals using geomagnetic intensity data: results from two lactating Weddell seals. Polar Biology, 26(5), 311-317

Moll RJ, Millspaugh JJ, Beringer J et al (2007) A new "view" of ecology and conservation through animal-borne video systems. Trends Ecol Evol 22:660–668. https://doi.org/10.1016/j.tree.2007.09.007

Montevecchi W, Fifield D, Burke C et al (2012a) Tracking long-distance migration to assess marine pollution impact. Biol Lett 8:218–221. https://doi.org/10.1098/rsbl.2011.0880

Montevecchi W, Hedd A, McFarlane Tranquilla L et al (2012b) Tracking seabirds to identify ecologically important and high risk marine areas in the western North Atlantic. Biol Conserv 156:62–71. https://doi.org/10.1016/j.biocon.2011.12.001

Muheim R, Phillips JB, Akesson S (2006) Polarized light cues underlie compass calibration in migratory songbirds. Science 313:837–839. https://doi.org/10.1126/science.1129709

Nachtsheim DA, Jerosch K, Hagen W et al (2017) Habitat modelling of crabeater seals (*Lobodon carcinophaga*) in the Weddell Sea using the multivariate approach Maxent. Polar Biol 40:961–976. https://doi.org/10.1007/s00300-016-2020-0

Naito Y, Bornemann H, Takahashi A et al (2010) Fine-scale feeding behavior of Weddell seals revealed by a mandible accelerometer. Polar Sci 4:309–316. https://doi.org/10.1016/j.polar.2010.05.009

NOAA (2017) National Oceanographic and Atmospheric Administration. http://www.noaa.gov/. Accessed 11 Nov 2017

Norman SA, Flynn KR, Zerbini AN et al (2017) Assessment of wound healing of tagged gray (*Eschrichtius robustus*) and blue whales (*Balaenoptera musculus*) in the eastern North Pacific using long-term series of photographs. Mar Mammal Sci. https://doi.org/10.1111/mms.12443

Nowacek DP, Johnson MP, Tyack PL et al (2001) Buoyant balaenids: the ups and downs of buoyancy in right whales. Proc R Soc B 268:1811–1816. https://doi.org/10.1098/rspb.2001.1730

O'Dwyer TW, Buttemer WA, Priddel DM (2007) Differential rates of offspring provisioning in Gould's petrels: are better feeders better breeders? Aust J Zool 55:155–160. https://doi.org/10.1071/ZO07005

Orians G, Pearson N (1979) On the theory of central place foraging. In: Horn DJ, Mitchell R, Stair G (eds) Analysis of ecological systems. Ohio State University Press, Columbus, pp 155–177

Paleczny M, Hammill E, Karpouzi V et al (2015) Population trend of the world's monitored seabirds, 1950–2010. PLoS One 10:e0129342. https://doi.org/10.1371/journal.pone.0129342

Patrick SC, Weimerskirch H (2014) Personality, foraging and fitness consequences in a long lived seabird. PLoS One 9:e87269. https://doi.org/10.1371/journal.pone.0087269

Patrick SC, Bearhop S, Grémillet D et al (2013) Individual differences in searching behaviour and spatial foraging consistency in a central place marine predator. Oikos 123:33–40. https://doi.org/10.1111/j.1600-0706.2013.00406.x

Patrick SC, Bearhop S, Bodey TW et al (2015) Individual seabirds show consistent foraging strategies in response to predictable fisheries discards. J Avian Biol 46:431–440. https://doi.org/10.1111/jav.00660

Patterson TA, McConnell BJ, Fedak MA et al (2010) Using GPS data to evaluate the accuracy of state-space methods for correction of Argos satellite telemetry error. Ecology 91:273–285. https://doi.org/10.1890/08-1480.1

Pavlov VV, Wilson RP, Lucke K (2007) A new approach to tag design in dolphin telemetry: computer simulations to minimise deleterious effects. Deep Res Part II 54:404–414. https://doi.org/10.1016/j.dsr2.2006.11.010

Pendoley KL, Schofield G, Whittock PA et al (2014) Protected species use of a coastal marine migratory corridor connecting marine protected areas. Mar Biol 161:1455–1466. https://doi.org/10.1007/s00227-014-2433-7

Phillips RA, Xavier JC, Croxall JP (2003) Effects of satellite transmitters on albatrosses and petrels. Auk 120:1082–1090

Phillips RA, Silk JRD, Croxall JP et al (2004) Accuracy of geolocation estimates for flying seabirds. Mar Ecol Prog Ser 266:265–272. https://doi.org/10.3354/meps266265

Piatt JF, Sydeman W, Wiese F (2007) Introduction: a modern role for seabirds as indicators. Mar Ecol Prog Ser 352:199–204

Pinaud D, Weimerskirch H (2007) At-sea distribution and scale-dependent foraging behaviour of petrels and albatrosses: a comparative study. J Anim Ecol 76:9–19. https://doi.org/10.1111/j.1365-2656.2006.01186.x

Pinaud D, Cherel Y, Weimerskirch H (2005) Effect of environmental variability on habitat selection, diet, provisioning behaviour and chick growth in yellow-nosed albatrosses. Mar Ecol Prog Ser 298:295–304. https://doi.org/10.3354/meps298295

Pistorius P, Hindell M, Crawford R et al (2017) At-sea distribution and habitat use in king penguins at sub-Antarctic Marion Island. Ecol Evol 7:3894–3903. https://doi.org/10.1002/ece3.2833

Ponganis PJ (2013) Diving physiology. J Exp Biol 2016:3381–3383. https://doi.org/10.1242/jeb.076455

Ponganis PJ, Van Dam RP, Marshall G et al (2000) Sub-ice foraging behavior of emperor penguins. J Exp Biol 203:3275–3278

Priede IG (1984) A basking shark (*Cetorhinus maximus*) tracked by satellite together with simultaneous remote sensing. Fish Res 2:201–216. https://doi.org/10.1016/0165-7836(84)90003-1

Priede IG, Miller PI (2009) A basking shark (*Cetorhinus maximus*) tracked by satellite together with simultaneous remote sensing II: new analysis reveals orientation to a thermal front. Fish Res 95:370–372. https://doi.org/10.1016/j.fishres.2008.09.038

Putman NF, Scanlan MM, Billman EJ et al (2014) An inherited magnetic map guides ocean navigation in juvenile pacific salmon. Curr Biol 24:446–450. https://doi.org/10.1016/j.cub.2014.01.017

Pyke GH (1984) Optimal foraging theory : a critical review. Annu Rev Ecol Syst 15:523–575

Quinn TP (2005) The behavior and ecology of Pacific salmon and trout. University of Washington Press, Seattle

Raymond B, Lea MA, Patterson T et al (2015) Important marine habitat off east Antarctica revealed by two decades of multi-species predator tracking. Ecography 38:121–129. https://doi.org/10.1111/ecog.01021

Reisinger RR, Oosthuizen WC, Péron G et al (2014) Satellite tagging and biopsy sampling of killer whales at subantarctic Marion Island: effectiveness, immediate reactions and long-term responses. PLoS One 9:e111835. https://doi.org/10.1371/journal.pone.0111835

Reisinger RR, Keith M, Andrews RD et al (2015) Movement and diving of killer whales (*Orcinus orca*) at a Southern Ocean archipelago. J Exp Mar Bio Ecol 473:90–102. https://doi.org/10.1016/j.jembe.2015.08.008

Reisinger RR, Raymond B, Hindell MA et al (2018) Habitat modelling of tracking data from multiple marine predators identifies important areas in the Southern Indian Ocean. Divers Distrib. https://doi.org/10.1111/ddi.12702

Rishworth GM, Tremblay Y, Green DB et al (2014) Drivers of time-activity budget variability during breeding in a pelagic seabird. PLoS One 9:e116544. https://doi.org/10.1371/journal.pone.0116544

Robinson RA, Learmonth JA, Hutson AM et al (2005) Climate change and migratory species. BTO Research Report 414

Robinson PW, Costa DP, Crocker DE et al (2012) Foraging behavior and success of a mesopelagic predator in the northeast Pacific Ocean: insights from a data-rich species, the northern elephant seal. PLoS One 7:e36728. https://doi.org/10.1371/journal.pone.0036728

Ropert-Coudert Y, Kato A (2006) Are stomach temperature recorders a useful tool to determine feeding activity. Polar Biosci 20:63–72

Ropert-Coudert Y, Wilson R (2005) Trends and perspectives in animal-attached remote sensing. Front Ecol Environ 3:437–444. https://doi.org/10.1890/1540-9295(2005)003

Ropert-Coudert Y, Kato A, Baudat J et al (2001) Feeding strategies of free-ranging Adélie penguins *Pygoscelis adeliae* analysed by multiple data recording. Polar Biol 24:460–466. https://doi.org/10.1007/s003000100234

Ropert-Coudert Y, Kato A, Liebsch N et al (2004) Monitoring jaw movements: a cue to feeding activity. Game Wildl Sci 20:1–19

Roquet F, Wunsch C, Forget G et al (2013) Estimates of the Southern Ocean general circulation improved by animal-borne instruments. Geophys Res Lett 40:6176–6180. https://doi.org/10.1002/2013GL058304

Rosen DAS, Gerlinsky CG, Trites AW (2017) Telemetry tags increase the costs of swimming in northern fur seals, *Callorhinus ursinus*. Mar Mammal Sci. https://doi.org/10.1111/mms.12460

Rosenbaum HC, Maxwell SM, Kershaw F et al (2014) Long-range movement of humpback whales and their overlap with anthropogenic activity in the South Atlantic Ocean. Conserv Biol 28:604–615. https://doi.org/10.1111/cobi.12225

Ryan PG, Petersen SL, Peters G et al (2004) GPS tracking a marine predator: the effects of precision, resolution and sampling rate on foraging tracks of African Penguins. Mar Biol 145:215–223. https://doi.org/10.1007/s00227-004-1328-4

Sakamoto KQ, Sato K, Ishizuka M et al (2009a) Can ethograms be automatically generated using body acceleration data from free-ranging birds? PLoS One 4:e5379. https://doi.org/10.1371/journal.pone.0005379

Sakamoto KQ, Takahashi A, Iwata T et al (2009b) From the eye of the albatrosses: a bird-borne camera shows an association between albatrosses and a killer whale in the Southern Ocean. PLoS One 4:e7322. https://doi.org/10.1371/journal.pone.0007322

Sato K, Mitani Y, Cameron MF et al (2003) Factors affecting stroking patterns and body angle in diving Weddell seals under natural conditions. J Exp Biol 206:1461–1470. https://doi.org/10.1242/jeb.00265

Schoener TW (1971) Theory of feeding strategies. Annu Rev Ecol Syst 2:369–404

Scholander PF (1940) Experimental investigations on the respiratory function in diving mammals and birds. Hvalrådets Skr 22:1–131

Shaffer SA, Tremblay Y, Awkerman JA et al (2005) Comparison of light- and SST-based geolocation with satellite telemetry in free-ranging albatrosses. Mar Biol 147:833–843. https://doi.org/10.1007/s00227-005-1631-8

Shamoun-Baranes J, Bouten W, van Loon EE et al (2016) Flap or soar? how a flight generalist responds to its aerial environment. Philos Trans R Soc Lond B 371:415–422. https://doi.org/10.1098/rstb.2015.0395

Shepard ELC, Wilson RP, Quintana F et al (2008) Identification of animal movement patterns using tri-axial accelerometry. Endanger Species Res 10:47–60. https://doi.org/10.3354/esr00084

Shoji A, Aris-Brosou S, Fayet A et al (2015) Dual foraging and pair coordination during chick provisioning by Manx shearwaters: empirical evidence supported by a simple model. J Exp Biol 218:2116–2123. https://doi.org/10.1242/jeb.120626

Shoji A, Aris-Brosou S, Owen E et al (2016) Foraging flexibility and search patterns are unlinked during breeding in a free-ranging seabird. Mar Biol 163:72. https://doi.org/10.1007/s00227-016-2826-x

Skórka P, Wójcik JD (2008) Habitat utilisation, feeding tactics and age related feeding efficiency in the Caspian Gull *Larus cachinnans*. J Ornithol 149:31–39. https://doi.org/10.1007/s10336-007-0208-3

Sommerfeld J, Kato A, Ropert-Coudert Y et al (2015) Flexible foraging behaviour in a marine predator, the Masked booby (*Sula dactylatra*), according to foraging locations and environmental conditions. J Exp Mar Bio Ecol 463:79–86. https://doi.org/10.1016/j.jembe.2014.11.005

Sommerfeld J, Mendel B, Fock HO et al (2016) Combining bird-borne tracking and vessel monitoring system data to assess discard use by a scavenging marine predator, the lesser black-backed gull *Larus fuscus*. Mar Biol 163:116. https://doi.org/10.1007/s00227-016-2889-8

Sonne C, Teilmann J, Wright AJ et al (2012) Tissue healing in two harbor porpoises (*Phocoena phocoena*) following long-term satellite transmitter attachment. Mar Mammal Sci 28:316–324. https://doi.org/10.1111/j.1748-7692.2011.00513.x

Spencer NC, Gilchrist HG, Mallory ML (2014) Annual movement patterns of endangered ivory gulls: the importance of sea ice. PLoS One 9:e115231. https://doi.org/10.1371/journal.pone.0115231

Stienen EWM, Desmet P, Aelterman B et al (2016) GPS tracking data of Lesser Black-backed Gulls and Herring Gulls breeding at the southern North Sea coast. Zookeys 555:115–124. https://doi.org/10.3897/zookeys.555.6173

Stillman RA, West AD, Goss-Custard JD et al (2003) An individual behaviour-based model can predict shorebird mortality using routinely collected shellfishery data. J Appl Ecol 40:1090–1101. https://doi.org/10.1111/j.1365-2664.2003.00853.x

Stillman RA, Railsback SF, Giske J et al (2015) Making predictions in a changing world: the benefits of individual-based ecology. Bioscience 65:140–150. https://doi.org/10.1093/biosci/biu192

Sugishita J, Torres LG, Seddon PJ (2015) A new approach to study of seabird-fishery overlap: connecting chick feeding with parental foraging and overlap with fishing vessels. Glob Ecol Conserv 4:632–644. https://doi.org/10.1016/j.gecco.2015.11.001

Takahashi A, Sato K, Naito Y et al (2004) Penguin-mounted cameras glimpse underwater group behaviour. Proc R Soc B 271:S281–S282. https://doi.org/10.1098/rsbl.2004.0182

Teilmann J, Larsen F, Desportes G (2007) Time allocation and diving behaviour of harbour porpoises (*Phocoena phocoena*) in Danish and adjacent waters. J Cetacean Res Manag 9:201–210

Thaxter CB, Lascelles B, Sugar K et al (2012) Seabird foraging ranges as a preliminary tool for identifying candidate marine protected areas. Biol Conserv 156:53–61. https://doi.org/10.1016/j.biocon.2011.12.009

Thaxter CB, Ross-Smith VH, Clark JA et al (2016) Contrasting effects of GPS device and harness attachment on adult survival of Lesser Black-backed Gulls *Larus fuscus* and Great Skuas *Stercorarius skua*. Ibis 158:279–290. https://doi.org/10.1111/ibi.12340

Thorrold SR, Afonso P, Fontes J, Braun CD, Santos RS, Skomal GB, Berumen ML (2014) Extreme diving behaviour in devil rays links surface waters and the deep ocean. Nat Commun 5:4274

Tomkiewicz SM, Fuller MR, Kie JG et al (2010) Global positioning system and associated technologies in animal behaviour and ecological research. Philos Trans R Soc Lond B 365:2163–2176. https://doi.org/10.1098/rstb.2010.0090

Tremblay Y, Bertrand S, Henry RW et al (2009) Analytical approaches to investigating seabird-environment interactions: a review. Mar Ecol Prog Ser 391:153–163. https://doi.org/10.3354/meps08146

van der Hoop JM, Fahlman A, Hurst T et al (2014) Bottlenose dolphins modify behavior to reduce metabolic effect of tag attachment. J Exp Biol 217:4229–4236. https://doi.org/10.1242/jeb.108225

van der Vaart E, Johnston ASA, Sibly RM (2016) Predicting how many animals will be where: how to build, calibrate and evaluate individual-based models. Ecol Modell 326:113–123. https://doi.org/10.1016/j.ecolmodel.2015.08.012

Vandenabeele SP, Wilson RP, Grogan A (2011) Tags on seabirds: how seriously are instrument-induced behaviours considered? Anim Welf 20:559–571

Vandenabeele SP, Shepard EL, Grogan A et al (2012) When three per cent may not be three percent; device-equipped seabirds experience variable flight constraints. Mar Biol 159:1–14. https://doi.org/10.1007/s00227-011-1784-6

Vandenabeele SP, Grundy E, Friswell MI et al (2014) Excess baggage for birds: inappropriate placement of tags on gannets changes flight patterns. PLoS One 9:e92657. https://doi.org/10.1371/journal.pone.0092657

Villegas-Amtmann S, Jeglinski JWE, Costa DP et al (2013) Individual foraging strategies reveal niche overlap between endangered Galapagos pinnipeds. PLoS One 8:e70748. https://doi.org/10.1371/journal.pone.0070748

Vincent C, Mcconnell BJ, Ridoux V et al (2002) Assessment of Argos location accuracy from satellite tags deployed on captive grey seals. Mar Mammal Sci 18:156–166. https://doi.org/10.1111/j.1748-7692.2002.tb01025.x

Volpov BL, Hoskins AJ, Battaile BC et al (2015) Identification of prey captures in Australian fur seals (*Arctocephalus pusillus doriferus*) using head-mounted accelerometers: field validation with animal-borne video cameras. PLoS One 10:e0128789. https://doi.org/10.1371/journal.pone.0128789

Volpov BL, Rosen DAS, Hoskins AJ et al (2016) Dive characteristics can predict foraging success in Australian fur seals (*Arctocephalus pusillus doriferus*) as validated by animal-borne video. Biol Open 5:262–271. https://doi.org/10.1242/bio.016659

Votier SC, Bearhop S, Ratcliffe N et al (2004) Reproductive consequences for great skuas specializing as seabird predators. Condor 106:275–287. https://doi.org/10.1650/7261

Votier SC, Bearhop S, Witt MJ et al (2010) Individual responses of seabirds to commercial fisheries revealed using GPS tracking, stable isotopes and vessel monitoring systems. J Appl Ecol 47:487–497. https://doi.org/10.1111/j.1365-2664.2010.01790.x

Votier SC, Bicknell A, Cox SL et al (2013) A bird's eye view of discard reforms: bird-borne cameras reveal seabird/fishery interactions. PLoS One 8:e57376. https://doi.org/10.1371/journal.pone.0057376

Walker BG, Boveng PL (1995) Effects of time-depth recorders on maternal foraging and attendance behavior of Antarctic fur seals (*Arctocephalus gazella*). Can J Zool 73:1538–1544. https://doi.org/10.1139/z95-182

Watanabe Y, Takahashi A (2013) Linking animal-borne video to accelerometers reveals prey capture variability. Proc Natl Acad Sci U S A 110:2199–2204. https://doi.org/10.1073/pnas.1216244110

Watanabe Y, Mitani Y, Sato K et al (2003) Dive depths of Weddell seals in relation to vertical prey distribution as estimated by image data. Mar Ecol Prog Ser 252:283–288. https://doi.org/10.3354/meps252283

Watanabe Y, Bornemann H, Liebsch N et al (2006) Seal-mounted cameras detect invertebrate fauna on the underside of an Antarctic ice shelf. Mar Ecol Prog Ser 309:297–300. https://doi.org/10.3354/meps309297

Watanuki Y, Daunt F, Takahashi A et al (2008) Microhabitat use and prey capture of a bottom-feeding top predator, the European shag, shown by camera loggers. Mar Ecol Prog Ser 356:283–293. https://doi.org/10.3354/meps07266

Weimerskirch H, Åkesson S, Pinaud D (2006) Postnatal dispersal of wandering albatrosses *Diomedea exulans*: implications for the conservation of the species. J Avian Biol 37:23–28. https://doi.org/10.1111/j.2006.0908-8857.03675.x

Weimerskirch H, Pinaud D, Pawlowski F et al (2007) Does prey capture induce area-restricted search? a fine-scale study using GPS in a marine predator, the wandering albatross. Am Nat 170:734–743. https://doi.org/10.1086/522059

Weimerskirch H, Louzao M, de Grissac S et al (2012) Changes in wind pattern alter albatross distribution and life-history traits. Science 335:211–214. https://doi.org/10.1126/science.1210270

Weimerskirch H, Cherel Y, Delord K et al (2014) Lifetime foraging patterns of the wandering albatross: life on the move! J Exp Mar Bio Ecol 450:68–78. https://doi.org/10.1016/j.jembe.2013.10.021

Weinrich M (1998) Early experience in habitat choice by Humpback whales (*Megaptera novaeangliae*). J Mamm 79:163–170

Weller D (2008) Report of the large whale tagging workshop. U.S. Marine Mammal Commission, San Diego

Weng KC, Castilho PC, Morrissette JM et al (2005) Satellite tagging and cardiac physiology reveal niche expansion in salmon sharks. Science 310:104–106. https://doi.org/10.1126/science.1114616

Wikelski M, Tertitski G (2016) Living sentinels for climate change effects. Science 352:775–776. https://doi.org/10.1126/science.aaf6544

Wikelski M, Kays RW, Kasdin NJ et al (2007) Going wild: what a global small-animal tracking system could do for experimental biologists. J Exp Biol 210:181–186. https://doi.org/10.1242/jeb.02629

Williams HJ, Holton MD, Shepard ELC et al (2017) Identification of animal movement patterns using tri-axial magnetometry. Mov Ecol 5:6. https://doi.org/10.1186/s40462-017-0097-x

Wilson RP (1997) A method for restraining penguins. Mar Ornithol 25:72–73

Wilson RP (2010) Resource partitioning and niche hyper-volume overlap in free-living Pygoscelid penguins. Funct Ecol 24:646–657. https://doi.org/10.1111/j.1365-2435.2009.01654.x

Wilson RP, McMahon CR (2006) Measuring devices on wild animals: what contitutes acceptable practice? Front Ecol Environ 4:147–154

Wilson RP, Grant WS, Duffy DC (1986) Recording devices on free-ranging marine animals: does measurement affect foraging performance? Ecol Soc Am 67:1091–1093

Wilson RP, Cooper J, Plötz J (1992) Can we determine when marine endotherms feed? a case study with seabirds. J Exp Biol 167:267–275

Wilson RP, Steinfurth A, Ropert-Coudert Y et al (2002) Lip-reading in remote subjects: an attempt to quantify and separate ingestion, breathing and vocalisation in free-living animals using penguins as a model. Mar Biol 140:17–27. https://doi.org/10.1007/s002270100659

Wilson RP, White CR, Quintana F et al (2006) Moving towards acceleration for estimates of activity-specific metabolic rate in free-living animals: the case of the cormorant. J Anim Ecol 75:1081–1090. https://doi.org/10.1111/j.1365-2656.2006.01127.x

Wilson RP, Liebsch N, Davies IM et al (2007) All at sea with animal tracks; methodological and analytical solutions for the resolution of movement. Deep Res Part II 54:193–210. https://doi.org/10.1016/j.dsr2.2006.11.017

Wilson RP, Shepard ELC, Liebsch N (2008) Prying into the intimate details of animal lives: use of a daily diary on animals. Endanger Species Res 4:123–137. https://doi.org/10.3354/esr00064

Wilson ADM, Wikelski M, Wilson RP et al (2015a) Utility of biological sensor tags in animal conservation. Conserv Biol 29:1065–1075. https://doi.org/10.1111/cobi.12486

Wilson RP, Liebsch N, Gómez-Laich A et al (2015b) Options for modulating intra-specific competition in colonial pinnipeds: the case of harbour seals (*Phoca vitulina*) in the Wadden Sea. PeerJ 3:e957. https://doi.org/10.7717/peerj.957

Wilson K, Littnan C, Halpin P et al (2017) Integrating multiple technologies to understand the foraging behavior of Hawaiian monk seals. R Soc Open Sci 4:160703

Wisniewska DMM, Johnson M, Teilmann J et al (2016) Ultra-high foraging rates of harbor porpoises make them vulnerable to anthropogenic disturbance. Curr Biol 26:1441–1446. https://doi.org/10.1016/j.cub.2016.03.069

Woakes AJ, Butler PJ, Bevan RM (1995) Implantable data logging system for heart rate and body temperature: its application to the estimation of field metabolic rates in Antarctic predators. Med Biol Eng Comput 33:145–151

Zamon JE, Greene CH, Meir E et al (1996) Acoustic characterization of the three-dimensional prey field of foraging chinstrap penguins. Mar Ecol Prog Ser 131:1–10. https://doi.org/10.3354/meps131001

Zooniverse (2017) People-powered research. https://www.zooniverse.org/. Accessed 15 Nov 2017

How Do They Do It? – Understanding the Success of Marine Invasive Species

Jonas C. Geburzi and Morgan L. McCarthy

Abstract

From the depths of the oceans to the shallow estuaries and wetlands of our coasts, organisms of the marine environment are teeming with unique adaptations to cope with a multitude of varying environmental conditions. With millions of years and a vast volume of water to call their home, they have become quite adept at developing specialized and unique techniques for survival and – given increasing human mediated transport – biological invasions. A growing world human population and a global economy drives the transportation of goods across the oceans and with them invasive species via ballast water and attached to ship hulls. In any given 24-hour period, there are about 10,000 species being transported across different biogeographic regions. If any of them manage to take hold and establish a range in an exotic habitat, the implications for local ecosystems can be costly. This review on marine invasions highlights trends among successful non-indigenous species (NIS), from vectors of transport to ecological and physiological plasticity. Apart from summarizing patterns of successful invasions, it discusses the implications of how successfully established NIS impact the local environment, economy and human health. Finally, it looks to the future and discusses what questions need to be addressed and what models can tell us about what the outlook on future marine invasions is.

J. C. Geburzi (✉)
Zoological Institute and Museum, Kiel University, Kiel, Germany

Alfred Wegener Institute, Helmholtz Centre for Polar and Marine Research, Wadden Sea Station, List/Sylt, Germany
e-mail: jonas.geburzi@zoolmuseum.uni-kiel.de

M. L. McCarthy
School of Biological Sciences, The University of Queensland, St. Lucia, QLD, Australia

Marine Biology, Vrije Universiteit Brussel (VUB), Brussels, Belgium
e-mail: m.l.mccarthy@uq.net.au

Introduction

The continuously rising numbers and extending ranges of non-indigenous species (NIS) are today widely seen as a major biological aspect of global change, affecting invaded ecosystems, economy and even human health (Vitousek et al. 1996; Ruiz et al. 2000; Simberloff et al. 2013). Marine species have been anthropogenically introduced into new habitats since humans travel overseas. However, only in the past 150 years, and especially the latter half of the 20th century, technical advances and extreme increases in global marine trade led to the exponential increase of marine species introductions (Carlton and Geller 1993; Bax et al. 2003).

Of the hundreds of species that get introduced to habitats out of their native range, only a small fraction actually establishes permanently in their new environment. An even smaller fraction reaches high population densities and/or successfully disperses over wider ranges with adverse impacts on the recipient system – being consequently termed 'invasive species' (Sakai et al. 2001; Colautti and MacIsaac 2004). The growing field of invasion biology uses various approaches, e.g., ecology, physiology, evolution, and genetics, to investigate mechanisms and consequences of the establishment of NIS. Finding answers to the questions what makes certain species successful invaders and how invasion processes actually happen is a main focus of invasion biology. These often include aspects that predict impacts of invasive species on the invaded communities and may disclose starting points for possible management strategies (e.g., Bremner 2008; Williams and Grosholz 2008). Furthermore, the study of biological invasions offers model systems to better understand general biological processes such as species interactions, physiological and ecological adaptations, and evolutionary processes (Ruiz et al. 2000; Stachowicz et al. 2002; Facon et al. 2006). While marine systems globally are amongst the most heavily invaded ones, they have long been underrepresented in invasion biology studies compared to terrestrial and limnic systems. A main reason for this might

S. Jungblut et al. (eds.), *YOUMARES 8 – Oceans Across Boundaries: Learning from each other*,
https://doi.org/10.1007/978-3-319-93284-2_8

be the vastness and open character of marine systems, which require higher (technical) efforts and make it generally more difficult to detect, investigate and manage marine invasions. However, this discrepancy is reduced by a quickly growing body of literature in recent years (Grosholz and Ruiz 1996; Ruiz et al. 2000; Chan and Briski 2017).

Even though it is difficult to identify universal factors and traits that lead to high invasion success due to its apparent dependency on the individual conditions of each invasion event (Sakai et al. 2001), some general patterns regularly occur in this context. These include for example common invasion pathways and vectors (Katsanevakis et al. 2013), as well as anthropogenic alterations or perturbations of recipient habitats (Bax et al. 2000; Briggs 2012; Mineur et al. 2012). Additional factors are high ecological and physiological plasticity of successful invaders (Hänfling et al. 2011; Parker et al. 2013; Tepolt and Somero 2014), and the general nature of interactions between native and non-native as well as among non-native species (Snyder and Evans 2006; Johnson et al. 2009; Briggs 2010). This review aims to give an introductory overview of important aspects of successful marine invasions, including human impacts, species' traits and interactions, and invasion genetics. The second part of this review copes with ecological and socio-economic consequences of marine invasions and their implications for policy and management, and closes with an outlook on future developments of the phenomenon under the perspectives of ongoing global (esp. climate) change. For clarification purposes, a glossary defining the most important terms can be found in Box 1.

Box 1: Glossary

Cryptogenic species: (Crypt-Greek, kryptos, secret; -genic, New Latin, genic, origin) as a species that is not demonstrably native or introduced (Carlton 1996).

Dispersal pathway: The combination of processes and opportunities resulting in the movement of propagules from one area to another, including aspects of the vectors involved, features of the original and recipient environments, and the nature and timing of what exactly is moved (Wilson et al. 2009).

Hybridization: The interbreeding of individuals of morphologically and presumably genetically distinct populations, regardless of the taxonomic status of such populations (Short 1969).

Native/indigenous/original: An organism occurring within its natural past or present range and dispersal potential (organisms whose dispersal is independent of human intervention) (Falk-Petersen et al. 2006, modified from IUCN 2000).

Non-native/alien/exo-tic/foreign/intro-duced/non-indigenous: An organism occurring outside its natural past or present range and dispersal potential including any parts of the organism that might survive and subsequently reproduce (organisms whose dispersal is caused by human action) (Falk-Petersen et al. 2006, modified from IUCN 2000).

Vector: The physical means or agent by which a species is transported, such as ballast water, ships' hulls, boats, hiking boats, cars, vehicles, packing material, or soil in nursery stock (Carlton 2001).

Promoters of Successful Spread and Establishment

Vectors, Pathways and Altered Habitats – Human Impacts

Anthropogenic activities are, by definition, major prerequisites for the occurrence of marine NIS, as only they allow species to reach regions beyond their natural range and dispersal limits. Besides obvious examples of direct species transportation, either intentional or unintentional, human impacts on marine habitats can also indirectly act as strong promoters of the spread and establishment of marine NIS.

Ship traffic is the most important vector of species' introductions (Fig. 1). Ships act as vectors in two ways. First, their hulls provide a habitat for fouling communities of sessile species, which are transported between ports and may eventually get removed or detached, or release offspring into a new environment (Ruiz et al. 1997; Gollasch 2002). If the fouling layer is thick enough, mobile species may survive transoceanic transport in sheltered cavities, as for example the Asian crab *Hemigrapsus takanoi*, which was first recorded in Europe in 1993 on a ship's hull (Gollasch 1999, then identified as *H. penicillatus*). Second, the exchange of huge amounts of ballast water holds the potential for all species with (at least temporal) planktonic or swimming lifestyle to be taken up in one and be released in another port. Since the 1880s, when seawater started to replace solid ballast, the number of marine NIS and the frequency of introductions has been constantly increasing (Carlton and Geller 1993; Ruiz et al. 1997; Ruiz and Smith 2005; Wolff 2005). The ongoing trend to ever more, bigger, and faster vessels fuels this trend by increasing ballast water volume and thus the number of transported organisms, as well as their survival probability. The importance of international ship traffic for the dispersal of marine NIS is also underlined by the fact that especially international ports and their surroundings have often turned into hot-spots for exotic species and that the dispersal routes of many species follow the main transoceanic shipping routes (Briggs 2012; Seebens et al. 2013). While the big container vessels and other large trading ships account to a large extent for primary species introductions across continents, regional traffic of smaller ships are important vectors for the secondary spread (range-expansion) of marine NIS. Recent studies showed that recreational boating is a particularly important driver of regional dispersal of non-native species (e.g. Clarke Murray et al. 2011; Hänfling et al. 2011).

Aquaculture is another important vector for marine NIS, which also accounts for a rising number of introductions parallel to the global growth of this economy during the last decades (Naylor et al. 2001). Organisms with a planktonic larval stage are especially prone to 'spill over' from their culture areas into the surrounding habitats. This introduction pathway led, for example, to the invasion of the Pacific oyster *Magallana gigas* along the southeastern coast of the European North Sea. The species was initially believed to not be able to reproduce in the cold climate of the North Sea, but a series of warm summers following the introduction of *M. gigas* promoted their dispersal. The case of *M. gigas* highlights how a combination of human actions, environmental change and species' traits can lead to a successful invasion (Diederich et al. 2005; Smaal et al. 2009). Aquaculture is not only a vector for the cultured target spe-

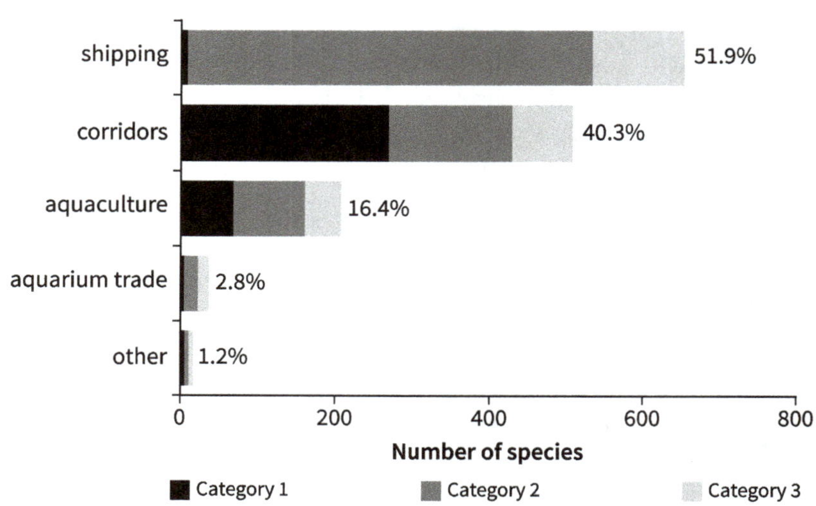

Fig. 1 Number of marine NIS in Europe, known or likely to be introduced by each of the main vectors. Percentages add to more than 100% as 147 out of the 1,264 species are linked to more than one vector. Categories refer to the certainty by which a species can be linked to a vector: (1) there is direct evidence of a vector; (2) a most likely vector can be inferred; (3) one or more possible vectors can be inferred. Redrawn and modified from Katsanevakis et al. (2013) with permission from Elsevier.

cies, but often also unintentionally introduces organisms, which are associated with them if they are not vigorously cleaned before transportation. Worldwide, the introductions of more than 40 marine species can be directly linked to the translocation of bivalves used in aquaculture (Padilla et al. 2011), and in total 206 NIS have been linked to this vector in Europe alone (Katsanevakis et al. 2013; Fig. 1). Particularly invasive ecosystem-engineers like reef-building oysters (esp. *M. gigas* and *Crassostrea virginica*) promote the establishment of NIS they brought along by providing favorable habitats, which eventually further enhances community shifts in the invaded systems (e.g., Ruiz et al. 2000; Markert et al. 2010; Padilla et al. 2011). Another important taxon in this context are macroalgae, which are regularly introduced as 'blind passengers' with aquaculture organisms. They can likewise change existing or form new habitats, thus affecting both native and other alien species (e.g., Jones and Thornber 2010; Salvaterra et al. 2013; Thomsen et al. 2016).

Floating (plastic) litter is a vector recently gaining attention. While the marine litter problem is mostly discussed under the aspect of pollution and the hazardous effects of microplastic accumulation, larger pieces of litter are also a possible habitat for fouling organisms, which might then be transported over large distances by oceanic currents. Recent studies found a variety of species from different taxonomic groups (including bryozoans, barnacles and mollusks) settling on macroplastic, with a considerable proportion of marine NIS among them (Barnes and Milner 2005; Gregory 2009; Gil and Pfaller 2016). While driftwood and other debris may already historically have played a role in the cosmopolitan distribution of species like *Teredo navalis* (Bivalvia, Myoida) or *Lepas anatifera* (Crustacea, Pedunculata), the recent extreme increase in amounts of marine litter may lead to a future increase in numbers on marine NIS dispersed by this vector (Gregory 2009).

Trade of ornamental and aquarium-kept organisms has been widely neglected by scientists and policy makers as an introduction pathway, although it bears a high potential for species invasions (Padilla and Williams 2004). Introductions of aquarium organisms to natural environments may occur accidentally, when organisms escape during transport or, for example, from public aquaria with in-/outflow from/to natural water bodies, or intentionally, when hobbyists or traders release single individuals or discard the contents of whole aquaria into the wild. Fish and macroalgae are the taxa with the highest numbers of species (potentially) introduced by aquarium trade. Zenetos et al. (2016) list 19 introduced fish species with a potential link to this vector in the Mediterranean Sea alone, and Vranken et al. (2018) identified at least 23 seaweed species commonly found in aquaria across Europe, which have the potential to thrive European natural waters, with the highly invasive *Caulerpa taxifolia* as the most striking example (compare also Padilla and Williams 2004; Fig. 1). Besides the usually ornamental target species, aquarium trade may also account for unintentional introductions of associated species, especially epibionts on seaweeds and live rock used for aquarium decoration, such as macro- and microalgae, (hemi)sessile cnidarians, crustaceans, polychaetes or mollusks (Padilla and Williams 2004). Aquarium trade is a strongly growing economy, and commercial and private online retailers make exotic species easily available worldwide via the internet, rendering this vector extremely difficult to control and regulate (Padilla and Williams 2004; Mazza et al. 2015; Vranken et al. 2018). Today, the Mediterranean and southern European Atlantic are the regions within Europe which are most affected by this introduction pathway, due to the (sub)tropical origin of most of the traded species. In the light of ongoing ocean warming, also more temperate regions might get invaded by these species in the future (Vranken et al. 2018). Thus, the number of marine invasions promoted by aquarium trade are very likely to increase in the future.

Although not thought of as a 'classical' vector, canals are a major introduction pathway for marine NIS. The best known example is the Suez Canal, connecting the Mediterranean Sea to the Red Sea and Indian Ocean, which accounts for the vast majority of species invasions to the Mediterranean by migration through the canal (Lessepsian migration) (Galil 2009). The Baltic Sea, as another example, was invaded by numerous ponto-caspian species since it is connected to the Black Sea by a system of canals and rivers (Leppäkoski et al. 2002; Katsanevakis et al. 2013). Additionally, the Kiel Canal provides a shortcut route between the southwestern Baltic and the southeastern North Sea. It likely served as an invasion pathway for numerous species native or invasive to the Atlantic, like the crabs *Rhithropanopeus harrisii* and *H. takanoi* (Fowler et al. 2013; Geburzi et al. 2015). An interesting case in this context is the shrimp *Palaemon elegans*, of which an Atlantic type invaded the Baltic Sea from the west, and a Mediterranean/Black Sea-type invaded from the southeast (Reuschel et al. 2010). Besides opening routes for the active migration or natural (e.g., larval) dispersal processes, canals also increase the probability for successful ship-mediated introductions, as they shorten transportation times, thus increasing survival probabilities, e.g., for organisms in ballast water tanks.

Human activities not only provide vectors and pathways for species' introductions, but they also impact the environment in ways that can promote the establishment success of marine NIS, in particular by changing natural habitats. The 'invasibility' of a community or habitat, i.e., its receptivity towards invasive species, can be strongly influenced by human activities. Apart from the propagule pressure of invaders, it largely depends on the availability of suitable niches and resources. Anthropogenic habitat changes (addition or depletion of different niches) and disturbances leading to a reduction in native diversity (increasing resource

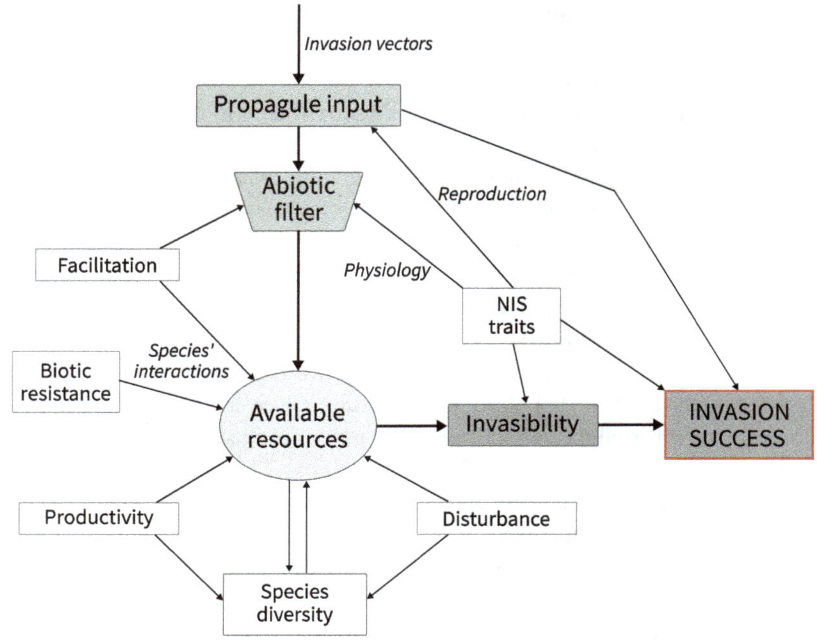

Fig. 2 Factors that have been shown to affect invasions of non-native species into marine communities. Propagules of NIS may be supplied by various vectors and propagule pressure is influenced by reproductive traits of these species. When they are able to survive under the environmental conditions of the new habitat ('Abiotic filter'), their ability to invade the community is influenced by several, potentially interacting factors that affect resource availability, as well as the NIS' biological traits (modified after Olyarnik et al. 2009, with permission from Springer). See text for details and examples.

availability for invaders) are thus important determinants of invasibility (Colautti et al. 2006; Fridley et al. 2007; Fig. 2). The construction of harbors and coastal defense structures such as groins or seawalls on sedimentary coasts or in estuaries, for example, adds artificial rocky habitats to naturally soft-bottomed environments. Such new habitats are often rapidly colonized by non-native species as native species are less adapted to their conditions (Mineur et al. 2012). They may also serve as 'stepping stones' for the dispersal of rocky-shore species (Landschoff et al. 2013), a function which is also currently investigated with regard to the increasing number of offshore wind farms (De Mesel et al. 2015). Water pollution and eutrophication are discussed as additional anthropogenic impacts which lead to disturbances of marine ecosystems and communities, making them more receptive to invasions (Reise et al. 2006; Briggs 2007). Several experimental studies revealed in fact a higher tolerance of NIS towards anthropogenic abiotic stress (i.e., water pollution) compared to related native taxa in marine communities (e.g., Piola and Johnston 2008; Crooks et al. 2010; Lenz et al. 2011). However, observational studies often show that invasions occur likewise in disturbed and pristine habitats, and that NIS are affected by disturbances as well (Klein et al. 2005; Boudouresque and Verlaque 2012, and references therein), indicating a more complex relationship between disturbance, native diversity, and invasibility. Clark and Johnston (2011) showed that the relationship between disturbance and invader success changes over time, being highly positive for initial invasions (due to increased resource availability by the reduction in native diversity), but turning to a negative relationship for later stages of establishment. The high prevalence of marine invasions in disturbed habitats might rather be due to the fact that these are often at the same time heavily affected by introduction vectors such as shipping and aquaculture (compare Colautti et al. 2006; Boudouresque and Verlaque 2012).

Life History, Ecology and Physiology – Species' Traits

Of the high numbers of transported and introduced species, only a small proportion successfully establishes and becomes invasive. Several ecological and life-history traits regularly occur in marine invasive species from different taxa and can, therefore, be associated with their success. Overall, the following traits and examples show that plasticity, for example in life-history strategies, behavior, and physiology, is a key feature of successful invaders.

Many of these traits are associated with reproduction, as in the end reproductive success is the one factor determining whether a species successfully establishes and spreads (Fig. 2). Invaders are often characterized by an *r*-selected breeding strategy (early maturity, short generation time, high fecundity, rapid growth rates) or the ability to switch between *r*- and *K*-selected strategies (reviewed in Sakai et al. 2001), enabling them to develop a high propagule pressure even from small founder populations. Likewise, the ability of females to produce several broods per season when environmental conditions allow for it has been often observed (Hines 1986; Anderson and Epifanio 2010; van den Brink et al. 2013). For crabs in particular, Zeng et al. (2014) discuss the ability to switch between, or combine, two strategies of resource allocation for reproduction as a potential promoter of invasiveness. The authors suggest that successful invaders might be able to use both, internal repositories as well as external energy sources, during reproduction, allowing for higher fecundity or extended breeding periods (Zeng et al. 2014). In some cases, seasonality also seems to have facilitated the establishment of NIS. Temporally shifted breeding periods are for example believed to reduce competition between early juveniles of native crabs and the invasive European shore crab *Carcinus maenas* in Australia (Garside et al. 2015).

Having a planktonic larval stage is a common feature of many marine taxa, but it provides particular benefits for establishing NIS by increasing their dispersal abilities. This trait mostly affects the secondary regional spread after the initial introduction of a species. Some marine NIS possess a significantly longer duration of larval development compared to native species of the same taxonomic group, which is regarded as a mean of further enhancing the dispersal potential (Roman and Palumbi 2004; Viard et al. 2006; Delaney et al. 2012; Katsanevakis et al. 2013). The choice of recruitment sites by the last larval stages is often positively influenced by chemical signals released by conspecifics (gregarious settlement), but for some successful invaders cues from suitable habitat act as strong as recruitment enhancer as conspecific cues. This mechanism has for example been shown for the crab *Hemigrapsus sanguineus* and is believed to enhance the species' ability to colonize new habitats (O'Connor 2007; Anderson and Epifanio 2009).

Resource utilization and food preference is another set of traits, where successful invaders often show high levels of plasticity. Omnivory (in animals) and the ability to quickly adapt to a changed food supply is often observed (e.g., Blasi and O'Connor 2016) and allows NIS to avoid food competition with native species of the same guild. On the other hand, newly arrived species can also cause dietary shifts in native species, being both, beneficial for themselves but detrimental for the native competitor (Griffen et al. 2011).

A species physiology is an important component in determining its ability to take hold in a new habitat (Fig. 2). In order for an invader to take hold and remain, it must pass biotic and abiotic conditions, known as the "ecological filter" of the environment (Crowl et al. 2008). Abiotic factors are the first part of the filter that NIS must endure. They must be able to survive physiological adjustments and abiotic stressors such as temperature, desiccation, and disturbance (Olyarnik et al. 2009). Granted a species is able to endure the abiotic factors, they must also be able to maintain performance and fitness through competition (Levine et al. 2004) and predation (deRivera et al. 2005) with other species in the environment. Failing to survive through this filter can prevent establishment or range expansion (Kelley 2014).

Temperature and salinity are two factors highly regarded in limiting an organism's ability to expand its range as a NIS, as is the case in *Mytilus* studies (Pickens 1965; Helm and Trueman 1967; Coleman and Trueman 1971; Stickle and Sabourin 1979; Nicholson 2002; Braby and Somero 2006a). *Mytilus trossulus* is native to the North Pacific, however, has been replaced along the California coastline from the Mexican border to Monterey Bay after the introduction of *Mytilus galloprovincialis*, a Mediterranean native, to Southern California via shipping in the 1900s (McDonald and Koehn 1988; Geller 1999). A habitat mosaic exists in San Francisco and Monterey Bay, which are both characterized by varying abiotic environmental conditions (Braby and Somero 2006b). While the invasive mussel (*Mytilus galloprovincialis*) is genetically inclined to high temperature thermal tolerance, *Mytilus trossulus* is well adapted to areas achieving a critical salinity level, making the matrix of habitats in Monterey and San Francisco Bay a mixed mosaic, where otherwise *Mytilus galloprovincialis* had displaced it along the southern coast given its thermal tolerance acclimation advantage (Braby and Somero 2006a).

In some cases, NIS are able to sustain populations in new ranges despite not being physiologically capable of reproduction in the surrounding environment. This is a leading hypothesis for the Chinese Mitten Crab *Eriocheir sinensis* in the Baltic Sea. Though other theories exist, evidence suggests that the Elbe River estuary is a donor area for *E. sinensis*, and that individuals are migrating to the Baltic via the Kiel Canal, traveling distances up to 1,500 km (Ojaveer et al. 2007). Under unfavourable combinations of temperature and salinity, it has been shown that additional larval stages may occur in *E. sinensis* (Ojaveer et al. 2007), which is a phenomenon unique among brachyuran crabs (Montú et al. 1996). Other crab species have also shown regional adaptations to physiological parameters. Populations of the European shore crab *C. maenas* in the Baltic Sea (salinity 15) have shown a higher capacity for hyper-regulation than populations of *C. maenas* in the North Sea (salinity 30) (Theede 1969).

Competition, Facilitation and Parasitism – Species' Interactions

Wherever NIS are introduced, they develop interactions with both native and other non-native species in their new habitat. At the same time, detrimental interactions with species of their native range, such as predators, parasites or pathogens, may fall away in the invaded range ('enemy-release-hypothesis', see e.g., Jeschke et al. 2012; Papacostas et al. 2017). Either way, shifts in the interaction regimes of NIS during the invasion process are probably among the most important factors determining the long-term potential for a successful establishment after initial introduction.

Many invasive species are known as strong competitors, having negative effects on native species occupying the same niche, which, in turn, facilitates their own or their offspring's establishment. Common mechanisms are superiority in the competition for food and shelter or for optimal settlement space in the case of sessile animals and plants, respectively (Ruiz et al. 1999; Jensen et al. 2002; Levin et al. 2002; van den Brink et al. 2012; Katsanevakis et al. 2013). Several studies also reported direct predation pressure by invaders on native species within the same guild (Ruiz et al. 1999, and references therein). Some authors (Briggs 2010) relate the strong competitiveness of many NIS to a regularly observed biogeographical pattern of marine invasions: They often originate from regions with high biodiversity (e.g., the western and central Indo-Pacific and the NW-Pacific for NIS in Europe, see Tsiamis et al. 2018) and are, therefore, well adapted to strong competition. This makes them superior over native species of their recipient regions, which are often characterized by lower biodiversity. This pattern is also incorporated in the 'enemy-release-hypothesis' (Bax et al. 2001; Brockerhoff and McLay 2011), and assumed to significantly contribute to the observation that successful invaders often 'perform better' (grow bigger, reproduce more) in their invaded compared to their native ranges (Parker et al. 2013).

The success of marine NIS may also be enhanced by positive interactions which benefit the invader. They have been described to occur among species invading the same region, where the establishment of a first species (often an ecosystem-engineer, see section "Vectors, Pathways and Altered Habitats – Human Impacts") facilitates subsequent invasions of further species (Fridley et al. 2007; Altieri and Irving 2017). The initial invader might either directly provide beneficial effects for subsequent invaders (e.g., habitat or food) or exert detrimental effects for native competitors of subsequent invaders (e.g., predation, pathogens, structural habitat changes, Fig. 2). Such cascading effects have led to the assumption that increasingly invaded systems become more susceptible to further introductions, cumulating in 'invasional meltdown' scenarios (Simberloff and Von Holle 1999; Grosholz 2005). Empirical evidence for 'invasional meltdown' is however scarce (Simberloff 2006; Briggs 2012). At the same time, an increasing number of studies report both negative interactions between NIS (Lohrer and Whitlatch 2002; Griffen et al. 2008; Griffen 2016) and positive effects of NIS on native species (Rodriguez 2006, and references therein). In summary, these studies underline the complexity of species interactions in the context of NIS establishment, making predictions on general interaction patterns and long-term invasion success extremely difficult.

Parasitism is another type of species' interaction with the potential to strongly affect invasion success. Just like being released from enemies, a release from parasites often occurs during the translocation process of many species, resulting in a much lower parasite load of introduced compared to native populations (Snyder and Evans 2006; McDermott 2011; Fowler et al. 2013). Direct positive effects of reduced parasite load include, for example, increased survival and fecundity (especially, when released from sterilizing parasites). Even more important are the indirect effects by the reduced need to invest in parasite defense, allowing organisms to reallocate those resources to traits like growth or reproduction (Goedknegt et al. 2016). Reduced investment in parasite defense, however, results in higher susceptibility to parasite infections, which may in turn negatively impact establishment success (Keogh et al. 2016). Introduced non-native parasites, on the other hand, can reach extreme invasion success when they are able to infect native species which are closely related to their original host, but have only weak defensive traits due to the lack of coevolution (examples in Ruiz et al. 1999; Feis et al. 2016). This could theoretically even promote the invasion success of the original host, which may gain competitive advantages over its native relative by being better adapted to infections.

Selection, Multiple Introductions and Hybridization – Invasion Genetics

Species introductions have the potential to trigger rapid evolutionary changes and adaptation processes acting on the genetic level. Invasion genetics, therefore, play an important role in determining long-term success of species introductions and their evolutionary consequences for the respective species (Holland 2000; Geller et al. 2010). Furthermore, invasion genetics is a tool to determine the origin of invasive species and potential pathways of introduction. The Veined rapa whelk (*Rapana venosa*), for example, is genetically highly diverse in native Chinese populations (Yang et al. 2008), but genetically monomorphic in all introduced populations in Europe and the Americas. This implies that all introduced populations originate from one single introduction, which has been localized in the Black Sea (Chandler et al. 2008). Similarly, all invasive populations of the seaweed *Caulerpa taxifolia* in the Mediterranean, Australia and North America could be traced back genetically to a strain that was released or escaped from a European aquarium (Wiedenmann et al. 2001; Padilla and Williams 2004).

Usually, introduction and colonization processes of species into new habitats are associated with a considerable reduction of genetic diversity by strong genetic drift or bottleneck effects. One would, therefore, expect to regularly observe negative effects of genetic depletion in newly establishing populations, especially a reduced ability to adapt to changing environmental conditions. This seems, however, often not to be the case (a terrestrial example in Tsutsui et al. 2000; Hänfling 2007). Possible reasons are for example co-segregation of fixed loci or changes in frequencies of rare (recessive) alleles caused by the reduction of population size, leading to an actual increase in additive genetic diversity (Hänfling 2007, and references therein; Facon et al. 2008). If, by chance events, advantageous genotypes develop under these conditions, they can rapidly become fixed in a small founder population due to the strong selective forces. Multiple introductions of the same species can further mitigate possibly negative effects of small founder populations. They will be often not recognized as long as no genetic studies are performed ('cryptic invasions'), but are likely to occur in many introduced species. If repeated introductions originate from different source populations, this leads to an admixture of genotypes, holding the potential to strongly increase the adaptive abilities of the species by novel combination of alleles (Hänfling 2007; Herborg et al. 2007; Chan and Briski 2017). This is believed to considerably contribute to the invasiveness of global invaders like the European shore crab *Carcinus maenas* (Geller et al. 1997; Roman 2006).

Hybridization between native species and NIS regularly occurs in animals with external fertilization like mollusks and fish, and especially in plants. From the invaders perspective, it increases the chances to successfully establish despite small founder population sizes either by introgression of native alleles which enhance adaptive evolution, or by the development of new hybrid lineages combining beneficial traits from both parental lineages (Sakai et al. 2001; Hänfling 2007). The latter can occasionally lead to hybrid superiority and eventually result in the displacement of native species by newly evolved hybrids. This has been for example observed for cordgrass, *Spartina sp.*, in Great Britain and North America, where hybrids between native and invasive species disperse more successfully than their parent species (Huxel 1999; Williams and Grosholz 2008, and references therein).

Why Does It Matter?

Ecological Impacts

The evidence is overwhelming that NIS invasions are a significant stressor to marine communities and has been observed in invasions by plants, fish, crabs, snails, clams, mussels, bryozoans, and nudibranchs (Ruiz et al. 1999). Furthermore, anthropogenic derived disturbances and the introduction of new species are skewing food webs towards a loss of higher trophic groups and a gain in lower order consumers (Byrnes et al. 2007). On the other hand, the invasion of marine NIS may increase local biodiversity, as marine invaders often appear to accommodate besides native species rather than replacing them (Briggs 2007; an example in Reise et al. 2017). As a higher biodiversity stabilizes communities, invaders may also have overall positive effects, especially in otherwise disturbed habitats. This can also include the resistance against further invasions (Stachowicz et al. 2002; Marraffini and Geller 2015). Species most likely to have wide-reaching ecosystem impacts are those that alter the biotic and abiotic factors of the environment, namely ecosystem engineers (Vitousek et al. 1996).

An ecosystem engineer is an organism that alters the availability of resources to other species. Jones et al. (1994) described ecosystem engineers as falling into two categories, autogenic and allogenic. Autogenic engineers change the environment through their own physical structure. Corals for example, provide habitats for many reef dwelling species. Allogenic engineers alter the environment by transforming living or non-living materials between physical states, as is the case for sea urchins which alter the environment by eating the kelp that would otherwise be providing a habitat for organisms as autogenic engineers (Jones et al. 1994). Broadly speaking, NIS as ecosystem engineers can provide both positive and negative impacts on their environments. As a prominent example, Pacific oysters (*Magallana gigas*) have been introduced globally for aquaculture purposes and have in some cases established wild oyster beds among its introduced ranges (Lejart and Hily 2011). The impact of Pacific oysters has varied from displacement of *Sabellaria* reefs, a species of conservation importance, to increases in sessile invertebrate diversity via secondary settlement on oyster shells (Olyarnik et al. 2009; Herbert et al. 2016).

Apart from the introduction of ecosystem engineers themselves, the introduction of pathogens can indirectly cause a significant alteration to the physical environment by infecting ecosystem engineers. The introduced protistan pathogens *Haplosporidium nelsoni* and *Perkinsus marinus* were partly responsible for the decline of the Virginia oyster (*Crassostrea virginica*) (Crooks 2002), a historically important ecosystem engineer in the Chesapeake Bay. The Chesapeake Bay has seen a decrease in over 90% of its oyster population in the last century and the pathogen introduction has been recorded as a dominant factor of mortality. Additionally, results of the pathogen introduction have limited the physical structure of oysters as a habitat and as a filter feeder, altering the benthic and planktonic food webs (Ruiz et al. 1999).

Positive ecological impacts of NIS also occur outside the group of ecosystem engineers. For example, the mitten crab *E. sinensis* is able to transfer native and non-native invertebrates to new habitats (Ojaveer et al. 2007). The large carapace acts as a substrate for flora and fauna (e.g., algae and barnacles) to inhabit. Furthermore, the 'hairy' patches on the crabs' claws could also provide a habitat for nematodes, bivalves, crustaceans, oligocheates, and gastropods (Normant et al. 2007). Other ecological advantages include new food sources for fish, novel habitats, and increased biofiltration. In a recent meta-analysis, Katsanevakis et al. (2014) found that among the assessed NIS, 35% had been reported to have a positive impact on other species.

The Economy and Human Health

Social and economic impacts are linked to invasive species altering fisheries, aquaculture, tourism, and marine infrastructure activities. Human health is also impacted, when the consequence of these alterations results in lost revenue and, potentially, a direct decrease in human health (Bax et al. 2001).

The economy drives the exchange of goods across the globe via shipping routes and trade and with it come new NIS. In some instances, NIS have negative economic impacts by altering ecosystems and reducing the stocks of exportable fish and shellfish through competition and disease. Few studies focus on the economic impacts of aquatic species alone and even fewer separate out marine from freshwater species impacts. In a recent case study, Ünal and Bodur (2017) investigated the negative impacts of Silver-cheeked toadfish *Lagocephalus scleratus* on small-scale fisheries in the eastern Mediterranean Sea. Marine invasions pose an additional challenge because of the widely dispersing planktonic larvae of some marine species (Thresher and Kuris 2004).

The European green crab *Carcinus maenas*, ranked in the IUCN list of the 'world's worst invasive alien species' (Lowe et al. 2004) has had quite an economic impact in North America since emerging from its native European range over 200 years ago (Carlton and Cohen 2003). The estimated annual losses to shellfisheries on the East Coast of the United States due to predation by *C. maenas* alone range from \$14.7 to \$18.7 million a year and sum up to \$805.9 million during the period from 1975 to 2005 (Abt Associates Inc. 2008). In addition to loss of profit from shellfish sales, green crabs are also responsible for the loss of eelgrass in restoration projects through bioturbation activities such as foraging and burrowing (Davis et al. 1998). The associated costs from these activities range from \$60,150 to \$77,433 as an estimate for the year 2006 (Abt Associates Inc. 2008). Apart from the costs associated with direct shellfish predation and eelgrass restoration projects, there are also projected costs for handling further losses from the NIS. Expenditures for a proposed monitoring and control program to the US Environmental Protection Agency would cost \$285,000 per year (Abt Associates Inc. 2008). The European shore crab is just one NIS, in one country and calculations are based on only the known impacts. Carlton (1999) and others demonstrated through the European shore crab that the economic cost of a single NIS can be quite significant, highlighting the need for effective control and management implementation. Keller et al. (2011) highlight the possibility that there may be economic benefits in some cases. The release of the Red king crab *Paralithodes camtschaticus* into the Barents Sea and subsequent expansion into the Norwegian coast provided an income of over 9 million EUR for fisherman (Galil et al. 2009). Other edible NIS used in fisheries or aquaculture include the fish *Planiliza haematocheila*, *Saurida lessepsianus*, *Siganus luridus*, *S. rivulatus*, the molluscs *Ensis directus*, *Mercenaria mercenaria* and *Aerococcus viridans* var. *homari* (Katsanevakis et al. 2014).

While known impacts of NIS can be calculated to the dollar, other factors may not have a monetary label, especially those concerning human health. Lafferty and Kuris (1996) describe the risk that the Chinese mitten crab in California presents as a second intermediate host for the Oriental lung fluke *Paragonimus westermani*, which can cause paralysis in humans. If the mitten crab becomes widely abundant, it may serve as a suitable host for the native North American lung flukes and increase the potential for infection in humans (Lafferty and Kuris 1996).

Management and Policy

As Thresher and Kuris (2004) summarize, there are management efforts in outbreaks across the globe, from the marine alga *Caulerpa taxifolia* in California, the Mediterranean Sea and Australia (Meinesz et al. 2001; Cheshire et al. 2002; Williams and Grosholz 2002) to the Asian whelk *Rapana venosa* in the Chesapeake Bay (Mann and Harding 2000) and the Asian mussel *Perna viridis* in Cairns, Australia (Thresher and Kuris 2004). These examples illustrate that marine invasions are truly a global challenge. In meeting this challenge, the study highlights four key differences between management in marine and terrestrial invasion approaches. The first is that the ocean is perceived as an open system and that, due to global patterns of circulation, pelagic larvae and large-scale migrations, local eradication efforts are insignificant. The open nature of the ocean establishes a defeatist attitude among public managers. Another challenge to marine invasions is who should bear the burden of cost. The benefits of management actions can be widespread and, therefore, which parties should be involved in paying for them can become convoluted. Furthermore, the public perceives the ocean and open coastline to be pristine, allowing invasions to go largely unnoticed by the public. Thus, public awareness typically does not arise until the later stages of the invasion process (expansion and persistence), while at the same time containment and other management efforts grow increasingly difficult (Fig. 3). Finally, scientific literature on the biology of most marine taxa is limited, making decisions and predicting outcomes of management practices difficult (Thresher and Kuris 2004). Despite these additional challenges in managing marine invasions, there have been approaches developed which have produced mixed results.

Lovell et al. (2006) highlights some of the policies developed to limit the spread of NIS. Two main approaches to international policy have been to focus on shipping vectors as a means of distribution, and by limiting the amount

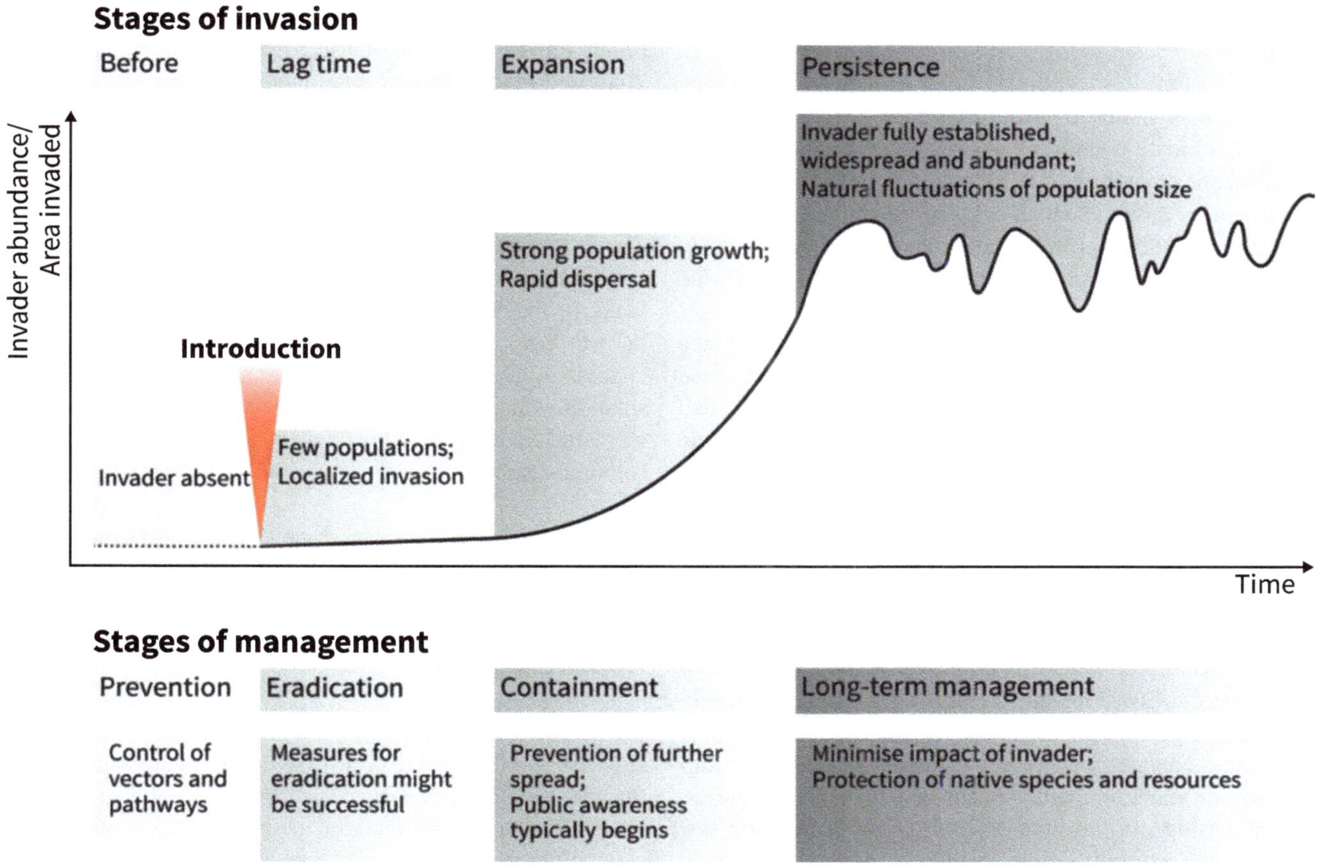

Fig. 3 Theoretical invasion curve (black line), stages of the invasion process and according stages of invasive species management (after Boudouresque et al. 2005; State of Victoria 2010; Gothland et al. 2014). Preventive measures before the introduction are the most effective and least disturbing to the environment. Once the expansion stage is reached, management is limited to containment of the invader and minimization of its impacts.

of imports via quarantine bans or tariffs (Lovell et al. 2006). Despite regulations, trade among the North Atlantic Free Trade Area (NAFTA) countries spreads invasive species that were introduced as a result of trade with countries outside of NAFTA (Perrault et al. 2003). It is in this manner that NIS can be introduced and widely distributed among trade partners who themselves adhere to stringent treaty regulations.

A mere handful of marine focused management recommendations have been suggested and are summarized by Secord (2003). Recommendations are centered around prevention and control. The least disturbing to the environment is the prevention of invasions in the first place (Fig. 3). This may be manifested through public education and outreach programs, ballast water exchange and treatment options, and regulation in the aquaculture industry and aquarium trade (Secord 2003). Second is the eradication of small invasive populations, while they are still susceptible to localized chemical or mechanical control methods (Fig. 3). This may take form through shading algal species, chlorine treatment of marinas, or the physical smashing of individual snail hosts of an invasive parasite (Bax 1999; Culver and Kuris 2000; Dalton 2000). Invasions can further be augmented through biocontrol using native species. The advantage of this method is that it introduces no further NIS, however, the implications for affected community dynamics may be put at risk. The most risky method to local ecosystem dynamics is through the introduction of other NIS to eat, parasitize, infect or compete with the invader (Secord 2003). Finally, public education, supporting research, and monitoring all help to inform, when management actions should be implemented or when new invaders have arrived.

Returning to the example of *C. maenas,* one considered bioinvasion control for this species was the introduction of the castrating barnacle *Sacculina carcini*. After infiltrating the crab body, the parasite castrates both male and females, effectively prohibiting its hosts from reproducing (Lafferty and Kuris 1996). The danger in this form of bioinvasion control is how it would impact native as well as commercially valuable crabs or shrimp (Secord 2003).

Future Implications in Light of Climate Change

Future studies should consider marine invasions in light of changing climate conditions. As sea level rises, how will it influence the expansion of NIS ranges? If bodies of water are connected by larger channels and increased water flow, how will this impact the spread of future invasions? Will prone to acclimatizing NIS thrive, when new areas of land sink beneath the rising oceans?

Educating the public on marine NIS is an important step towards keeping invasions in check. The ocean is large and looking out at it from on land, it can seem undisturbed and peaceful. Beneath the surface, however, NIS are thriving under rocks and floating through the water column as microscopic plankton. Using molecular genetic techniques, we can first study the past and use it to understand how established NIS came to be. Through further monitoring, we can keep invasions in check and observe, whether new invasions are underway. Finally, by carefully predicting climatic conditions of the future, we can hypothesize about the course of future marine invasions and begin to plan future management campaigns in light of global climate change.

Climate change is expected to impact aquatic invasions by warming water temperatures, altering water flow patterns, and increasing storm events (Poff et al. 2002). Aquatic systems that are naturally saline will likely increase in salinity, though, whether this will necessarily allow marine species to invade inland waters is still largely unknown (Rahel and Olden 2008). Climate change will also result in physiological changes, which will become apparent at the population level and as seen by shifts in abundance, timing of annually recurring events and distribution and dispersion of organisms (Doney et al. 2012). For example, invasive ectotherms have a greater ability to acclimate their thermal tolerance and can achieve a higher upper thermal tolerance threshold than native ectotherms (Kelley 2014). As Occhipinti-Ambrogi (2007) highlights, the range expansions of populations of NIS has already been observed to be coupled with increasing water temperatures. The Mediterranean Sea has witnessed the establishment of alien microalgae species, an increase that is largely attributed to increased water temperatures (Gómez and Claustre 2003). Other microalgae, whose spread is also thought to be linked to increasing water temperatures, has impacted human health. The NIS *Ostreopsis cf. ovata*, which bloomed in the Ligurian sea, caused respiratory illness in tourists exposed to it (Brescianini et al. 2006; Durando et al. 2007; Vila et al. 2016).

One of the most comprehensive models for predicting the fate of marine invasions found that overall there would be a high species turnover rate attributed to invasions and extinctions by the mid-21st century. In considering the distributional ranges of 1,066 marine fish and invertebrates for 2050 in a bioclimate envelope model, Cheung et al. (2009) found that patterns of species invasion as well as turnover (accounting for invading and locally extinct species) were predicted in high latitude regions of the Arctic and Southern Ocean and that, combined with global extinctions, invasions and extinctions will amount to a turnover of 60% of the present biodiversity. The potential disruptions in ecosystem services in the future based on this model remain yet to be known. With a growing world population and in light of a changing global climate, studies are needed to better understand how marine invasions will further impact our environment and economy, and how managers can better prepare for future invasions.

Acknowledgements The authors thank A. Zenetos and an anonymous reviewer for their comments, which greatly improved the manuscript.

Appendix

This article is related to the YOUMARES 8 conference session no. 11: "How Do They Do It? – Understanding the Success of Marine Invasive Species". The original Call for Abstracts and the abstracts of the presentations within this session can be found in the appendix "Conference sessions and Abstracts", chapter "7 How Do They Do It? – Understanding the Success of Marine Invasive Species", of this book.

References

Abt Associates Inc (2008) Ecological and economic impacts and invasion management strategies for the European Green Crab. Cambridge, MA

Altieri AH, Irving AD (2017) Species coexistence and the superior ability of an invasive species to exploit a facilitation cascade habitat. Peer J 5:e2848. https://doi.org/10.7717/peerj.2848

Anderson JA, Epifanio CE (2009) Induction of metamorphosis in the Asian shore crab *Hemigrapsus sanguineus*: characterization of the cue associated with biofilm from adult habitat. J Exp Mar Bio Ecol 382:34–39. https://doi.org/10.1016/j.jembe.2009.10.006

Anderson JA, Epifanio CE (2010) Mating and sperm storage of the Asian shore crab *Hemigrapsus sanguineus*. J Shellfish Res 29:497–501. https://doi.org/10.2983/035.029.0228

Barnes DKA, Milner P (2005) Drifting plastic and its consequences for sessile organism dispersal in the Atlantic Ocean. Mar Biol 146:815–825. https://doi.org/10.1007/s00227-004-1474-8

Bax NJ (1999) Eradicating a dreissenid from Australia. Dreissena! 10:1–5

Bax NJ, Hayes K, Marshall A et al (2000) Man-made marinas as sheltered islands for alien marine organisms: establishment and eradication of an alien invasive marine species. In: Veitch CR, Clout MN (eds) Turning the tide: the eradication of invasive species. IUCN SSC Invasive Specialist Group, Gland, pp 26–39

Bax NJ, Carlton JT, Mathews-Amos A et al (2001) The control of biological invasions in the world's oceans. Conserv Biol 15:1234–1246. https://doi.org/10.1046/j.1523-1739.2001.99487.x

Bax NJ, Williamson A, Aguero M et al (2003) Marine invasive alien species: a threat to global biodiversity. Mar Policy 27:313–323

Blasi JC, O'Connor NJ (2016) Amphipods as potential prey of the Asian shore crab *Hemigrapsus sanguineus*: Laboratory and field experiments. J Exp Mar Bio Ecol 474:18–22. https://doi.org/10.1016/j.jembe.2015.09.011

Boudouresque C-F, Verlaque M (2012) An overview of species introduction and invasion processes in marine and coastal lagoon habitats. Cah Biol Mar 53:309–317

Boudouresque C-F, Ruitton S, Verlaque M (2005) Large-scale disturbances, regime shift and recovery in littoral systems subject to biological invasions. In: Velikova V, Chipev N (eds) UNESCO-ROSTE/BAS workshop on regime shifts. Varna, pp 85–101

Braby CE, Somero GN (2006a) Following the heart: temperature and salinity effects on heart rate in native and invasive species of blue mussels (genus *Mytilus*). J Exp Biol 209:2554–2566. https://doi.org/10.1242/jeb.02259

Braby CE, Somero GN (2006b) Ecological gradients and relative abundance of native (*Mytilus trossulus*) and invasive (*Mytilus galloprovincialis*) blue mussels in the California hybrid zone. Mar Biol 148:1249–1262. https://doi.org/10.1007/s00227-005-0177-0

Bremner J (2008) Species' traits and ecological functioning in marine conservation and management. J Exp Mar Bio Ecol 366:37–47. https://doi.org/10.1016/j.jembe.2008.07.007

Brescianini C, Grillo C, Melchiorre N et al (2006) *Ostreopsis ovata* algal blooms affecting human health in Genova, Italy, 2005 and 2006. Euro Surveillance2 11:3040

Briggs JC (2007) Marine biogeography and ecology: invasions and introductions. J Biogeogr 34:193–198. https://doi.org/10.1111/j.1365-2699.2006.01632.x

Briggs JC (2010) Marine biology: the role of accommodation in shaping marine biodiversity. Mar Biol 157:2117–2126. https://doi.org/10.1007/s00227-010-1490-9

Briggs JC (2012) Marine species invasions in estuaries and harbors. Mar Ecol Prog Ser 449:297–302. https://doi.org/10.3354/meps09553

Brockerhoff AM, McLay CL (2011) Human-Mediated Spread of Alien Crabs. In: Galil BS, Clark PF, Carlton JT (eds) In the wrong place – alien marine crustaceans: distribution, biology and impacts. Springer, Dordrecht, pp 27–106

Byrnes JE, Reynolds PL, Stachowicz JJ (2007) Invasions and extinctions reshape coastal marine food webs. PLoS ONE 2:e295. https://doi.org/10.1371/journal.pone.0000295

Carlton JT (1996) Biological Invasions and Cryptogenic Species. Ecology 77:1653–1655

Carlton JT (1999) The scale and ecological consequences of biological invasions in the World's oceans. In: Sandlund OT, Schei PJ, Viken Å (eds) Invasive species and biodiversity management. Kluwer Academic Publishers, Dordrecht, pp 195–212

Carlton JT (2001) Introduced species in US coastal waters: environmental impacts and management priorities. Pew Oceans Commission, Arlington, Virginia

Carlton JT, Cohen AN (2003) Episodic global dispersal in shallow water marine organisms: the case history of the European shore crabs *Carcinus maenas* and *C. aestuarii*. J Biogeogr 30:1809–1820. https://doi.org/10.1111/j.1365-2699.2003.00962.x

Carlton JT, Geller JB (1993) Ecological roulette: the global transport of nonindigenous marine organisms. Science 261:78–82. https://doi.org/10.1126/science.261.5117.78

Chan FT, Briski E (2017) An overview of recent research in marine biological invasions. Mar Biol 164:121. https://doi.org/10.1007/s00227-017-3155-4

Chandler EA, McDowell JR, Graves JE (2008) Genetically monomorphic invasive populations of the rapa whelk, *Rapana venosa*. Mol Ecol 17:4079–4091. https://doi.org/10.1111/j.1365-294X.2008.03897.x

Cheshire A, Westphalen G, Boxall V et al (2002) *Caulerpa taxifolia* in West Lakes and the Port River: distribution and eradication options. Adelaide, Australia

Cheung WWL, Lam VWY, Sarmiento JL et al (2009) Projecting global marine biodiversity impacts under climate change scenarios. Fish Fish 10:235–251. https://doi.org/10.1111/j.1467-2979.2008.00315.x

Clark GF, Johnston EL (2011) Temporal change in the diversity – invasibility relationship in the presence of a disturbance regime. Ecol Lett 14:52–57. https://doi.org/10.1111/j.1461-0248.2010.01550.x

Clarke Murray C, Pakhomov EA, Therriault TW (2011) Recreational boating: A large unregulated vector transporting marine invasive species. Divers Distrib 17:1161–1172. https://doi.org/10.1111/j.1472-4642.2011.00798.x

Colautti RI, MacIsaac HJ (2004) A neutral terminology to define "invasive" species. Divers Distrib 10:135–141. https://doi.org/10.1111/j.1366-9516.2004.00061.x

Colautti RI, Grigorovich IA, MacIsaac HJ (2006) Propagule pressure: a null model for biological invasions. Biol Invasions 8:1023–1037. https://doi.org/10.1007/s10530-005-3735-y

Coleman N, Trueman ER (1971) The effect of aerial exposure on the activity of the mussels *Mytilus edulis* L. and *Modiolus modiolus* (L.). J Exp Mar Bio Ecol 7:295–304. https://doi.org/10.1016/0022-0981(71)90011-6

Crooks JA (2002) Characterizing ecosystem-level consequences of biological invasions: the role of ecosystem engineers. Oikos 97:153–166

Crooks JA, Chang AL, Ruiz GM (2010) Aquatic pollution increases the relative success of invasive species. Biol Invasions 13:165–176. https://doi.org/10.1007/s10530-010-9799-3

Crowl TA, Crist TO, Parmenter RR et al (2008) The spread of invasive species and infectious disease as drivers of ecosystem change. Front Ecol Environ 6:238–246. https://doi.org/10.1890/070151

Culver CS, Kuris AM (2000) The apparent eradication of a locally established introduced marine pest. Biol Invasions 2:245–253. https://doi.org/10.1023/A:1010082407254

Dalton R (2000) Researchers criticize response to killer algae. Nature 406:447–447

Davis RC, Short FT, Burdick DM (1998) Quantifying the effects of green crab damage to eelgrass transplants. Restor Ecol 6:297–302. https://doi.org/10.1046/j.1526-100X.1998.00634.x

De Mesel I, Kerckhof F, Norro A et al (2015) Succession and seasonal dynamics of the epifauna community on offshore wind farm foundations and their role as stepping stones for non-indigenous species. Hydrobiologia 756:37–50. https://doi.org/10.1007/s10750-014-2157-1

Delaney DG, Edwards PK, Leung B (2012) Predicting regional spread of non-native species using oceanographic models: validation and identification of gaps. Mar Biol 159:269–282. https://doi.org/10.1007/s00227-011-1805-5

deRivera CE, Ruiz GM, Hines AH et al (2005) Biotic resistance to invasion: native predator limits abundance and distribution of an introduced crab. Ecology 86:3364–3376. https://doi.org/10.1890/05-0479

Diederich S, Nehls G, van Beusekom JE et al (2005) Introduced Pacific oysters (*Crassostrea gigas*) in the northern Wadden Sea: Invasion accelerated by warm summers? Helgol Mar Res 59:97–106. https://doi.org/10.1007/s10152-004-0195-1

Doney SC, Ruckelshaus M, Emmett Duffy J et al (2012) Climate change impacts on marine ecosystems. Ann Rev Mar Sci 4:11–37. https://doi.org/10.1146/annurev-marine-041911-111611

Durando P, Ansaldi F, Oreste P et al (2007) *Ostreopsis ovata* and human health: epidemiological and clinical features of respiratory syndrome outbreaks from a two-year syndromic surveillance, 2005–06, in north-west Italy. Euro Surveill 12:3212. https://doi.org/10.2807/esw.12.23.03212-en

Facon B, Genton BJ, Shykoff J et al (2006) A general eco-evolutionary framework for understanding bioinvasions. Trends Ecol Evol 21:130–135. https://doi.org/10.1016/j.tree.2005.10.012

Facon B, Pointier JP, Jarne P et al (2008) High genetic variance in life-history strategies within invasive populations by way of multiple introductions. Curr Biol 18:363–367. https://doi.org/10.1016/j.cub.2008.01.063

Falk-Petersen J, Bøhn T, Sandlund OT (2006) On numerous concepts in invasion biology. Biol Invasions 8:1409–1424. https://doi.org/10.1007/s10530-005-0710-6

Feis ME, Goedknegt MA, Thieltges DW et al (2016) Biological invasions and host-parasite coevolution: different coevolutionary trajectories along separate parasite invasion fronts. Zoology 119:366–374. https://doi.org/10.1016/j.zool.2016.05.012

Fowler AE, Forsström T, von Numers M et al (2013) The North American mud crab *Rhithropanopeus harrisii* (Gould, 1841) in newly colonized Northern Baltic Sea: distribution and ecology. Aquat Invasions 8:89–96. https://doi.org/10.3391/ai.2013.8.1.10

Fridley JD, Stachowitz TJ, Naeem S et al (2007) The invasion paradox: reconciling pattern and process in species invasions. Ecology 88:3–17

Galil BS (2009) Taking stock: inventory of alien species in the Mediterranean sea. Biol Invasions 11:359–372. https://doi.org/10.1007/s10530-008-9253-y

Galil BS, Gollasch S, Minchin D et al (2009) Alien marine biota of Europe. In: Hulme PE, Nentwig W, Pyšek P et al (eds) Handbook of Alien Species in Europe. Springer, Dordrecht, pp 93–104

Garside CJ, Glasby TM, Stone LJ et al (2015) The timing of *Carcinus maenas* recruitment to a south-east Australian estuary differs to that of native crabs. Hydrobiologia 762:41–53. https://doi.org/10.1007/s10750-015-2332-z

Geburzi JC, Graumann G, Köhnk S et al (2015) First record of the Asian crab *Hemigrapsus takanoi* Asakura & Watanabe, 2005 (Decapoda, Brachyura, Varunidae) in the Baltic Sea. BioInvasions Rec 4:103–107. https://doi.org/10.3391/bir.2015.4.2.06

Geller JB (1999) Decline of a native mussel masked by sibling species invasion. Conserv Biol 13:661–664. https://doi.org/10.1046/j.1523-1739.1999.97470.x

Geller JB, Walton ED, Grosholz ED et al (1997) Cryptic invasions of the crab *Carcinus* detected by molecular phylogeography. Mol Ecol 6:901–906. https://doi.org/10.1046/j.1365-294X.1997.00256.x

Geller JB, Darling JA, Carlton JT (2010) Genetic perspectives on marine biological invasions. Ann Rev Mar Sci 2:367–393. https://doi.org/10.1146/annurev.marine.010908.163745

Gil MA, Pfaller JB (2016) Oceanic barnacles act as foundation species on plastic debris: implications for marine dispersal. Sci Rep 6:19987. https://doi.org/10.1038/srep19987

Goedknegt MA, Feis ME, Wegner KM et al (2016) Parasites and marine invasions: ecological and evolutionary perspectives. J Sea Res 113:11–27. https://doi.org/10.1016/j.seares.2015.12.003

Gollasch S (1999) The Asian decapod *Hemigrapsus penicillatus* (De Haan, 1835) (Grapsidae, Decapoda) introduced in european waters: status quo and future perspective. Helgoländer Meeresun 52:359–366. https://doi.org/10.1007/BF02908909

Gollasch S (2002) The importance of Ship Hull Fouling as a vector of species introductions into the North Sea. Biofouling 18:105–121

Gómez F, Claustre H (2003) The genus *Asterodinium* (Dinophyceae) as a possible biological indicator of warming in the western Mediterranean Sea. J Mar Biol Assoc UK 83:173–174. https://doi.org/10.1017/S0025315403006945h

Gothland M, Dauvin J-C, Denis L et al (2014) Biological traits explain the distribution and colonisation ability of the invasive shore crab *Hemigrapsus takanoi*. Estuar Coast Shelf Sci 142:41–49. https://doi.org/10.1016/j.ecss.2014.03.012

Gregory MR (2009) Environmental implications of plastic debris in marine settings – entanglement, ingestion, smothering, hangers-on, hitch-hiking and alien invasions. Philos Trans R Soc Lond B 364:2013–2025. https://doi.org/10.1098/rstb.2008.0265

Griffen BD (2016) Scaling the consequences of interactions between invaders from the individual to the population level. Ecol Evol 6:1769–1777. https://doi.org/10.1002/ece3.2008

Griffen BD, Guy T, Buck JC (2008) Inhibition between invasives: a newly introduced predator moderates the impacts of a previously established invasive predator. J Anim Ecol 77:32–40. https://doi.org/10.1111/j.1365-2656.2007.01304.x

Griffen BD, Altman I, Hurley J et al (2011) Reduced fecundity by one invader in the presence of another: a potential mechanism leading to species replacement. J Exp Mar Bio Ecol 406:6–13. https://doi.org/10.1016/j.jembe.2011.06.005

Grosholz ED (2005) Recent biological invasion may hasten invasional meltdown by accelerating historical introductions. Proc Natl Acad Sci USA 102:1088–1091. https://doi.org/10.1073/pnas.0308547102

Grosholz ED, Ruiz GM (1996) Predicting the impact of introduced marine species: lessons from the multiple invasions of the european green crab *Carcinus maenas*. Biol Conserv 78:59–66

Hänfling B (2007) Understanding the establishment success of non-indigenous fishes: lessons from population genetics. J Fish Biol 71:115–135. https://doi.org/10.1111/j.1095-8649.2007.01474.x

Hänfling B, Edwards F, Gherardi F (2011) Invasive alien Crustacea: dispersal, establishment, impact and control. BioControl 56:573–595. https://doi.org/10.1007/s10526-011-9380-8

Helm MM, Trueman ER (1967) The effect of exposure on the heart rate of the mussel, *Mytilus edulis* L. Comp Biochem Physiol 21:171–177. https://doi.org/10.1016/0010-406X(67)90126-0

Herbert RJH, Humphreys J, Davies CJ et al (2016) Ecological impacts of non-native Pacific oysters (*Crassostrea gigas*) and management measures for protected areas in Europe. Biodivers Conserv 25:2835–2865. https://doi.org/10.1007/s10531-016-1209-4

Herborg L-M, Weetman D, van Oosterhout C et al (2007) Genetic population structure and contemporary dispersal patterns of a recent European invader, the Chinese mitten crab, *Eriocheir sinensis*. Mol Ecol 16:231–242. https://doi.org/10.1111/j.1365-294X.2006.03133.x

Hines AH (1986) Larval patterns in the life histories of Brachyuran crabs (Crustacea, Decapoda, Brachyura). Bull Mar Sci 39:444–466

Holland BS (2000) Genetics of marine bioinvasions. Hydrobiologia 420:63–71. https://doi.org/10.1023/A:1003929519809

Huxel GR (1999) Rapid displacement of native species by invasive species: effects of hybridization. Biol Conserv 89:143–152. https://doi.org/10.1016/S0006-3207(98)00153-0

IUCN (2000) IUCN guidelines for the prevention of biodiversity loss caused by alien invasive species. IUCN, Gland

Jensen GC, McDonald PS, Armstrong DA (2002) East meets west: competitive interactions between green crab *Carcinus maenas*, and native and introduced shore crab *Hemigrapsus spp*. Mar Ecol Prog Ser 225:251–262. https://doi.org/10.3354/meps225251

Jeschke JM, Gómez Aparicio L, Haider S et al (2012) Support for major hypotheses in invasion biology is uneven and declining. NeoBiota 14:1–20. https://doi.org/10.3897/neobiota.14.3435

Johnson PTJ, Olden JD, Solomon CT, Vander Zanden MJ (2009) Interactions among invaders: Community and ecosystem effects of multiple invasive species in an experimental aquatic system. Oecologia 159:161–170. https://doi.org/10.1007/s00442-008-1176-x

Jones E, Thornber CS (2010) Effects of habitat-modifying invasive macroalgae on epiphytic algal communities. Mar Ecol Prog Ser 400:87–100. https://doi.org/10.3354/meps08391

Jones CG, Lawton JH, Shachak M (1994) Organisms as ecosystem engineers. Oikos 69:373–386

Katsanevakis S, Zenetos A, Belchior C et al (2013) Invading European Seas: assessing pathways of introduction of marine aliens. Ocean Coast Manag 76:64–74. https://doi.org/10.1016/j.ocecoaman.2013.02.024

Katsanevakis S, Wallentinus I, Zenetos A et al (2014) Impacts of invasive alien marine species on ecosystem services and biodiversity: a pan-European review. Aquat Invasions 9:391–423. https://doi.org/10.3391/ai.2014.9.4.01

Keller RP, Geist J, Jeschke JM et al (2011) Invasive species in Europe: ecology, status, and policy. Environ Sci Eur 23:23. https://doi.org/10.1186/2190-4715-23-23

Kelley AL (2014) The role thermal physiology plays in species invasion. Conserv Physiol 2:1–14. https://doi.org/10.1093/conphys/cou045

Keogh CL, Miura O, Nishimura T et al (2016) The double edge to parasite escape: invasive host is less infected but more infectable. Ecology 38:42–49. https://doi.org/10.1002/ecy.1953

Klein J, Ruitton S, Verlaque M et al (2005) Species introductions, diversity and disturbances in marine macrophyte assemblages of the northwestern Mediterranean Sea. Mar Ecol Prog Ser 290:79–88. https://doi.org/10.3354/meps290079

Lafferty KD, Kuris AM (1996) Biological control of marine pests. Ecology 77:1989–2000

Landschoff J, Lackschewitz D, Kesy K et al (2013) Globalization pressure and habitat change: Pacific rocky shore crabs invade armored shorelines in the Atlantic Wadden Sea. Aquat Invasions 8:77–87. https://doi.org/10.3391/ai.2013.8.1.09

Lejart M, Hily C (2011) Differential response of benthic macrofauna to the formation of novel oyster reefs (*Crassostrea gigas*, Thunberg) on soft and rocky substrate in the intertidal of the Bay of Brest, France. J Sea Res 65:84–93

Lenz M, da Gama BAP, Gerner NV et al (2011) Non-native marine invertebrates are more tolerant towards environmental stress than taxonomically related native species: results from a globally replicated study. Environ Res 111:943–952. https://doi.org/10.1016/j.envres.2011.05.001

Leppäkoski E, Gollasch S, Gruszka P et al (2002) The Baltic – a sea of invaders. Can J Fish Aquat Sci 59:1175–1188. https://doi.org/10.1139/f02-089

Levin PS, Coyer JA, Petrik R et al (2002) Community-wide effects of nonindigenous species on temperate rocky reefs. Ecology 83:3182–3193

Levine JM, Adler PB, Yelenik SG (2004) A meta-analysis of biotic resistance to exotic plant invasions. Ecol Lett 7:975–989. https://doi.org/10.1111/j.1461-0248.2004.00657.x

Lohrer AM, Whitlatch RB (2002) Interactions among aliens: apparent replacement of one exotic species by another. Ecology 83:719–732

Lovell SJ, Stone SF, Fernandez L (2006) The economic impacts of aquatic invasive species: a review of the literature. Agric Resour Econ Rev 35:195–208. https://doi.org/10.1017/S1068280500010157

Lowe S, Browne M, Boudjelas S et al (2004) 100 of the world's worst invasive species: a selection from the global invasive species database. Invasive Species Specialist Group, Auckland

Mann R, Harding JM (2000) Invasion of the North American Atlantic coast by a large predatory Asian mollusc. Biol Invasions 2:7–22

Markert A, Wehrmann A, Kröncke I (2010) Recently established *Crassostrea*-reefs versus native *Mytilus*-beds: differences in ecosystem engineering affects the macrofaunal communities (Wadden Sea of Lower Saxony, southern German Bight). Biol Invasions 12:15–32. https://doi.org/10.1007/s10530-009-9425-4

Marraffini ML, Geller JB (2015) Species richness and interacting factors control invasibility of a marine community. Proc R Soc B 282:20150439. https://doi.org/10.1098/rspb.2015.0439

Mazza G, Aquiloni L, Inghilesi A et al (2015) Aliens just a click away: the online aquarium trade in Italy. Manag Biol Invasions 6:253–261. https://doi.org/10.3391/mbi.2015.6.3.04

McDermott JJ (2011) Parasites of shore crabs in the genus *Hemigrapsus* (Decapoda: Brachyura: Varunidae) and their status in crabs geographically displaced: a review. J Nat Hist 45:2419–2441. https://doi.org/10.1080/00222933.2011.596636

McDonald JH, Koehn RK (1988) The mussels *Mytilus galloprovincialis* and *M. trossulus* on the Pacific coast of North America. Mar Biol 99:111–118. https://doi.org/10.1007/BF00644984

Meinesz A, Belsher T, Thibaut T et al (2001) The introduced green alga *Caulerpa taxifolia* continues to spread in the Mediterranean. Biol Invasions 3:201–210

Mineur F, Cook EJ, Minchin D et al (2012) Changing coasts: marine aliens and artificial structures. Oceanogr Mar Biol An Annu Rev 50:189–234

Montú M, Anger K, de Bakker C (1996) Larval development of the Chinese mitten crab *Eriocheir sinensis* H. Milne-Edwards (Decapoda: Grapsidae) reared in the laboratory. Helgoländer Meeresun 50:223–252

Naylor RL, Williams SL, Strong DR (2001) Aquaculture – a gateway for exotic species. Science 294:1655–1656

Nicholson S (2002) Ecophysiological aspects of cardiac activity in the subtropical mussel *Perna viridis* (L.) (Bivalvia: Mytilidae). J Exp Mar Bio Ecol 267:207–222. https://doi.org/10.1016/S0022-0981(01)00362-8

Normant M, Korthals J, Szaniawska A (2007) Epibiota associated with setae on Chinese mitten crab claws (*Eriocheir sinensis* H. Milne-Edwards, 1853): a first record. Oceanologia 49:137–143

O'Connor NJ (2007) Stimulation of molting in megalopae of the Asian shore crab *Hemigrapsus sanguineus*: physical and chemical cues. Mar Ecol Prog Ser 352:1–8. https://doi.org/10.3354/meps07315

Occhipinti-Ambrogi A (2007) Global change and marine communities: alien species and climate change. Mar Pollut Bull 55:342–352. https://doi.org/10.1016/j.marpolbul.2006.11.014

Ojaveer H, Gollasch S, Jaanus A et al (2007) Chinese mitten crab *Eriocheir sinensis* in the Baltic Sea – a supply-side invader? Biol Invasions 9:409–418. https://doi.org/10.1007/s10530-006-9047-z

Olyarnik SV, Bracken MES, Byrnes JE et al (2009) Ecological factors affecting community invasibility. In: Rilov G, Crooks JA (eds) Biological invasions in marine ecosystems. Springer, Berlin, pp 215–238

Padilla DK, Williams SL (2004) Beyond ballast water: aquarium and ornamental trades as sources of invasive species in aquatic ecosystems. Front Ecol Environ 2:131–138. https://doi.org/10.1890/1540-9295(2004)002[0131:BBWAAO]2.0.CO;2

Padilla DK, Mccann MJ, Shumway SE (2011) Marine invaders and bivalve aquaculture: sources, impacts, and consequences. In: Shumway SE (ed) Shellfish aquaculture and the environment. Wiley, Chichester, pp 395–424

Papacostas KJ, Rielly-Carroll EW, Georgian SE et al (2017) Biological mechanisms of marine invasions. Mar Ecol Prog Ser 565:251–268. https://doi.org/10.3354/meps12001

Parker JD, Torchin ME, Hufbauer RA et al (2013) Do invasive species perform better in their new ranges? Ecology 94:985–994

Perrault A, Bennett M, Burgiel S et al (2003) Invasive species, agriculture and trade: case studies from the NAFTA context. In: Second North American symposium on assessing the environmental effects of trade, Mexico City, p 55

Pickens PE (1965) Heart rate of mussels as a function of latitude, intertidal height, and acclimation temperature. Physiol Zool 38:390–405. https://doi.org/10.1086/physzool.38.4.30152416

Piola RF, Johnston EL (2008) Pollution reduces native diversity and increases invader dominance in marine hard-substrate communities. Divers Distrib 14:329–342. https://doi.org/10.1111/j.1472-4642.2007.00430.x

Poff NL, Brinson MM, Day JW (2002) Aquatic ecosystems & global climate change. Pew Center on Global Climate Change, Arlington

Rahel FJ, Olden JD (2008) Assessing the effects of climate change on aquatic invasive species. Conserv Biol 22:521–533. https://doi.org/10.1111/j.1523-1739.2008.00950.x

Reise K, Olenin S, Thieltges DW (2006) Are aliens threatening European aquatic coastal ecosystems? Helgol Mar Res 60:77–83. https://doi.org/10.1007/s10152-006-0024-9

Reise K, Buschbaum C, Büttger H et al (2017) Invading oysters and native mussels: from hostile takeover to compatible bedfellows. Ecosphere. https://doi.org/10.1002/ecs2.1949

Reuschel S, Cuesta JA, Schubart CD (2010) Marine biogeographic boundaries and human introduction along the European coast revealed by phylogeography of the prawn *Palaemon elegans*. Mol Phylogenet Evol 55:765–775. https://doi.org/10.1016/j.ympev.2010.03.021

Rodriguez LF (2006) Can invasive species facilitate native species? Evidence of how, when, and why these impacts occur. Biol Invasions 8:927–939. https://doi.org/10.1007/s10530-005-5103-3

Roman J (2006) Diluting the founder effect: cryptic invasions expand a marine invader's range. Proc R Soc B 273:2453–2459. https://doi.org/10.1098/rspb.2006.3597

Roman J, Palumbi SR (2004) A global invader at home: population structure of the green crab, *Carcinus maenas*, in Europe. Mol Ecol 13:2891–2898. https://doi.org/10.1111/j.1365-294X.2004.02255.x

Ruiz GM, Smith G (2005) Biological study of container vessels at the port of Oakland. Final report. Submitted to the Port of Oakland, 155 pp

Ruiz GM, Carlton JT, Grosholz ED et al (1997) Global invasions of marine and estuarine habitats by non-indigenous species: mechanisms, extent, and consequences. Am Zool 37:621–632. https://doi.org/10.1093/icb/37.6.621

Ruiz GM, Fofonoff PW, Hines AH et al (1999) Non-indigenous species as stressors in estuarine and marine communities: assessing invasion impacts and interactions. Limnol Oceanogr 44:950–972. https://doi.org/10.4319/lo.1999.44.3_part_2.0950

Ruiz GM, Fofonoff PW, Carlton JT et al (2000) Invasion of coastal marine communities in North America: apparent patterns, processes, and biases. Annu Rev Ecol Syst 31:481–531. https://doi.org/10.2307/annurev.ecolsys.37.091305.30000016

Sakai AK, Allendorf FW, Holt JS et al (2001) The population biology of invasive species. Annu Rev Ecol Syst 32:305–332. https://doi.org/10.1146/annurev.ecolsys.32.081501.114037

Salvaterra T, Green DS, Crowe TP et al (2013) Impacts of the invasive alga *Sargassum muticum* on ecosystem functioning and food web structure. Biol Invasions 15:2563–2576. https://doi.org/10.1007/s10530-013-0473-4

Secord D (2003) Biological control of marine invasive species: cautionary tales and land-based lessons. Biol Invasions 5:117–131. https://doi.org/10.1023/A:1024054909052

Seebens H, Gastner MT, Blasius B (2013) The risk of marine bioinvasion caused by global shipping. Ecol Lett 16:782–790. https://doi.org/10.1111/ele.12111

Short LL (1969) Aspects of avian hybridization. The Auk 86:84–105

Simberloff D (2006) Invasional meltdown 6 years later: important phenomenon, unfortunate metaphor, or both? Ecol Lett 9:912–919. https://doi.org/10.1111/j.1461-0248.2006.00939.x

Simberloff D, Von Holle B (1999) Positive interactions of nonindigenous species: invasional meltdown? Biol Invasions 1:21–32

Simberloff D, Martin JL, Genovesi P et al (2013) Impacts of biological invasions: what's what and the way forward. Trends Ecol Evol 28:58–66. https://doi.org/10.1016/j.tree.2012.07.013

Smaal AC, Kater BJ, Wijsman J (2009) Introduction, establishment and expansion of the Pacific oyster *Crassostrea gigas* in the Oosterschelde (SW Netherlands). Helgol Mar Res 63:75–83. https://doi.org/10.1007/s10152-008-0138-3

Snyder WE, Evans EW (2006) Ecological effects of invasive arthropod generalist predators. Annu Rev Ecol Evol Syst 37:95–122. https://doi.org/10.2307/annurev.ecolsys.37.091305.30000006

Stachowicz JJ, Fried H, Osman RW et al (2002) Biodiversity, invasion resistance, and marine ecosystem function: reconciling pattern and process. Ecology 83:2575–2590. https://doi.org/10.2307/3071816

State of Victoria (2010) Invasive plants and animals policy framework. In: Department of economic development, jobs, transport and resources. http://agriculture.vic.gov.au/agriculture/pests-diseases-and-weeds/protecting-victoria-from-pest-animals-and-weeds/invasive-plants-and-animals/invasive-plants-and-animals-policy-framework. Accessed 11 Mar 2018

Stickle WB, Sabourin TD (1979) Effects of salinity on the respiration and heart rate of the common mussel, *Mytilus edulis* L., and the black chiton, *Katherina tunicata* (Wood). J Exp Mar Bio Ecol 41:257–268. https://doi.org/10.1016/0022-0981(79)90135-7

Tepolt CK, Somero GN (2014) Master of all trades: thermal acclimation and adaptation of cardiac function in a broadly distributed marine invasive species, the European green crab, *Carcinus maenas*. J Exp Biol 217:1129–1138. https://doi.org/10.1242/jeb.093849

Theede H (1969) Einige neue Aspekte bei der Osmoregulation von *Carcinus maenas*. Mar Biol 2:114–120

Thomsen MS, Wernberg T, South PM et al (2016) Non-native Seaweeds Drive Changes in Marine Coastal Communities Around the World. In: Hu Z-M, Fraser C (eds) Seaweed phylogeography – adaptation and evolution of seaweeds under environmental change. Springer, Dordrecht, pp 147–185

Thresher RE, Kuris AM (2004) Options for managing invasive marine species. Biol Invasions 6:295–300. https://doi.org/10.1023/B:BINV.0000034598.28718.2e

Tsiamis K, Zenetos A, Deriu I et al (2018) The native distribution range of the European marine non-indigenous species. https://doi.org/10.3391/ai.2018.13.2.01

Tsutsui ND, Suarez AV, Holway DA et al (2000) Reduced genetic variation and the success of an invasive species. Proc Natl Acad Sci USA 97:5948–5953. https://doi.org/10.1073/pnas.100110397

Ünal V, Bodur HG (2017) The socio-economic impacts of the silver-cheeked toadfish on small-scale fishers: a comparative study from the Turkish coast. Ege J Fish Aquat Sci 34:119–127. https://doi.org/10.12714/egejfas.2017.34.2.01

van den Brink AM, Wijnhoven S, McLay CL (2012) Competition and niche segregation following the arrival of *Hemigrapsus takanoi* in the formerly *Carcinus maenas* dominated Dutch delta. J Sea Res 73:126–136. https://doi.org/10.1016/j.seares.2012.07.006

van den Brink AM, Godschalk M, Smaal AC et al (2013) Some like it hot: the effect of temperature on brood development in the invasive crab *Hemigrapsus takanoi* (Decapoda: Brachyura: Varunidae). J Mar Biol Assoc United Kingdom 93:189–196. https://doi.org/10.1017/S0025315412000446

Viard F, Ellien C, Dupont L (2006) Dispersal ability and invasion success of *Crepidula fornicata* in a single gulf: Insights from genetic markers and larval-dispersal model. Helgol Mar Res 60:144–152. https://doi.org/10.1007/s10152-006-0033-8

Vila M, Abós-Herràndiz R, Isern-Fontanet J et al (2016) Establishing the link between *Ostreopsis cf. ovata* blooms and human health impacts using ecology and epidemiology. Sci Mar 80:107–115. https://doi.org/10.3989/scimar.04395.08A

Vitousek PM, D'Antonio CM, Loope LL et al (1996) Biological invasions as global environmental change. Am Nat 84:468–478

Vranken S, Bosch S, Peña V et al (2018) A risk assessment of aquarium trade introductions of seaweed in European waters. Biol Invasions:1–17. https://doi.org/10.1007/s10530-017-1618-7

Wiedenmann J, Baumstark A, Pillen TL et al (2001) DNA fingerprints of *Caulerpa taxifolia* provide evidence for the introduction of an aquarium strain into the Mediterranean Sea and its close relationship to an Australian population. Mar Biol 138:229–234. https://doi.org/10.1007/s002270000456

Williams SL, Grosholz ED (2002) Preliminary reports from the *Caulerpa taxifolia* invasion in southern California. Mar Ecol Prog Ser 233:307–310. https://doi.org/10.3354/meps233307

Williams SL, Grosholz ED (2008) The invasive species challenge in estuarine and coastal environments: marrying management and science. Estuar Coast 31:3–20. https://doi.org/10.1007/s12237-007-9031-5

Wilson JRU, Dormontt EE, Prentis PJ et al (2009) Something in the way you move: dispersal pathways affect invasion success. Trends Ecol Evol 24:136–144. https://doi.org/10.1016/j.tree.2008.10.007

Wolff WJ (2005) Non-indigenous marine and estuarine species in The Netherlands. Zool Meded Leiden 79:1–116

Yang J, Li Q, Kong L et al (2008) Genetic structure of the veined rapa whelk (*Rapana venosa*) populations along the coast of China. Biochem Genet 46:539–548. https://doi.org/10.1007/s10528-008-9168-4

Zenetos A, Apostolopoulos G, Crocetta F (2016) Aquaria kept marine fish species possibly released in the Mediterranean Sea: first confirmation of intentional release in the wild. Acta Ichthyol Piscat 46:255–262. https://doi.org/10.3750/AIP2016.46.3.10

Zeng Y, McLay CL, Yeo DCJ (2014) Capital or income breeding crabs: who are the better invaders? Crustaceana 87:1648–1656. https://doi.org/10.1163/15685403-00003385

For a World Without Boundaries: Connectivity Between Marine Tropical Ecosystems in Times of Change

Hannah S. Earp, Natalie Prinz, Maha J. Cziesielski, and Mona Andskog

Abstract

Tropical mangrove forests, seagrass beds, and coral reefs are among the most diverse and productive ecosystems on Earth. Their evolution in dynamic, and ever-changing environments means they have developed a capacity to withstand and recover (i.e., are resilient) from disturbances caused by anthropogenic activities and climatic perturbations. Their resilience can be attributed, in part, to a range of cross-ecosystem interactions whereby one ecosystem creates favorable conditions for the maintenance of its neighbors. However, in recent decades, expanding human populations have augmented anthropogenic activities and driven changes in global climate, resulting in increased frequencies and intensities of disturbances to these ecosystems. Many contemporary environments are failing to regenerate following these disturbances and consequently, large-scale degradation and losses of ecosystems on the tropical seascape are being observed. This chapter reviews the wealth of available literature focused on the tropical marine seascape to investigate the degree of connectivity between its ecosystems and how cross-ecosystem interactions may be impacted by ever-increasing anthropogenic activities and human-induced climate change. Furthermore, it investigates how disruption and/or loss of these cross-ecosystem interactions may impact the success of neighboring ecosystems and consequently, the highly-valued ecosystem services to which these ecosystems give rise. The findings from this review highlight the degree of connectivity between mangroves, seagrasses and coral reefs, and emphasizes the need for a holistic, seascape-wide research approach to successfully protect and preserve these critically important ecosystems and their associated services for future generations.

H. S. Earp (✉)
Faculty of Biology and Chemistry, University of Bremen, Bremen, Germany
e-mail: hannahsearp@hotmail.com

Leibniz Centre for Tropical Marine Research (ZMT), Bremen, Germany

School of Ocean Sciences, Bangor University, Menai Bridge, Wales, UK

N. Prinz · M. Andskog
Faculty of Biology and Chemistry, University of Bremen, Bremen, Germany

Leibniz Centre for Tropical Marine Research (ZMT), Bremen, Germany
e-mail: nprinz@uni-bremen.de; andskog.mona@gmail.com

M. J. Cziesielski
Red Sea Research Centre, King Abdullah University of Science and Technology, Thuwal, Kingdom of Saudi Arabia
e-mail: maha.olschowsky@kaust.edu.sa

Introduction

Within the tropical zone, cartographically defined as the area between the Tropics of Cancer and Capricorn (~23.5 °N and S) (Gnanadesikan and Stouffer 2006), three ecologically distinct marine ecosystems; mangroves forests, seagrass beds and coral reefs, can be found (Fig. 1). These ecosystems have long been known for; their rich biodiversity, with coral reefs alone hosting 25% of known marine species (McAllister 1995; Plaisance et al. 2011), their high levels of gross productivity (for coral reefs it is estimated at ca. 0.4–5.5 kg C m^{-2} $year^{-1}$ (Douglas 2001)), which rival those of terrestrial ecosystems, and the array of ecosystem goods and services which they provide (Costanza et al. 1997). Across the land-sea boundary, mangroves have an annual economic value of approximately US$ 200,000–900,000 per square kilometer (UNEP-WCMC 2006), and their extent is closely correlated to the success of adjacent fisheries (Manson et al. 2005; Aburto-Oropeza et al. 2008). A step further into the ocean, nutrient cycling by seagrasses has been valued at US$ 19,000 ha^{-1} $year^{-1}$ (Costanza et al. 1997). Finally, in the most

S. Jungblut et al. (eds.), *YOUMARES 8 – Oceans Across Boundaries: Learning from each other*,
https://doi.org/10.1007/978-3-319-93284-2_9

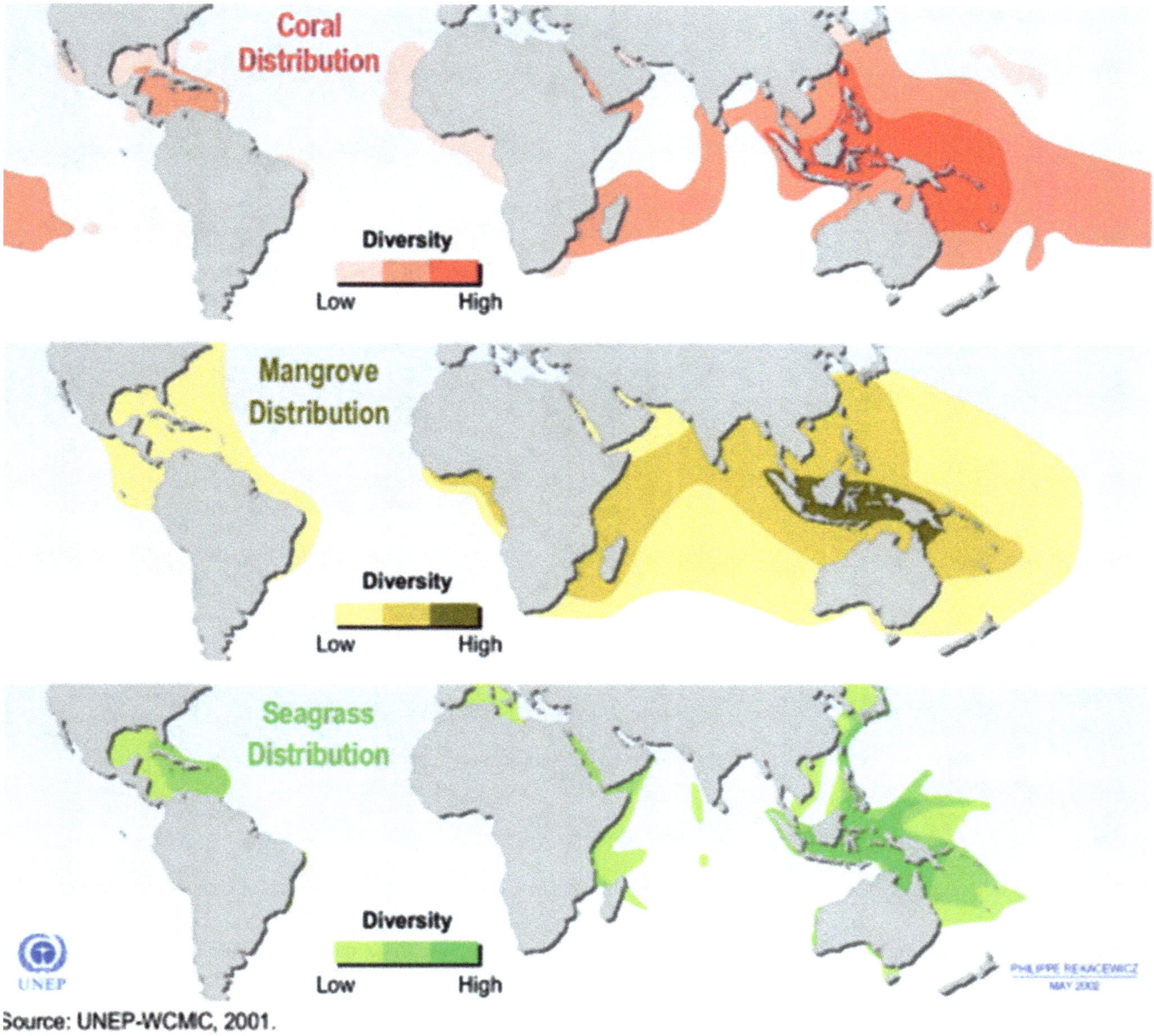

Fig. 1 Global distribution of coral, mangrove, and seagrass diversity. (Image created by Philippe Rekacewicz in May 2002 from data compiled by UNEP-WCMC, 2001. Reproduced from Jennerjahn 2012 with permission from Elsevier)

offshore habitat, coral reefs provide coastal protection, sustain fisheries, and drive tourism activities which are valued globally at US$ 30 billion per annum (Stone 2007; Khan and Larrosa 2008).

Although these ecosystems can thrive in isolation (Parrish 1989), in regions where they occur together, the value of the services provided is enhanced. For example, each ecosystem alone provides a form of coastal protection (Koch et al. 2009), but together the three have been shown to supply more protection compared to any one ecosystem alone, or any combination of two ecosystems (Guannel et al. 2016). These services are, in part, the result of highly complex cross-ecosystem interactions occurring on physical, chemical and biological levels (Ogden 1980; Nagelkerken 2009; Gillis et al. 2014). Early research, such as the mangrove 'outwelling' hypothesis proposed by Odum (1968), and Odum and Heald (1972) highlighted the importance of research beyond ecosystem boundaries. This hypothesis postulated that a fraction of organic matter, ~50%, produced by mangroves is exported to the coastal ocean (Dittmar et al. 2006), where it is either stored as carbon in marine sediments (Jennerjahn and Ittekkot 2002), or provides essential habitat and food resources to adjacent ecosystems including coral reefs (Bouillon et al. 2008; Granek et al. 2009). Since then, a wealth of studies have investigated individual connectivity pathways between tropical marine ecosystems. Biologically,

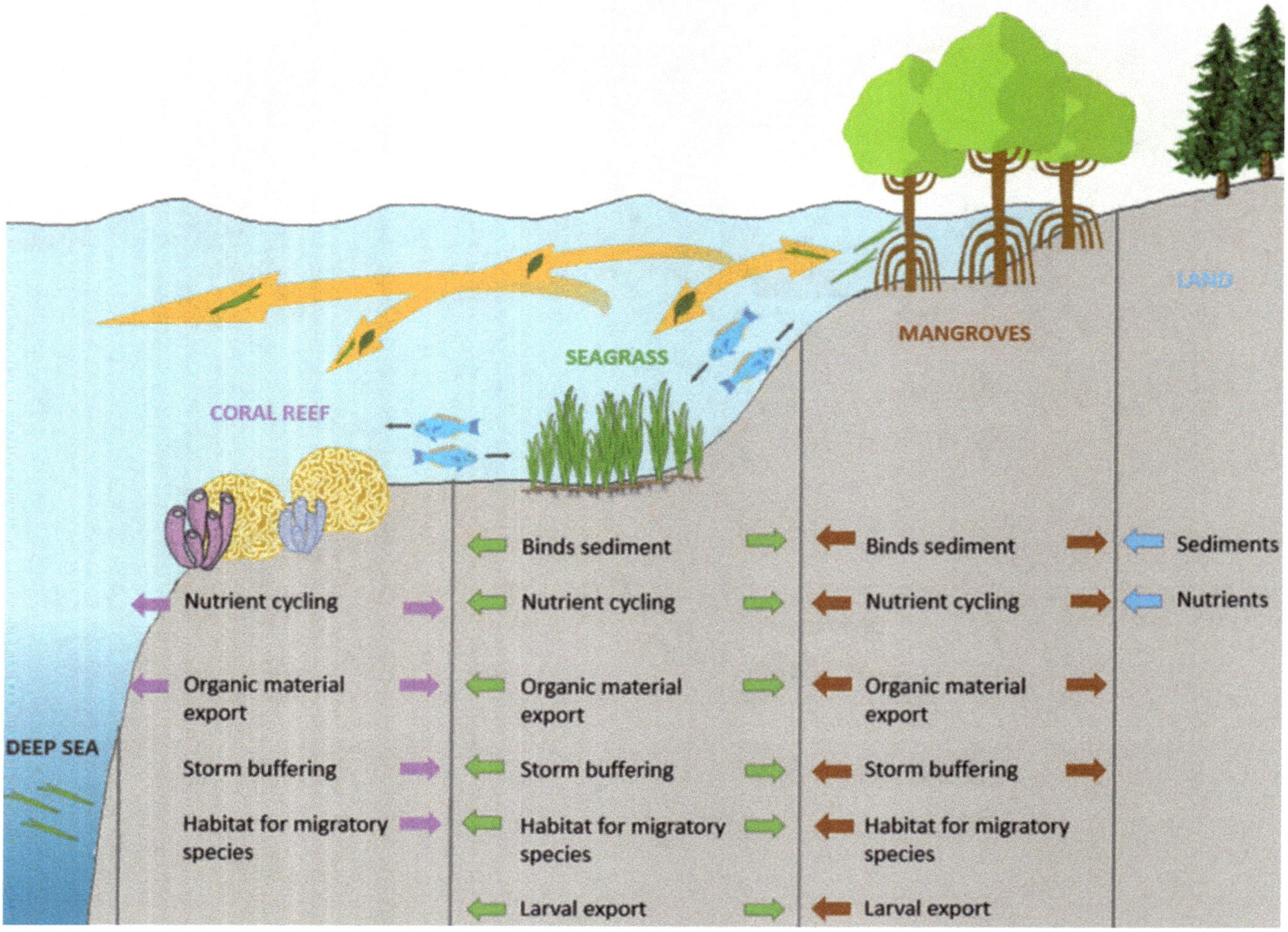

Fig. 2 Interdependencies of ecosystems along the tropical seascape. (Based on Moberg and Folke 1999; Heck et al. 2008; Berkström et al. 2012)

mobile organisms transition between ecosystems to forage, spawn, as part of seasonal migrations, or through ontogenesis (Parrish 1989; Cocheret de la Morinière et al. 2002; Mumby 2006). Water movement, including tidal regimes and currents further connect these systems by facilitating the exchange of organic matter, sediments, nutrients and pollutants (Fig. 2) (Grober-Dunsmore et al. 2009).

Although we are just beginning to uncover and understand the extent of these connectivity pathways, in most cases they are involved in creating optimal conditions for the successful maintenance of neighboring ecosystems. Coral reefs dissipate the energy of waves and currents, providing calm environments for seagrass and mangroves, whilst they in return stabilize the sediment and trap nutrients, creating the oligotrophic waters in which coral reefs thrive (Kitheka 1997; McGlathery et al. 2007; Mumby et al. 2011). Inevitably, the success of one ecosystem is directly linked to the success of the others, meaning the response of one ecosystem to change could result in profound consequences for neighboring systems (Grober-Dunsmore et al. 2009; Saunders et al. 2014).

Change is however, a natural attribute of global ecosystems (Alongi 2002), and tropical marine ecosystems have evolved under a regime of natural disturbances (Lamy et al. 2015). Consequently, they have developed a capacity to withstand and recover (i.e., are resilient) from periodic disruptions to their ecological equilibrium or 'steady-state' and readily regenerate (Connell 1997). However, in recent decades, disruptions in the form of anthropogenic activities (i.e., pollution and exploitation), human-induced climate change (i.e., temperature rise, ocean acidification, sea level rise, expansion of oxygen minimum zones, and severe weather events), and a combination of the two, have increased in intensity, duration, and extent (Vitousek et al. 1997). These disruptions pose significant challenges to tropical marine ecosystems and their associated cross-ecosystem interactions.

A lack of empirical data for tropical environments, compared to temperate regions, has resulted in conflicting predictions regarding the impact of future anthropogenic and climatic perturbations on tropical marine ecosystems (Alongi 2002). However, field studies have shown that many of these contemporary ecosystems are failing to regenerate following disturbances (Bellwood et al. 2004). The global loss and fragmentation of mangrove forests equates to a loss of ecosystem services worth US$ 7.2 trillion year^{-1} (Costanza et al. 2014). Other studies have noted significant global declines in seagrass areas, at rates of 110 km^2 year^{-1} since 1980, meaning at least 29% of their known areal extent has been lost (Green and Short 2003; Waycott et al. 2009; Short et al. 2014). In terms of coral reefs, estimates show ~19% of the world's reefs have been lost (Wilkinson 2008), with 75% of present day reefs considered threatened when climatic and anthropogenic threats are combined (Burke et al. 2011), and 20% of these are expected to disappear within 20 years (Wilkinson 2008). Furthermore, 55% of coral reef fisheries in 49 island countries are considered as unsustainable (Newton et al. 2007). Overfishing threatens reef health by causing trophic cascades that may induce phase-shifts to macroalgal dominated environments, which subsequently impacts adjacent ecosystems and cross-ecosystem interactions (Jackson et al. 2001). Modelling studies compliment this research, revealing that impacts on one ecosystem can have profound impacts on neighboring ecosystems, and in turn, the ecosystem services they provide (Saunders et al. 2014).

Despite the overwhelming evidence that loss and degradation of these vital marine ecosystems will have far reaching ecological and economic impacts, significant gaps in our knowledge regarding the interconnectivity between these ecosystems remain (Duarte et al. 2008). Appealing to scientific research efforts, this review provides an overview of the impacts of augmenting anthropogenic activities, and human induced climate on the known interconnectivity pathways amongst tropical marine ecosystems, as opposed to each ecosystem in isolation. Sections "A Nutritious Ocean" and "An Empty Ocean" explore the response of cross-ecosystem interactions and ecosystems services to anthropogenic activities in the form of eutrophication and exploitation successively, whilst sections "A Warmer Ocean" and "A Sour Ocean" investigate their responses to ocean warming and acidification consecutively. Understanding the threats facing interdependencies between these ecosystems is suggested to be an opportunity for science to prevent large-scale losses of these critical environments in the face of disturbances in the years to come.

A Nutritious Ocean

Mangroves, seagrasses, and coral reefs are located either on, or near land masses (Spalding et al. 2001; Green and Short 2003), exposing them to local anthropogenic threats including periodic fertilizer runoff and sewage discharge, which are delivered to coastal waters (Fabricius 2005; Burke et al. 2011). This process, known as eutrophication, can stimulate phytoplankton blooms and algal growth in coastal ecosystems (McGlathery et al. 2007), which can lead to anoxia and toxic sulphide production due to increased microbial activity degrading this additional biomass (Flindt et al. 1999; Herbeck et al. 2014). These periodic enrichment events have become more prevalent within the last five decades, as annual global usage of nitrogen fertilizers has increased 14-fold, and is expected to increase even further (Matson et al. 1997; FAO 2016a).

Eutrophication can impact coastal ecosystems either directly, by affecting the fitness of organisms, or indirectly, by affecting processes within the ecosystem or altering the connectivity between ecosystems. In tropical regions, such as the Great Barrier Reef, seagrasses and mangroves are nutrient limited (Schaffelke et al. 2005). Therefore, the most common direct effect of nutrient enrichment is an increase in productivity and growth of these marine plants (Schaffelke et al. 2005), which alone is a positive effect. These ecosystems can therefore buffer eutrophication to a certain extent, and protect the oligotrophic waters of their vulnerable neighbors, coral reefs, from nutrient enrichment (Kitheka 1997; McGlathery et al. 2007). Indirect effects of eutrophication on marine plant communities are more commonplace, as excess nutrients also increase the productivity of other competing autotrophs, namely algae (Schaffelke and Klumpp 1998; McGlathery et al. 2007). In mangroves, there is little evidence of the direct effects of excess nutrients, however indirect links to mangrove dieback and damage do exist. The dieback of *Avicennia marina* in southern Australia was indirectly linked to eutrophication through the increased proliferation of the green macroalgae, *Ulva* sp., which smothered and killed the aerial roots of established mangroves, as well as smothering and inhibiting the growth of mangrove seedlings (Fig. 3) (Schaffelke et al. 2005). As mangroves have a high nutrient uptake capacity, they are also at risk of taking up herbicides and heavy metals which run-off agricultural land together with the nutrients. The build-up of these toxic substance has also been linked to mangrove dieback and damage in downstream estuarine habitats (Schaffelke et al. 2005).

Algal blooms may be even more detrimental for seagrasses as they are completely submerged in water and are highly dependent on light availability and water quality (McGlathery 2001). Light availability can be reduced by both biotic and abiotic factors. Biotic factors are primarily based on the abundance of phytoplankton, epiphytic algae, and seaweed wracks (Fig. 3) (McGlathery 2001; Herbeck et al. 2014; van Tussenbroek et al. 2017), whilst abiotic factors include increased particle loads from sewage effluent which settle on seagrass leaves or attenuate light within the water column (Herbeck et al. 2014). The increase in algal biomass will be followed by faster decomposition rates (Flindt et al. 1999), and therefore increased sulphide production in the sediment (Herbeck et al. 2014) (Fig. 3). Sulphide is toxic for seagrasses and leads to a decrease in shoot density, rhizome extension, and growth (Díaz-Almela et al. 2008; Herbeck et al. 2014;

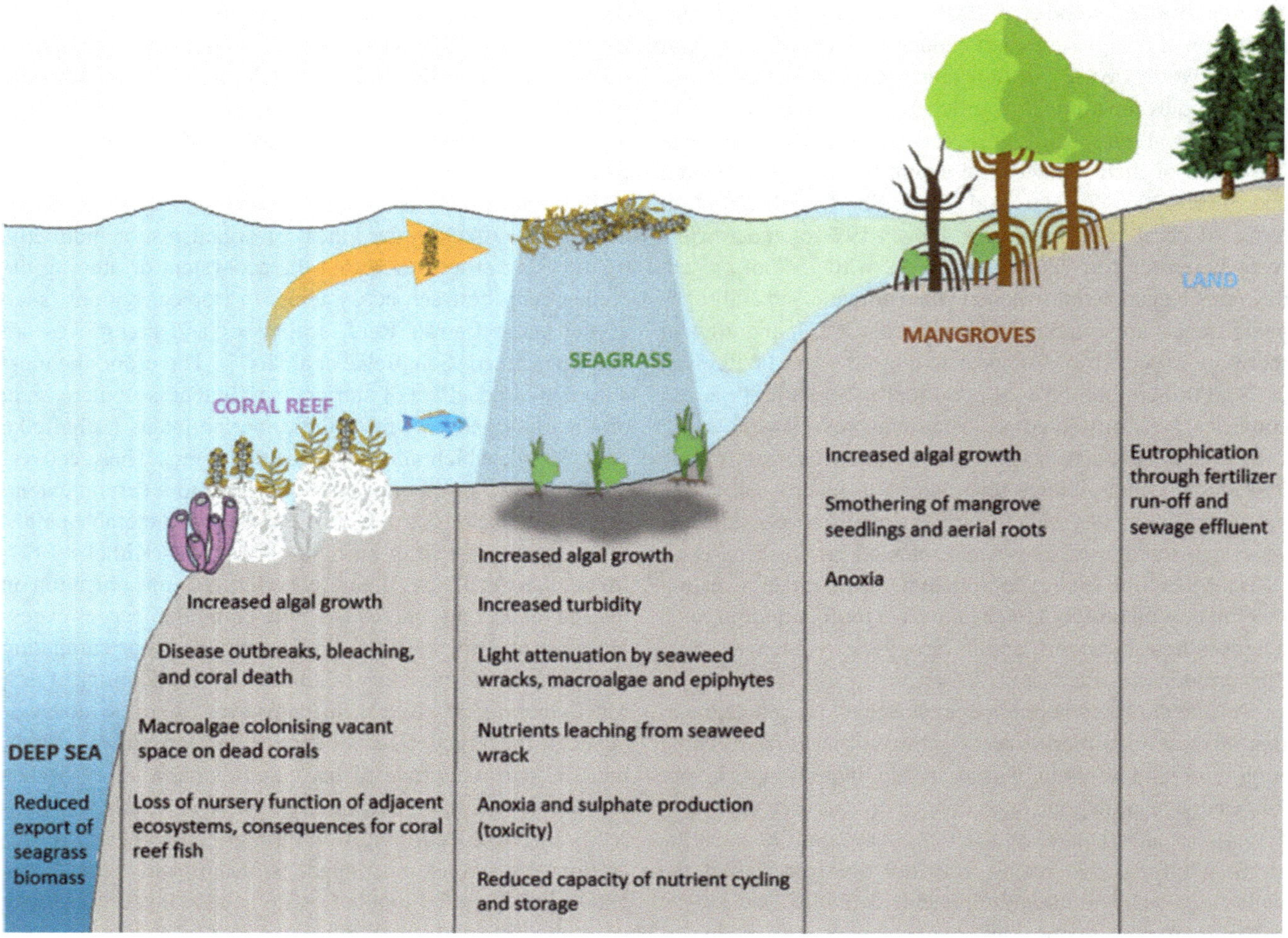

Fig. 3 Conceptual diagram detailing the possible consequences of eutrophication on each individual ecosystem as well as on the connectivity between them. Effects are not isolated within each system, but changes in one system may cause changes in others, either directly (e.g., transport of seaweed biomass) or indirectly (e.g., loss of ecosystem function such as nutrient retention or nursery areas). (Based on Moberg and Folke 1999; Heck et al. 2008; Berkström et al. 2012)

Maxwell et al. 2017). One way in which seagrasses counteract sulphide toxicity is by transporting oxygen from the leaves to the roots, thereby oxygenating the sediment (Borum et al. 2005). However, oxygen production is highly dependent on photosynthesis and light availability, which will be decreased during algal bloom conditions. This negative feedback loop can ultimately result in increased losses of seagrass and subsequent replacement by algae, which in turn stimulates further seagrass loss through elevated decomposition rates (Maxwell et al. 2017).

In regards to the connectivity between seagrass meadows and other ecosystems, we can expect multiple disruptions of important processes. For example, there are more than 50 records of seagrass shoots laying on the deep-sea floor (>1000 m), and there is evidence that seagrass detritus is an important subsidy to the deep-sea food web (Heck et al. 2008; Duarte and Krause-Jensen 2017). Seagrass has a relatively slow decomposition time compared to macroalgae (degradation rate constants range from 0.001 to 0.018 d^{-1} in seagrass and from 0.02 to 0.26 d^{-1} for *Ulva* spp.), enabling seagrass detritus to reach the deep-sea floor (Flindt et al. 1999; Heck et al. 2008). Algae are also exported from seagrass communities, but they have more labile organic matter than seagrasses and decompose before reaching the seafloor (Flindt et al. 1999; Heck et al. 2008). This important organic matter subsidy will be lost as seagrasses areas decline or shift to macroalgal meadows (Fig. 3). Furthermore, continuous, non-patchy seagrass beds with mangrove neighbors have a higher beta-diversity compared to patchy seagrass beds with greater distances to mangrove forests (Henderson et al. 2017). Proximity to mangroves is also positively related to parrotfish grazing of seagrass, which is crucial for the removal of epiphytes and leaf turnover (Swindells et al. 2017). We can expect that fragmentation, patchiness, and seagrass loss will further increase the distance to nearby mangroves and thereby affect these important ecosystem processes and diversity (Fig. 3). Furthermore, a recent study found that the prevalence of coral disease was 50% less on reefs with adjacent seagrass meadows, compared to reefs without (Lamb et al. 2017). Coral tissue mortality caused by bleaching and sedimentation was also significantly less on reefs with neighboring seagrass beds (Lamb et al. 2017). This is due to the role seagrasses play in filtering our pathogenic bacteria, and thus, in the face of continued sewage outflow in some regions, we can expect elevated bacterial and disease prevalence on neighboring coral reefs (Lamb et al. 2017) (Fig. 3). However, in other regions, seagrass meadow functions are not as vulnerable or important to ecosystem connectivity. In Brazil, evidence shows that macroalgal beds serve as a better nursery area for juvenile fish than adjacent seagrass beds, and that these seagrass beds had much lower juvenile fish abundances compared similar beds in the Indo-Pacific or Caribbean (Eggertsen et al. 2017). That being said, the negative effects of eutrophication across the rest of the world far outweigh this sole positive scenario.

Eutrophication of coral reefs environments, can promote phytoplankton blooms and thereby increase developmental success of the coral-consuming crown-of-thorns starfish (*Acanthaster planci*) larvae, which feed on phytoplankton (Fabricius et al. 2010; De'ath et al. 2012). Nutrient enrichment can also, as previously mentioned, enhance the growth and productivity of macroalgae (Schaffelke and Klumpp 1998). Once established on a coral reef, macroalgae may continue to proliferate if nutrients are available and herbivory is limited. Ultimately, this can result in a phase-shift (Lapointe 1997; McCook 1999) (Fig. 3). In the case of some South Pacific islands, the range and abundance of two native macroalgal species, *Turbinaria ornata* and *Sargassum pacificum*, have increased noticeably throughout the reefs since the 1980's (Payri and Naim 1982; Stewart 2008). These algae are primarily found on dead patches of corals on top of *Porites* heads, where they form dense aggregations (Stewart 2006). High swells frequently remove these macroalgae from their substrate, resulting in masses of floating algae which aggregate in currents to form large compact seaweed wracks (Zubia et al. 2015). These wracks are also seen as communities of drift algae such as *Sargassum fluitans* and *Sargassum natans* in the Caribbean. However, when exceptionally large blooms of these wracks reach nearby ecosystems such as seagrasses, they eutrophicate them through decomposition, reduction of light availability, increased hypoxia/anoxia, and reduction of pH (van Tussenbroek et al. 2017) (Fig. 3). We can therefore expect that these wracks stimulate algal growth inside the meadows, leading to seagrass loss, and inhibiting seagrass recovery (van Tussenbroek et al. 2017).

To summarize the effects of eutrophication on ecosystems within the tropical seascape, the effects on mangroves are not well documented, probably because they have a large capacity to absorb nutrients and are not dependent on water clarity to survive. Seagrass meadows can also buffer nutrient enrichment, but may be more vulnerable than mangroves as they are completely submerged and their survival depends on light availability. The greatest impacts of eutrophication are seen on coral reefs, which are the most vulnerable to excess nutrients, but may not be as exposed when seagrass meadows and mangrove forest are in the vicinity. With a loss of ecosystem functions in these ecosystems as a result of nutrient enrichment, we can expect that they will no longer be able to provide important services for each other, such as; nursery grounds, habitats, feeding grounds for mobile fauna, nutrient and sediment retention, and export of biomass (Fig. 3).

Nutrients and herbivory are two well-connected concepts in marine ecosystems, and in most cases, phase-shifts are not attributed to one or the other but rather a combination of the two (Adam et al. 2015). That is, eutrophication by itself is rarely the only reason why a system experiences algal blooms, as it is also highly dependent on grazing pressure from consumers (Hughes 1994). The health of these ecosystems is therefore not only dependent on what we add to them, but also what we remove, through the harvesting mangroves, corals, fish, and invertebrates.

An Empty Ocean

Human existence is directly and indirectly dependent on marine ecosystems (Halpern et al. 2008). In the tropics, millions of people rely directly on marine ecosystems to harvest food (e.g., fishes, clams, crabs) and raw materials (e.g., timber, curio artefacts, medicinal products), either for subsistence purposes, or for their livelihood (Hoegh-Guldberg 2014). As a consequence, these ecosystems are experiencing accelerating losses of biodiversity with largely unknown consequences (Worm et al. 2006). With over 1.3 billion people residing along tropical coastlines, primarily in developing countries (Sale et al. 2014), it is important to understand the impacts of harvesting activities within the tropical seascape, and its subsequent consequences for ecosystem services upon which so many people rely. This section will investigate the impact of mangrove harvesting for raw materials and the impact of fishing on the functioning and connectivity amongst tropical marine ecosystems.

Mangrove use by humans has a long history, extending back over 7000 years (Spalding et al. 2010; Tomlinson 2016), as a diverse array of goods can be harvested from them, including; tannins, honey, medicinal products, thatch, timber, and firewood (Hamilton and Snedaker 1984; Ellison 1994; Kathiresan and Bingham 2001; Spalding et al. 2010). The physical properties of mangroves differ amongst species; *Rhizophora* is most widely harvested due to its hard, dense, easily-splitting wood which makes it an ideal material for poles, firewood, and more recently wood-chips (for conversion to rayon), whilst *Xylocarpus* is more suitable for furniture/carving, and *Aviennia* is considered too soft to have any real harvesting value (Ewel et al. 1998). As a result of expanding human populations in recent decades, anthropogenic activities and our exploitation of these ecosystems has resulted in large-scale degradation and destruction of mangroves forests (Kathiresan and Bingham 2001; Spalding et al. 2010). Monospecific stands are exceptionally threatened by harvesting (Ewel et al. 1998), and when coupled with forest clearance to make way for development, vast areas of mangrove are being removed. It is estimated that one quarter of original mangrove cover (>200,000 km^2), has been lost due to human activities, at a rate of 0.66–2% per annum (Duke et al. 2007; Spalding et al. 2010). This exceeds the loss rates reported for other threatened ecosystems (Stone 2007; Kathiresan 2008). For instance, of coral reefs, 10% have already been lost (Wilkinson 1992) and rainforests are lost at a rate of 0.8% per annum (Valiela et al. 2001). Consequently, mangroves are considered critically endangered or approaching extinction in 26 of the 120 countries in which they exist (Kathiresan 2008). Clearance and fragmentation of mangroves is of global concern due to its impact on ecosystem services like coastal protection, sediment trapping, nutrient cycling, and loss of habitats for commercially important species.

Mangroves provide coastal protection by mitigating the impact of tidal surges and waves caused by hurricanes and tsunamis (Duke et al. 2007). Estimates show that per kilometer of mangrove width, surges reduce in height by 5–50 cm, and surface wind waves reduce by up to 75% (McIvor et al. 2012). During the super cyclone, which hit Orissa (India) in 1999, mangroves significantly reduced the number of deaths and damage to property (Badola and Hussain 2005). Evidence from the Indian Ocean tsunami in December 2004 showed that villages in India with mangrove buffers were damaged to a lesser extent compared to nearby villages with no mangroves (Kathiresan and Rajendran 2005; Vermaat and Thampanya 2006). The degree of protection provided by mangroves is attributed to several factors: forest width and slope, tree and root density, and tree height (Alongi 2002). Yet in many regions, clear-cutting and felling of mangroves significantly reduces the forest width as well as tree and root densities, and consequently lessenes the buffering capacity of mangrove ecosystems to the threats posed by hurricanes and tsunamis (Ellison 1994; Kathiresan and Bingham 2001; Spalding et al. 2010). This buffering capacity is cited as one of the most severely undervalued ecosystem services provided by mangroves (Barbier et al. 2011). More recently, studies have shown that the value of this service is augmented at sites where other foreshore ecosystems (i.e., seagrasses and coral reefs) are present. Guannel et al. (2016) concluded that mangroves in combination with a second foreshore ecosystem attenuate significantly more wave energy compared to any one ecosystem alone.

Sediment trapping and nutrient cycling pathways further connect mangroves to adjacent ecosystems (Ewel et al. 1998). Riverine transport and terrestrial runoff are important pathways to coastal environments and provide loads rich in sediments, nutrients, organic matter, and at times, pollutants, to coastal environments (Ramos et al. 2004). These terrestrially derived components are caught and slowed by the complex aerial root structure of mangroves, and become immobilized and sequestered within mangrove systems before they reach the clear, nutrient-limited waters of often adjacent seagrass and coral reefs (Morell and Corredor 1993; Valiela and Cole 2002). On Pohnpei (Federal States of Micronesia), reduction of forest width to make way for a road, led to the death of the remaining downstream mangroves which could not withstand the increased sediment loads that buried lenticels on pneumatophores, prop roots and young stems (Ewel et al. 1998). In regions where seagrass beds and coral reefs neighbor mangroves, loss and degradation of the mangrove forest due to harvesting activities can be seen to reduce sediment and nutrient trapping capacities, thus increasing the risk of sedimentation and eutrophication (see section "A Nutritious Ocean" for a review) in neighboring ecosystems. Despite several mentions of the important role mangroves play in protecting adjacent systems from sedimentation (Morell and Corredor 1993; Valiela and Cole 2002; Schaffelke et al. 2005), limited case studies exist showing the impact of mangrove harvesting on sedimentation of adjacent ecosystems.

In terms of carbon, mangroves have a dual capacity as both a sink of atmospheric CO_2, and a source of oceanic carbon (Singh et al. 2005; Duke et al. 2007). Their high levels of productivity, which reached 26.70 t ha^{-1} year^{-1} for *Rizophora apiculata* in Thailand (Christensen 1978), shows that their role in atmospheric CO_2 assimilation to build biomass is of considerable importance (Spalding et al. 2010). However, it is hypothesized that net primary production of mangroves may be in excess of the carbon utilized in the system, consequently an estimated 20–30% is 'outwelled' to adjacent ecosystems (Bouillon et al. 2008; Granek et al. 2009), corresponding well to the 50% organic matter export estimate proposed by Dittmar et al. (2006). Although accurate quantification of the mangrove carbon budget remains limited (Bouillon et al. 2008; Alongi 2009), research has shown that clearing of mangroves could result in carbon emissions of up to 112–392 Mg ha^{-1} (Donato et al. 2011). These emissions would significantly influence global CO_2 concentrations, which in turn drive climate change (see sections "A Warmer Ocean" and "A Sour Ocean" for reviews). Although their impact on carbon export to the coastal ocean remains unknown (Donato et al. 2011), what is known is that alterations to these fluxes could impact habitats and food resources for organisms which depend upon them, e.g., marine fishes and invertebrates.

Marine organisms including fishes and invertebrates have long been consumed by humans, with the earliest evidence extending back some 140,000 years to South Africa where shellfish and shallow-water fishes were consumed (Marean et al. 2007). Yet the development of fishing equipment is believed to have arisen 40,000 years later, based upon the oldest known fishing hooks found in East Timor (O'Connor et al. 2011). Since then, the evolution of fishing gears and vessels have supported a transition from small-scale subsistence fishing to modern-day commercial fishing, making seafood one of the most traded food commodities worldwide (FAO 2016b). Fishing is now considered to be the most widespread, unsustainable human impact on the oceans (Pauly et al. 2002; Halpern et al. 2008; Ricard et al. 2012), with 31.4% of fish stocks estimated to be fished at biologically unsustainable levels and therefore overfished in 2016 (FAO 2016b). More recently, the Food and Agricultural Organization of the United Nations (FAO) estimated that 89% of global fish stocks are exploited or overexploited (Zhou 2017).

On tropical coasts, fishing occurs at both subsistence and commercial levels, and targets an array of vertebrate (i.e., fishes such as snapper, parrotfish and grouper), and invertebrate (e.g., penaeid shrimp and mud crab) species. Many of these target species are considered 'mobile links' (Moberg and Folke 1999) due to the roles they play in connecting ecosystems across the tropical seascape through diel, seasonal and/or ontogenetic migrations (Parrish 1989; Cocheret de la Morinière et al. 2002; Mumby 2006). The larvae of the grey snapper (*Lutjanus griseus*) migrate towards their nursery area among the mangroves where they develop into juveniles, which later migrate to seagrass beds, and finally to coral reefs as adults, where they reproduce and the cycle repeats (Fig. 4) (Luo et al. 2009). Penaeid prawns also undergo a number of habitat shifts during their development, with the eggs released by adults on offshore waters undergoing two post-larval stages before they migrate to mangrove areas as juveniles. Late stage juveniles then move towards alternative habitats such as seagrasses before they transfer to their offshore adult habitat (Fig. 5) (Robertson and Duke 1987). Several studies have indicated that the abundance and diversity of fish communities in particular, are higher in regions where three tropical ecosystems were in close proximity, compared to those where they were a significant distance apart (Unsworth et al. 2008). The transition of organisms is not only a biological link between tropical ecosystems, it also results in a substantial transfer of organic matter, nutrients, and energy across ecosystems (Deegan 1993). It can therefore be postulated that the exploitation of certain species, would have knock-on effects for connectivity pathways among tropical marine ecosystems.

Gulf menhaden (*Brevoortia patronus*), small euryhaline clupeid fish found in the waters of the Gulf of Mexico, play an important role in exporting nutrients and energy between estuaries and offshore waters (Deegan 1993). They feed on phytoplankton and detritus and are in turn an important prey item for larger predatory fishes. They also support the second largest commercial fishery (by weight) in North America (Vaughan et al. 2007). When combined, their ecological and economic values mean this species, along with other *Brevoortia* species have been described as "*the most important fish in the sea*" (Franklin 2007). Although not currently considered overfished, exploitation of this species correlates to reduced production of larger pelagic fishes, and may lead to considerable effects on the trophic structure of ecosystems in the Gulf of Mexico (Robinson et al. 2015). Further research is essential to understand the impact of harvesting *B. patronus* on nutrient and energy export to adjacent systems. However, it can be seen that the exploitation of organisms with key ecological roles could have adverse effects on resource transfer among ecosystems and trophic levels. This theory could be applied to multiple exploited organisms transitioning between tropical ecosystems, however, research into the ecological roles of many of these organisms remains, at present, uninvestigated, thus the impact of their exploitation unknown.

One organism, whose role is known, and of vital importance to the health of coral reefs is the Caribbean rainbow parrotfish (*Scarus guacamaia*). Adults of this species play a pivotal role in regulating algal cover on reefs, and consequently preventing phase-shifts (Heenan and Williams 2013). There is evidence that the success of this species is dependent on the success of nearby mangroves. The juveniles of *S. guacamaia* are dependent on mangroves as nursery areas, and in Belizean coral reefs it was found that the density of adult parrotfish was significantly higher in mangrove-rich regions compared to mangrove-scarce regions (Mumby et al. 2004). Similar findings were made in Aruba, where recruitment of juvenile parrotfish from mangroves to coral reefs was dependent on the maximum distance (10 km) between these two habitats. *S. guacamaia* were therefore not be able to be recruited to coral reefs situated at a greater distance from mangroves (Dorenbosch et al. 2006). Coral reefs with adjacent mangrove nurseries exhibit increased parrotfish grazing (Mumby et al. 2007), and are consequently considered more resilient to perturbations. However, parrotfish are highly sensitive to exploitation, and several species including *S. guacamaia* are currently classified by the International Union for the Conservation of Nature as 'near threatened'. Exploited populations can only maintain 5% of a reef in a permanently grazed state compared to 40% in unexploited populations (Mumby 2006), which has implications concerning increased algal proliferation and its effect on adjacent ecosystems (see section "A Nutritious Ocean"

Fig. 4 The grey snapper uses many habitats throughout its life. Open water, mangroves, seagrass and coral reefs are important for the growth and survival during different stages of this fish. Art by Ryan Kleiner. (Reproduced from www.piscoweb.org, with permission from Kristen Milligan)

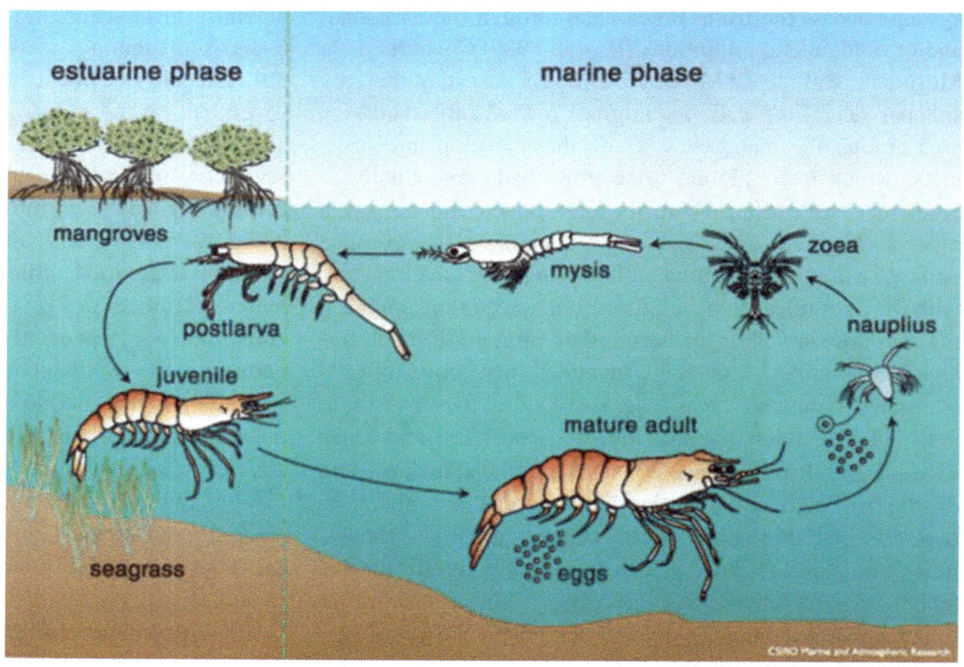

Fig. 5 The lifecycle of a penaeid prawn involves several habitats within the tropical seascape. (Reproduced from www.csiropedia.csiro.au, with permission from Pamela Tyres)

for a review). Bozec et al. (2016) discussed the trade-off between fisheries harvest and the resilience of coral reefs concluding that reefs will only remain resilient if <10% of parrotfish (>30 cm in length) are harvested. Highlighting harvesting once more, the clear-cutting and fragmentation of mangroves reduces nursery areas, and increases the migration distance to reefs and thereby further threatening parrotfish populations. In the Caribbean, the combination of historical overexploitation of parrotfish and mangrove deforestation synergistically reduced herbivory and secondary production (Mumby et al. 2004). This highlights once again that research into the effect of exploitation on connectivity between tropical ecosystems is limited, and that there is a need for seascape-wide and cross-disciplinary research in the tropics.

A Warmer Ocean

Anthropogenic activities have had demonstrated localized impacts in tropical marine ecosystems, however, anthropogenic-induced stress in the form of global climate change is having impacts on these ecosystems on a global scale. Climate change occurs in several forms, but one of the most studied is the increase of atmospheric and oceanic temperatures. These temperature elevations can be attributed to the post-industrialization increase in atmospheric carbon dioxide (CO_2), a greenhouse gas, which is responsible for trapping the Earth's outgoing radiation within the atmosphere, and consequently allowing the planet to warm. By the end of the century, global surface temperature is estimated to have increased by 4 °C (IPCC 2013). The ocean has been heating up by absorbing 90% of incoming solar radiation since 1971 (Riser et al. 2016). This temperature increase is not the only impact of global warming, a number of indirect impacts are also expected including; the melting of glaciers and ice sheets resulting in sea level rise, increased precipitation resulting in elevated terrestrial runoff, and increased frequency and intensity of storms (Knutson et al. 2010; Trenberth 2011; Godoy and De Lacerda 2015).

At present, the distribution of some mangroves and seagrass species is confined by minimum and maximum air and sea temperatures (Short et al. 2007; Bjork et al. 2008; Ward et al. 2016). For the mangroves and seagrass systems not living towards the edge of their tolerance limits, an increase in temperature could initially result in positive responses (Alongi 2015). However, once their tolerance limits are surpassed the consequences are severe. A decrease in productivity and growth leads to a shift in community composition, favoring those better adapted to cope with the elevated temperatures which could ultimately lead to the disappearance of mangrove and seagrasses species with low thermal tolerances (Pernetta 1993; Short et al. 2011). Furthermore, temperature increase has been shown to cause changes in reproduction patterns and altered metabolism (Short and Neckles 1999; Gilman et al. 2008; Yeragi and Yeragi 2014; Arunparasath and Gomathinayagam 2015). These examples highlight the direct consequences of elevated atmospheric and oceanic temperatures, however, some of the most threatening impacts come from the collateral impacts of climate change.

If sea level rises due to glacial melting, it will not only result in changes to flooding duration and frequency, but also of salinity. Although mangroves are sensitive to such changes (Friess et al. 2012; Ward et al. 2016), they also exhibit exceptional resilience through their ability to actively modify their environment and migrate both inland and seawards (Fig. 6) (Krauss et al. 2014; Ward et al. 2016). Roots of mangrove trees trap sediment, allowing it to settle in the surrounding area. However, the ability of mangrove forests to respond to sea level rise will depend on sediment type and, importantly, the rate of sediment accretion (Ward et al. 2016). If sedimentation rates remain higher than the rate of sea level rise, then mangrove forests will respond by raising the seafloor and progressively moving inland (Alongi 2008; Godoy and De Lacerda 2015; Lovelock et al. 2015). By contrast, if sea level rise exceeds the sedimentation rate, then the forest will drown (Godoy and De Lacerda 2015; Ward et al. 2016).

The exceptional migratory capabilities of mangroves have enabled some forests to, contrary to what one might expect, have a positive response to climate change scenarios. Due to changes in global temperature regimes, mangroves are expanding their inland and poleward limits. The decreases in cold and arid conditions are enabling mangrove expansion into new territories (Godoy and De Lacerda 2015). Whilst the polar shift of mangrove forests essentially represents a re-distribution but potential survival of the ecosystem as a whole, inland migration may have severe impacts on other ecosystems, such as seagrass meadows, begging the question: will the survival of mangrove vegetation come at the cost of other ecosystems, or will these ecosystems also adapt by migrating?

Although also having distributional limits defined by sea and air temperatures, seagrass meadows can be seen as more sensitive than mangroves due to the fact they are exposed to elevated temperatures during the day, and can be subject to desiccation during low tides and consequently high ultraviolet and photosynthetically active radiation (Dawson and Dennison 1996; Durako and Kunzelman 2002; Campbell et al. 2007). Whilst direct impacts of temperature increase on seagrasses are similar to mangroves, changes in the carbon balance have been observed as an additional consequence. Carbon balances, particularly in the substrate, are affected by the increase in photosynthesis, which initially has benefits, but also disadvantages as the increased labile carbon (a product of photosynthesis), is transferred to the substrate, and in turn may alter microbial communities important for the maintenance of soil nutrients (Cotner et al. 2004; Koch et al. 2007). Thus, increasing water temperature in seagrass meadows directly affects nutrient composition not only of the sediment, but also the water column.

However, similarly to mangroves, the collateral impacts of global warming will also have substantial impacts on seagrass meadows. Light, nutrients, and turbidity, which influ-

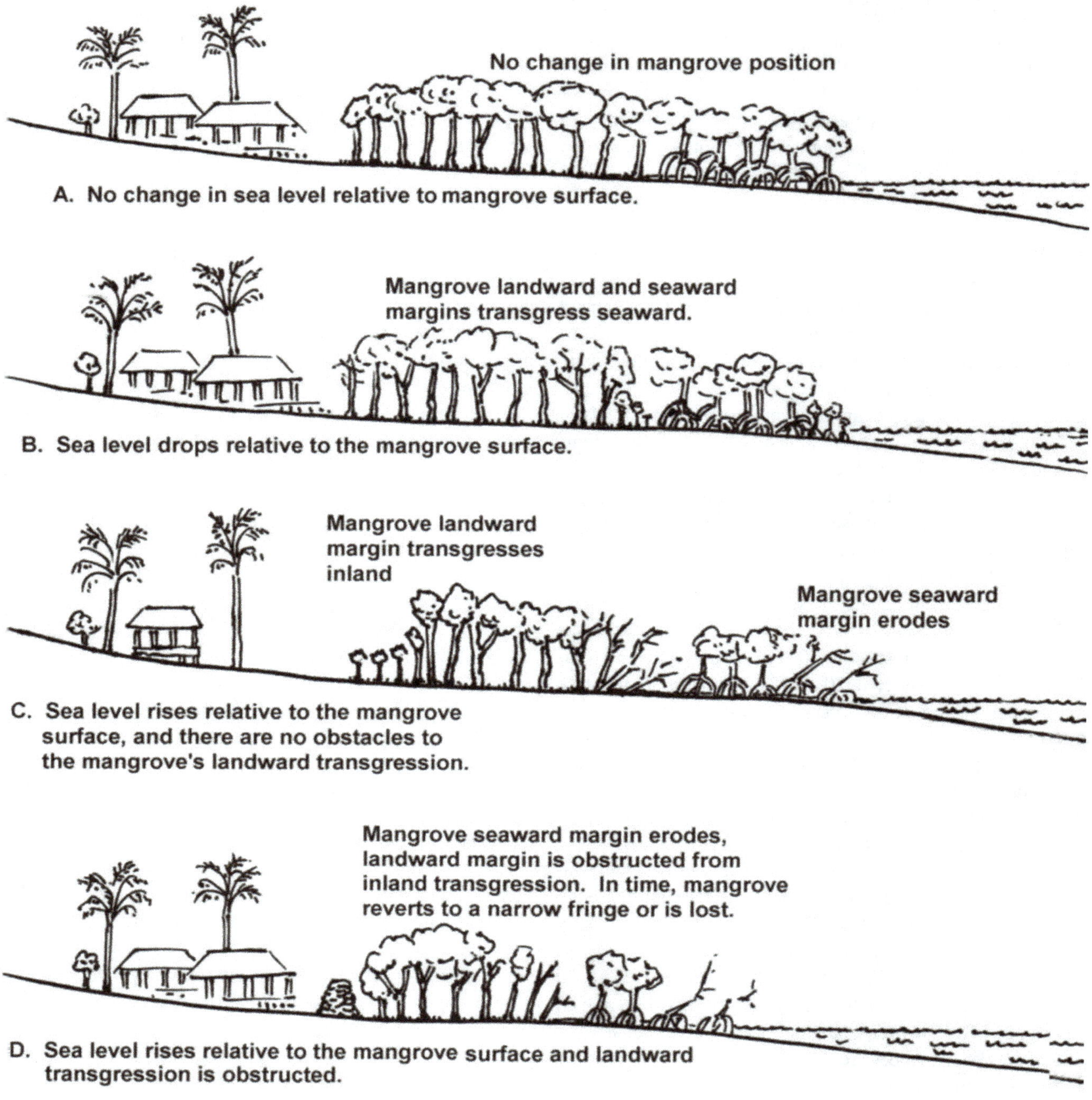

Fig. 6 Sea level rise generalized responses of mangroves. (**a**) Stable sea level where mangrove margins remain the same. (**b**) Sea level falls leading to a seaward shift of margins. (**c**) Sea level rises, no landward obstruction and high sedimentation rate allows margins to move inwards. (**d**) Sea level rises, but landward obstruction and/or lack in sedimentation rate prevent migration of mangroves resulting in eroding margins and eventual loss. (Reproduced from Gilman et al. 2006 with permission from Eric Gilman)

ence biochemical processes of seagrasses, will be altered due to changes in the atmosphere. Increased precipitation and sea level rise will result in changes in nutrient fluxes, sedimentation and salinity (Lee et al. 2007). Smothering of seagrasses due to sedimentation (potentially due to the loss of neighboring mangrove systems) and elevated sea levels, further limits sunlight to the meadows and consequently decreases their productivity. Furthermore, climate change may also affect the frequency and strength of tropical storms, which carry their own set of consequences for seagrass meadows (Trenberth 2005), including sediment movement burying the seagrasses and increased turbulence caused by strong storms, that can last for long after the storm has passed could uproot and completely flatten the meadow (Bjork et al. 2008).

Coral reefs act as natural barriers protecting the coastline from currents, waves and storms by dissipating their force and hence overall destructive impact (Moberg and Folke 1999). Calcite skeletons allow coral reefs to be robust structures that can withstand strong currents. However, studies have shown that increasing temperature, along with ocean acidification, are causing changes in skeletal growth and robustness (Tambutté et al. 2015). To understand how, one must first look at the physiology of corals.

Symbiodinium (i.e., dinoflagellates) residing in coral tissue provide the majority of the coral hosts' energy demands, allowing them to successfully thrive in oligotrophic waters (Yellowlees et al. 2008; Baker et al. 2015). The distribution of coral reefs is heavily limited by sea temperature, where the sensitivity to temperature stress depends on the physiological tolerance limits of both endosymbiotic partners (Putnam et al. 2012). When temperatures exceed tolerance limits, the most common response is the expulsion of symbionts from the tissue of the coral hosts, an event known as coral bleaching. The response to the loss of symbionts and their photosynthetic by-products, the largest energy source of corals, is a decline in coral productivity and skeletal growth (Langdon and Atkinson 2005; Pandolfi et al. 2011). Without *Symbiodinium,* corals can only survive for a limited time before the onset of tissue necrosis and ultimately their death. Hope for the survival of the reef exists in two forms; some species of symbionts are more thermotolerant than others, and if temperatures return to base-level the coral can recover by re-establishing symbiosis with *Symbiodinium*.

Some species of corals with different symbiont associations have managed to inhabit areas with extreme temperatures, highlighting their temperature resilient capacity (van Oppen et al. 2015). These symbionts have been accredited with a particularly important role in the overall thermal tolerance of corals, where temperature adapted symbiont clades can reduce the overall bleaching response (Berkelmans and van Oppen 2006; Sampayo et al. 2008; Howells et al. 2016). In recent years, advances in technology have made it possible to uncover some of the underlying mechanisms allowing for thermotolerance of corals and their symbionts, which potentially play important roles in their adaptation and acclimatization potential (Richier et al. 2008; DeSalvo et al. 2010; Kvitt et al. 2011; Bellantuono et al. 2012; Kenkel et al. 2013). By understanding these mechanisms it is hoped that we can aid corals in future response and survival, in light of global warming (Magris et al. 2015; van Oppen et al. 2015). Thus, it can be seen that temperature drastically impacts the survival of corals and if temperatures continue to increase as predicted, corals will struggle to recover from bleaching events and subsequently large-scale losses of coral reefs will be observed.

Whilst the direct effects of elevated temperatures on corals are frightening enough, they extend, as with seagrass and mangroves, into related atmospheric changes. Increased precipitation and sea level rise have two primary impacts, a decrease in salinity and alteration of nutrient fluxes. In the case of salinity, a decrease may actually directly influence thermotolerance of corals, whereas, recent studies have indicated that more saline environments could increase coral holobiont temperature resilience (Gegner et al. 2017). Whilst not directly impacting thermotolerance, eutrophication of the water column (as the result of declines in seagrass and mangroves areas) can cause imbalances in coral-symbiont relationships and ultimately leads to the breakdown of symbiosis (D'Angelo and Wiedenmann 2014). If the symbiotic relationship is already compromised by imbalances in nutrient exchange between the two partners, it is less likely to withstand additional temperature stress. Consequently, if corals undergo repeated long-term bleaching, calcification rates will be continuously affected. Since calcification is a costly process that requires a lot of energy, the lack of sufficient nutrients, due to symbiont absence, not only reduces growth but also increases the porosity of the skeleton (Tambutté et al. 2015). As storm frequency and intensity increase, coral skeletons will be less resilient to withstand turbulent waters, potentially leading to fractures and breaks, effectively destroying the reef. Thus, the existence of coral reefs is currently threatened not only by temperature increase, but also by most associated atmospheric changes which accompany global warming.

We can clearly see that the existence of mangroves, seagrass and coral reefs is currently threatened not only by temperature increase, but also by many associated changes in abiotic factors. However, the extent to which certain changes affect the ecosystem depends wholly on the system in question, as some are more resilient to certain abiotic stressors than others (Guannel et al. 2016). Sea level rise has shown contrasting impacts on mangrove, seagrasses and coral reefs, whereby seagrass showed the most resilience over longer periods of time (Albert et al. 2017). On the other hand, temperature increase of the water column is having the most detrimental effects on corals whose algal partners can escape suboptimal conditions, whilst the animal and its skeleton are left behind. Although mangroves, seagrass and coral reefs may respond differently to temperature and associated changes, the end point appears to be a decline in all three either through migration to new locations or permanent loss. As each ecosystem provides a service to help mitigate global warming impacts, the slow disappearance of one could increase the stress experienced by its neighbors. This is particularly important in terms of increased storm intensities and frequencies which have the potential to significantly impact sedimentation and nutrient enrichment, especially in regions where losses of ecosystems and their associated buffering, trapping, and absorbing capacities have occurred (Golbuu et al. 2003; Unsworth et al. 2012). Another global stressor that is tightly linked to temperature is ocean acidification. If temperature itself will have such extensive effects on tropical marine ecosystems, then the combination with ocean acidification will be unfathomable.

A Sour Ocean

Within the literature, ocean acidification has been closely coupled with the rise in atmospheric and oceanic temperatures, with all three being attributed primarily to rising levels of CO_2 in the Earth's atmosphere (Caldeira and Wickett 2003; Hoegh-Guldberg and Bruno 2010). CO_2 is not only a key player in climate change due to its ability to trap heat, but also a vital component of biological mechanisms (e.g., photosynthesis), which are important in sustaining life. The oceans play an important role in the global carbon cycle, acting as a 'carbon sink' by taking up about one third of CO_2 from the atmosphere and transporting it around the globe (Le Quéré et al. 2013). The acidification of our oceans alters ocean chemistry, which poses significant challenges to these already threatened tropical marine ecosystems (Kroeker et al. 2010; Gaylord et al. 2015).

Since the industrial revolution our, on average, slightly alkaline ocean, with a pH of 8.2, has experienced a decrease in pH of 0.1 units, which represents a 30% increase in acidity. The pH is predicted to drop a further 0.3 units by the end of the century (IPCC 2013). The trajectory towards an ocean with a lower pH will have both positive and negative consequences for marine organisms (Garrard et al. 2014). This was demonstrated by research conducted near oceanic vents, which emit large quantities of CO_2 and consequently create areas of seabed with a lower than usual pH (Frankignoulle and Distèche 1984; Hall-Spencer et al. 2008; Fabricius et al. 2011; Scartazza et al. 2017). However, research on ocean acidification primarily focuses on the impacts to calcifying organisms such as corals, molluscs, echinoderms, crustaceans, coccolithophores, foraminifera, pteropods, and some species of algae. Increasing atmospheric CO_2 alters the dissolved inorganic carbon distribution in seawater, reducing its pH and with it the availability of carbonate ions (CO_3^{2-}) (Cohen and Holcomb 2009; Findlay et al. 2010). This impacts the energy-costly process of calcification as the rate at which calcifying organisms produce calcium carbonate ($CaCO_3$), is slowed to a point where rates of erosion exceed those of skeletal accretion (Cohen and Holcomb 2009; Gerber et al. 2014). In terms of non-calcifying species, acidification is believed to disturb acid–based (metabolic) physiology and impact their survival, growth, and reproduction (Fabry et al. 2008; Kroeker et al. 2010). Research into the response of non-calcifying organisms including jellyfish, fish, fleshy algae, and seagrasses to acidification is becoming more commonplace, yet, in the case of mangroves, the impacts remain vastly understudied (Guinotte and Fabry 2008; Kroeker et al. 2010).

Calcifying species appear to be the 'losers' in the case of a more acidic ocean, and exhibit a range of negative responses, especially when acidification is combined with other stressors (Hoegh-Guldberg et al. 2007; Fabricius et al. 2011). For coral calcification, studies have indicated that the extracellular calcifying medium is maintained at a higher pH than that of the surrounding seawater in order to facilitate $CaCO_3$ precipitation (McCulloch et al. 2012). However, how changes in seawater pH would affect this internal biological control is currently unknown. More recently, studies have revealed that instead of decreasing their growth rate, corals are acclimatizing by decreasing their skeletal density and increasing their porosity (Tambutté et al. 2015). Although this morphological plasticity ensures slower, but continuous growth of the colony, it weakens the overall reef structure, making it more susceptible to damage resulting from anthropogenic or climatic perturbations (Hoegh-Guldberg et al. 2007). This weakening of skeletons also affects reef-building gastropods of the family Vermetidae, which provide coastal protection to neighboring ecosystems such as mangroves and seagrasses (Milazzo et al. 2014). The decreased reef resilience can be attributed to reduced structural complexity and coral species diversity (Anthony et al. 2011; Fabricius et al. 2011). A pH of 7.7 has been shown to cease reef development due to a shift in coral species dominance, away from structural corals (branching, foliose, and tabulate growth forms) towards massive growth forms such as *Porites* corals (Fabricius et al. 2011). These reductions in reef complexity can in turn impact the biodiversity of reef-associated species as well as trophic interactions, and other ecosystem processes (Raven et al. 2005; Kleypas et al. 2006).

For non-calcifying marine consumers, elevated oceanic CO_2 and the accompanying change in pH will have negative effects as it will require additional energy for metabolic acid–based regulation (Pörtner 2008). Ocean acidification slows larval development in fishes, molluscs and echinoderms (Kurihara 2008; Miller et al. 2009; Dupont et al. 2010; Talmage and Gobler 2010; Dineshram et al. 2013; Gazeau et al. 2013). The early life stages of fish are impacted by a reduction of their oxygen consumption capacity and hence their activity, along with olfactory cues for predation, settlement, and reproduction (Munday et al. 2009). However, these highly mobile organisms have developed intra- or extracellular pH regulatory mechanisms that may aid them to be more resilient to ocean acidification (Kroeker et al. 2010). An additional option for fishes to escape a more acidic environment is finding refuge in seagrass meadows (Hendriks et al. 2014), and potentially mangrove roots (Chakraborty et al. 2013).

Primary producers including seagrasses and macroalgae appear to be the winners in the face of elevated oceanic CO_2 concentrations and lower seawater pH. Increased CO_2 concentrations in seawater are a resource for these primary producers (Palacios and Zimmerman 2007; Fabricius et al. 2011; Hepburn et al. 2011) which allow them to enhance their productivity and growth (Harley et al. 2012; Koch et al. 2013). Seagrasses are known to alter the carbonate chemistry in the water column, which is of particular importance in regions where they neighbor coral reef environments (Dorenbosch et al. 2005; Hendriks et al. 2014). In the Mediterranean, *Posidonia oceanica* diurnally modify the water column pH by as much as 0.2–0.7 units through photosynthesis and respiration (Frankignoulle and Distèche 1984; Invers et al. 1997; Hall-Spencer et al. 2008; Scartazza et al. 2017), and a similar process is also exhibited by macroalgae dominated reef-tops (Russell et al. 2009). Additionally, ocean acidification results in decreased carbon to nitrogen (C:N) ratios in *P. oceanica* tissues, which increases shoot density, leaf proteins, and asparagine accumulation in the rhizomes (Scartazza et al. 2017). This in turn provides a positive contribution to associated food-webs given the nutritional quality of organic matter available for herbivores and consequently an increase in the grazing rate is observed (Kroeker et al. 2010; Arnold et al. 2012; Rossoll et al. 2012; Scartazza et al. 2017). However, the spatial scale of these processes, ranging from millimeters to entire water layers, must be kept in mind when extrapolating their impacts to an ecosystem extent (Hendriks et al. 2015).

This enhanced productivity of seagrass meadows is likely to contribute to enhanced productivity in neighboring coral reef ecosystems on the tropical seascape. Modelling studies suggests that calcification on coral reefs with seagrass neighbors may be up to 18% greater compared to reefs without neighboring seagrasses (Unsworth et al. 2012). Their role in enhancing calcification rates will continue and possibly even increase (Zimmerman et al. 1997), allowing coral and invertebrate communities to persist (Unsworth et al. 2012; Garrard et al. 2014). The term connectivity is primarily used in the context of ocean acidification to discuss the disruption to organismal reproduction, dispersal and hence, the connectivity of populations in a more acidic ocean (Cowen et al. 2006; Gerber et al. 2014). Ocean acidification appears to exhibit an especially strong capacity to drive ecological change and hence its impacts are not straight forward in the bigger picture (Gaylord et al. 2015). The coupled responses create a complex interplay among the physiological susceptibility of organisms to ocean acidification, the availability of resources, and the intensity of competition (Gaylord et al. 2015). Models suggest that a decreasing ocean pH will impose additional physiological stresses to the global distribution of species, narrowing the breadth of the thermal performance curve (Pörtner 2008). Ocean acidification effects would lead to smaller overall ranges, and ranges for which equatorward boundaries shift more dramatically towards poleward ones (Gaylord et al. 2015). How species will respond within the context of their communities is yet to be investigated. However, it is almost certain that many of the most striking consequences of acidification will arise through altered biotic interactions (Fabricius et al. 2011; Falkenberg et al. 2013; Kroeker et al. 2013; McCormick et al. 2013).

In summary, primary producers like seagrass beds are a crucial buffer zone of potential stressors for the calcifying fauna of coral reefs, with which interactions seem to be key for the resilience of many different species and even ecosystems in the face of environmental perturbations. With this in mind a more interconnected approach needs to be taken into consideration for tropical ecology under ocean acidification (Fabry et al. 2008; Garrard et al. 2014). Similar to Gaylord et al. (2015) and their argumentation that ocean acidification needs to be seen not only in the individual but ecosystem context, we argue that ecosystems need to be investigated in a connected manner. It is unequivocal that this issue requires global human action (Kennedy et al. 2013).

Summary

The evolution of mangroves, seagrasses, and coral reefs in ever-changing environments has allowed them to form highly-adapted, and for the most part, resilient ecosystems. This resilience, however, is often facilitated by their connectivity to adjacent ecosystems. But within one generation, anthropogenic activities and human-induced climate change have exerted such pressures on these connectivity pathways that a decline in ecosystem resilience and services has been observed. Consequently, places on Earth previously considered refugia for a range of species may cease to exist. Perhaps one of the most significant examples of these combined stressors on tropical marine ecosystems occurred in the Red Sea, where in the 1960s 98% of its coasts were considered to be "*in practically virgin condition*" (Ormond 1987). However, rapid development in this area, as a result of expansion in petroleum-based economies, meant the 'virgin' status of many regions was lost (Gladstone 2008). Over 75% of mangrove forests were degraded by activities including felling, cutting, sewage, root burial or overgrazing by camels (Gladstone 2008), and coral reefs, especially those by industrializing areas were impacted by dredging, sewage, and tourism (Gladstone 2008). Further to these threats, industrial trawling depleted economically important species (Gladstone 2008).

The underlying cause of many of these activities, both in the Red Sea and around the globe are: expanding populations, rapid urbanization, and weak governance, coupled with a lack of baseline information on tropical marine ecosystems, limited awareness of the consequences of human activities, and, most importantly, a lack of perspective regarding the connectivity among ecosystems on the tropical seascape (Duda and Sherman 2002). This review has highlighted the scale and importance of connectivity between tropical marine ecosystems, and investigated the impact of select anthropogenic activities and climatic perturbations on their associated connectivity pathways and ecosystem services. Only by progressing our understanding of these environments can the impact of human activities and changes in environmental conditions on nature be better elucidated. It is concluded that in order to effectively protect and preserve these critically important ecosystems and their associated services for future generations, we can no longer consider each ecosystem as a separate entity, and instead a holistic, seascape-wide approach is paramount. This means, the static, 'boundary-based' norm of scientific thinking must be overcome, and instead a more flexible, inter-ecosystem and interdisciplinary approach employed, which may, in turn, lead to strategies, which balance environmental change whilst allowing human subsistence to be ensured.

Acknowledgements The authors are grateful to constructive input by Mirco Wölfelschneider, Timothy Thomson and Dr. Siobhan Vye. Thanks also to Dr. Simon Jungblut for his continuous support and to two reviewers who contributed significantly to the final version of this chapter.

Appendix

This article is related to the YOUMARES 8 conference session no. 14: "Tropical Aquatic Ecosystems Across Time, Space and Disciplines". The original Call for Abstracts and the abstracts of the presentations within this session can be found in the appendix "Conference Sessions and Abstracts", chapter "8 Tropical Aquatic Ecosystems Across Time, Space and Disciplines", of this book.

References

Aburto-Oropeza O, Ezcurra E, Danemann G et al (2008) Mangroves in the Gulf of California increase fishery yields. Proc Natl Acad Sci USA 105:10456–10459. https://doi.org/10.1073/pnas.0804601105

Adam TC, Burkepile DE, Ruttenberg BI et al (2015) Herbivory and the resilience of Caribbean coral reefs: knowledge gaps and implications for management. Mar Ecol Prog Ser 520:1–20. https://doi.org/10.3354/meps11170

Albert S, Saunders MI, Roelfsema CM et al (2017) Winners and losers as mangrove, coral and seagrass ecosystems respond to sea-level rise in Solomon Islands. Environ Res Lett 12(9):094009. https://doi.org/10.1088/1748-9326/aa7e68

Alongi DM (2002) Present state and future of the world's mangrove forests. Environ Conserv 29:331–349. https://doi.org/10.1017/S0376892902000231

Alongi DM (2008) Mangrove forests: resilience, protection from tsunamis, and responses to global climate change. Estuar Coast Shelf Sci 76:1–13

Alongi DM (2009) The energetics of mangrove forests. Springer Science/Business Media BV, New York

Alongi DM (2015) The impact of climate change on mangrove forests. Curr Clim Chang Rep 1:30–39. https://doi.org/10.1007/s40641-015-0002-x

Anthony KRN, Maynard JA, Diaz-Pulido G et al (2011) Ocean acidification and warming will lower coral reef resilience. Glob Chang Biol 17:1798–1808. https://doi.org/10.1111/j.1365-2486.2010.02364.x

Arnold TM, Mealey C, Leahey H et al (2012) Ocean acidification and the loss of phenolic substances in marine plants. PLoS ONE 7:e35107. https://doi.org/10.1371/journal.pone.0035107

Arunparasath A, Gomathinayagam M (2015) Reproductive phenology of true mangrove species in Pichavaram mangrove forests, Tamilnadu, India – a comparative account. J Environ Treat Tech 15:17–21

Badola R, Hussain SA (2005) Valuing ecosystem functions: an empirical study on the storm protection function of Bhitarkanika mangrove ecosystem, India. Environ Conserv 32:85–92. https://doi.org/10.1017/S0376892905001967

Baker DM, Freeman CJ, Knowlton N et al (2015) Productivity links morphology, symbiont specificity and bleaching in the evolution of Caribbean octocoral symbioses. ISME J 9:2620–2629. https://doi.org/10.1038/ismej.2015.71

Barbier EB, Hacker SD, Kennedy C et al (2011) The value of estuarine and coastal ecosystem services. Ecol Monogr 81:169–193. https://doi.org/10.1890/10-1510.1

Bellantuono AJ, Granados-Cifuentes C, Miller DJ et al (2012) Coral thermal tolerance: tuning gene expression to resist thermal stress. PLoS ONE 7:e50685. https://doi.org/10.1371/journal.pone.0050685

Bellwood DR, Hughes TP, Folke C et al (2004) Confronting the coral reef crisis. Nature 429:827–833. https://doi.org/10.1038/nature02691

Berkelmans R, van Oppen MJ (2006) The role of zooxanthellae in the thermal tolerance of corals: a "nugget of hope" for coral reefs in an era of climate change. Proc R Soc B 273:2305–2312. https://doi.org/10.1098/rspb.2006.3567

Berkström C, Gullström M, Lindborg R et al (2012) Exploring "knowns" and "unknowns" in tropical seascape connectivity with insights from East African coral reefs. Estuar Coast Shelf Sci 107:1–21. https://doi.org/10.1016/j.ecss.2012.03.020

Bjork M, Short F, Mcleod E et al (2008) Managing seagrasses for resilience to climate change. IUCN, Gland. https://doi.org/10.1017/CBO9781107415324.004

Borum J, Pedersen O, Greve TM et al (2005) The potential role of plant oxygen and sulphide dynamics in die-off events of the tropical seagrass, *Thalassia testudinum*. J Ecol 93:148–158. https://doi.org/10.1111/j.1365-2745.2004.00943.x

Bouillon S, Borges AV, Castañeda-Moya E et al (2008) Mangrove production and carbon sinks: a revision of global budget estimates. Glob Biogeochem Cycles 22:GB2013. https://doi.org/10.1029/2007GB003052

Bozec Y-M, O'Farrell S, Bruggemann JH et al (2016) Tradeoffs between fisheries harvest and the resilience of coral reefs. Proc Natl Acad Sci USA 113:4536–4541. https://doi.org/10.1073/pnas.1601529113

Burke L, Reytar K, Spalding MD et al (2011) Reefs at risk revisited. World Resources Institute, Washington, DC

Caldeira K, Wickett ME (2003) Oceanography: anthropogenic carbon and ocean pH. Nature 425:365. https://doi.org/10.1038/425365a

Campbell PKE, Middleton EM, McMurtrey JE et al (2007) Assessment of vegetation stress using reflectance or fluorescence measurements. J Environ Qual 36:832–845. https://doi.org/10.2134/jeq2005.0396

Chakraborty S, Zaman S, Pramanick P et al (2013) Acidification of Sundarbans mangrove estuarine system. Discov Nat 6:14–20

Christensen B (1978) Biomass and primary production of *Rhizophora apiculata* Bl. in a mangrove in southern Thailand. Aquat Bot 4:43–52. https://doi.org/10.1016/0304-3770(78)90005-0

Cocheret de la Morinière E, BJA P, Nagelkerken I et al (2002) Post-settlement life cycle migration patterns and habitat preference of coral reef fish that use seagrass and mangrove habitats as nurseries. Estuar Coast Shelf Sci 55:309–321. https://doi.org/10.1006/ecss.2001.0907

Cohen A, Holcomb M (2009) Why corals care about ocean acidification: uncovering the mechanism. Oceanography 22:118–127. https://doi.org/10.5670/oceanog.2009.102

Connell JH (1997) Disturbance and recovery of coral assemblages. Coral Reefs 16:S101–S113. https://doi.org/10.1007/s003380050246

Costanza R, Arge R, de Groot R et al (1997) The value of the world's ecosystem services and natural capital. Nature 387:253–260. https://doi.org/10.1038/387253a0

Costanza R, de Groot R, Sutton P et al (2014) Changes in the global value of ecosystem services. Glob Environ Chang 26:152–158. https://doi.org/10.1016/j.gloenvcha.2014.04.002

Cotner JB, Suplee MW, Chen NW et al (2004) Nutrient, sulfur and carbon dynamics in a hypersaline lagoon. Estuar Coast Shelf Sci 59:639–652. https://doi.org/10.1016/j.ecss.2003.11.008

Cowen RK, Paris CB, Srinivasan A (2006) Scaling of connectivity in marine populations. Science 311:522–527. https://doi.org/10.1126/science.1122039

D'Angelo C, Wiedenmann J (2014) Impacts of nutrient enrichment on coral reefs: new perspectives and implications for coastal management and reef survival. Curr Opin Environ Sustain 7:82–93. https://doi.org/10.1016/j.cosust.2013.11.029

Dawson SP, Dennison WC (1996) Effects of ultraviolet and photosynthetically active radiation on five seagrass species. Mar Biol 125:629–638. https://doi.org/10.1007/BF00349244

De'ath G, Fabricius KE, Sweatman H et al (2012) The 27-year decline of coral cover on the Great Barrier Reef and its causes. Proc Natl Acad Sci USA 109:17995–17999. https://doi.org/10.1073/pnas.1208909109

Deegan LA (1993) Nutrient and energy transport between estuaries and coastal marine ecosystems by fish migration. Can J Fish Aquat Sci 50:74–79. https://doi.org/10.1139/f93-009

DeSalvo MK, Sunagawa S, Fisher PL et al (2010) Coral host transcriptomic states are correlated with Symbiodinium genotypes. Mol Ecol 19:1174–1186. https://doi.org/10.1111/j.1365-294X.2010.04534.x

Díaz-Almela E, Marbà N, Álvarez E et al (2008) Benthic input rates predict seagrass (*Posidonia oceanica*) fish farm-induced decline. Mar Pollut Bull 56:1332–1342. https://doi.org/10.1016/j.marpolbul.2008.03.022

Dineshram R, Thiyagarajan V, Lane A et al (2013) Elevated CO_2 alters larval proteome and its phosphorylation status in the commercial oyster, *Crassostrea hongkongensis*. Mar Biol 160:2189–2205. https://doi.org/10.1007/s00227-013-2176-x

Dittmar T, Hertkorn N, Kattner G et al (2006) Mangroves, a major source of dissolved organic carbon to the oceans. Glob Biogeochem Cycles 20:GB1012. https://doi.org/10.1029/2005GB002570

Donato DC, Kauffman JB, Murdiyarso D et al (2011) Mangroves among the most carbon-rich forests in the tropics. Nat Geosci 4:293–297. https://doi.org/10.1038/ngeo1123

Dorenbosch M, Grol MGG, Christianen MJA et al (2005) Indo-Pacific seagrass beds and mangroves contribute to fish density and diversity on adjacent coral reefs. Mar Ecol Prog Ser 302:63–76. https://doi.org/10.3354/meps302063

Dorenbosch M, Grol MGG, Nagelkerken I et al (2006) Seagrass beds and mangroves as potential nurseries for the threatened Indo-Pacific humphead wrasse, *Cheilinus undulatus* and Caribbean rainbow parrotfish, *Scarus guacamaia*. Biol Conserv 129:277–282. https://doi.org/10.1016/j.biocon.2005.10.032

Douglas AE (2001) The productivity of corals. Encyclopedia of life support systems 2. Developed under the Auspices of the UNESCO, Eolss Publishers, Paris, France, [http://www.eolss.net]

Duarte CM, Krause-Jensen D (2017) Export from seagrass meadows contributes to marine carbon sequestration. Front Mar Sci 4:13. https://doi.org/10.3389/fmars.2017.00013

Duarte CM, Borum J, Short FT et al (2008) Seagrass ecosystem: their global status and prospects. In: NVC P (ed) Aquatic ecosystems: trend and global prospects. Cambridge University Press, Cambridge, pp 281–294

Duda AM, Sherman K (2002) A new imperative for improving management of large marine ecosystems. Ocean Coast Manag 45:797–833. https://doi.org/10.1016/S0964-5691(02)00107-2

Duke NC, Meynecke J-O, Dittmann S et al (2007) A world without mangroves? Science 317:41b–42b. https://doi.org/10.1126/science.317.5834.41b

Dupont S, Ortega-Martínez O, Thorndyke M (2010) Impact of near-future ocean acidification on echinoderms. Ecotoxicology 19:449–462

Durako MJ, Kunzelman JI (2002) Photosynthetic characteristics of *Thalassia testudinum* measured in situ by pulse-amplitude modulated (PAM) fluorometry: methodological and scale-based considerations. Aquat Bot 73:173–185. https://doi.org/10.1016/S0304-3770(02)00020-7

Eggertsen L, Ferreira CEL, Fontoura L et al (2017) Seaweed beds support more juvenile reef fish than seagrass beds in a south-western Atlantic tropical seascape. Estuar Coast Shelf Sci 196:97–108. https://doi.org/10.1016/j.ecss.2017.06.041

Ellison J (1994) Climate change and sea level rise impacts on mangrove ecosystems. In: Pernetta J, Leemans R, Elder D et al (eds) Impacts of climate change on ecosystems and species. IUCN, Gland, pp 11–30

Ewel KC, Twilley RR, Ong JE (1998) Different kinds of mangrove forests provide different goods and services. Glob Ecol Biogeogr Lett 7:83–94. https://doi.org/10.2307/2997700

Fabricius KE (2005) Effects of terrestrial runoff on the ecology of corals and coral reefs: review and synthesis. Mar Pollut Bull 50:125–146. https://doi.org/10.1016/j.marpolbul.2004.11.028

Fabricius KE, Okaji K, De'ath G (2010) Three lines of evidence to link outbreaks of the crown-of-thorns seastar *Acanthaster planci* to the release of larval food limitation. Coral Reefs 29:593–605. https://doi.org/10.1007/s00338-010-0628-z

Fabricius KE, Langdon C, Uthicke S et al (2011) Losers and winners in coral reefs acclimatized to elevated carbon dioxide concentrations. Nat Clim Chang 1:165–169. https://doi.org/10.1038/nclimate1122

Fabry VJ, Seibel BA, Feely RA et al (2008) Impacts of ocean acidification on marine fauna and ecosystem processes. ICES J Mar Sci 65:414–432

Falkenberg LJ, Russell BD, Connell SD (2013) Future herbivory: the indirect effects of enriched CO_2 may rival its direct effects. Mar Ecol Prog Ser 492:85–95. https://doi.org/10.3354/meps10491

FAO (2016a) World fertilizer trends and outlook to 2019. FAO, Rome

FAO (2016b) The state of world fisheries and aquaculture. Contributing to food security and nutrition for all. FAO, Rome

Findlay HS, Kendall MA, Spicer JI et al (2010) Post-larval development of two intertidal barnacles at elevated CO_2 and temperature. Mar Biol 157:725–735. https://doi.org/10.1007/s00227-009-1356-1

Flindt MR, Pardal MÂ, Lillebø AI et al (1999) Nutrient cycling and plant dynamics in estuaries : a brief review. Acta Oecol 20:237–248

Frankignoulle M, Distèche A (1984) CO_2 chemistry in the water column above a *Posidonia* seagrass bed and related air-sea exchanges. Oceanol Acta 7:209–219

Franklin HB (2007) The most important fish in the sea: menhaden and America. Island Press, Washington, DC

Friess DA, Krauss KW, Horstman EM et al (2012) Are all intertidal wetlands naturally created equal? Bottlenecks, thresholds and knowledge gaps to mangrove and saltmarsh ecosystems. Biol Rev 87:346–366. https://doi.org/10.1111/j.1469-185X.2011.00198.x

Garrard SL, Gambi MC, Scipione MB et al (2014) Indirect effects may buffer negative responses of seagrass invertebrate communities to ocean acidification. J Exp Mar Biol Ecol 461:31–38. https://doi.org/10.1016/j.jembe.2014.07.011

Gaylord B, Kroeker KJ, Sunday JM et al (2015) Ocean acidification through the lens of ecological theory. Ecology 96:3–15. https://doi.org/10.1890/14-0802.1

Gazeau F, Parker LM, Comeau S et al (2013) Impacts of ocean acidification on marine shelled molluscs. Mar Biol 160:2207–2245. https://doi.org/10.1007/s00227-013-2219-3

Gegner HM, Ziegler M, Rädecker N et al (2017) High salinity conveys thermotolerance in the coral model *Aiptasia*. Biol Open 6:1943–1948. https://doi.org/10.1242/bio.028878

Gerber LR, Mancha-Cisneros MDM, O'Connor M et al (2014) Climate change impacts on connectivity in the ocean: implications for conservation. Ecosphere 5:1–18. https://doi.org/10.1890/ES13-00336.1

Gillis LG, Bouma TJ, Jones CG et al (2014) Potential for landscape-scale positive interactions among tropical marine ecosystems. Mar Ecol Prog Ser 503:289–303. https://doi.org/10.3354/meps10716

Gilman E, van Lavieren J, Ellison V (2006) Pacific Island mangroves in a changing climate and rising sea. United Nations Environment Programme, Regional Seas Programme, Nairobi

Gilman EL, Ellison J, Duke NC et al (2008) Threats to mangroves from climate change and adaptation options: a review. Aquat Bot 89:237–250

Gladstone W (2008) Towards conservation of a globally significant ecosystem: the Red Sea and Gulf of Aden. Aquat Conserv Mar Freshwat Ecosyst 18:1–5

Gnanadesikan A, Stouffer RJ (2006) Diagnosing atmosphere-ocean general circulation model errors relevant to the terrestrial biosphere using the Köppen climate classification. Geophys Res Lett 33:L22701. https://doi.org/10.1029/2006GL028098

Godoy MDP, De Lacerda LD (2015) Mangroves response to climate change: a review of recent findings on mangrove extension and distribution. An Acad Bras Cienc Ann Braz Acad Sci 87:651–667. https://doi.org/10.1590/0001-3765201520150055

Golbuu Y, Victor S, Wolanski E et al (2003) Trapping of fine sediment in a semi-enclosed bay, Palau, Micronesia. Estuar Coast Shelf Sci 57:941–949. https://doi.org/10.1016/S0272-7714(02)00424-9

Granek EF, Compton JE, Phillips DL (2009) Mangrove-exported nutrient incorporation by sessile coral reef invertebrates. Ecosystems 12:462–472. https://doi.org/10.1007/s10021-009-9235-7

Green EP, Short FT (2003) World atlas of seagrasses. Prepared by the UNEP World Conservation Monitoring Centre. University of California Press, Berkeley

Grober-Dunsmore R, Pittman SJ, Caldow C et al (2009) A landscape ecology approach for the study of ecological connectivity across tropical marine seascapes. In: Nagelkerken I (ed) Ecological connectivity among tropical coastal ecosystems. Springer, Dordrecht, pp 493–530

Guannel G, Arkema K, Ruggiero P et al (2016) The power of three: coral reefs, seagrasses and mangroves protect coastal regions and increase their resilience. PLoS ONE 11:e0158094. https://doi.org/10.1371/journal.pone.0158094

Guinotte JM, Fabry VJ (2008) Ocean acidification and its potential effects on marine ecosystems. Ann NY Acad Sci 1134:320–342

Hall-Spencer JM, Rodolfo-Metalpa R, Martin S et al (2008) Volcanic carbon dioxide vents show ecosystem effects of ocean acidification. Nature 454:96–99. https://doi.org/10.1038/nature07051

Halpern BS, Walbridge S, Selkoe KA et al (2008) A global map of human impact on marine ecosystems. Science 319:948–952. https://doi.org/10.1126/science.1149345

Hamilton LS, Snedaker SC (1984) Handbook for mangrove area management. East-West Environment and Policy Institute, Honolulu

Harley CDG, Anderson KM, Demes KW et al (2012) Effects of climate change on global seaweed communities. J Phycol 48:1064–1078. https://doi.org/10.1111/j.1529-8817.2012.01224.x

Heck KL, Carruthers TJB, Duarte CM et al (2008) Trophic transfers from seagrass meadows subsidize diverse marine and terrestrial consumers. Ecosystems 11:1198–1210. https://doi.org/10.1007/s10021-008-9155-y

Heenan A, Williams ID (2013) Monitoring herbivorous fishes as indicators of coral reef resilience in American Samoa. PLoS ONE 8:e79604. https://doi.org/10.1371/journal.pone.0079604

Henderson CJ, Gilby BL, Lee SY et al (2017) Contrasting effects of habitat complexity and connectivity on biodiversity in seagrass meadows. Mar Biol 164:117. https://doi.org/10.1007/s00227-017-3149-2

Hendriks IE, Olsen YS, Ramajo L et al (2014) Photosynthetic activity buffers ocean acidification in seagrass meadows. Biogeosciences 11:333–346. https://doi.org/10.5194/bg-11-333-2014

Hendriks IE, Duarte CM, Olsen YS et al (2015) Biological mechanisms supporting adaptation to ocean acidification in coastal ecosystems. Estuar Coast Shelf Sci 152:A1–A8. https://doi.org/10.1016/j.ecss.2014.07.019

Hepburn CD, Pritchard DW, Cornwall CE et al (2011) Diversity of carbon use strategies in a kelp forest community: implications for a high CO_2 ocean. Glob Chang Biol 17:2488–2497. https://doi.org/10.1111/j.1365-2486.2011.02411.x

Herbeck LS, Sollich M, Unger D et al (2014) Impact of pond aquaculture effluents on seagrass performance in NE Hainan, tropical China. Mar Pollut Bull 85:190–203. https://doi.org/10.1016/j.marpolbul.2014.05.050

Hoegh-Guldberg O (2014) Tropical marine ecosystems. In: Lindenmayer D, Morton S, Dovers S (eds) Ten commitments revisited: securing Australia's future environment. CSIRO Publishing, Collingwood, pp 59–70

Hoegh-Guldberg O, Bruno JF (2010) The impact of climate change on the world's marine ecosystems. Science 328:1523–1528. https://doi.org/10.1126/science.1189930

Hoegh-Guldberg O, Mumby PJ, Hooten AJ et al (2007) Coral reefs under rapid climate change and ocean acidification. Science 318:1737–1742

Howells EJ, Abrego D, Meyer E et al (2016) Host adaptation and unexpected symbiont partners enable reef-building corals to tolerate extreme temperatures. Glob Chang Biol 22:2702–2714. https://doi.org/10.1111/gcb.13250

Hughes TP (1994) Catastrophes, phase shifts, and large-scale degradation of a Caribbean coral reef. Science 265:1547–1551. https://doi.org/10.1126/science.265.5178.1547

Invers O, Romero J, Pérez M (1997) Effects of pH on seagrass photosynthesis: a laboratory and field assessment. Aquat Bot 59:185–194. https://doi.org/10.1016/S0304-3770(97)00072-7

IPCC (2013) Climate change 2013 – the physical science basis. Contribution of working group I to the fifth assessment report of the intergovernmental panel on climate change. Cambridge University Press, Cambridge, New York. https://doi.org/10.1017/CBO9781107415324

Jackson JBC, Kirby MX, Berger WH et al (2001) Historical overfishing and the recent collapse of coastal ecosystems. Science 293:629–637

Jennerjahn TC (2012) Biogeochemical response of tropical coastal systems to present and past environmental change. Earth Sci Rev 114:19–41

Jennerjahn TC, Ittekkot V (2002) Relevance of mangroves for the production and deposition of organic matter along tropical continental margins. Naturwissenschaften 89:23–30

Kathiresan K (2008) Threats to mangroves – degradation and destruction of mangroves. Centre of Advanced Study in Marine Biology. Annamalai University

Kathiresan K, Bingham BL (2001) Biology of mangroves and mangrove ecosystems. Adv Mar Biol 40:81–251. https://doi.org/10.1016/S0065-2881(01)40003-4

Kathiresan K, Rajendran N (2005) Coastal mangrove forests mitigated tsunami. Estuar Coast Shelf Sci 65:601–606. https://doi.org/10.1016/j.ecss.2005.06.022

Kenkel CD, Meyer E, Matz MV (2013) Gene expression under chronic heat stress in populations of the mustard hill coral (*Porites astreoides*) from different thermal environments. Mol Ecol 22:4322–4334. https://doi.org/10.1111/mec.12390

Kennedy EV, Perry CT, Halloran PR et al (2013) Avoiding coral reef functional collapse requires local and global action. Curr Biol 23:912–918. https://doi.org/10.1016/j.cub.2013.04.020

Khan S, Larrosa C (2008) Economic values of coral reefs, mangroves, and seagrasses: a global compilation. Center for Applied Biodiversity Science, Conservation International, Arlington

Kitheka JU (1997) Coastal tidally-driven circulation and the role of water exchange in the linkage beween tropical coastal ecosystems. Estuar Coast Shelf Sci 45:177–187. https://doi.org/10.1006/ecss.1996.0189

Kleypas JA, Langdon C, Phinney JT et al (2006) Coral reefs and changing seawater carbonate chemistry. In: Phinney JT, Hoegh-Guldberg O, Kleypas J et al (eds) Coral reefs and climate change: science and management. American Geophysical Union, Washington, DC, pp 73–110

Knutson TR, McBride JL, Chan J et al (2010) Tropical cyclones and climate change. Nat Geosci 3:157–163

Koch MS, Schopmeyer S, Kyhn-Hansen C et al (2007) Synergistic effects of high temperature and sulfide on tropical seagrass. J Exp Mar Biol Ecol 341:91–101. https://doi.org/10.1016/j.jembe.2006.10.004

Koch EW, Barbier EB, Silliman BR et al (2009) Non-linearity in ecosystem services: temporal and spatial variability in coastal protection. Front Ecol Environ 7:29–37

Koch M, Bowes G, Ross C et al (2013) Climate change and ocean acidification effects on seagrasses and marine macroalgae. Glob Chang Biol 19:103–132

Krauss KW, McKee KL, Lovelock CE et al (2014) How mangrove forests adjust to rising sea level. New Phytol 202:19–34. https://doi.org/10.1111/nph.12605

Kroeker KJ, Kordas RL, Crim RN et al (2010) Meta-analysis reveals negative yet variable effects of ocean acidification on marine organisms. Ecol Lett 13:1419–1434

Kroeker KJ, Kordas RL, Crim R et al (2013) Impacts of ocean acidification on marine organisms: quantifying sensitivities and interaction with warming. Glob Chang Biol 19:1884–1896. https://doi.org/10.1111/gcb.12179

Kurihara H (2008) Effects of CO_2-driven ocean acidification on the early developmental stages of invertebrates. Mar Ecol Prog Ser 373:275–284

Kvitt H, Rosenfeld H, Zandbank K et al (2011) Regulation of apoptotic pathways by Stylophora pistillata (anthozoa, pocilloporidae) to survive thermal stress and bleaching. PLoS ONE 6:e28665. https://doi.org/10.1371/journal.pone.0028665

Lamb JB, van de Water JAJM, Bourne DG et al (2017) Seagrass ecosystems reduce exposure to bacterial pathogens of humans, fishes, and invertebrates. Science 355:731–733. https://doi.org/10.1126/science.aal1956

Lamy T, Legendre P, Chancerelle Y et al (2015) Understanding the spatio-temporal response of coral reef fish communities to natural disturbances: insights from beta-diversity decomposition. PLoS ONE 10:e0138696. https://doi.org/10.1371/journal.pone.0138696

Langdon C, Atkinson MJ (2005) Effect of elevated pCO_2 on photosynthesis and calcification of corals and interactions with seasonal change in temperature/irradiance and nutrient enrichment. J Geophys Res 110:C09S07. https://doi.org/10.1029/2004JC002576

Lapointe BE (1997) Nutrient thresholds for bottom-up control of macroalgal blooms on coral reefs in Jamaica and southeast Florida. Limnol Oceanogr 42:1119–1131. https://doi.org/10.4319/lo.1997.42.5

Le Quéré C, Andres RJ, Boden T et al (2013) The global carbon budget 1959–2011. Earth Syst Sci Data 5:165–185. https://doi.org/10.5194/essd-5-165-2013

Lee KS, Park SR, Kim YK (2007) Effects of irradiance, temperature, and nutrients on growth dynamics of seagrasses: a review. J Exp Mar Biol Ecol 350:144–175

Lovelock CE, Cahoon DR, Friess DA et al (2015) The vulnerability of Indo-Pacific mangrove forests to sea-level rise. Nature 526:559–563. https://doi.org/10.1038/nature15538

Luo J, Serafy JE, Sponaugle S et al (2009) Movement of gray snapper *Lutjanus griseus* among subtropical seagrass, mangrove, and coral reef habitats. Mar Ecol Prog Ser 380:255–269. https://doi.org/10.3354/meps07911

Magris RA, Heron SF, Pressey RL (2015) Conservation planning for coral reefs accounting for climate warming disturbances. PLoS ONE 10:e0140828. https://doi.org/10.1371/journal.pone.0140828

Manson FJ, Loneragan NR, Harch BD et al (2005) A broad-scale analysis of links between coastal fisheries production and mangrove extent: a case-study for northeastern Australia. Fish Res 74:69–85. https://doi.org/10.1016/j.fishres.2005.04.001

Marean CW, Bar-Matthews M, Bernatchez J et al (2007) Early human use of marine resources and pigment in South Africa during the Middle Pleistocene. Nature 449:905–908. https://doi.org/10.1038/nature06204

Matson PA, Parton WJ, Power AG et al (1997) Agricultural intensification and ecosystem properties. Science 227:504–509. https://doi.org/10.1126/science.277.5325.504

Maxwell PS, Eklöf JS, van Katwijk MM et al (2017) The fundamental role of ecological feedback mechanisms for the adaptive management of seagrass ecosystems – a review. Biol Rev 92:1521–1538. https://doi.org/10.1111/brv.12294

McAllister DE (1995) Status of the World Ocean and its biodiversity. Sea Wind 9:1–72

McCook LJ (1999) Macroalgae, nutrients and phase shifts on coral reefs: scientific issues and management consequences for the Great Barrier Reef. Coral Reefs 367:357–367

McCormick MI, Watson SA, Munday PL (2013) Ocean acidification reverses competition for space as habitats degrade. Sci Rep 3:3280. https://doi.org/10.1038/srep03280

McCulloch M, Falter J, Trotter J et al (2012) Coral resilience to ocean acidification and global warming through pH up-regulation. Nat Clim Chang 2:623–627. https://doi.org/10.1038/nclimate1473

McGlathery KJ (2001) Macroalgal blooms contribute to the decline of seagrass in nutrient-enriched coastal waters. J Phycol 37:453–456. https://doi.org/10.1046/j.1529-8817.2001.037004453.x

McGlathery KJ, Sundbäck K, Anderson IC (2007) Eutrophication in shallow coastal bays and lagoons: the role of plants in the coastal filter. Mar Ecol Prog Ser 348:1–18. https://doi.org/10.3354/meps07132

McIvor A, Spencer T, Möller I et al (2012) Storm surge reduction by mangroves. Natural coastal protection series: report 2. Cambridge Coastal Research Unit working paper 41. Published by The Nature Conservancy and Wetlands International

Milazzo M, Rodolfo-Metalpa R, Chan VBS et al (2014) Ocean acidification impairs vermetid reef recruitment. Sci Rep 4:4189. https://doi.org/10.1038/srep04189

Miller AW, Reynolds AC, Sobrino C et al (2009) Shellfish face uncertain future in high CO_2 world: influence of acidification on oyster larvae calcification and growth in estuaries. PLoS ONE 4:e5661. https://doi.org/10.1371/journal.pone.0005661

Moberg F, Folke C (1999) Ecological goods and services of coral reef ecosystems. Ecol Econ 29:215–233. https://doi.org/10.1016/S0921-8009(99)00009-9

Morell JM, Corredor JE (1993) Sediment nitrogen trapping in a mangrove lagoon. Estuar Coast Shelf Sci 37:203–212. https://doi.org/10.1006/ecss.1993.1051

Mumby PJ (2006) The impact of exploiting grazers (Scaridae) on the dynamics of caribbean coral reefs. Ecol Appl 16:747–769. https://doi.org/10.1890/1051-0761(2006)016[0747:TIOEGS]2.0.CO;2

Mumby PJ, Edwards AJ, Arias-González JE et al (2004) Mangroves enhance the biomass of coral reef fish communities in the Caribbean. Nature 427:533–536. https://doi.org/10.1038/nature02286

Mumby PJ, Hastings A, Edwards HJ (2007) Thresholds and the resilience of Caribbean coral reefs. Nature 450:98–101. https://doi.org/10.1038/nature06252

Mumby PJ, Iglesias-Prieto R, Hooten AJ et al (2011) Revisiting climate thresholds and ecosystem collapse. Front Ecol Environ 9:94–96

Munday PL, Crawley NE, Nilsson GE (2009) Interacting effects of elevated temperature and ocean acidification on the aerobic performance of coral reef fishes. Mar Ecol Prog Ser 388:235–242. https://doi.org/10.3354/meps08137

Nagelkerken I (2009) Ecological connectivity among tropical coastal ecosystems. Springer, Dordrecht

Newton K, Côté IM, Pilling GM et al (2007) Current and future sustainability of island coral reef fisheries. Curr Biol 17:655–658. https://doi.org/10.1016/j.cub.2007.02.054

O'Connor S, Ono R, Clarkson C (2011) Pelagic fishing at 42,000 years before the present and the maritime skills of modern humans. Science 334:1117–1121. https://doi.org/10.1126/science.1207703

Odum EP (1968) Energy flow in ecosystems: a historical review. Integr Comp Biol 8:11–18. https://doi.org/10.1093/icb/8.1.11

Odum WE, Heald EJ (1972) Trophic analyses of an estuarine mangrove community. Bull Mar Sci 22:671–738

Ogden JC (1980) Faunal relationships in Caribbean seagrass beds. In: Phillips RC, CP MR (eds) Handbook of seagrass biologi: an ecosystem perspective. Garland STPM Press, New York, pp 173–198

Ormond R (1987) Conservation and management. In: Edwards A, Head S (eds) Key environments: Red Sea. Pergamon Press, Oxford, pp 405–423

Palacios S, Zimmerman R (2007) Response of eelgrass *Zostera marina* to CO_2 enrichment: possible impacts of climate change and potential for remediation of coastal habitats. Mar Ecol Prog Ser 344:1–13. https://doi.org/10.3354/meps07084

Pandolfi JM, Connolly SR, Marshall DJ et al (2011) Projecting coral reef futures under global warming and ocean acidification. Science 333:418–422. https://doi.org/10.1126/science.1204794

Parrish JD (1989) Fish communities of interacting shallow-water habitats in tropical oceanic regions. Mar Ecol Prog Ser 58:143–160. https://doi.org/10.3354/meps058143

Pauly D, Christensen V, Guénette S et al (2002) Towards sustainability in world fisheries. Nature 418:689–695. https://doi.org/10.1038/nature01017

Payri CE, Naim O (1982) Variations entre 1971 et 1980 de la biomasse et de la composition des populations de macroalgues sur le ṛecif corallien de Tiahura (le de Moorea, Polyṇesie francaise). Cryptogam Algol 3:229–240

Pernetta JC (1993) Mangrove forests, climate change and sea level rise: hydrological influences on community structure and survival, with examples from the Indo-West Pacific. IUCN, Gland

Plaisance L, Caley MJ, Brainard RE et al (2011) The diversity of coral reefs: what are we missing? PLoS ONE 6:e25026. https://doi.org/10.1371/journal.pone.0025026

Pörtner HO (2008) Ecosystem effects of ocean acidification in times of ocean warming: a physiologist's view. Mar Ecol Prog Ser 373:203–217

Putnam HM, Stat M, Pochon X et al (2012) Endosymbiotic flexibility associates with environmental sensitivity in scleractinian corals. Proc R Soc B 279:4352–4361. https://doi.org/10.1098/rspb.2012.1454

Ramos AA, Inoue Y, Ohde S (2004) Metal contents in Porites corals: anthropogenic input of river run-off into a coral reef from an urbanized area, Okinawa. Mar Pollut Bull 48:281–294. https://doi.org/10.1016/j.marpolbul.2003.08.003

Raven J, Caldeira K et al (2005) Ocean acidification due to increasing atmospheric carbon dioxide. R Soc Rep 12:1–68

Ricard D, Minto C, Jensen OP et al (2012) Examining the knowledge base and status of commercially exploited marine species with the RAM legacy stock assessment database. Fish Fish 13:380–398. https://doi.org/10.1111/j.1467-2979.2011.00435.x

Richier S, Rodriguez-Lanetty M, Schnitzler CE et al (2008) Response of the symbiotic cnidarian *Anthopleura elegantissima* transcriptome to temperature and UV increase. Comp Biochem Physiol D 3:283–289. https://doi.org/10.1016/j.cbd.2008.08.001

Riser SC, Freeland HJ, Roemmich D et al (2016) Fifteen years of ocean observations with the global argo array. Nat Clim Chang 6:145–153

Robertson AI, Duke NC (1987) Mangroves as nursery sites: comparisons of the abundance and species composition of fish and crustaceans in mangroves and other nearshore habitats in tropical Australia. Mar Biol 96:193–205. https://doi.org/10.1007/BF00427019

Robinson KL, Ruzicka JJ, Hernandez FJ et al (2015) Evaluating energy flows through jellyfish and gulf menhaden (*Brevoortia patronus*) and the effects of fishing on the northern Gulf of Mexico ecosystem. ICES J Mar Sci 72:2301–2312. https://doi.org/10.1093/icesjms/fsv088

Rossoll D, Bermúdez R, Hauss H et al (2012) Ocean acidification-induced food quality deterioration constrains trophic transfer. PLoS ONE 7:e34737. https://doi.org/10.1371/journal.pone.0034737

Russell BD, Thompson JAI, Falkenberg LJ et al (2009) Synergistic effects of climate change and local stressors: CO_2 and nutrient-driven change in subtidal rocky habitats. Glob Chang Biol 15:2153–2162. https://doi.org/10.1111/j.1365-2486.2009.01886.x

Sale PF, Agardy T, Ainsworth CH et al (2014) Transforming management of tropical coastal seas to cope with challenges of the 21st century. Mar Pollut Bull 85:8–23. https://doi.org/10.1016/j.marpolbul.2014.06.005

Sampayo EM, Ridgway T, Bongaerts P et al (2008) Bleaching susceptibility and mortality of corals are determined by fine-scale differences in symbiont type. Proc Natl Acad Sci U S A 105:10444–10449. https://doi.org/10.1073/pnas.0708049105

Saunders MI, Leon JX, Callaghan DP et al (2014) Interdependency of tropical marine ecosystems in response to climate change. Nat Clim Chang 4:724–729. https://doi.org/10.1038/NCLIMATE2274

Scartazza A, Moscatello S, Gavrichkova O et al (2017) Carbon and nitrogen allocation strategy in *Posidonia oceanica* is altered by seawater acidification. Sci Total Environ 607–608:954–964. https://doi.org/10.1016/j.scitotenv.2017.06.084

Schaffelke B, Klumpp DW (1998) Short-term nutrient pulses enhance growth and photosynthesis of the coral reef macroalga *Sargassum baccularia*. Mar Ecol Prog Ser 170:95–105. https://doi.org/10.3354/meps170095

Schaffelke B, Mellors J, Duke NC (2005) Water quality in the Great Barrier Reef region: responses of mangrove, seagrass and macroalgal communities. Mar Pollut Bull 51:279–296. https://doi.org/10.1016/j.marpolbul.2004.10.025

Short FT, Neckles HA (1999) The effects of global climate change on seagrasses. Aquat Bot 63:169–196. https://doi.org/10.1016/S0304-3770(98)00117-X

Short F, Carruthers T, Dennison W et al (2007) Global seagrass distribution and diversity: a bioregional model. J Exp Mar Biol Ecol 350:3–20. https://doi.org/10.1016/j.jembe.2007.06.012

Short FT, Polidoro B, Livingstone SR et al (2011) Extinction risk assessment of the world's seagrass species. Biol Conserv 144:1961–1971. https://doi.org/10.1016/j.biocon.2011.04.010

Short FT, Coles R, Fortes MD et al (2014) Monitoring in the Western Pacific region shows evidence of seagrass decline in line with global

trends. Mar Pollut Bull 83:408–416. https://doi.org/10.1016/j.marpolbul.2014.03.036
Singh G, Ramanathan AL, Prasad MBK (2005) Nutrient cycling in mangrove ecosystem: a brief overview. Int J Ecol Environ Sci 31:231–244
Spalding MD, Green EP, Ravilious C (2001) World atlas of coral reefs. University of California Press/UNEP-WCMC, Berkeley
Spalding M, Kainuma M, Collins L (2010) World atlas of mangroves. Earthscan, London
Stewart HL (2006) Morphological variation and phenotypic plasticity of buoyancy in the macroalga Turbinaria ornata across a barrier reef. Mar Biol 149:721–730. https://doi.org/10.1007/s00227-005-0186-z
Stewart HL (2008) The role of spatial and ontogenetic morphological variation in the expansion of the geographic range of the tropical brown alga, Turbinaria ornata. Integr Comp Biol 48:713–719. https://doi.org/10.1093/icb/icn028
Stone R (2007) Ecology. A world without corals? Science 316:678–681. https://doi.org/10.1126/science.316.5825.678
Swindells KL, Murdoch RJ, Bazen WD et al (2017) Habitat configuration alters herbivory across the tropical seascape. Front Mar Sci 4:48. https://doi.org/10.3389/fmars.2017.00048
Talmage SC, Gobler CJ (2010) Effects of past, present, and future ocean carbon dioxide concentrations on the growth and survival of larval shellfish. Proc Natl Acad Sci U S A 107:17246–17251. https://doi.org/10.1073/pnas.0913804107
Tambutté E, Venn AA, Holcomb M et al (2015) Morphological plasticity of the coral skeleton under CO_2-driven seawater acidification. Nat Commun 6:7368. https://doi.org/10.1038/ncomms8368
Tomlinson PB (2016) The botany of mangroves. Cambridge University Press, Cambridge
Trenberth KE (2005) Uncertainty in hurricanes and global warming. Science 308:1753–1754
Trenberth KE (2011) Changes in precipitation with climate change. Clim Res 47:123–138. https://doi.org/10.3354/cr00953
UNEP-WCMC (2006) In the front line: shoreline protection and other ecosystem services from mangroves and coral reefs. UNEP-WCMC, Cambridge
Unsworth RKF, De León PS, Garrard SL et al (2008) High connectivity of Indo-Pacific seagrass fish assemblages with mangrove and coral reef habitats. Mar Ecol Prog Ser 353:213–224. https://doi.org/10.3354/meps07199
Unsworth RKF, Collier CJ, Henderson GM et al (2012) Tropical seagrass meadows modify seawater carbon chemistry: implications for coral reefs impacted by ocean acidification. Environ Res Lett 7:024026. https://doi.org/10.1088/1748-9326/7/2/024026
Valiela I, Cole ML (2002) Comparative evidence that salt marshes and mangroves may protect seagrass meadows from land-derived nitrogen loads. Ecosystems 5:92–102. https://doi.org/10.1007/s10021-001-0058-4
Valiela I, Bowen JL, York JK (2001) Mangrove forests: one of the world's threatened major tropical environments. Bioscience 51:807–815. https://doi.org/10.1641/0006-3568(2001)051[0807:MFOOTW]2.0.CO;2
van Oppen MJH, Oliver JK, Putnam HM et al (2015) Building coral reef resilience through assisted evolution. Proc Natl Acad Sci U S A 112:2307–2313. https://doi.org/10.1073/pnas.1422301112
van Tussenbroek BI, Hernández Arana HA, Rodríguez-Martínez RE et al (2017) Severe impacts of brown tides caused by *Sargassum* spp. on near-shore Caribbean seagrass communities. Mar Pollut Bull 122:272–281. https://doi.org/10.1016/j.marpolbul.2017.06.057
Vaughan DS, Shertzer KW, Smith JW (2007) Gulf menhaden (*Brevoortia patronus*) in the U.S. Gulf of Mexico: fishery characteristics and biological reference points for management. Fish Res 83:263–275. https://doi.org/10.1016/j.fishres.2006.10.002
Vermaat JE, Thampanya U (2006) Mangroves mitigate tsunami damage: a further response. Estuar Coast Shelf Sci 69:1–3. https://doi.org/10.1016/j.ecss.2006.04.019
Vitousek PM, Mooney HA, Lubchenco J et al (1997) Human domination of Earth's ecosystems. Science 277:494–499. https://doi.org/10.1126/science.277.5325.494
Ward RD, Friess DA, Day RH et al (2016) Impacts of climate change on mangrove ecosystems: a region by region overview. Ecosyst Heal Sustain 2:e01211. https://doi.org/10.1002/ehs2.1211
Waycott M, Duarte CM, Carruthers TJB et al (2009) Accelerating loss of seagrasses across the globe threatens coastal ecosystems. Proc Natl Acad Sci U S A 106:12377–12381. https://doi.org/10.1073/pnas.0905620106
Wilkinson CR (1992) Coral reefs of the world are facing widespread devastation: can we prevent this through sustainable management practices? Proc Seventh Int Coral Reef Symp 1:11–21
Wilkinson C (2008) Status of coral reefs of the world: 2008. Global Coral Reef Monitoring Network and Reef and Rainforest Research Centre. The Australian Institute of Marine Science (AIMS), Townsville
Worm B, Barbier EB, Beaumont N et al (2006) Impacts of biodiversity loss on ocean ecosystem services. Science 314:787–790. https://doi.org/10.1126/science.1132294
Yellowlees D, Rees TAV, Leggat W (2008) Metabolic interactions between algal symbionts and invertebrate hosts. Plant Cell Environ 31:679–694
Yeragi SS, Yeragi SG (2014) Status, biodiversity and distribution of mangroves in South Konkan, Sindhudurg District, Maharashtra State India: an overview. Int J Life Sci 2:67–69
Zhou X (2017) An overview of recently published global aquaculture statistics. FAO Aquact Newslett 56:6–8
Zimmerman RC, Kohrs DG, Steller DL et al (1997) Impacts of CO_2 enrichment on productivity and light requirements of eelgrass. Plant Physiol 115:599–607. https://doi.org/10.1104/pp.115.2.599
Zubia M, Andréfouët S, Payri CE (2015) Distribution and biomass evaluation of drifting brown algae from Moorea lagoon (French Polynesia) for eco-friendly agricultural use. J Appl Phycol 27:1277–1287. https://doi.org/10.1007/s10811-014-0400-9

Arctic Ocean Biodiversity and DNA Barcoding – A Climate Change Perspective

Katarzyna S. Walczyńska, Maciej K. Mańko, and Agata Weydmann

Abstract

Global changes are initiating a cascade of complex processes, which result, among other things, in global climate warming. Effects of global climate change are most pronounced in the Arctic, where the associate processes are progressing at a more rapid pace than in the rest of the world. Intensified transport of warmer water masses into the Arctic is causing shifts in species distributions and efforts to understand and track these change are currently intensified. However, Arctic marine fauna is the result of different recurring colonization events by Atlantic and Pacific Ocean populations, producing a very confounding evolutionary signal and making species identification by traditional morphological taxonomic analysis extremely challenging. In addition, many marine species are too small or too similar to identify reliably, even with profound taxonomic expertise. Nevertheless, the majority of current research focusing on artic marine communities still relies on the analysis of samples with traditional taxonomic methods, which tends to lack the necessary taxonomic, spatial and temporal resolution needed to understand the drastic ecosystem shifts underway. However, molecular methods are providing new opportunities to the field and their continuous development can accelerate and facilitate ecological research in the Arctic. Here, we discuss molecular methods currently available to study marine Arctic biodiversity, encouraging the DNA barcoding for improved descriptions, inventory and providing examples of DNA barcoding utilization in Arctic diversity research and investigations into ecosystem drivers.

K. S. Walczyńska (✉) · M. K. Mańko · A. Weydmann
Department of Marine Plankton Research, Institute of Oceanography, University of Gdańsk, Gdynia, Poland
e-mail: katarzyna.walczynska@phdstud.ug.edu.pl; mmanko@ug.edu.pl; agataw@ug.edu.pl

Biodiversity of the Arctic Ocean

Today's Artic marine biodiversity is highly impacted by newly formed current systems that bring warmer waters and their boreal inhabitants from the Atlantic and Pacific Oceans through the Fram and Bering Straits, respectively (Piepenburg et al. 2011). In the past, the resident diversity was primarily shaped by recurrent invasions, habitat fragmentation, and processes associated with glacial and interglacial periods, like bathymetric changes (e.g., Hewitt 2000, 2004; Ronowicz et al. 2015; Weydmann et al. 2017).

The Quaternary glaciation and deglaciation events were associated with global sea level fluctuations often exceeding 100 m, which lead to recurrent eradication of shelf biota and favored the survival of bathyal species and those thriving in isolated refugia, with subsequent recolonizations from the Atlantic and Pacific Oceans (Golikov and Scarlato 1989). In addition, the presence of ice sheets covering the open ocean further limited the dispersal of planktonic organisms (including larval stages of the benthic fauna) in the transarctic perspective (Hardy et al. 2011). The relatively recent, dynamic glacial history of the area have created complex evolutionary patterns, often blurring species delineations and hampering traditional morphological taxonomic methods, whereby, e.g., cryptic taxa can be easily overlooked (Hardy et al. 2011). Evidence for the underlying processes can also be gleaned from paleoceanographic data (Gladenkov and Gladenkov 2004). The geology of the Bering Strait, for example, revealed that, since its first opening at the Miocene-Pliocene boundary, this gateway between the Pacific and Arctic Oceans has been opened and closed repeatedly, providing opportunities for multiple invasions (Gladenkov and Gladenkov 2004; Hardy et al. 2011) from both sides (during the first 0.9–1.0 Ma after opening the prevailing currents flowed southward; Haug and Tiedemann 1998).

The five oceanic basins of the Arctic Ocean (Canada, Makarov, Amundsen, Nansen and Eurasian Basin) are separated by mid-oceanic ridges that limit dispersal of the

S. Jungblut et al. (eds.), *YOUMARES 8 – Oceans Across Boundaries: Learning from each other*,
https://doi.org/10.1007/978-3-319-93284-2_10

deep-sea species within the Arctic (Bluhm et al. 2011a), but also their inflow of waters from the adjoining oceanic regions (Carmack and Wassmann 2006). These dispersal barriers, together with the glacial history of the area, have resulted in isolated assemblages of distinctive marine biota, while maintaining the close relatedness to species found in neighboring oceanic regions (Bucklin et al. 2010).

Once thought to be relatively poor, the biodiversity of the Arctic Ocean is now considered to be at an intermediate level (Hardy et al. 2011), with the number of extant species estimated to about 8000 (Bluhm et al. 2011b). However, this number is dynamically increasing, with new taxa described ever more frequently (see e.g. Matsuyama et al. 2017) and estimates of several thousand yet undescribed species (Bluhm et al. 2011b; Appeltans et al. 2012). The ecologically harsh, but diverse setting of the Arctic Ocean underlies the local biodiversity (see Table 1). Sea ice, for example, aside from aforementioned dispersal limitation, constitutes a unique ecosystem where sympagic (ice-associated) organisms thrive (Bluhm et al. 2009a). This group includes many endemic taxa and those of panarctic distribution (Bluhm et al. 2009a), but remains largely unstudied with many taxa still awaiting descriptions (see Piraino et al. 2008).

The diversity level of each Artic marine ecological group is also tightly coupled with the highly specific ecosystem functioning of the Arctic. Seasonality, with light and dark periods lasting for large parts of the year (polar day and night, respectively), and the variable sea ice extent, govern the phenology of the whole ecosystem. Algal blooms, as main energy source for secondary producers and thus higher trophic levels, follow a two-part succession. The first ice algae bloom appears towards the end of winter, which is succeeded by a second bloom of planktonic algae, once the sea-ice melts (Leu et al. 2015). Both phases are significantly restricted in duration, due to light availability and water stratification (Sakshaug 2004). When the sea ice melts, surface waters warm up and, together with the presence of the fresh melt water, limit water mixing and consequently the amount of nutrients available to autotrophs, thus terminating the bloom (Sakshaug 2004). In spite of limited primary production, the trophic web of the marine Arctic is relatively rich and diverse. It can probably be explained by lower metabolic rates of organisms from higher trophic levels, resulting from permanently low temperatures in the Arctic Ocean (Bluhm et al. 2011b).

Table 1 Species diversity of marine Arctic biota of different ecological groups

Ecological group	Number of species
Unicellular eukaryotes	2106 (1027 sympagic; 1875 planktonic)
Sea ice fauna	At least 50
Zooplankton	354
Seaweeds	c. 160
Zoobenthos	c. 4600
Fish	243
Seabirds	64
Marine mammals	16

Modified after Bluhm et al. (2011b)

Most of the primary production is spatially restricted to shelves, and thus the most diverse community of consumers can be found there (Piepenburg et al. 2011; Wei et al. 2010). Availability of concentrated organic matter attracts primary consumers (zooplankton), which later become easy prey for secondary consumers (e.g. macrozooplankton, fish, sea birds) at shallow depths. Ungrazed organic matter, metabolic products and remains of the organisms sink to the bottom, where they fuel the complex benthic community. This concentration of biomass in the shelf regions draws the attention of top predators, like sea birds and marine mammals, for whom the Arctic shelves constitute the main forage areas (Wei et al. 2010).

The tight coupling between the functioning of the diverse marine Arctic ecosystems and environmental drivers renders them particularly susceptible to changes. The most detrimental anthropogenic impacts affecting the state of the Arctic Ocean usually include enterprises like shipping (including tourism), oil and gas exploration and fisheries related damages (ACIA 2004). However, the factor with the most obvious impact on the future of the marine Arctic is clearly climate change (IPCC 2014). An increase in sea surface temperatures reduces the geographic extent and thickness of the sea-ice cover directly, inducing a habitat loss for sympagic organisms, but also initiating regional shifts in species distributions or declines in primary production on a larger scale (Bluhm et al. 2011a; IPPC 2014).

In spite of insufficient amounts of decadal biodiversity studies encompassing the broad range of Arctic ecosystems, rapid (year-to-year) changes in different aspects of species biology have already been detected. On the autecological scale, these changes included e.g., biomass, diet or fitness (see review by Wassmann et al. 2011). On a broader view, the climate change driven modifications in Arctic communities are leading to a northward extension of the distribution ranges of boreal species (see examples in Hegseth and Sundfjord (2008) for phytoplankton; Weydmann et al. (2014) for zooplankton; Bluhm et al. (2009b) for zoobenthos; Mueter and Litzow (2008) for fish; Piatt and Kitaysky (2002) and CAFF (2010) for sea birds; Moore (2008) for marine mammals), replacing the long-lived and slow growing Arctic organisms with their smaller and short-lived boreal counterparts (e.g., Berge et al. 2005; Węsławski et al. 2010), while population of more susceptible, and usually less plastic species decline (e.g., Gilchrist and Mallory 2005).

DNA Barcoding

Biodiversity studies represent the first step to provide a baseline for detecting the effect of climate change on marine biota. A precise identification of all ecosystem components will allow to analyze interspecific interactions and will enable to determine factors, which influence its functioning. Until recently, most of the biodiversity research has been based on morphological analyses, which have many limitations, what might result in underestimation of diversity. In the marine environment, cryptic speciation is common, resulting in genetically differentiated lineages that are undistinguishable morphologically (Bickford et al. 2006). Nonetheless, their recognition is important, as they can have different functions in ecosystems (Fišer et al. 2015). Similarly, the identification of very small organisms or early life stages may be problematic, resulting in identification restricted to the phylum or family level.

A promising auxiliary approach is the use of molecular methods for identification and discrimination of species, known as DNA barcoding, which enables not only the assignment of unknown species, but it also enhances the discovery of new species (Bucklin et al. 2011), by matching their genetic fingerprint to a known barcode reference. Its development in recent years enabled more accurate species identification (Hebert et al. 2003), and the effectiveness of this approach has been established for several large groups of organisms (Bucklin et al. 2011), due to contribution of big, international projects, like Barcode of Life (www.barcodeoflife.org). Here, species identification is achieved by the analysis of a short DNA sequence from a specific gene region, called "the barcode", by comparing it with the library of reference barcode sequences derived from species of known identity (Hajibabaei et al. 2007). The method is based on the assumption that genetic differences between sequences within a species (intraspecific variability) are smaller than genetic differences among species (interspecific variability), reflected in the so-called "barcoding gap" (a min. % difference between intra- and interspecific variability), can be used to match the specimen's barcode in the database, if an appropriate reference sequence is available. The presence, extent, and "position" of the barcoding gap differs between species, and hence there is a need to use different markers for different groups of organisms. One of the most commonly used markers in animals is a 648-base fragment at the 5′ end of mitochondrial gene cytochrome *c* oxidase I (COI), as it has no introns (in some groups of animals), limited recombination and many copies per cell (Hajibabaei et al. 2007). Other popular markers include the genomic ITS (internal transcribed spacer I and II), 18S and the mitochondrial 16S rDNA. The number of sequences in databases like GenBank or BOLD are constantly increasing at a very fast rate. Hajibabaei et al. (2007) summarized the number of available sequences in public databases, and in only few years these numbers have increased several times. Information on popular markers used for DNA barcoding and the corresponding number of available sequences per organism group are presented in Table 2.

Like all identification methods, DNA barcoding has its flaws, as it requires a reference sequence in the database based on accurately identified organisms. Even though the development of Gen Bank is very dynamic – new sequences are submitted every day – sequences from many organisms are lacking whilst other sequences may be present under a wrongly identified species name. Nevertheless, molecular methods may have advantages over morphological methods in species identification as there is a lack of unique diagnostic morphological or morphometric characteristics separating species, but it can also be performed by a person without specialized taxonomic knowledge. An integrative approach using both molecular and morphological analyses, has been shown to strengthen species identification in previous polar taxonomic studies and provided the most reliable taxonomic resolution (Heimeier et al. 2010) as compared to using either method alone.

Indeed, identification of organisms based on nucleotide sequences it is not always 100% accurate, which has led to the use of the term Operational Taxonomic Unit (OTU) or – in case of barcoding – Molecular Operational Taxonomic Unit (MOTU), instead of "species". Studies have been carried out where the function of particular organisms in the ecosystem have been attributed to MOTUs (Ryberg 2015).

In the following sections we will provide examples to illustrate the use of DNA barcoding in Arctic diversity research and how can it be useful for detecting and monitoring of different processes in several important groups of marine organisms.

Plankton

Plankton is a very diverse group, containing very small organisms like viruses, heterotrophic single-cell organisms (bacterioplankton), autotrophic organisms (phytoplankton) and bigger animals (zooplankton). The diverse planktonic communities encompass both the tiniest autotrophs, like unicellular algae *Synechoccocus* and *Prochloroccocus,* which are responsible for the production of approximately 60% of the atmospheric oxygen, as well as the siphonophores, which can grow to about 40 m in length (Robison 1995). Yet another important component of the plankton are pelagic copepod crustaceans, which in many regions of the World's Ocean are the key species of the pelagic food webs, constituting up to 70% of the whole plankton biomass (Søreide et al. 2008). Their relatively short life cycles, high reproductive outputs, lack of direct antropogenic pressure and distributions depen-

Table 2 Common molecular markers. Numbers of available sequences in GenBank on 01.02.2017

Marker	Region	Numbers of sequences			
		Animals	Plants	Protists	Fungi
COI	Mitochondrial	2,219,762	30,511	1162	2043
18S	Genomic	161,263	25,130	9264	583,384
16S	Mitochondrial	345,915	4072	5221	382,418
ITS1	Genomic	47,842	82,880	33,235	481,840
ITS2	Genomic	61,956	88,157	14,535	236,705
CYTB	Mitochondrial	413,039	619		15,090
rbcL	Plastid	–	45,737	31,463	–

dent on the local hydrography make the plankton ideal for monitoring climate related changes in biodiversity (Hays et al. 2005). However, uncertainty in the taxonomic identification impedes further reasoning on climate-driven alterations of pelagic ecosystems.

Arctic zooplankton is characterized by a high seasonality and a strong spatial diversification resulting from distinct biogeographic origins of species (Błachowiak-Samołyk et al. 2008; Weydmann et al. 2014). A good example of such structuring of the plankton, comes from the analysis of the *Calanus* species complex. Three species of *Calanus* copepods coexist in the European Arctic: *C. finmarchicus*, *C. glacialis* and *C. hyperboreus*. In spite of similarities in their morphology and life cycles, there are some striking differences such as the type of lipids that characterize these congenerics, what should be taken into account, as they play a role in the lipid-based energy flux in the Arctic (Falk-Petersen et al. 2008). So far, *C. finmarchicus* was considered a boreal species, *C. glacialis* a typical Arctic shelf species, and *C. hyperboreus* the Arctic open-water species (Falk-Petersen et al. 2008). Their distribution ranges were clearly established, and in areas where they coexisted, species identification just followed the size criterion (Unstad and Tande 1991). However, the accuracy of this method, has been questioned, because of the potential interspecific hybridization and growth plasticity (Gabrielsen et al. 2012; Nielsen et al. 2014), which already has been documented by Parent et al. (2012) in the Arctic and Northwest Atlantic.

Hence, the distribution records of these three key planktonic species may have to be revised whilst knowledge on exact distribution ranges is crucial for the understanding of ecosystem functioning. In the Arctic, little auks (*Alle alle*), an ecologically important sea bird species, mainly feed on *Calanus glacialis*. With the observable increase of Atlantic water inflow to the Arctic (Polyakov et al. 2011), the distribution of this Arctic copepod is predicted to decline, while a northward range expansion is expected for its boreal sister-species *C. finmarchicus*. This comparatively much smaller Atlantic counterpart, *C. finmarchicus,* is an undesirable food source for little auks since it is not as energy rich as *C. glacialis,* and thus capture of a sufficient amount of *C. finmarchicus* comes with more energy expenses (Wojczulanis-Jakubas et al. 2013). In order to validate the hypothesis of distribution shifts between those two species, Lindeque et al. (2004) employed both morphological (based on the prosome length) and molecular (barcoding of the 16S rDNA gene) methods for species identification. Results obtained by molecular techniques proved that *Calanus* species co-occur and have wider distribution than it was established based on morphological analysis.

Another example illustrating the efficiency of molecular methods for plankton species identification is a study on pandeid hydromedusae. Four morphologically similar genera are currently co-existing in the Arctic: *Catablema, Halitholus, Leuckartiara* and *Neoturris.* The taxonomic features used for species delineation are often inconspicuous and in some cases assumed to be growth-dependent, and thus variable within the species (see comments in Schuchert 2007). Besides the need to thoroughly re-examine the life cycle of some of these species, molecular methods can be a solution for the identification problems. In the case of Hydrozoa, the use of 16S rDNA as barcode marker has certain advantages over COI (Lindsay et al. 2015), and therefore initiatives aiming at supplementing sequence data, using this particular gene should be encouraged (see project HYPNO, Dr. Aino Hosia, https://artsdatabanken.no/Pages/168312).

Microorganisms

Microorganisms, are particularly important as primary producers for the functioning of marine ecosystems, but they also play an important role in all biogeochemical processes (Sogin et al. 2006). Nonetheless, knowledge is limited due to the difficulties associated with the investigation of small organisms like pico- (0.2–2 µm), and nanoplankton (2–20 µm). Previous research in the Arctic has shown strong seasonal variations in microorganism communities, related to changes in irradiation. However the development of molecular techniques in recent years enabled further investigation of their diversity (Marquardt et al. 2016). Genetic analyses proved that microorganisms in Arctic waters are of greater importance than previously believed. Furthermore, they are also widely spread during polar night: in fjords and open ocean, deep and shallow water (Vader et al. 2015), which is particularly interesting as our knowledge regarding processes during the dark season was limited for a long time due to logistic difficulties with conducting research in winter. It should be taken into account that temperature increase and decrease in sea ice cover may influence the community structure of microorganisms and this effect has the potential to be translated to all upper trophic levels (Berge et al. 2015).

One of the most common methods used in the analysis of microorganisms, is barcoding based on the comparison of DNA and RNA derived OTU. While DNA is a very stable molecule and able to persist outside of the source organism for a long time, RNA is less stable and degrades rapidly. RNA analysis is therefore useful in informing about the current situation in the water column. In Svalbard waters, 4000 OTUs were differentiated based on DNA and only 2000 OTUs based on RNA (Marquardt et al. 2016). Differences can be explained by the fact that DNA is stable and may be present in the water column even after the death of an organism, but may also be caused by the high number copies of rRNA genes (Gong et al. 2013). The result of this research based on molecular data, has shown a high activity of heterotrophic groups during the polar night. It also revealed that species considered as autotrophic can become mixotrophic during winter. Based on a seasonal analysis of DNA and RNA, a succession of different microbial groups was demonstrated and their presence explained by particular environmental preferences, which may suggest that increasing temperatures will significantly influence community composition (Marquardt et al. 2016). Another study, in which microorganism communities were compared before and after the Record Sea Ice Minimum in the Arctic in 2007 (next were observed in 2012 and 2016), the genetic diversity of microorganisms appeared to be much lower (Comeau et al. 2011). This may be the result of particular adaptations to the sea-ice environment, as some are known to belong to the sympagic community. Differences in the community composition of Bacteria and Archaea, responsible for carbon and nutrient cycles, may influence productivity, but also the release of CO_2 from the Arctic Ocean (Legendre and Le Fèvre 1995). These findings underline the importance of future research focusing on the ecology and functions of microorganisms to predict consequences of forthcoming changes.

Benthos

Some areas of the Arctic Ocean, especially the continental shelves, are well-recognized for their tight bentho-pelagic coupling, inferred from the high amount of carbon fixed near the oceans' surface that sinks ungrazed to the seafloor, where it fuels benthic communities (Ambrose Jr. and Renaud 1995; Renaud et al. 2008). In the Arctic, biogenic sedimentation is far greater than at lower latitudes, thus explaining the high biomass of benthos thriving there (Petersen and Curtis 1980; Ambrose Jr. and Renaud 1995). Even with winter-limited primary production these benthic communities are relatively stable (Dunton et al. 2005).

The Arctic benthos is composed of a relatively young community that acquired lot of its current form during Quaternary glaciations (Zenkevitch 1963). The ice-mediated inflow and -outflow of mature organisms and their offspring, the isolation in refugia, and species extinctions led to the present day state of the Arctic benthic biodiversity (Hardy et al. 2011; Ronowicz et al. 2015). Although much is known about the current state of the Arctic benthos, a higher spatial and taxonomic resolution for biodiversity data is needed for an improved inferring of its future.

High phenotypic plasticity (e.g., in body pigmentation) further impedes species identification and hence the understanding of environmentally-dependent spatial diversification of benthic communities (Hardy et al. 2011). Bottom-dwelling polychaetes of the Arctic properly portray this trend. Until recently, this speciose group was perceived as lacking geographic structure on the global scale (Fauchald 1984). However, the use of molecular methods revealed numerous phenotypically indistinctive sibling species whilst it confirmed the presumed cosmopolitism of others (Carr et al. 2011). Hence, morphology-based taxonomy coupled with COI barcoding better resolved the diversity of Arctic polychaetes, showing that almost 25% of the over 300 "species" examined, were in fact complexes of two or more divergent lineages (Carr et al. 2011). Using COI sequences, Carr et al. (2011), were also able to retrace possible historical changes in distribution ranges of polychaetes found on Canadian coast of the Arctic, suggesting the Pleistocene glaciation as the main factor responsible of the increased diversification observed in this taxon.

Similar studies were conducted on echinoderms of the Canadian Arctic (Layton et al. 2016). Out of 141 taxa examined, 118 constituted morphologically distinctive species, while the remaining 23 were taxa assigned to different genera but not representing recognized species (Layton et al. 2016). It may suggest that in this area 23 morphologically indistinctive species new to science, or new for this region, may exist. Interestingly, with the sole usage of COI sequences, these authors also discussed various aspects of the phylogeography of echinoderms. For example, they pointed out that all species, where no pronounced spatial genetic structure could be observed between specimens collected in two or three oceanic regions of Canada, possessed a planktonic larval stage, which may justify the high levels of gene flow (Layton et al. 2016).

The above examples illustrate the utility of barcoding in delineation of the species composing the benthic communities of shallow shelf areas of the Arctic. Unfortunately, similar studies, focusing on the deep ocean assemblage remain uncommon, mostly because of the obvious difficulties of sampling below certain depths (Layton et al. 2016). One of the few examples of such studies, is Song et al. (2016), who used combined morphological and molecular approach to investigate the collections of Chinese National Arctic Research Expeditions in the Bering Sea. By means of 16S rDNA sequences, a new species, *Sertularia xuelongi*, was described and the potential biogeographic origin of this species discussed. By comparing 16 sequences of *S. xuelongi* and of other congeneric species from the northwest of France, Iceland, and the Chukchi Sea, they suggested that these species are of Pacific origin, but may in fact constitute a significant part of the deep-sea benthic fauna of the Arctic (Song et al. 2016).

As mentioned earlier, important factors in shaping nowadays Arctic diversity were glaciation processes, during which species were forced into refugia in order to survive, what caused long-term isolation and thus differentiation of the species. After glaciation ceased, some of the expanding species went in secondary contact, however, undergoing processes were much more complicated (Maggs et al. 2008). One of the interesting examples is blue mussel, *Mytilus edulis,* which was gone for a long time from Svalbard waters, however warming of the Arctic enabled its re-appearance (Berge et al. 2005). It has been proven that blue mussels can create hybrids with other species, like *Mytilus trossulus* and *Mytilus galloprovancialis* in different Arctic regions, what leads to local adaptations (Mathiesen et al. 2017). This topic has not been investigated well yet, nonetheless it requires more insight as *Mytilus* spp. are ecosystem engineers and global warming opens new paths for invasions of boreal species in the Arctic.

Nekton

The benthic and planktonic organisms discussed above constitute food sources for higher trophic levels, which in the Arctic are primarily nektonic vertebrates. Aside from marine mammals and sea birds, this group is represented by a speciose community of fish. In the Arctic, there are 243 species of fish (Bluhm et al. 2011a), comprising several key species like polar cod and capelin (Hop and Gjøsæter 2013) as well as species with unique traits including the longest living vertebrate, the Greenland shark (Nielsen et al. 2016).

The biogeography of this ecologically and economically important group remained, unfortunately, largely unknown. Only recently, Mecklenburg et al. (2011) have improved the taxonomic identification of all Arctic species, thereby improving the resolution available for the spatial structure of their diversity. COI barcoding, combined with morphological analyses, allowed them to revise the biogeographic origin of species, showing that some of the past fish records from Arctic waters were misidentified. They found that a majority Arctic fish species (59%) are cosmopolitan species with boreal distribution, while the remaining 41% are Arctic, mainly-Arctic, and boreo-Arctic species (Mecklenburg et al. 2011). Such detailed knowledge on the biodiversity is required to trace the climate-change derived alteration of, for example, species distribution. The study also shows the hidden potential of the simultaneous morphological-molecular approach to taxonomy. In this particular case, it could be used to resolve the cod mother identity, or to acquire data of unprecedented species-resolution (Carr and Marshall 2008). For some fish species like *Arctogadus glacialis, Boreogadus saida* complete genomes are available (Breines et al. 2008). There is a high interest in postglacial colonization of fishes like *Salvelinus fontinalis* (Pilgrim et al. 2012), *Coregonus nasus* (Harris and Taylor 2010) or *Coregonu lavaretus* (Østbye et al. 2006). We can also find lots of studies about genetic diversity of different species (Kai et al. 2011; Kovpak et al. 2011), as in the future it might be crucial for adaptations to a changing environment.

The overall low number of species and distinctive morphology allow a relatively easy acquisition of high-resolution data on marine mammal diversity by means of classic taxonomic methods. Furthermore, such approaches have already revealed pronounced modifications in species ecology and biology, by detecting shifts in distribution ranges, decrease in body size and size of the separate populations, as well as alterations of food migrations (Kovacs et al. 2010). All of these changes might affect marine mammal species populations. Even though molecular research does not focus on biodiversity, it may cover a wide range of other aspects, like evolution, population genetics or phylogeography.

Future Perspectives

The Arctic Ocean is warming three times faster than the global average (IPCC 2014), thus further changes in species composition and entire ecosystem functioning are inevitable. Temperature has an impact on many aspects of physiological processes and it can affect reproduction, growth, and survival. Changes in single species distributions can effect entire ecosystems through all trophic levels, as was shown for the case of a potential mismatch between phytoplankton blooms and reproduction of *Calanus glacialis* in Arctic waters (Søreide et al. 2010). Hence, using only traditional methods might not be enough to timely observe what is going on in this fragile ecosystem. Kędra et al. (2015) emphasized the lack of biodiversity research in some Arctic areas, especially in the deep-sea region, but also the lack of research predicting direction of changes in species distribution resulting from global change.

DNA barcoding has been proven a useful tool in biodiversity assessment, however, the evolution of molecular methods is very fast, including the development of new approaches such as metabarcoding. This method involves the extraction of DNA from an entire sample, without the need of picking out single individuals, like larvae or other targeted groups of mesozooplankton. It is based on the New Generation Sequencing (NGS) technology, where millions of short sequences (reads) are produced allowing to screen entire genomes or transcriptomes in order to obtain a higher resolution of spatio-temporal patterns of species distribution (Bucklin et al. 2016). This technique is becoming increasingly available as sequencing is getting cheaper. Commonly used genetic markers for metabarcoding are 16S, 18S and 28S, while COI is not often used as it requires specific primers (Deagle et al. 2014). So far, metabarcoding has mainly been used for microorganism research, however, it might also be used for monitoring of zooplankton for which the dynamic changes may not be detected with other tools. It is now also possible to obtain DNA from environmental samples (environmental DNA, eDNA), like water or soil, without prior isolation of target organisms, as they continuously expel DNA into their surroundings from where it can be collected (Thomsen and Willerslev 2015). This approach can provide information about the presence and type of organisms which were in a particular location in the recent past, like fishes or whales (Sigsgaard et al. 2016). Metagenomics represent an even more advanced method, for which entire genomes present in environmental samples are analyzed. Yet it is mostly applied on microorganisms, since not enough reference genomes exist for metazoans (Wooley et al. 2010). Nevertheless, as mentioned at the beginning of this chapter, databases are growing at an enormous speed and new genomes are published every day, what means that analyses of metagenomes of different ecological groups will become possible in the nearest future. Metagenomics significantly exceeds beyond species identification, in biodiversity research it allows for investigation of uncultured microbial populations. It is a very powerful tool, which enables exploration of metabolic diversity, isolation and identification of enzymes and it may be an effective way to produce novel bioresources (Kodzius and Gojobori 2015).

Currently, the analysis of high-throughput sequence (NGS) data requires an in-depth knowledge in bioinformatics. Moreover, the obtained results are rather qualitative than quantitative, e.g. based on presence/absence of DNA in a sample, however, this is currently being improved. Nonetheless, until these methods are not optimized for converting number of sequences into abundances of organisms in the field, the best method remains the integrative taxonomic approach, which combines molecular with morphological data.

In conclusion, molecular data are a promising tool for detecting the influence and consequences of global warming on different communities. Standard molecular methods have been successfully applied in Arctic research and their fast development will render analysis even more feasible and cost-effective. The use of DNA barcoding should be emphasized for long-time monitoring studies. Considering the opportunity of acquiring fast results, however, caution should be taken with regard to the choice of an adequate molecular marker, a careful analysis of the data and if possible, the application of an integrative approach by supporting these results with morphological analyses.

Knowledge on the existing biodiversity is the baseline for many studies, e.g. on ecological and physiological aspects. In order to investigate the future of the Arctic ecosystems, further research should focus on combining data obtained from biodiversity assessments with modelling and experiments, in which molecular tools can be used as well.

Appendix

This article is related to the YOUMARES 8 conference session no. 8: "Polar Ecosystems in the Age of Climate Change". The original Call for Abstracts and the abstracts of the presentations within this session can be found in the appendix "Conference Sessions and Abstracts", chapter "9 Polar Ecosystems in the Age of Climate Change", of this book.

References

ACIA (Arctic Climate Impact Assessment) (2004) Impacts of warming Arctic. Cambridge University Press, Cambridge

Ambrose WG Jr, Renaud PE (1995) Benthic response to water column productivity patterns: evidence for benthic-pelagic coupling in the Northeast Water Polynya. J Geophys Res 100:4411–4421

Appeltans W, Ahyong ST, Anderson G et al (2012) The magnitude of global marine species diversity. Curr Biol 22:2189–2202

Berge J, Johnsen G, Nilsen F et al (2005) Ocean temperature oscillations enable reappearance of blue mussel *Mytilus edulis* in Svalbard after 1000 years of absence. Mar Ecol Prog Ser 303:167–175

Berge J, Renaud PE, Darnis G et al (2015) In the dark: a review of ecosystem processes during the Arctic polar night. Prog Oceanogr 139:258–271

Bickford D, Lohman DJ, Sodhi NS et al (2006) Cryptic species as a window on diversity and conservation. Trends Ecol Evol 22(3):148–155

Błachowiak-Samołyk K, Søreide J, Kwasniewski S et al (2008) Hydrodynamic control of mesozooplankton abundance and biomass in northern Svalbard waters (79–81°N). Deep Sea Res Part 2 55:2210–2224

Bluhm BA, Gradinger RR, Schnack-Schiel SB (2009a) Sea ice meio- and macrofauna. In: Thomas DN, Dieckmann GS (eds) Sea ice, 2nd edn. Wiley-Blackwell, Oxford, pp 357–393

Bluhm BA, Iken SL, Mincks BI et al (2009b) Community structure of epibenthic megafauna in the Chukchi Sea. Aquatic Biol 7:269–293

Bluhm BA, Gebruk AV, Gradinger R et al (2011a) Arctic marine biodiversity: an update of species richness and examples of biodiversity change. Oceanography 24(3):232–248

Bluhm BA, Gradinger R, Hopcroft RR (2011b) Editorial – Arctic Ocean diversity: synthesis. Mar Biodivers 41:1–4

Breines R, Ursvik A, Nymark M et al (2008) Complete mitochondrial genome sequences of the Arctic Ocean codfishes *Arctogadus glacialis* and *Boreogadus saida* reveal oriL and tRNA gene duplications. Polar Biol 31:1245–1252

Bucklin A, Hopcroft RR, Kosobokova KN et al (2010) DNA barcoding of Arctic Ocean holozooplankton for species identification and recognition. Deep Sea Res Part 2 57:40–48

Bucklin A, Steinke D, Blanco-Bercial L (2011) DNA barcoding of marine Metazoa. Annu Rev Mar Sci 3:471–508

Bucklin A, Lindeque PK, Rodriguez-Ezpeleta N et al (2016) Metabarcoding of marine zooplankton: prospects, progress and pitfalls. J Plankton Res 38:393–400

CAFF (2010) Arctic biodiversity trends 2010: selected indicators of change. CAFF International Secretariat, Akureyri

Carmack E, Wassmann P (2006) Food webs and physical–biological coupling on pan-Arctic shelves: unifying concepts and comprehensive perspectives. Prog Oceanogr 71:446–477

Carr SM, Marshall HD (2008) Intraspecific phylogeographic genomics from multiple complete mtDNA genomes in Atlantic Cod (*Gadus morhua*): origins of the "Codmother" transatlantic vicariance and midglacial population expansion. Genetics 180:381–389

Carr CM, Hardy SM, Brown TM et al (2011) A tri-oceanic perspective: DNA barcoding reveals geographic structure and cryptic diversity in Canadian polychaetes. PLoS One 6(7):e22232. https://doi.org/10.1371/journal.pone.0022232

Comeau AM, Li WKW, Tremblay J-E et al (2011) Arctic Ocean microbial community structure before and after the 2007 record sear ice minimum. PLoS One 6(11):e27492. https://doi.org/10.1371/journal.pone.0027492

Deagle BE, Jarmen SN, Coissac E et al (2014) DNA metabarcoding and the cytochrome *c* oxidase subunit I marker: not a perfect match. Biol Lett 10:140562. https://doi.org/10.1098/rsbl.2014.0562

Dunton KH, Goodall JL, Schonberg SV et al (2005) Multi-decadal synthesis of benthic-pelagic coupling in the western Arctic: role of cross-shelf advective processes. Deep Sea Res Part 2 52:3462–3477

Falk-Petersen S, Mayzaud P, Kattner G et al (2008) Lipids and life strategy of Arctic *Calanus*. Mar Biol Res 5:18–39

Fauchald K (1984) Polychaete distribution patterns, or: can animals with Palaeozoic cousins show large-scale geographical patterns? In: Hutchings PA (ed) Proceedings of the first international polychaete conference, Sydney, The Linnean Society of New South Wales, pp 1–6

Fišer Ž, Altermatt F, Zakšek V et al (2015) Morphologically cryptic amphipod species are "ecological clones" at regional but not at local scale: a case study of four *Niphargus* species. PLoS One 10(7):e0134384. https://doi.org/10.1371/journal.pone.0134384

Gabrielsen TM, Merkel B, Søreide J et al (2012) Potential misidentifications of two climate indicator species of the marine arctic ecosystem: *Calanus glacialis* and *C. finmarchicus*. Polar Biol 35:1621–1628

Gilchrist HG, Mallory ML (2005) Declines in abundance and distribution of the ivory gull (Pagophila eburnea) in Arctic Canada. Biol Conserv 121:303–309

Gladenkov AY, Gladenkov YB (2004) Onset of connections between the Pacific and Arctic Oceans through the Bering Strait in the Neogene. Stratigr Geol Correl 12:175–187

Golikov AN, Scarlato OA (1989) Evolution of Arctic ecosystems during the Neogene period. In: Herman Y (ed) The Arctic Seas climatology, oceanography and biology. Van Nostrand Reinhold, New York, pp 257–279

Gong J, Dong J, Liu X et al (2013) Extremely high copy numbers and polymorphisms of the rDNA operon estimated from single cell analysis of Oligotrich and Peritrich ciliates. Protist 164:369–379

Hajibabaei M, Singer GAC, Hebert PDN et al (2007) DNA barcoding: how it complements taxonomy, molecular phylogenetics and population genetics. Trends Genet 23:167–172

Hardy SM, Carr CM, Hardman M et al (2011) Biodiversity and phylogeography of Arctic marine fauna: insights from molecular tools. Mar Biodivers 41:195–210

Harris LN, Taylor EB (2010) Pleistocene glaciations and contemporary genetic diversity in a Beringian fish, the broad whitefish, *Coregonus nasus* (Pallas): inferences from microsatellite DNA variation. J Evol Biol 23:72–86

Haug GH, Tiedemann R (1998) Effect of the formation of the Isthmus of Panama on Atlantic Ocean thermocline circulation. Nature 393:673–676

Hays GC, Richardson AJ, Robinson C (2005) Climate change and marine plankton. Trends Ecol Evol 20:337–344

Hebert PDN, Cywinska A, Ball SL et al (2003) Biological identifications through DNA barcodes. Proc R Soc Lond B 270:313–321

Hegseth EN, Sundfjord A (2008) Intrusion and blooming of Atlantic phytoplankton species in the high Arctic. J Mar Syst 74:108–119

Heimeier D, Lavery S, Sewell MA (2010) Using DNA barcoding and phylogenetics to identify Antarctic invertebrate larvae: lesson from a large scale study. Mar Gen 3:165–177

Hewitt GM (2000) The genetic legacy of the quaternary ice ages. Nature 45:907–913. https://doi.org/10.1038/35016000

Hewitt GM (2004) Genetic consequences of climatic oscillations in the quaternary. Philos Trans R Soc Lond Ser B Biol Sci 359:183–195

Hop H, Gjøsæter H (2013) Polar cod (*Boerogadus saida*) and capelin (*Mallotus villosus*) as key species in marine food webs of the Arctic and Barents Sea. Mar Biol Res 9:878–894

IPCC (Intergovernmental Panel on Climate Change) (2014) Climate change 2014: synthesis report. Contribution of working groups I, II and III to the fifth assessment report of the IPCC. IPCC, Switzerland, Genève

Kai Y, Orr JW, Sakai K et al (2011) Genetic and morphological evidence for cryptic diversity in the *Careproctus rastrinus* species complex (Liparidae) of the North Pacific. Ichthyol Res 58:143–154

Kędra M, Moritz C, Choy ES et al (2015) Status and trends in the structure of Arctic benthic food webs. Pol Res 34:23775. https://doi.org/10.3402/polar.v34.23775

Kodzius R, Gojobori T (2015) Marine metagenomics as a source for bioprospecting. Mar Gen 24:21–30

Kovacs KM, Moore S, Lydersen C et al (2010) Impacts of changing sea ice conditions on Arctic marine mammals. Mar Biodivers 41:181–194

Kovpak NE, Skurikhina LA, Kukhlevsky AD et al (2011) Genetic divergence and relationships among smelts of the genus *Osmerus* from the Russian waters. Russ J Genet 47:958–972

Layton KKS, Corstorphine EA, Hebert PDN (2016) Exploring Canadian echinoderm diversity through DNA barcodes. PLoS One 11(11):e0166118. https://doi.org/10.1371/journal.pone.0166118

Legendre L, Le Fèvre J (1995) Microbial food webs and the export of biogenic carbon in oceans. Aquat Microb Ecol 9:69–77

Leu E, Mundy CJ, Assmy P et al (2015) Arctic spring awakening – steering principles behind the phenology of vernal ice algal blooms. Prog Oceanogr 139:151–170

Lindeque PK, Harris R, Jones MB et al (2004) Distribution of *Calanus* spp. as determined using a genetic identification system. Sci Mar 68:121–128

Lindsay DJ, Grossmann MM, Nishikawa J et al (2015) DNA barcoding of pelagic cnidarians: current status and future prospects. Bull Plankton Soc Jpn 62:39–43

Maggs CA, Castihlo R, Foltz D et al (2008) Evaluating signatures of glacial refugia for North Atlanthic benthic marine taxa. Ecology 89:108–122

Marquardt M, Vader A, Stübner EI et al (2016) Strong seasonality of marine microbial eukaryotes in a High-Arctic Fjord (Isfjorden, in West Spitsbergen, Norway). Appl Environ Microbiol 82:1868–1880

Mathiesen SS, Thyrring J, Hemmer-Hansen J et al (2017) Genetic diversity and connectivity within *Mytilus* spp. in the subarctic and Arctic. Evol Appl 10:39–55

Matsuyama K, Martha SO, Scholz J et al (2017) *Ristedtia vestiflua* n. gen. et sp., a new bryoazon genus and species (Gymnolaemata:

Cheilostomata) from an Arctic seamount in the Central Greenland Sea. Mar Biodivers. https://doi.org/10.1007/s12526-017-0645-z

Mecklenburg CW, Møller PR, Steinke D (2011) Biodiversity of arctic marine fishes: taxonomy and zoogeography. Mar Biodivers 41:109–140

Moore SE (2008) Marine mammals as ecosystem sentinels. J Mammal 89:534–540

Mueter FJ, Litzow MA (2008) Sea ice retreat alters the biogeography of the Bering Sea continental shelf. Ecol Appl 18:309–320

Nielsen TG, Kjellerup S, Smolina I et al (2014) Live discrimination of *Calanus glacialis* and *C. finmarchicus* females: can we trust phenological differences? Mar Biol 161:1299–1306

Nielsen J, Hedeholm RB, Heinemeier J (2016) Eye lens radiocarbon reveals centuries of longevity in the Greenland shark (*Somniosus microcephalus*). Science 353:702–704

Østbye K, Amundsen P, Bernatchez L et al (2006) Parallel evolution of ecomorphological traits in the European whitefish *Coregonus lavaretus* (L.) species complex during postglacial times. Mol Ecol 15:3983–4001

Parent GJ, Plourde S, Turgeon J (2012) Natural hybridization between *Calanus finmarchicus* and *C. glzcialis* (Copepoda) in the Arctic and Northwest Atlantic. Limnol Oceanogr 57:1057–1066

Petersen GH, Curtis MA (1980) Differences in energy flow through major components of subarctic, temperate and tropical marine shelf ecosystems. Dana 1:53–64

Piatt JF, Kitaysky AS (2002) Tufted puffin: *Fratercula cirrata*. In: Poole A (ed) The birds of North America online, Cornell Laboratory of Ornithology, Ithaca. Available online at: http://bna.birds.cornell.edu/bna/species/708

Piepenburg D, Archambault P, Ambrose WG Jr (2011) Towards a pan-Arctic inventory of the species diversity of the macro- and megabenthic fauna of the Arctic shelf seas. Mar Biodivers 41:51–70

Pilgrim BL, Perry RC, Barron JL et al (2012) Nucleotide variation in the mitochondrial genome provides evidence for dual routes of postglacial recolonization and genetic recombination in the northeastern brook trout (*Salvelinus fontinalis*). Gen Mol Res 11:3466–3481

Piraino S, Bluhm BA, Gradinger R et al (2008) *Sympagohydra tuuli* gen. nov. and sp. nov. (Cnidaria: Hydrozoa) a cool hydroid from the Arctic Sea ice. J Mar Biol Assoc UK 88:1637–1641

Polyakov IV, Alexeev VA, Ashik IM et al (2011) Fate of early 2000s Arctic warm water pulse. Bull Am Meteorol Soc 92:561–566

Renaud PE, Morata N, Carroll ML et al (2008) Pelagic-benthic coupling in the western Barents Sea: processes and time scales. Deep Sea Res 2(55):2372–2380

Robison BH (1995) Light in the ocean's midwaters. Sci Am 273:60–64

Ronowicz M, Kukliński P, Mapstone GM (2015) Trends in the diversity, distribution and life history strategy of Arctic Hydrozoa (Cnidaria). PLoS One 10:e0120204

Ryberg M (2015) Molecular operational taxonomic units as approximations of species in the light of evolutionary models and empirical data from Fungi. Mol Ecol 24:5770–5777

Sakshaug E (2004) Primary and secondary production in the Arctic Seas. In: Stein R, MacDonald R (eds) The organic carbon cycle in the Arctic Ocean. Springer, Berlin, pp 57–82

Schuchert P (2007) The European athecate hydroids and their medusae (Hydrozoa, Cnidaria): Filifera Part 2. Rev Suisse Zool 114:195–396

Sigsgaard EE, Nielsen IB, Bach SS et al (2016) Population characteristics of a large whale shark aggregation inferred from seawater environmental DNA. Nat Ecol Evol 1:0004. https://doi.org/10.1038/s41559-016-0004

Sogin ML, Morrison HG, Huber JA et al (2006) Microbial diversity in the deep sea and the underexplored "rare biosphere". Proc Natl Acad Sci U S A 103:12115–12120

Song X, Gravili C, Wang J et al (2016) A new deep-sea hydroid (Cnidaria: Hydrozoa) from the Bering Sea Basin reveals high genetic relevance to Arctic and adjacent shallow-water species. Pol Biol 39:461–471

Søreide J, Falk-Petersen S, Hegseth EN et al (2008) Saesonal feeding startegies of *Calanus* in the high-Arctic Svalbard region. Deep Sea Res Part 2 55:2225–2244

Søreide J, Leu E, Berge J et al (2010) Timing of blooms, algal food quality and *Calanus glacialis* reproduction and growth in a changing Arctic. Glob Chang Biol 16:3154–3163

Thomsen PF, Willerslev E (2015) Environmental DNA – an emerging tool in conservation for monitoring past and present biodiversity. Biol Conserv 183:4–18

Unstad KM, Tande KS (1991) Depth distribution of *Calanus finmarchicus* and *C. glacialis* in relation to environmental conditions in the Barents Sea. Polar Res 10:409–420

Vader A, Marquardt M, Meshram AR et al (2015) Key Arctic phototrophs are widespread in the polar night. Polar Biol 38:13–21

Wassmann P, Duarte CM, Agusti S et al (2011) Footprints of climate change in the Arctic marine ecosystem. Glob Chang Biol 17:1235–1249

Wei CK, Rowe GT, Escobar-Briones E et al (2010) Global patterns and predictions of seafloor biomass using random forests. PLoS One 5(12):e15323. https://doi.org/10.1371/journal.pone.0015323

Wesławski JM, Wiktor J Jr, Kotwicki L (2010) Increase in biodiversity in the arctic rocky littoral, Sorkappland, Svalbard, after 20 years of climate warming. Mar Biodivers 40:123–130

Weydmann A, Carstensen J, Goszczko I et al (2014) Shift towards the dominance of boreal species in the Arctic: inter-annual and spatial zooplankton variability in the West Spitsbergen current. Mar Ecol Prog Ser 501:41–52

Weydmann A, Przyłucka A, Lubośny M et al (2017) Postglacial expansion of the Arctic keystone copepod Calanus glacialis. Mar Biodivers. https://doi.org/10.1007/s12526-017-0774-4

Wojczulanis-Jakubas K, Jakubas D, Stempniewicz L (2013) Alczyk – sztandarowy gatunek Arktyki. Kosmos 62:401–407

Wooley JC, Godzik A, Friedberg I (2010) A primer on metagenomics. PLoS One 6(2):e1000667

Zenkevitch L (1963) Biology of the seas of the U.S.S.R. George Allen & Unwin Ltd, London

Regime Shifts – A Global Challenge for the Sustainable Use of Our Marine Resources

Camilla Sguotti and Xochitl Cormon

Abstract

Over the last decades many marine systems have undergone drastic changes often resulting in new ecologically structured and sometimes economically less valuable states. In particular, the additive effects of anthropogenic stressors (e.g., fishing, climate change) seem to play a fundamental role in causing unexpected and sudden shifts between system states, generally termed regime shifts. Recently, many examples of regime shifts have been documented worldwide and their mechanisms and consequences have been vigorously discussed. Understanding causes and mechanisms of regime shifts is of great importance for the sustainable use of natural resources and their management, especially in marine ecosystems. Hence, we conducted a session entitled "Ecosystem dynamics in a changing world, regime shifts and resilience in marine communities" during the 8th YOUMARES conference (Kiel, 13–15th September 2017) to present regime shifts concepts and examples to a broad range of marine scientists (e.g., biologists and/or ecologists, physicists, climatologists, sociologists) and highlight their importance for the marine ecosystems worldwide.

In this chapter, we first provide examples of regime shifts which have occurred over the last decades in our oceans and discuss their potential implications for the sustainable use of marine resources; then we review regime shift theory and associated concepts. Finally, we review recent advances and future challenges to integrate regime shift theory into holistic marine ecosystem-based management approaches.

C. Sguotti (✉) · X. Cormon (✉)
Institute for Marine Ecosystem and Fishery Science, Centre for Earth System Research and Sustainability (CEN), University of Hamburg, Hamburg, Germany
e-mail: camilla.sguotti@uni-hamburg.de; xochitl.cormon@uni-hamburg.de

Introduction

Today, living marine resources represent a primary source of proteins for more than 2.6 billion people and support the livelihoods of about 11% of the world's population (UNESCO 2012; FAO 2014). Oceans worldwide concentrate dense and diversified human activities, e.g., fishing, tourism, shipping, offshore energy production, while experiencing a range of environmental pressures, e.g., increase of water temperature, acidification (Halpern et al. 2008; Boyd et al. 2014). Together anthropogenic and environmental pressures may threaten the integrity of marine systems and their sustainable use, altering their different components in many ways. These ecosystem changes may have great impacts for the social-ecological systems they are a part of, particularly when associated with changes in ecological keystone, cultural and/or commercial species (Garibaldi and Turner 2004; Casini et al. 2008a; Möllmann et al. 2008; Llope et al. 2011; Blenckner et al. 2015b).

The World Summit on Sustainable Development in Johannesburg (2002) provided a legally binding framework to implement the Ecosystem Approach to Fisheries Management (EAFM). This holistic approach aims (i) to conserve the structure, diversity and functioning of marine ecosystems and (ii) to provide the economic benefits of a sustainable exploitation of marine ecosystems. Scientific activities supporting approaches such as the EAFM are hence highly encouraged (FAO 2003). However, the insufficient knowledge on the diversity and entanglement of interactions between the ecological system components (deYoung et al. 2008), as well as their vulnerability to increasing anthropogenic and environmental pressures, may hinder successful management.

Even if systems may react to stressors in a non-linear way shifting suddenly to a different state and losing important ecosystem services, management is indeed still more based on continuous dynamics (Scheffer et al. 2001; Sugihara et al. 2012; Glaser et al. 2014; Travis et al. 2014; Levin and

S. Jungblut et al. (eds.), *YOUMARES 8 – Oceans Across Boundaries: Learning from each other*,
https://doi.org/10.1007/978-3-319-93284-2_11

Möllmann 2015). Some ecosystems may be able to absorb stronger disturbances than others depending on their characteristics, but in general, marine ecosystems are known to be particularly vulnerable to drastic and unexpected shifts, referred in ecology as regime shifts (deYoung et al. 2008). Such non-linear dynamics may have positive or negative outcomes for the sustainable use of natural resources and their management, therefore they should be taken into account and dealt with great precaution when taking environmental policy decisions (Holling 1973; Carpenter 2001; Scheffer 2009; Rocha et al. 2014a).

In this chapter, we first present some examples of marine ecosystems which have exhibited non-linear dynamics in response to external changes. These examples allow us to highlight different mechanisms potentially involved in regime shifts from an empirical point of view, as well as their potential implications for the sustainable use of marine resources. Secondly, we review the regime shift theory and associated concepts to finally consider recent advances and future challenges of integrating regime shift theory into holistic marine ecosystem-based management approaches.

Marine Ecosystems Regime Shifts All Over the World

Although the regime shift concept is still vigorously discussed, an increasing number of studies provide evidence for the potential of abrupt changes and surprises in marine ecosystems worldwide (Steneck et al. 2002; Beaugrand 2004; Mumby et al. 2007; Möllmann et al. 2008, 2009; Mumby 2009; Bestelmeyer et al. 2011; Frank et al. 2011, 2016; Llope et al. 2011; Beaugrand et al. 2015; Gårdmark et al. 2015; Ling et al. 2015; Vasilakopoulos and Marshall 2015; Auber et al. 2015). These studies, based on empirical observations, highlight mechanisms of regime shifts, firstly formulated by theoretical studies (Holling 1973; May 1977; Scheffer et al. 2001).

The Atlantic Cod Trophic Cascade

Surprises in natural systems are relatively common and can happen even in well-studied systems, due to different drivers. One driver of non-linear dynamics is the overfishing of top-predators. Top-predator overfishing may cause the depletion and collapse of their population resulting in unexpected ecosystem structure reorganizations through trophic cascades (Myers and Worm 2005; Fauchald 2010; Llope et al. 2011; Möllmann and Diekmann 2012; Steneck and Wahle 2013). Atlantic cod (*Gadus morhua*) is an important top-predator fish species, which can regulate marine ecosystems through top-down control, and has supported entire human communities through fisheries for centuries (Haedrich and Hamilton 2000; Myers and Worm 2005). After the industrial revolution and the increase of fishing power and capacity around the 1980s–1890s, many cod stocks collapsed bringing high economic losses (Myers et al. 1997; Frank et al. 2016). Multiple analyses conducted in different basins such as in the Baltic Sea or in the Eastern Scotian Shelf, showed that the collapse of cod stocks was caused by a combination of increased fishing pressure and unfavorable climatic conditions (Frank et al. 2005, 2016; Casini et al. 2008b; Möllmann et al. 2008, 2009). The high economic loss and social issues induced, led governments to adopt a range of management measures, such as drastic quota reductions and, in some cases, even fishing moratoria. Nevertheless, despite all the management measures and plans adopted, cod stocks failed to recover (Hutchings 2000; Frank et al. 2011; Hutchings and Rangeley 2011).

One of the reasons advanced to explain these management failures is the undergoing non-linear dynamics known as trophic cascades (Casini et al. 2008a; Star et al. 2011). Indeed, the collapse of this top-predator resulted in a shift from a cod-dominated to a forage fishes-dominated system (Frank et al. 2005; Gårdmark et al. 2015). Before overfishing, adult cod biomass level was high and cod controlled forage fish populations through predation. This hindered the forage fish from negatively impacting younger cod (through predation and/or competition), thus enhancing its biologically sustainable biomass. However, when cod biomass became depleted, the consequently increased forage fish abundance caused a further decline of cod population by increasing their negative direct (predation) or indirect (competition) impacts on younger cod. This feedback loop is then very difficult to reverse (Walters and Kitchell 2001; Möllmann et al. 2009; Nyström et al. 2012). Based on this example, it is clear how such systems can show two distinct configurations depending on their level of top-predator biomass. Of course, changes in mid-trophic levels will also reflect in lower ones, for instance high abundance of forage fishes will likely reduce plankton abundance. Under this new configuration with low cod biomass, a reduction in fishing pressure would likely lead to a very delayed or even none cod recovery, since new mechanisms would keep its population in the new depleted state. To summarize, both Baltic Sea and Scotian Shelf regime shifts were caused by a combination of overfishing and climate variation, and characterized by a trophic cascade (top-down mechanism) due to the depletion of Atlantic cod stocks (Frank et al. 2005; Casini et al. 2008b; Llope et al. 2011; Möllmann and Diekmann 2012). This led to immediate high social and economic losses for cod fishery, followed by a fisheries reorganization in order to adapt to the new ecosystem configuration. Finally, this regime shift led to a considerable increase of fisheries profits due to an outburst of lobster and crustaceans productivity.

Regime Shifts – A Global Challenge for the Sustainable Use of Our Marine Resources

Camilla Sguotti and Xochitl Cormon

Abstract

Over the last decades many marine systems have undergone drastic changes often resulting in new ecologically structured and sometimes economically less valuable states. In particular, the additive effects of anthropogenic stressors (e.g., fishing, climate change) seem to play a fundamental role in causing unexpected and sudden shifts between system states, generally termed regime shifts. Recently, many examples of regime shifts have been documented worldwide and their mechanisms and consequences have been vigorously discussed. Understanding causes and mechanisms of regime shifts is of great importance for the sustainable use of natural resources and their management, especially in marine ecosystems. Hence, we conducted a session entitled "Ecosystem dynamics in a changing world, regime shifts and resilience in marine communities" during the 8th YOUMARES conference (Kiel, 13–15th September 2017) to present regime shifts concepts and examples to a broad range of marine scientists (e.g., biologists and/or ecologists, physicists, climatologists, sociologists) and highlight their importance for the marine ecosystems worldwide.

In this chapter, we first provide examples of regime shifts which have occurred over the last decades in our oceans and discuss their potential implications for the sustainable use of marine resources; then we review regime shift theory and associated concepts. Finally, we review recent advances and future challenges to integrate regime shift theory into holistic marine ecosystem-based management approaches.

C. Sguotti (✉) · X. Cormon (✉)
Institute for Marine Ecosystem and Fishery Science, Centre for Earth System Research and Sustainability (CEN), University of Hamburg, Hamburg, Germany
e-mail: camilla.sguotti@uni-hamburg.de; xochitl.cormon@uni-hamburg.de

S. Jungblut et al. (eds.), *YOUMARES 8 – Oceans Across Boundaries: Learning from each other*,
https://doi.org/10.1007/978-3-319-93284-2_11

Introduction

Today, living marine resources represent a primary source of proteins for more than 2.6 billion people and support the livelihoods of about 11% of the world's population (UNESCO 2012; FAO 2014). Oceans worldwide concentrate dense and diversified human activities, e.g., fishing, tourism, shipping, offshore energy production, while experiencing a range of environmental pressures, e.g., increase of water temperature, acidification (Halpern et al. 2008; Boyd et al. 2014). Together anthropogenic and environmental pressures may threaten the integrity of marine systems and their sustainable use, altering their different components in many ways. These ecosystem changes may have great impacts for the social-ecological systems they are a part of, particularly when associated with changes in ecological keystone, cultural and/or commercial species (Garibaldi and Turner 2004; Casini et al. 2008a; Möllmann et al. 2008; Llope et al. 2011; Blenckner et al. 2015b).

The World Summit on Sustainable Development in Johannesburg (2002) provided a legally binding framework to implement the Ecosystem Approach to Fisheries Management (EAFM). This holistic approach aims (i) to conserve the structure, diversity and functioning of marine ecosystems and (ii) to provide the economic benefits of a sustainable exploitation of marine ecosystems. Scientific activities supporting approaches such as the EAFM are hence highly encouraged (FAO 2003). However, the insufficient knowledge on the diversity and entanglement of interactions between the ecological system components (deYoung et al. 2008), as well as their vulnerability to increasing anthropogenic and environmental pressures, may hinder successful management.

Even if systems may react to stressors in a non-linear way shifting suddenly to a different state and losing important ecosystem services, management is indeed still more based on continuous dynamics (Scheffer et al. 2001; Sugihara et al. 2012; Glaser et al. 2014; Travis et al. 2014; Levin and

Möllmann 2015). Some ecosystems may be able to absorb stronger disturbances than others depending on their characteristics, but in general, marine ecosystems are known to be particularly vulnerable to drastic and unexpected shifts, referred in ecology as regime shifts (deYoung et al. 2008). Such non-linear dynamics may have positive or negative outcomes for the sustainable use of natural resources and their management, therefore they should be taken into account and dealt with great precaution when taking environmental policy decisions (Holling 1973; Carpenter 2001; Scheffer 2009; Rocha et al. 2014a).

In this chapter, we first present some examples of marine ecosystems which have exhibited non-linear dynamics in response to external changes. These examples allow us to highlight different mechanisms potentially involved in regime shifts from an empirical point of view, as well as their potential implications for the sustainable use of marine resources. Secondly, we review the regime shift theory and associated concepts to finally consider recent advances and future challenges of integrating regime shift theory into holistic marine ecosystem-based management approaches.

Marine Ecosystems Regime Shifts All Over the World

Although the regime shift concept is still vigorously discussed, an increasing number of studies provide evidence for the potential of abrupt changes and surprises in marine ecosystems worldwide (Steneck et al. 2002; Beaugrand 2004; Mumby et al. 2007; Möllmann et al. 2008, 2009; Mumby 2009; Bestelmeyer et al. 2011; Frank et al. 2011, 2016; Llope et al. 2011; Beaugrand et al. 2015; Gårdmark et al. 2015; Ling et al. 2015; Vasilakopoulos and Marshall 2015; Auber et al. 2015). These studies, based on empirical observations, highlight mechanisms of regime shifts, firstly formulated by theoretical studies (Holling 1973; May 1977; Scheffer et al. 2001).

The Atlantic Cod Trophic Cascade

Surprises in natural systems are relatively common and can happen even in well-studied systems, due to different drivers. One driver of non-linear dynamics is the overfishing of top-predators. Top-predator overfishing may cause the depletion and collapse of their population resulting in unexpected ecosystem structure reorganizations through trophic cascades (Myers and Worm 2005; Fauchald 2010; Llope et al. 2011; Möllmann and Diekmann 2012; Steneck and Wahle 2013). Atlantic cod (*Gadus morhua*) is an important top-predator fish species, which can regulate marine ecosystems through top-down control, and has supported entire human communities through fisheries for centuries (Haedrich and Hamilton 2000; Myers and Worm 2005). After the industrial revolution and the increase of fishing power and capacity around the 1980s–1890s, many cod stocks collapsed bringing high economic losses (Myers et al. 1997; Frank et al. 2016). Multiple analyses conducted in different basins such as in the Baltic Sea or in the Eastern Scotian Shelf, showed that the collapse of cod stocks was caused by a combination of increased fishing pressure and unfavorable climatic conditions (Frank et al. 2005, 2016; Casini et al. 2008b; Möllmann et al. 2008, 2009). The high economic loss and social issues induced, led governments to adopt a range of management measures, such as drastic quota reductions and, in some cases, even fishing moratoria. Nevertheless, despite all the management measures and plans adopted, cod stocks failed to recover (Hutchings 2000; Frank et al. 2011; Hutchings and Rangeley 2011).

One of the reasons advanced to explain these management failures is the undergoing non-linear dynamics known as trophic cascades (Casini et al. 2008a; Star et al. 2011). Indeed, the collapse of this top-predator resulted in a shift from a cod-dominated to a forage fishes-dominated system (Frank et al. 2005; Gårdmark et al. 2015). Before overfishing, adult cod biomass level was high and cod controlled forage fish populations through predation. This hindered the forage fish from negatively impacting younger cod (through predation and/or competition), thus enhancing its biologically sustainable biomass. However, when cod biomass became depleted, the consequently increased forage fish abundance caused a further decline of cod population by increasing their negative direct (predation) or indirect (competition) impacts on younger cod. This feedback loop is then very difficult to reverse (Walters and Kitchell 2001; Möllmann et al. 2009; Nyström et al. 2012). Based on this example, it is clear how such systems can show two distinct configurations depending on their level of top-predator biomass. Of course, changes in mid-trophic levels will also reflect in lower ones, for instance high abundance of forage fishes will likely reduce plankton abundance. Under this new configuration with low cod biomass, a reduction in fishing pressure would likely lead to a very delayed or even none cod recovery, since new mechanisms would keep its population in the new depleted state. To summarize, both Baltic Sea and Scotian Shelf regime shifts were caused by a combination of overfishing and climate variation, and characterized by a trophic cascade (top-down mechanism) due to the depletion of Atlantic cod stocks (Frank et al. 2005; Casini et al. 2008b; Llope et al. 2011; Möllmann and Diekmann 2012). This led to immediate high social and economic losses for cod fishery, followed by a fisheries reorganization in order to adapt to the new ecosystem configuration. Finally, this regime shift led to a considerable increase of fisheries profits due to an outburst of lobster and crustaceans productivity.

The North Sea Regime Shift

The North Sea regime shift involved different mechanisms that induced changes which started at the bottom of the trophic chain and propagated up to higher trophic levels (Reid et al. 2001; Beaugrand 2004; deYoung et al. 2008; Conversi et al. 2010; Lynam et al. 2017). The North Sea regime shift occurred during the 1980s and was mainly induced by a combination of increased sea surface temperatures and changes in hydro-climatic forces (Beaugrand 2004). Due to the increase of sea surface temperature and changes in the water inflows, phytoplankton biomass increased. As a consequence, the zooplankton assemblage, originally dominated by cold waters species, e.g., *Calanus finmarchicus*, shifted to an assemblage dominated by warmer water species, e.g., *Calanus helgolandicus* and gelatinous zooplankton such as jellyfish (Reid et al. 2001; Beaugrand 2004; Möllmann and Diekmann 2012). These changes in the zooplankton community, combined with hydro-climatic changes, propagated to higher trophic levels. Changes in temperature and/or salinity led to an increase of flatfish biomass (Möllmann and Diekmann 2012) while the decline of *C. finmarchicus,* which is the preferred prey of gadoids and especially of cod larvae, led to cod recruitment failures (Beaugrand et al. 2003; Beaugrand 2004) enhancing the negative sea warming effects. These changes in recruitment had a lagged impact on the adult gadoids biomass that, already stressed by overfishing, started to decline inexorably at the end of the 1980s (Hislop 1996). The changes in fish biomass and composition, together with warmer temperatures, favored the emergence of previously scarcely present species such as horse mackerel (*Trachurus trachurus*) and mackerel (*Scomber scombrus*), especially in the northern North Sea (Reid et al. 2001; Beaugrand et al. 2003; Beaugrand 2004).

This regime shift, induced by bottom-up processes, was more qualitative than quantitative in the sense that changes in assemblage and not in total biomass of trophic levels occurred (Beaugrand 2004). The dynamics of these changes highlighted different response time patterns depending on the organisms affected. Indeed, the phytoplankton and zooplankton communities, with their fast life cycles, responded to climatic changes faster than the fish community. Spatial patterns were also different: the coastal areas were less sensitive to change in hydrodynamic conditions, and the regime shift was stronger in the northern North Sea (Reid et al. 2001; Beaugrand 2004; Möllmann and Diekmann 2012). This regime shift completely changed the structure of the North Sea fish community and led to the decline of various commercial species like cod, while the abundance of other species like flatfishes and mackerel increased, consequently having impacts on fisheries (Reid et al. 2001).

Coral Reefs and Kelp Forests Transitions

Other examples of marine regime shifts are coral and kelps transitions (Rocha et al. 2014b). For instance, the Caribbean coral reefs were flourishing ecosystems providing many ecosystem services, sustaining large fish populations and associated human communities. The integrity of the reefs depended on the presence of sea urchins and grazing fishes, which, by eating the algae, maintained the coral reef structure. When the populations of grazing fish started to decrease due to overfishing, nothing seemed to change in the system. Indeed, sea urchins were still able to regulate algae population through predation, preserving the reef structure (Nyström 2006; Standish et al. 2014). However, the ability of the reef to absorb disturbances was already eroded by overfishing, when two concomitant and dramatic events occurred, leading to the total destruction of the reef (Mumby et al. 2007). Sea-urchin populations quickly collapsed due to an illness outbreak, while more nutrients, discarded from the islands, were added to the system, causing rapid eutrophication. In a short time, coral reefs were substituted by algae beds which were not regulated by any top-down (sea urchin predation) or bottom-up (limitation of nutrients) processes. This algae-dominated system is now difficult to reverse due to the feedback mechanisms maintaining the system in its new status (i.e., the number of new algae growing every year can impede the reintroduction of corals, Mumby et al. 2007; Mumby 2009; Kates et al. 2012).

Similarly, kelp forests are highly diverse ecosystems which can maintain flourishing fish populations and offer many services for humans such as fisheries and cultural values (Steneck et al. 2013; Ling et al. 2015). Kelp forests are mainly maintained by fish predation on sea urchins, which controls sea urchin populations. In Australia, overharvesting of predatory fish, coupled with diseases weakening the kelp, led to a boom of the sea urchin population and a shift from high biodiversity kelp forest to poorer urchin's barren (Ling et al. 2015). This state was then difficult to reverse due to various feedback mechanisms such as the increase of juvenile urchin abundance and facilitation of juvenile survival, but also because of the lack of efficient measures to recover the stocks of the sea urchin's predators (Ling et al. 2015). In these two examples, the regime shifts were caused by multiple stressors which altered the regulation (top-down and/or bottom-up) of previously highly productive ecosystems and led to huge economic, social and ecological losses. Similarly to the Atlantic cod example, management measures failed to reverse these unexpected regime shifts due to feedback loop mechanisms (Steneck et al. 2002; Ling et al. 2015).

From Examples to Theory

From these four examples, several conclusions can be drawn. Stressors potentially inducing regime shifts may affect a system gradually, e.g., decline of top-predator due to fishing (Baltic Sea and Scotian Shelf regime shifts), or abrupt and exceptionally, e.g., disease outbreak (Caribbean coral reef destruction). The examples of the Atlantic cod stock collapse and the North Sea regime shift showed that climate change may play and important role in such mechanisms (Beaugrand 2004; Conversi et al. 2015; Yletyinen et al. 2016). In addition, these examples showed the cumulative effects of different stressors and how they may act together in synergistic ways. The mechanisms and processes involved in regime shifts may be induced by top-down and/or bottom-up regulation (Holling 1973; Beisner et al. 2003; Conversi et al. 2015; Pershing et al. 2015). Finally, these examples highlight the importance and necessity to understand regime shifts mechanisms for a sustainable use of marine resources in order to provide ecosystem services and benefits for human communities (Doak et al. 2008). Also, they uncovered some fundamental properties of regime shifts, e.g., the abruptness of changes and their lack or low reversibility (Scheffer et al. 2001, 2015; Dakos et al. 2012). However, due to the complexity and entanglement of the mechanisms involved, defining regime shifts based on empirical evidences is challenging. A review of the concepts associated with regime shifts, which are mostly theoretical (Levin and Möllmann 2015), is essential to understand the non-linear mechanisms potentially involved in complex systems dynamics, particularly in a time of pronounced environmental changes.

The Regime Shift Theory

Different mathematical frameworks lead to the development of the regime shift theory (Jones 1975, 1977; Thom 1975; Crawford 1991), describing how changes in some controlling factors can lead to huge and abrupt changes in various systems (e.g., biological, physical, behavioral; Jones 1975; Carpenter 2001; Scheffer et al. 2001). Marine regime shifts can be defined as dramatic and abrupt changes in the system structure and function that are persistent in time, where the system can range from a single cell to a population or an ecosystem (Beisner et al. 2003; Scheffer and Carpenter 2003). Due to the high number of terminologies and definitions used in the literature, a glossary was added to this chapter in order to have consistent and clear definitions. All terms highlighted in italics in the following text can be found in the glossary section (Box 1).

Box 1: Glossary

Regime shift: dramatic and abrupt change in the structure and function of a system causing a shift between two alternate stable states following discontinuous non-linear dynamics and exhibiting three equilibria. There are some debates about the definition and *critical transition* or *phase shift* might be considered synonyms depending on the literature.

Resilience: capacity of the system to absorb disturbances and reorganize in a way that it retains the same functions, structure, identity and feedback mechanisms, potentially impeding a regime shift.

Regime: dynamic system configuration maintaining certain structures and functions. It is also known as *stable state, basin of attraction* or *domain of attraction.*

Tipping point: threshold separating two dynamics regimes. It is also known as *critical threshold* or *bifurcation point.*

Feedback mechanism: ecological mechanisms stabilizing a regime by amplifying (positive) or damping (negative) the response to a forcing. Positive feedbacks (reinforcing) move the system to an alternate stable state, out of equilibrium. Negative feedbacks (balancing) maintain the status of the system, close to the equilibrium dynamics.

Hysteresis: phenomenon for which the return path from regime B to regime A, is drastically different from the path that led from regime A to regime B.

The easiest way to understand and visualize *regime shifts* is the example of the ball-in-cup or ball-in-valley diagram developed from the pioneer work of Poincare in the 1800's in Crawford 1991; Fig. 1). The ball represents the study system, for instance the Caribbean coral reef. The system reef (our ball) has certain parameters such as coral abundance, coverage, and biodiversity. The system state is represented by the valley in which our ball (system) lies (regime 1 in Fig. 1). The dimension of the valley (width and height in our two dimensions' figure) corresponds to the

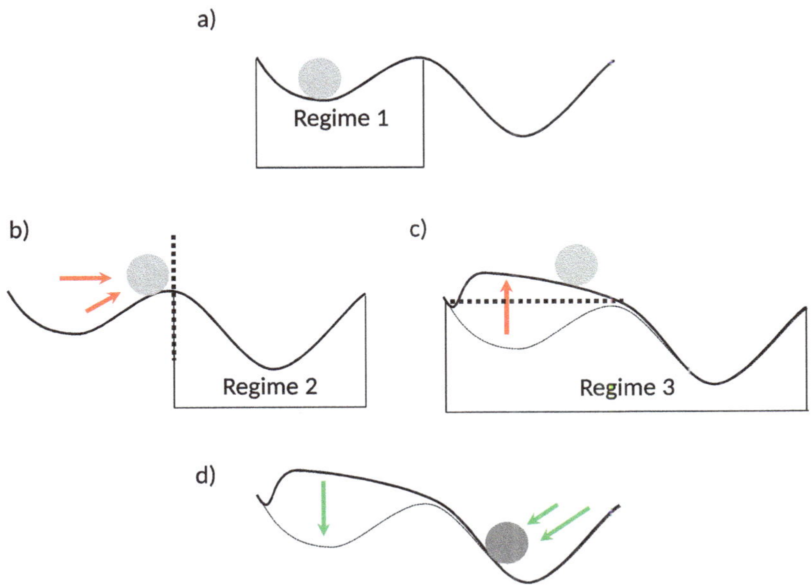

Fig. 1 Regime shift theory represented by ball-in-cup diagrams (Crawford 1991). The ball represents the system and the cups (or valleys) the system states (see text for more information). The thick dotted lines represent the tipping points. The arrows represent disturbances, red for disturbances inducing a shift and green for reversed disturbances having no effects. (**a**) System in its original state. (**b**) Regime shift induced by changes in system state variables. (**c**) Regime shift induced by change in system parameter variables. (**d**) System in its new state showing hysteresis. Referring to our Caribbean example (section "Coral reefs and kelp forests transitions") the light grey ball represents coral reef dominated system while the dark grey ball, the algae dominated system

resilience of the system state. For instance, even when the Caribbean coral reef system was stressed by intensive fishing on grazing fishes, the system maintained its original state and did not shift because its resilience was high (i.e., the sea urchins were able to maintain top-down regulation on algae, Mumby et al. 2007). Indeed, when the valley is large and deep, the ball/system remains in it, maintaining its structure, despite the disturbances. Repetitive disturbances such as overfishing and eutrophication did, however, reduce the system resilience (the valley became narrower and shallower) and when a strong disturbance occurred (here a disease outbreak), the system shifted abruptly to a new state (i.e., algae beds). This new state is now resilient, maintained by new *feedback mechanisms* that help its stabilization, e.g., the higher survival of algae and the non-recovery of grazer fishes (Beisner et al. 2003; Roe 2009; Conversi et al. 2015). Resilience is defined as the capacity of the system to absorb disturbances and reorganize, so as to still retain essentially the same functions, structure, identity and feedback mechanisms (Holling 1973; Beisner et al. 2003; Vasilakopoulos and Marshall 2015; Folke 2016).

Some perturbations may act either on the system state variables (pushing our ball from its valley into a new one, e.g., disease outbreak, Fig. 1b) or on the system parameter variables (modifying the shape of the valley, hence affecting system resilience, e.g., overfishing and eutrophication, Fig. 1c; Beisner et al. 2003). As highlighted by the Caribbean coral reefs example, it is the combination of multiple mechanisms that generally causes a system to shift from a stable state to another (Biggs et al. 2012). This shift of a system between two alternate stable states is the foundation of regime shift theory (Carpenter 2001; Scheffer et al. 2001). The separation point between two *regimes* (or alternate stable states) is the so-called *tipping point* (Selkoe et al. 2015). Once crossed, the system will shift to a new regime with new characterizing parameters. Clearly, once a tipping point is crossed, it is not easy to push the ball back in its original valley, since the new valley is deep and large, thus highly resilient, and/or the original valley might have disappeared. This can hinder a return of the system to the previous state even when disturbances stop (e.g., fishing ban, end of disease outbreak) or are reversed (Figs. 1d and 2; Beisner et al. 2003). This property of regime shifts is called *hysteresis* and can be defined as the phenomenon for which the return path of a system from the altered to the original state can be drastically different from the one which have led to this altered state (Beisner et al. 2003; Bestelmeyer et al. 2011). Hysteresis is a typical feature of discontinuous regime shifts and can be detected when the relationship between the stressors and the system differs depending on the regime (stable state) of the system (Scheffer and Carpenter 2003; Bestelmeyer et al. 2011).

Another way to visualize the regime shift is the fold bifurcation curve (Fig. 2; Scheffer et al. 2001). The system reacts in a smooth way to condition changes until a tipping point

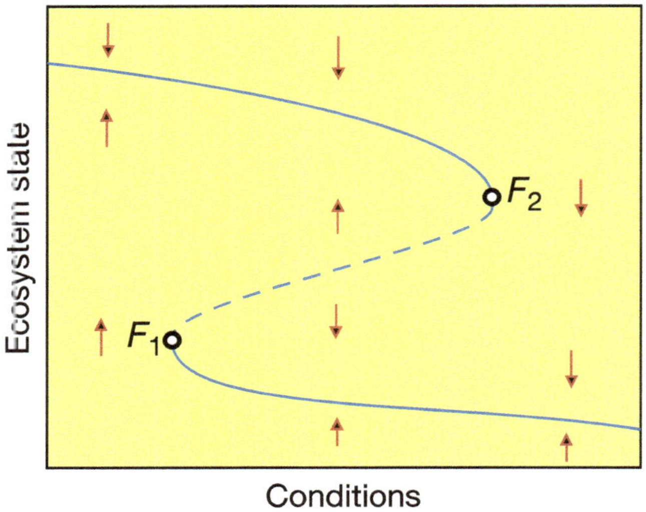

Fig. 2 Fold bifurcation curve (Reproduced from Scheffer et al. 2001). The dashed line represents the unstable equilibria and the border between the two alternate stable states represented by plain lines. F1 and F2 represent the tipping points

(F1 or F2) is reached and the system jumps from one state to another. In the area of discontinuity (Fig. 2, dashed blue line) the system can present three equilibria. As evidenced by this visualization, systems that show such behavior are difficult to reverse to previous state even when condition changes are reversed (hysteresis). Although some debates exist regarding the definition of regime shift we adopted the definition of Scheffer et al. (2001) and Selkoe et al. (2015) of an abrupt change over time with discontinuous dynamics exhibiting hysteresis. This is opposed to phase shifts *sensu* Selkoe et al. (2015), where system state's response to condition change is continuous, e.g., a logistic response, with two states but only one equilibrium.

Resilience, feedback mechanisms, tipping points and hysteresis are important attributes of regime shifts (van der Maas et al. 2003; Bestelmeyer et al. 2011). These properties make regime shifts extremely important in the real world and have profound implications for management (Travis et al. 2014; Selkoe et al. 2015; Angeler et al. 2016). Imagine having as system a fish population. When you start fishing, the population still manage to absorb the perturbation and might decline, but would remain in a state with high biomass, high recruitment, a certain growth rate, etc. At some point the fishing pressure, usually combined with other external stressors, increases so much that the population collapses and its internal mechanisms change. The exploited population is now in a new state at low abundance, possibly with different growth and mortality rates. Now suppose that we are the managers. We could assume that reducing the fishing pressure to precollapse levels would make the population quickly rebound. This could work in a context of linear dynamics but if the population has crossed a tipping point and it is now in a new alternate stable state, controlled by new mechanisms that cause hysteresis, recovery of the system may be slow and difficult, or even impossible. From this example, the importance of regime shift appears clear. In order to apply efficient and useful management measures, we should aim to detect regime shifts in advance or, at least, we should consider the possibility that an exploited system can show non-linear behaviors, and apply precautionary management approaches (Holling 1973; Carpenter 2001; Scheffer and Carpenter 2003; deYoung et al. 2008; Dakos et al. 2012; Punt et al. 2012; Levin and Möllmann 2015). Many marine ecosystems have undergone drastic shifts often resulting in new ecologically structured and/or economically less valuable states (Conversi et al. 2015; Möllmann et al. 2015). These regimes shifts have brought catastrophic ecological and social consequences (Rocha et al. 2015), such as economic losses, social issues and losses of ecosystem services (Casini et al. 2008a; Möllmann et al. 2008; Blenckner et al. 2015b). Thus, since several processes at several levels of the ecosystem are often involved, it appears evident from these examples that an ecosystem approach to management of marine ecosystems prone to regime shifts is essential (Long et al. 2015).

Challenges and Implications of Regime Shifts for Management Purpose

To include the concept of regime shift into management perspectives, multiple *a priori* steps have to be made to first identify the mechanisms and the drivers involved (feedback loops, interactions, etc.), and then integrate this information into suitable and adapted policy. The documentation of a broad range of regime shift examples, involving different mechanisms applied to different ecosystems may be very useful to compare the various processes involved, to understand potential implications in a better way (Rocha et al. 2015) and therefore to adapt management to local characteristics (deYoung et al. 2008). In this context, the Regime Shift Database (Rocha et al. 2014b), based on a participatory approach, aims to review regime shifts of social-ecological systems worldwide with a particular focus on regime shifts having a potential large impact on human well-being and ecosystem services. This database, available online (www.regimeshifts.org), is an initiative led by the Stockholm Resilience Centre to increase general knowledge and understanding of regime shifts and associated concepts and to help managers and policy makers to take these concepts into account in their future decisions.

Knowledge of different mechanisms and local characteristics of regime shifts may facilitate their detection. Indeed, the first step and challenge to consider regime shifts in management, is to actually detect them (Carpenter 2001; deYoung et al. 2008; Rocha et al. 2015). For instance, regime shifts in the North Sea and English Channel communities were only detected 10 years after they occurred (Beaugrand 2004; Auber et al. 2015). This late detection may partly be explained by the very large scale at which these shifts occurred and highlights the need of studying different spatial scales when wanting to understand ecosystems processes and dynamics. Similarly, temporal scales of changes might be different depending on the lifespan of the affected organisms and might lead to temporal lags in system responses to stressors (Holling 1973; deYoung et al. 2008) as it was the case in the North Sea. These differences in spatio-temporal patterns need to be addressed and disentangled as they might hinder or delay regime shift detection and exacerbate social and economic consequences (Levin 1992; Scheffer and Carpenter 2003; Kerkhoff and Enquist 2007; Levin and Möllmann 2015). It might also be necessary to disentangle regime shifts (*sensu* Selkoe et al. 2015) from simple logistic dynamics and highlight hysteresis (which requires additional observations in time). For these reasons, regime shift detection requires long and extensive observation datasets of the system which is generally costly in time and money (Carpenter 2001; Scheffer et al. 2009; Levin and Möllmann 2015). Moreover, the required time to obtain time series of suitable length might prove too long, particularly when such shifts strongly impact ecosystems services and human well-being. For these reasons, experimental studies are necessary to enhance the understanding of systems responses to disturbances (Angeler et al. 2016). Particularly, experiments may help to understand multi-causality and dual relationships between stressors and systems which generally participate in hindering detection of regimes shifts (Scheffer and Carpenter 2003; Conversi et al. 2015; Levin and Möllmann 2015).

While regime shifts detection may be delayed, their unexpected and abrupt behavior hinders regime shift prediction, which is necessary to ensure effective management measures. In addition, a post-regime shift detection may result in increased management challenges, particularly due to hysteresis, as described in the previous section for coral reefs (Mumby et al. 2007; Mumby 2009), kelp forests (Steneck et al. 2002) and various fish stock shifts (Myers et al. 1997; Hutchings 2000; Myers and Worm 2005; Hutchings and Rangeley 2011). Challenges in prediction may be partly related to the common use of linear relationships to statistically describe natural processes which need to be overcome in favor of more realistic (thus more complex) models (Holling 1973; Ludwig et al. 1997; Scheffer and Carpenter 2003). Indeed, the non-linear relationships between stressors and system variables need to be understood to be able to correctly predict system responses. Also, a new branch of science has been currently developing regime shift indicators, the so-called early-warning signals, to anticipate regimes shifts. These signals are generally based on the fact that the recovery of a highly disturbed system to an equilibrium is slow, i.e., critically slowing down (Scheffer et al. 2001, 2015; Dakos et al. 2012; Lindegren et al. 2012). Indeed, when systems are close to tipping points, their stability decreases, generally leading to an increased variability, and autocorrelation of the data describing them. These indicators work well with simulation models but still they have some limitations in predicting shifts using empirical data (Dakos et al. 2008, 2017; Scheffer et al. 2009; Dai et al. 2013). They may be constrained by the length of the times series available and/or the limited amount of data, by methodological assumptions and/or sampling errors (deYoung et al. 2008; Lindegren et al. 2012). Moreover, they are not suitable to predict stochastically driven shifts. To overcome these limitations Lade and Gross (2012) developed a new approach to detect early warning signals with reduced time-series. Lindegren et al. (2012) recommended a multiple approach based on knowledge of the system and its local characteristics (key ecological thresholds, relationships with drivers), data availability, sensitivity and bias of the analysis carried out. Such advances need to be followed by the scientific community to develop more approaches overcoming these limitations. Alternative sources of data, e.g., public records and narratives, must be found and used, particularly when ecological data are not available, and systems must be monitored at an appropriate time scale to ensure shift detection as early as possible.

Because prediction of regime shifts is so challenging, and because the potential consequences for ecosystem services and human well-being may be abrupt and very difficult (or even impossible) to reverse, precautionary approaches are recommended (Holling 1973; Carpenter 2001; Scheffer and Carpenter 2003; Selkoe et al. 2015). When managing systems prone to regime shifts, risks and uncertainties must be assessed before any management action is taken (Levin and Möllmann 2015; DePiper et al. 2017). Risk assessment requires a clear definition of the system of interest, its potential tipping points, as well as suitable indicators. However, all the challenges already mentioned (multiple-causality, dual relationships to drivers, spatio-temporal different patterns, limitation of data, etc.) may impede the definition of appropriate indicators (Kelly et al. 2015; Selkoe et al. 2015). For instance, Vasilakopoulos and Marshall (2015) showed that the spawning stock biomass (SSB) of Barents Sea cod did not suffice to detect a regime shift of this population, while SSB levels are generally the reference points used in current fishery management plans (single- or multi-species advices), and sometimes the only ones. These results evidence the need to base scientific advice to fishery managers on the monitoring of several ecosystem (community/population) parameters, particularly when suspecting potential impending shifts. Similarly, stressors effects may be unclear when studied individually, while their importance may appear only when combined with other stressors (Rocha et al. 2015; Vasilakopoulos and Marshall 2015). The factors undermining resilience (eutrophication, global warming, species invasion, etc.) should be of prior concern as small variations in stressors might lead to large changes in ecosystem structure and/or functioning when resilience is eroded (Ricker 1963; Ludwig et al. 1997; Scheffer et al. 2001; Beisner et al. 2003; Scheffer and Carpenter 2003). The quantitative assessment of risk and associated resilience is difficult and challenging. Economic cost-benefit analysis might be useful when trying to quantify risks for ecosystem services (Carpenter 2001), however, it might totally underestimate them when too narrow-focused, e.g., focusing on yield in fisheries while neglecting age-structure of the stock (deYoung et al. 2008). Quantitative assessment of resilience may prove very useful but requires an extensive amount of data particularly in complex systems (Vasilakopoulos and Marshall 2015). Therefore, qualitative analysis and/or conceptual models may be preferred (DePiper et al. 2017), particularly when studying data-poor systems or when dealing with complex adaptive systems such as social-ecological ones.

Despite the increasing effort in scientific research, even when risk (or resilience) may be assessed, ecological uncertainties (about system evolution) and livelihood uncertainties (about impacts on human communities) related to regime shifts are high (Pindyck 2000). When managing social-ecological systems (SES) prone to regime shifts, policy makers must face these uncertainties and different management strategies might emerge: reducing or limiting system stressors (mitigation), building up system resilience (adaptation) and/or reversing a shift (restoration, Kates et al. 2012; Angeler et al. 2013). These strategies might have different outcomes, benefits, costs and efficiency depending of goals and focus of management as well as the status of the system (Selkoe et al. 2015; Lade et al. 2015; Fenichel and Horan 2016; Mathias et al. 2017). For example, because of hysteresis, building up resilience might be more effective and less costly than restoration measures (Selkoe et al. 2015). These measures might also require different levels of governance. For instance, the reduction of tuna fishing effort in the Pacific Ocean would require an international consortium for management to be efficient while similar measures applied to a coral reef fishery would be relevant at the local management scale. In addition, when mitigation generally requires international and global management (e.g., gas emissions reduction), building up systems resilience (adaptation) may succeed at local scales, countering global inaction (Rocha et al. 2015). While decreasing variance of a system may seem a good idea, Carpenter et al. (2015) highlighted the adverse effects for system resilience management. Staying within a safe-operating space (Rockström et al. 2009), including uncertainties around tipping points and using history as guideline (Fenichel and Horan 2016; Liski and Salanié 2016) might, however, prove effective and reduce risks of management failures. Adversely, managers might need to erode resilience of a system to tip it towards a preferable regime, i.e., more pristine or more valuable (Derissen et al. 2011). This so-called transformation would require intentional changes in the institutional framework in which the utilization of marine systems (e.g., including switch to a novel management system), as well as a transparent and equitable redistribution of benefits across stakeholders takes place (Selkoe et al. 2015). Uncertainties may as well increase immediate costs, and even if costs of inaction would be high in the future, they might hinder immediate decisions (Pindyck 2000; Selkoe et al. 2015).

Adaptive co-management might be ideal when cooperation between local and global stakeholders is possible (Plummer et al. 2017). However, it might slow down management processes opposed to the potential flexibility and responsiveness of local stakeholders required for a good management of regime shift effects (deYoung et al. 2008; Horan et al. 2011; Blenckner et al. 2015a; Rocha et al. 2015; Valman et al. 2016). Similarly, polycentric governance holds great potential at the international scale but is vulnerable to negative interactions between institutions and weak coordination (Galaz et al. 2012; Mathias et al. 2017). In both cases, the question of responsibility might be raised in case of management failures (Baumgärtner et al. 2006; Fenichel and Horan 2016). Local and/or global stakeholder cooperation, as well as responsiveness, may be improved by the knowledge of the stressors involved in regime shifts mechanisms, their shared interactions with the different components of the system, and the different scales at which they interact (Rocha et al. 2015). Such knowledge may also help policy makers to set suitable management targets otherwise challenged when uncertainties are high.

Finally, the integration of management and regime shift theory may prove quite complicated. The complex responses to stressors, the multiple, cross-disciplinary interactions between each system components, the high uncertainties and the different stakeholder perspectives and conflicts need to be understood and accounted for when considering regime shifts (and/or resilience) in social-ecological systems (SES) management decisions. This requires holistic and integrative approaches such as integrative ecosystem assessment (IEA, (Levin and Möllmann 2015). In this context, scientists have recently developed frameworks to conceptualize SES and assess their sustainability and uncertainties (Ostrom 2009; Leslie et al. 2015; Levin et al. 2016). Particularly, these frameworks allow the combination of classic scientific information and local stakeholders' ecological, cultural and/or social knowledge of the system. These conceptual models may be used to promote interdisciplinary research, discussions between stakeholders, and allow a holistic management strategy evaluation after their operationalization (Levin and Möllmann 2015; Levin et al. 2016; DePiper et al. 2017).

Conclusions

Regime shifts are abrupt changes that can happen in complex systems worldwide at different temporal and spatial scales, depending on the resilience of the systems (Scheffer et al. 2001; deYoung et al. 2008). It is extremely important to study and understand these mechanisms since many regime shifts have led to catastrophic changes including ecological, social and economic losses worldwide (Mumby 2009; Steneck and Wahle 2013; Blenckner et al. 2015b). Despite the fact that many studies and methods have focused on the detection of regime shifts, there is still a lot to be done to achieve marine ecosystem management integrating resilience and regime shifts (Travis et al. 2014; Selkoe et al. 2015; Angeler et al. 2016). New tools, such as early warning signals or new ways to assess the resilience of different systems, combined with an in-depth study of the mechanisms and stressors affecting natural systems are a good start to incorporate resilience and regime shift into policy-maker decisions (Carpenter and Brock 2006; Scheffer et al. 2009; Dakos et al. 2012, 2017; Ling et al. 2015; Vasilakopoulos and Marshall 2015). Since regime shifts often affect many components of an ecosystem in different ways, ecosystem-based management (EBM) is necessary to include effectively regime shifts into management considerations (Blenckner et al. 2015a; Levin and Möllmann 2015; Long et al. 2015; Rocha et al. 2015). To make this holistic approach effective and to preserve the natural environment and ecosystems in a more integrative way, there is a real need to translate regime shift and resilience concepts from theory to applications (Punt et al. 2012; Travis et al. 2014; Selkoe et al. 2015). Recently, the operationalization of social-ecological systems (SES) conceptual models have shown promising improvements in this sense (Leslie et al. 2015; DePiper et al. 2017). Due to the different spatial and temporal scales at which regime shifts can act, i.e., from extremely local to global, and the high degree of associated uncertainties, innovative and flexible management options need to be developed at different levels of governance. For instance, Rockström et al. (2009) suggested a management at the planetary boundaries. Such management would require, in addition to adaptive management and polycentric governance, a societal shift in order to achieve a fair use of global resources, and a transformed economy (Hughes et al. 2013; Lade et al. 2013; O'Brien et al. 2014). Finally, we can expect that the increasing awareness of the implications of regime shifts and associated concepts for human well-being worldwide will likely lead to more precautionary management approaches, while new tools and technics will be developed to achieve an integrative and efficient management of our natural resources.

Acknowledgments The authors gratefully acknowledge financial support from the MARmaED project (CS) and marEEshift project (XC) without which this study could not have been conducted. The MARmaED project has received funding from the European Union's Horizon 2020 research and innovation program under the Marie Skłodowska-Curie grant agreement No 675997. The results of this study reflect only the author's view and the Commission is not responsible for any use that may be made of the information it contains. The marEEshift - marine ecological economic systems in the Western Baltic Sea and beyond: shifting the baseline to a regime of sustainability - project has received funding by German Ministry for Education and Research.

We thank Christian Möllmann for his suggestions on the first draft of this manuscript. Finally, we are indebted to the thoughtful comments of three anonymous referees, who helped to improve the initial manuscript.

Appendix

This article is related to the YOUMARES 8 conference session no. 5: "Ecosystems Dynamics in a Changing World: Regime Shifts and Resilience in Marine Communities". The original Call for Abstracts the abstracts of the presentations, and the report of the session can be found in the appendix "Conference Sessions and Abstracts", chapter "10 Ecosystems Dynamics in a Changing World: Regime Shifts and Resilience in Marine Communities".

References

Angeler DG, Allen CR, Johnson RK (2013) Measuring the relative resilience of subarctic lakes to global change: redundancies of functions within and across temporal scales. J Appl Ecol 50:572–584. https://doi.org/10.1111/1365-2664.12092

Angeler DG, Allen CR, Barichievy C et al (2016) Management applications of discontinuity theory. J Appl Ecol 53:688–698. https://doi.org/10.1111/1365-2664.12494

Auber A, Travers-Trolet M, Villanueva MC et al (2015) Regime shift in an exploited fish community related to natural climate oscillations. PLoS One 10:e0129883. https://doi.org/10.1371/journal.pone.0129883

Baumgärtner S, Faber M, Schiller J (2006) Joint production and responsibility in ecological economics – on the foundations of environmental policy, advances. Edward Elgar Publishing, Cheltenham

Beaugrand G (2004) The North Sea regime shift: evidence, causes, mechanisms and consequences. Prog Oceanogr 60:245–262. https://doi.org/10.1016/j.pocean.2004.02.018

Beaugrand G, Brander KM, Alistair Lindley J et al (2003) Plankton effect on cod recruitment in the North Sea. Nature 426:661–664. https://doi.org/10.1038/nature02164

Beaugrand G, Conversi A, Chiba S et al (2015) Synchronous marine pelagic regime shifts in the Northern Hemisphere. Philos Trans R Soc B 370:20130272. https://doi.org/10.1098/rstb.2013.0272

Beisner BE, Haydon DT, Cuddington K (2003) Alternative stable states in ecology. Front Ecol Environ 1:376–382. https://doi.org/10.1890/1540-9295(2003)001[0376:ASSIE]2.0.CO;2

Bestelmeyer BT, Ellison AM, Fraser WR et al (2011) Analysis of abrupt transitions in ecological systems. Ecosphere 2:129. https://doi.org/10.1890/ES11-00216.1

Biggs R, Blenckner T, Folke C et al (2012) Regime shifts. In: Encyclopedia of theoretical ecology. University of California Press, Berkeley, pp 609–617

Blenckner T, Kannen A, Barausse A et al (2015a) Past and future challenges in managing European seas. Ecol Soc 20:40. https://doi.org/10.5751/ES-07246-200140

Blenckner T, Llope M, Möllmann C et al (2015b) Climate and fishing steer ecosystem regeneration to uncertain economic futures. Proc R Soc B 282:20142809. https://doi.org/10.1098/rspb.2014.2809

Boyd PW, Cheung W, Lluch-Cota SE et al (2014) Ocean systems. In: Climate change 2014: impacts, adaptation, and vulnerability. Part A: global and sectoral aspects. Contribution of working group II to the fifth assessment report of the intergovernmental panel on climate change. In: Field CB, Barros VR, Dokken DJ et al (eds) Fifth assesment report. Cambridge University Press, Cambridge/New York, pp 411–484

Carpenter SR (2001) Alternate states of ecosystems: evidence and some implications. In: Press MC, Huntly NJ, Levin S (eds) Ecology: achievement and challenge. Blackwell Science, Oxford, pp 357–383

Carpenter SR, Brock WA (2006) Rising variance: a leading indicator of ecological transition. Ecol Lett 9:311–318. https://doi.org/10.1111/j.1461-0248.2005.00877.x

Carpenter SR, Brock WA, Folke C et al (2015) Allowing variance may enlarge the safe operating space for exploited ecosystems. Proc Natl Acad Sci U S A 112:14384–14389. https://doi.org/10.1073/pnas.1511804112

Casini M, Hjelm J, Molinero JC et al (2008a) Trophic cascades promote threshold-like shifts. Proc Natl Acad Sci U S A 106:197–202. https://doi.org/10.1073/pnas.0806649105

Casini M, Lövgren J, Hjelm J et al (2008b) Multi-level trophic cascades in a heavily exploited open marine ecosystem. Proc R Soc B 275:1793–1801. https://doi.org/10.1098/rspb.2007.1752

Conversi A, Umani SF, Peluso T et al (2010) The mediterranean sea regime shift at the end of the 1980s, and intriguing parallelisms with other European basins. PLoS One 5:e10633. https://doi.org/10.1371/journal.pone.0010633

Conversi A, Dakos V, Gårdmark A et al (2015) A holistic view of marine regime shifts. Philos Trans R Soc B 370:20130279. https://doi.org/10.1098/rstb.2013.0279

Crawford JD (1991) Introduction to bifurcation theory. Rev Mod Phys 63:991–1037. https://doi.org/10.1103/RevModPhys.63.991

Dai L, Korolev KS, Gore J (2013) Slower recovery in space before collapse of connected populations. Nature 496:355–358. https://doi.org/10.1038/nature12071

Dakos V, Scheffer M, van Nes EH et al (2008) Slowing down as an early warning signal for abrupt climate change. Proc Natl Acad Sci U S A 105:14308–14312. https://doi.org/10.1073/pnas.0802430105

Dakos V, Carpenter SR, Brock WA et al (2012) Methods for detecting early warnings of critical transitions in time series illustrated using simulated ecological data. PLoS One 7:e41010. https://doi.org/10.1371/journal.pone.0041010

Dakos V, Glaser SM, Hsieh C-H et al (2017) Elevated nonlinearity as an indicator of shifts in the dynamics of populations under stress. J R Soc Interface 14:20160845. https://doi.org/10.1098/rsif.2016.0845

DePiper GS, Gaichas SK, Lucey SM et al (2017) Operationalizing integrated ecosystem assessments within a multidisciplinary team: lessons learned from a worked example. ICES J Mar Sci 74:2076–2086. https://doi.org/10.1093/icesjms/fsx038

Derissen S, Quaas MF, Baumgärtner S (2011) The relationship between resilience and sustainability of ecological-economic systems. Ecol Econ 70:1121–1128. https://doi.org/10.1016/J.ECOLECON.2011.01.003

deYoung B, Barange M, Beaugrand G et al (2008) Regime shifts in marine ecosystems: detection, prediction and management. Trends Ecol Evol 23:402–409. https://doi.org/10.1016/j.tree.2008.03.008

Doak DF, Estes JA, Halpern BS et al (2008) Understanding and predicting ecological dynamics: are major surprises inevitable. Ecology 89:952–961. https://doi.org/10.1890/07-0965.1

FAO (2003) The ecosystem approach to fisheries. In: FAO technical guidelines for responsible fisheries. FAO, Rome

FAO (2014) Changements climatiques et sécurité alimentaire. http://www.fao.org/climatechange/16651-044a7adbada949701 1c8e-3d4a4d32c692.pdf

Fauchald P (2010) Predator-prey reversal: a possible mechanism for ecosystem hysteresis in the North Sea? Ecology 91:2191–2197. https://doi.org/10.1890/10-1922.1

Fenichel EP, Horan RD (2016) Tinbergen and tipping points: could some thresholds be policy-induced? J Econ Behav Organ 132:137–152. https://doi.org/10.1016/j.jebo.2016.06.014

Folke C (2016) Resilience. In: Oxford research encyclopedia of environmental science, vol 1. Oxford University Press, New York, pp 1–63

Frank KT, Petrie B, Choi JS et al (2005) Trophic cascades in a formerly cod-dominated ecosystem. Science 308:1621–1623. https://doi.org/10.1126/science.1113075

Frank KT, Petrie B, Fisher JAD et al (2011) Transient dynamics of an altered large marine ecosystem. Nature 477:86–89. https://doi.org/10.1038/nature10285

Frank KT, Petrie B, Leggett WC et al (2016) Large scale, synchronous variability of marine fish populations driven by commercial exploitation. Proc Natl Acad Sci U S A 113:8248–8253. https://doi.org/10.1073/pnas.1602325113

Galaz V, Crona B, Österblom H et al (2012) Polycentric systems and interacting planetary boundaries - emerging governance of climate change-ocean acidification-marine biodiversity. Ecol Econ 81:21–32. https://doi.org/10.1016/j.ecolecon.2011.11.012

Gårdmark A, Casini M, Huss M et al (2015) Regime shifts in exploited marine food webs: detecting mechanisms underlying alternative stable states using size-structured community dynamics theory. Philos Trans R Soc B 370:2013026. https://doi.org/10.1098/rstb.2013.0262

Garibaldi A, Turner N (2004) Cultural keystone species: implications for ecological conservation and restoration. Ecol Soc 9:1. https://doi.org/10.5751/ES-00669-090301

Glaser SM, Fogarty MJ, Liu H et al (2014) Complex dynamics may limit prediction in marine fisheries. Fish Fish 15:616–633. https://doi.org/10.1111/faf.12037

Haedrich RL, Hamilton LC (2000) The fall and future of Newfoundland's cod fishery. Soc Nat Resour 13:359–372. https://doi.org/10.1080/089419200279018

Halpern BS, Walbridge S, Selkoe KA et al (2008) A global map of human impact on marine ecosystems. Science 319:948–952. https://doi.org/10.1126/science.1149345

Hislop JR (1996) Changes in North Sea gadoid stocks. ICES J Mar Sci 53:1146–1156

Holling CS (1973) Resilience and stability of ecological systems. Annu Rev Ecol Syst 4:1–23. https://doi.org/10.1146/annurev.es.04.110173.000245

Horan RD, Fenichel EP, Drury KLS et al (2011) Managing ecological thresholds in coupled environmental–human systems. Proc Natl Acad Sci U S A 108:7333–7338

Hughes TP, Carpenter S, Rockström J et al (2013) Multiscale regime shifts and planetary boundaries. Trends Ecol Evol 28:389–395. https://doi.org/10.1016/j.tree.2013.05.019

Hutchings JA (2000) Collapse and recovery of marine fishes. Nature 406:882–885

Hutchings JA, Rangeley RW (2011) Correlates of recovery for Canadian Atlantic cod (*Gadus morhua*). Can J Zool 89:386–400

Jones DD (1975) The application of catastrophe theory to ecological systems. IIASA research report. IIASA, Laxenburg, Austria: RR-75-015

Jones DD (1977) Catastrophe theory applied to ecological systems. Simulation 29:1–15

Kates RW, Travis WR, Wilbanks TJ (2012) Transformational adaptation when incremental adaptations to climate change are insufficient. Proc Natl Acad Sci U S A 109:7156–7161. https://doi.org/10.1073/pnas.1115521109

Kelly RP, Erickson AL, Mease LA et al (2015) Embracing thresholds for better environmental management. Philos Trans R Soc B 370:20130276. https://doi.org/10.1098/rstb.2013.0276

Kerkhoff AJ, Enquist BJ (2007) The implications of scaling approaches for understanding resilience and reorganization in ecosystems. Bioscience 57:489–499. https://doi.org/10.1641/b570606

Lade SJ, Gross T (2012) Early warning signals for critical transitions: a generalized modeling approach. PLoS Comput Biol 8:e1002360. https://doi.org/10.1371/journal.pcbi.1002360

Lade SJ, Tavoni A, Levin SA et al (2013) Regime shifts in a social-ecological system. Theor Ecol 6:359–372. https://doi.org/10.1007/s12080-013-0187-3

Lade SJ, Niiranen S, Hentati-Sundberg J et al (2015) An empirical model of the Baltic Sea reveals the importance of social dynamics for ecological regime shifts. Proc Natl Acad Sci U S A 112:11120–11125. https://doi.org/10.1073/pnas.1504954112

Leslie HM, Basurto X, Nenadovic M et al (2015) Operationalizing the social-ecological systems framework to assess sustainability. Proc Natl Acad Sci U S A 112:5979–5984. https://doi.org/10.1073/pnas.1414640112

Levin SA (1992) The problem of pattern and scale in ecology: The Robert H. MacArthur Award Lecture. Ecology 73:1943–1967. https://doi.org/10.2307/1941447

Levin PS, Möllmann C (2015) Marine ecosystem regime shifts: challenges and opportunities for ecosystem-based management. Philos Trans R Soc B 370:20130275. https://doi.org/10.1098/rstb.2013.0275

Levin PS, Breslow SJ, Harvey CJ et al (2016) Conceptualization of social-ecological systems of the California current : an examination of interdisciplinary science supporting ecosystem-based management. Coast Manag 44:397–408

Lindegren M, Dakos V, Gröger JP et al (2012) Early detection of ecosystem regime shifts: a multiple method evaluation for management application. PLoS One 7:e38410. https://doi.org/10.1371/journal.pone.0038410

Ling SD, Scheibling RE, Rassweiler A et al (2015) Global regime shift dynamics of catastrophic sea urchin overgrazing. Philos Trans R Soc B 370:20130269. https://doi.org/10.1098/rstb.2013.0269

Liski M, Salanié F (2016) Tipping points, delays, and the control of catastrophes. 19th annual bioecon conference: evidence based environmental policies and the optimal management of natural resources

Llope M, Daskalov GM, Rouyer TA et al (2011) Overfishing of top predators eroded the resilience of the Black Sea system regardless of the climate and anthropogenic conditions. Glob Chang Biol 17:1251–1265. https://doi.org/10.1111/j.1365-2486.2010.02331.x

Long RD, Charles A, Stephenson RL (2015) Key principles of marine ecosystem-based management. Mar Policy 57:53–60. https://doi.org/10.1016/j.marpol.2015.01.013

Ludwig D, Walker B, Holling CS (1997) Sustainability, stability, and resilience. Conserv Ecol 1:7

Lynam CP, Llope M, Möllmann C et al (2017) Interaction between top-down and bottom-up control in marine food webs. Proc Natl Acad Sci U S A 114:1952–1957. https://doi.org/10.1073/pnas.1621037114

Mathias J-D, Lade S, Galaz V (2017) Multi-level policies and adaptive social networks – a conceptual modeling study for maintaining a polycentric governance system. Int J Commons 11:220–247. https://doi.org/10.18352/ijc.695

May RM (1977) Thresholds and breakpoints in ecosystems with a multiplicity of stable states. Nature 269:471–477. https://doi.org/10.1038/269471a0

Möllmann C, Diekmann R (2012) Marine ecosystem regime shifts induced by climate and overfishing: a review for the Northern Hemisphere. In: Woodward G, Jacob U, O'Gorman EJ (eds) Global change in multispecies systems: Part II. Academic, London, pp 303–347

Möllmann C, Müller-Karulis B, Kornilovs G et al (2008) Effects of climate and overfishing on zooplankton dynamics and ecosystem structure: regime shifts, trophic cascade, and feedback loops in a simple ecosystem. ICES J Mar Sci 65:302–310. https://doi.org/10.1093/icesjms/fsm197

Möllmann C, Diekmann R, Müller-Karulis B et al (2009) Reorganization of a large marine ecosystem due to atmospheric and anthropogenic pressure: a discontinuous regime shift in the Central Baltic Sea. Glob Chang Biol 15:1377–1393. https://doi.org/10.1111/j.1365-2486.2008.01814.x

Möllmann C, Folke C, Edwards M et al (2015) Marine regime shifts around the globe: theory, drivers and impacts. Philos Trans R Soc B 370:20130260. https://doi.org/10.1098/rstb.2013.0260

Mumby PJ (2009) Phase shifts and the stability of macroalgal communities on Caribbean coral reefs. Coral Reefs 28:761–773. https://doi.org/10.1007/s00338-009-0506-8

Mumby PJ, Hastings A, Edwards HJ (2007) Thresholds and the resilience of Caribbean coral reefs. Nature 450:98–101. https://doi.org/10.1038/nature06252

Myers RA, Worm B (2005) Extinction, survival or recovery of large predatory fishes. Philos Trans R Soc B 360:13–20. https://doi.org/10.1098/rstb.2004.1573

Myers RA, Hutchings JA, Barrowman NJ (1997) Why do fish stocks collapse? The example of cod in Atlantic Canada. Ecol Appl 7:91–106

Nyström M (2006) Redundancy and response diversity of functional groups: implications for the resilience of coral reefs. Ambio 35:30–35. https://doi.org/10.1579/0044-7447-35.1.30

Nyström M, Norström AV, Blenckner T et al (2012) Confronting feedbacks of degraded marine ecosystems. Ecosystems 15:695–710. https://doi.org/10.1007/s10021-012-9530-6

O'Brien M, Hartwig F, Schanes K et al (2014) Living within the safe operating space: a vision for a resource efficient Europe. Eur J Futur Res 2:48. https://doi.org/10.1007/s40309-014-0048-3

Ostrom E (2009) A general framework for analyzing sustainability of social-ecological systems. Science 325:419–422. https://doi.org/10.1126/science.1172133

Pershing AJ, Mills KE, Record NR et al (2015) Evaluating trophic cascades as drivers of regime shifts in different ocean ecosystems. Philos Trans R Soc B 370:20130265. https://doi.org/10.1098/rstb.2013.0265

Pindyck RS (2000) Irreversibilities and the timing of environmental policy. Resour Energy Econ 22:233–259. https://doi.org/10.1016/S0928-7655(00)00033-6

Plummer R, Baird J, Dzyundzyak A et al (2017) Is adaptive co-management delivering? Examining relationships between collaboration, learning and outcomes in UNESCO biosphere reserves. Ecol Econ 140:79–88. https://doi.org/10.1016/j.ecolecon.2017.04.028

Punt AE, Siddeek MSM, Garber-Yonts B et al (2012) Evaluating the impact of buffers to account for scientific uncertainty when setting TACs: application to red king crab in Bristol Bay, Alaska. ICES J Mar Sci 69:624–634. https://doi.org/10.1093/icesjms/fsr174

Reid PC, De Fatima Borges M, Svendsen E (2001) A regime shift in the north sea circa 1988 linked to changes in the north sea horse mackerel fishery. Fish Res 50:163–171. https://doi.org/10.1016/S0165-7836(00)00249-6

Ricker WE (1963) Big effects from small causes: two examples from fish population dynamics. J Fish Res Board Can 20:257–264. https://doi.org/10.1139/f63-022

Rocha JC, Biggs R, Peterson G (2014a) Regime shifts: what are they and why do they matter? Regime Shifts Database 20:681–697

Rocha JC, Biggs R, Peterson G (2014b) Regime shifts database. www.regimeshifts.org

Rocha JC, Yletyinen J, Biggs R et al (2015) Marine regime shifts: drivers and impacts on ecosystems services. Philos Trans R Soc B 370:20130273. https://doi.org/10.1098/rstb.2013.0273

Rockström J, Steffen W, Noone K et al (2009) A safe operating space for humanity. Nature 461:472–475. https://doi.org/10.1038/461472a

Roe G (2009) Feedbacks, timescales, and seeing red. Annu Rev Earth Planet Sci 37:93–115. https://doi.org/10.1146/annurev.earth.061008.134734

Scheffer M (2009) Critical transitions in nature and society. Princeton University Press, Princeton

Scheffer M, Carpenter SR (2003) Catastrophic regime shifts in ecosystems: linking theory to observation. Trends Ecol Evol 18:648–656. https://doi.org/10.1016/j.tree.2003.09.002

Scheffer M, Carpenter SR, Foley JA et al (2001) Catastrophic shifts in ecosystems. Nature 413:591–596. https://doi.org/10.1038/35098000

Scheffer M, Bascompte J, Brock WA et al (2009) Early-warning signals for critical transitions. Nature 461:53–59. https://doi.org/10.1038/nature08227

Scheffer M, Carpenter SR, Dakos V et al (2015) Generic indicators of ecological resilience: inferring the change of a critical tranistion. Annu Rev Ecol Evol Syst 46:145–167. https://doi.org/10.1146/annurev-ecolsys-112414-054242

Selkoe KA, Blenckner T, Caldwell MR et al (2015) Principles for managing marine ecosystems prone to tipping points. Ecosyst Heal Sustain 1:17. https://doi.org/10.1890/EHS14-0024.1

Standish RJ, Hobbs RJ, Mayfield MM et al (2014) Resilience in ecology: abstraction, distraction, or where the action is? Biol Conserv 177:43–51. https://doi.org/10.1016/j.biocon.2014.06.008

Star B, Nederbragt AJ, Jentoft S et al (2011) The genome sequence of Atlantic cod reveals a unique immune system. Nature 477:207–210. https://doi.org/10.1038/nature10342

Steneck RS, Wahle RA (2013) American lobster dynamics in a brave new ocean. Can J Fish Aquat Sci 70:1612–1624. https://doi.org/10.1139/cjfas-2013-0094

Steneck RS, Graham MH, Bourque BJ et al (2002) Kelp forest ecosystems: biodiversity, stability, resilience and future. Environ Conserv 29:436–459. https://doi.org/10.1017/S0376892902000322

Steneck RS, Leland A, Mcnaught DC et al (2013) Ecosystem flips, locks, and feedbacks: the lasting effects of fisheries on Maine's Kelp Forest ecosystem. Bull Mar Sci 89:31–55. https://doi.org/10.5343/bms.2011.1148

Sugihara G, May R, Ye H et al (2012) Detecting causality in complex ecosystems. Science 338:496–500. https://doi.org/10.1126/science.1227079

Thom R (1975) Structural stability and morphogenesis: an outline of a general theory of models. Benjamin Cummings Inc, Reading

Travis J, Coleman FC, Auster PJ et al (2014) Integrating the invisible fabric of nature into fisheries management. Proc Natl Acad Sci U S A 111:581–584. https://doi.org/10.1073/pnas.1402460111

UNESCO (2012) Blueprint for the future we want, Oceans. http://www.un.org/en/sustainablefuture/pdf/Rio+20_FS_Oceans.pdf

Valman M, Duit A, Blenckner T (2016) Organizational responsiveness: the case of unfolding crises and problem detection within HELCOM. Mar Policy 70:49–57. https://doi.org/10.1016/j.marpol.2016.04.016

van der Maas HLJ, Kolstein R, van der Pligt J (2003) Sudden transitions in attitudes. Sociol Methods Res 32:125–152. https://doi.org/10.1177/0049124103253773

Vasilakopoulos P, Marshall CT (2015) Resilience and tipping points of an exploited fish population over six decades. Glob Chang Biol 21:1834–1847. https://doi.org/10.1111/gcb.12845

Walters C, Kitchell JF (2001) Cultivation/depensation effects on juvenile survival and recruitment: implications for the theory of fishing. Can J Fish Aquat Sci 58:39–50. https://doi.org/10.1139/cjfas-58-1-39

Yletyinen J, Bodin Ö, Weigel B et al (2016) Regime shifts in marine communities: a complex systems perspective on food web dynamics. Proc R Soc B 283:20152569. https://doi.org/10.1098/rspb.2015.2569

Biodiversity and the Functioning of Ecosystems in the Age of Global Change: Integrating Knowledge Across Scales

Francisco R. Barboza, Maysa Ito, and Markus Franz

Abstract

The dramatic decline of biodiversity worldwide has raised a general concern on the impacts this process could have for the well-being of humanity. Human societies strongly depend on the benefits provided by natural ecosystems, which are the result of biogeochemical processes governed by species activities and their interaction with abiotic compartments. After decades of experimental research on the biodiversity-functioning relationship, a relative agreement has been reached on the mechanisms underlying the impacts that biodiversity loss can have on ecosystem processes. However, a general consensus is still missing. We suggest that the reason preventing an integration of existing knowledge is the scale discrepancy between observations on global change impacts and biodiversity-functioning experiments. The present chapter provides an overview of global change impacts on biodiversity across various ecological scales and its consequences for ecosystem functioning, highlighting what is known and where knowledge gaps still persist. Furthermore, the reader will be introduced to a set of tools that allow a multi-scale analysis of how global change drivers impact ecosystem functioning.

What We Know and What We Do Not: Biodiversity and Functioning in the Anthropocene

Environmental changes have ruled the geological history of Earth and have been responsible for the shifts that life has undergone during the past 3.5 billion years (Hoegh-Guldberg and Bruno 2010). Alternations between glacial and interglacial episodes, tectonic activity, and abrupt changes in atmospheric and oceanic chemistry have promoted five massive extinctions in the last 500 million years (Barnosky et al. 2011 and citations therein). These catastrophic events, each of which killed more than three-quarters of existing biota in a period of less than 2 million years, erased or dramatically rearranged ecosystems worldwide (Hull 2015). The expansion of the human population since the beginning of the Industrial Revolution in the nineteenth century, and its acceleration between the 1940s and 1960s, is severely altering the biogeochemistry of our planet (Vitousek et al. 1997; Doney 2010). Imposed anthropogenic pressures on natural ecosystems are so extreme that the projected magnitude of their effects is only comparable with those observed during massive extinctions (Barnosky et al. 2011). Degradation and loss of habitats, biological invasions, overexploitation of natural resources, pollution, and climate change are driving an unprecedented loss of biodiversity at a global scale (Pimm et al. 2014).

Humans, being unique in terms of the scale of their impacts, are as vulnerable as any other species to changes in the ecosystems to which they belong. Human societies rely on the goods and services provided by the functioning of ecosystems, which depends on the cycling of matter and flux of energy that the interactions of living and non-living compartments make possible (Díaz et al. 2006). Thus, direct impacts of global change stressors on biogeochemical processes (e.g., excessive increase of nutrient loads in land and waters) or those mediated by the loss of biodiversity, alter the dynamics and functioning of ecosystems compromising the well-being of humans (Isbell et al. 2017). The consequences that the current rates of biodiversity loss could have on ecosystem services called for research on the role that biodiversity plays in determining the structure, functioning and stability of ecosystems (Cardinale et al. 2012). The extensive body of theoretical, observational, and experimental evidence generated in the last decades, has led to a certain

F. R. Barboza (✉) · M. Ito · M. Franz
GEOMAR Helmholtz Centre for Ocean Research, Kiel, Germany
e-mail: fbarboza@geomar.de; mito@geomar.de; mfranz@geomar.de

S. Jungblut et al. (eds.), *YOUMARES 8 – Oceans Across Boundaries: Learning from each other*,
https://doi.org/10.1007/978-3-319-93284-2_12

consensus on the following set of statements, trends and potential underlying mechanisms:

Biodiversity Increases Stability at the Ecosystem Level The diversity-stability debate is probably one of the most relevant — given its implications in light of the anthropogenic-induced loss of biodiversity — and long standing ones in Ecology (McCann 2000). The pioneering observational works of Odum (1953) and Elton (1958), awakened this discussion by acknowledging that simplified terrestrial communities (e.g., in agricultural systems) exhibit stronger fluctuations and are more vulnerable to biological invasions. Blindly accepted until the beginning of the 1970s, these statements were questioned by a series of thoughtful mathematical essays developed by Robert May (May 1971, 1972, 1973). The linear stability analysis of constructed random communities[1] showed that the higher complexity is (in terms of connectance, strength of interaction and number of interacting species) the more unstable[2] population dynamics will be. May's arguments, and beyond the unrealistic assumptions of the proposed models (i.e., communities are randomly structured and exhibit stable equilibrium dynamics, McCann 2000), highlighted the absence of a mechanistic understanding of existing empirical evidence. In other words, if more diverse natural ecosystems tend to be more stable but those randomly constructed are not, natural ecosystems must be structured by a set of non-random principles that determine their stability. The challenge raised by May's results triggered the search for a set of properties capable of conferring stability to complex ecological systems. The accumulated evidence by the analysis of empirical ecological networks highlighted, for example, the role of weak interactions and modularity as properties that prevent the spread of disturbances (Paine 1992; McCann et al. 1998; Neutel et al. 2002; Olesen et al. 2007; Gilarranz et al. 2017).[3] A large body of empirical evidence supporting the diversity-stability relationship has been generated in the last four decades (McNaughton 1977; Stachowicz et al. 2007; Tilman et al. 2014). The manipulation of species or functional richness has shown that diversity reduces the temporal variability in the structure and functioning of communities (e.g., measured as biomass production). A remarkable conclusion of the syntheses of these results is that the positive correlation between diversity and stability at the community level cannot necessarily be extended to single populations (Gross et al. 2014; Tilman et al. 2014). Alternative hypotheses have been proposed to account for these results (Yachi and Loreau 1999; Lehman and Tilman 2000). The averaging and covariance effects predict that the variability of the overall community will be dampened due to the balance between contrasting single species dynamics (Lehman and Tilman 2000). These hypotheses assume that the higher the diversity, the higher the probability of observing species that respond differentially to conditions and disturbances (McCann 2000). Furthermore, the insurance hypothesis added the idea that the higher the diversity, the higher the probability of having functionally redundant species. Thus, the loss of species with particular functions can be replaced by others, increasing the temporal stability of ecosystems' functioning (Yachi and Loreau 1999). All in all, existing theoretical and experimental evidence provided a potential solution to the diversity-stability debate: the stabilizing effects of biodiversity at the ecosystem level (i.e., the observations of Odum and Elton) can occur at the expenses of decreasing single species stability (i.e., the theoretical conclusions of May) (Lehman and Tilman 2000).

Biodiversity Increases the Efficiency and Productivity of Ecosystems The number of observational and experimental studies analyzing how changes in biodiversity impact the functioning of ecosystems has rapidly increased since the 1990s. Research across ecosystems (from terrestrial to marine) and considering diversity at different levels of biological organization (from genes to functional groups) has been developed worldwide. Recent meta-analyses have summarized available bibliography, obtaining conclusive evidence that, on average, the decrease of biodiversity is translated into altered ecosystem functions (e.g., a lower capacity of communities to use resources and produce biomass, see Cardinale et al. 2012 and citations therein). Regardless of the clarity of these findings, a consensus on the responsible mechanisms is still elusive. The selection effect (i.e., the prevalence of species with certain traits in the determination of ecosystem processes) and/or the complementarity effect (i.e., a better performance of the community due to an efficient partitioning of resources or facilitation among species) have been proposed for the explanation of biodiversity-functioning relationships (Loreau and Hector 2001). A sampling process[4] is involved in both mechanisms, which means that the higher the diversity, the higher the odds

[1]Theoretical communities where the type and magnitude of the interactions are defined using statistical distributions (see May 1972 for a brief but enlightening summary).

[2]Original works of Robert May define stability in terms of resilience, assuming that stable systems are those able to return to the equilibrium after a perturbation (see McCann 2000).

[3]The list of features mentioned for ecological networks is far from being exhaustive, but a detailed presentation of described topological patterns and underlying mechanisms is out of the scope of the present chapter. In this sense, we recommend Montoya et al. (2006) and Ronney and McCann (2012) for a general overview of the state of the art in food webs theory.

[4]In light of the existing literature, it is important to draw the attention of the readers on the fact that the sampling and selection effects, sometimes, are incorrectly used as interchangeable concepts. Please see Loreau and Hector (2001) for a clear explanation of the differences.

of sampling a dominant species with specific traits or a set of species with complementary traits (Loreau and Hector 2001; Fargione et al. 2007). In light of these mechanisms, most of the empirical research developed in the last 10 years focused on disentangling the relative contribution of community composition (i.e., role of the taxonomic and/or functional identity of species) and complementarity to the effect of biodiversity on ecosystem processes. Cardinale et al. (2012) estimated an even contribution of both mechanisms, but highlighted that available evidence is still fragmentary for solving this debate.

Functional Diversity Determines Ecosystem Processes and Services Changes in biodiversity at all levels of biological organization could affect, to a greater or lesser extent, the functioning of ecosystems (e.g., Reusch et al. 2005; Worm et al. 2006). Nevertheless, there is a general agreement that functional diversity is the dimension of biodiversity that contributes the most to the determination of ecosystem processes (Díaz and Cabido 2001). Traits determine how species capture and use different resources, and interact with the environment. Thus, the role of species in the flux of energy and cycling of matter is shaped by their traits, being the identity, abundance, and range of these traits what links species and ecosystems from a functional perspective (Fig. 1; Naeem 1996; Bengtsson 1998). The goods and services provided by ecosystems depend on the persistence of biogeochemical processes, which rely on functional groups (i.e., sets of species that exhibit certain functional traits). It is the loss of functional groups, beyond species,[5] that compromises the capacity of ecosystems to continue providing benefits to humanity (Díaz et al. 2006). During mass extinctions, and the current one is not the exception, the loss of species is driven by negative selection against certain traits. Thus, identifying traits that determine a greater extinction risk, and how they directly or indirectly (through the correlation with other traits) influence ecosystem processes, is essential to predict the consequences of extinctions on ecosystem services (Cardinale et al. 2012, Fig. 1).

The information gathered so far has certainly been valuable for describing the effects that biodiversity has on ecosystem functioning (among other ecosystem characteristics) and elucidating the underlying mechanisms that mediate these effects. Nevertheless, a scale discrepancy still persists between the local nature of the evidence on which the current understanding of the biodiversity-functioning relationship is held and the global scale at which the impacts of anthropogenic activities on biodiversity have usually been described (Isbell et al. 2017). The understanding of the potential cascading effects that large-scale changes in biodiversity might have on ecosystems at a local scale is a challenge that still needs to be addressed. In general, data have been generated in a fragmented way at different spatial, temporal and ecological scales. In addition, there are almost no attempts in the literature to integrate this knowledge (but see Isbell et al. 2017 for an example with a management background). In a context where current methodological constraints prevent "multi-scale" observational and experimental analyses of certain phenomena and processes, theoretical essays and modeling provide a powerful approach to bridge isolated empirical efforts. Thus, constructing on the existing bibliography, this chapter will give an integrated perspective of the impacts that global change drivers will have at different ecological scales — from regional species pools to the interaction between species in local communities — and their potential consequences on the functioning of ecosystems (Fig. 1). Beyond the literature review, we introduce a set of tools which allow a holistic analysis of the consequences that changes in biodiversity have on ecosystem processes under global change.

Regional Pools of Species Under Global Change: Is Biodiversity Decreasing?

Regional species pools are defined as the overall set of species that can colonize local communities.[6] The total number of species observed in these pools is the result of the balance between processes that increase (i.e., speciation and immigration) and decrease (i.e., extinction) species diversity (Cornell and Harrison 2014). Human activities have heavily altered these processes mainly by increasing the rates of extinction and immigration. On one hand, the overexploitation of species of economic interest, the rapid and in many cases irreversible loss of habitat and the reduction of distributional ranges due to changes in prevailing climatic conditions are responsible for the loss of species at a regional scale. On the other hand, the dissemination of species out of their native range has promoted the exchange of species among previously isolated regions and in consequence the introduction of exotic species (Sax and Gaines 2003). The arrival and establishment of new species could have two

[5]It is important to clarify that keystone species (i.e., species with a disproportionately effect on the functioning of the ecosystem in comparison to its abundance) can be considered as single-species functional groups, since they are fully non-redundant and non-replaceable (Bond 1994).

[6]Recent reviews and perspective articles have extensively discussed the regional species pool concept. We recommend Carstensen et al. (2013) and Cornell and Harrison (2014) for an overview on the topic.

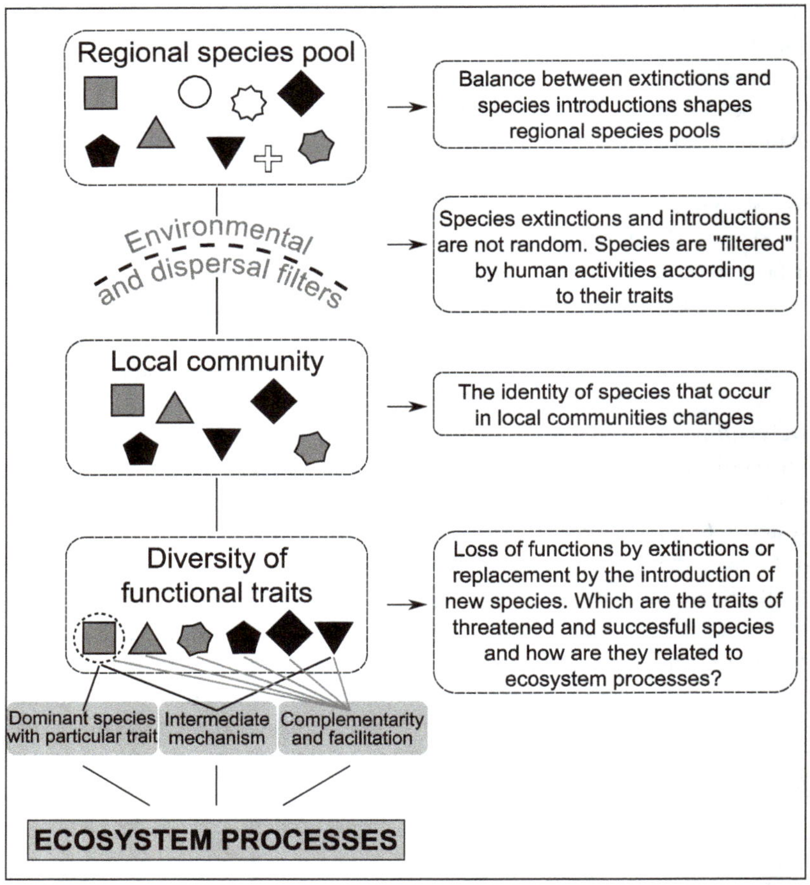

Fig. 1 Conceptual scheme integrating current knowledge on how biodiversity determines ecosystem functioning and expected cascading impacts of global change drivers.
The left side of the scheme (Adapted from Loreau et al. 2001) depicts a regional pool integrated by a set of species (represented by different shapes) with a range of functional traits (represented by different colors). From this initial set, only those species with particular traits can cope with experienced environmental and dispersal filters, occurring in a theoretical local community (i.e., only certain colors are observed in the community). The spectra of retained traits (functional diversity) determines the ecosystem processes and services provided by the community. A gradient of explanatory mechanisms, with selection and complementarity effects as extremes, have been suggested to explain how changes in functional diversity alter ecosystem processes (see details in the main text). The right side shows structuring mechanisms (species extinctions and introductions) that are being enhanced in the course of global change across ecological scales. Imposed anthropogenic pressures modify functional diversity in a non-random way, making it possible to predict how ecosystem processes will change during the Anthropocene

potential consequences on the diversity of a region: i) increase it due to the occurrence of a species that was not present within the original pool and that could even facilitate the arrival of other species or ii) diminish it by promoting the loss of native species through competition or predation (Sax and Gaines 2003, 2008), exceeding the gain that the introduction of a new species implies.[7] Even though the vast majority of articles have focused on the negative consequences of exotic species, some authors are discussing the introduction of species from a new perspective. Recent works showed that from those species classified as endangered or extinct by the IUCN, a small percentage have exotic species as the main or single cause of decline (the numbers increase if only island regions are considered; Gurevitch and Padilla 2004; Sax and Gaines 2008). Much of the evidence on the negative impacts of exotic species is correlational (or based on small scale experiments) and it cannot be discarded that the spread of the new species was favored by the impacts of other drivers on native communities. In addition, worldwide evidence suggests that the number of species introduced in a given region exceeds the number of extirpated ones, generating on average an increase of species richness at the regional scale (Thomas 2013a, b). Therefore, what at a global scale is only determined by the balance between speciation and extinction, at a regional scale it is

[7]An additional possibility will imply the generation of new species (and eventually new functional traits) by hybridization between native and non-native species. Please see Seehausen (2004) for a broad revision on the topic.

also shaped by the influx of new species that can compensate (regarding the number of species) extinctions or even generate an overall increase of regional diversity. But, as mentioned previously in this section, species richness is not the only dimension of biodiversity and the arrival of new species does not necessarily guarantee the functional replacement of extinct ones. In this context, it is crucial to better understand: (i) which are the traits of extirpated and introduced species, (ii) to what extent do they functionally overlap and (iii) if introduced species will be able to keep the functioning of ecosystems (Fig. 1).

Functional Diversity in Local Communities: Are Species Lost Functionally Replaced by Those Introduced?

As previously stated, human driven extinctions are not random, because certain species traits are favored or hampered by anthropogenic pressures, which act as environmental filters (Hillebrand and Blenckner 2002; Fig. 1). Traits like body size, fecundity, motility and physiological tolerance, among others, have been identified as potential predictors of both species' extinction risk and capacity to spread and colonize new environments. In this sense, it has been suggested that large body size, low fecundity, slow dispersal and resource specialization are generally filtered out, while small, fast reproducing, wide spreading, and generalist species are favored (McKinney and Lockwood 1999). According to these observations, it has been proposed that in the spectrum of variability of these traits, threatened and successful species must be in opposite extremes. Thus, those traits positively correlated with extinction risk must be negatively correlated with the probability of a species to get established and successfully spread (Blackburn and Jeschke 2009). This hypothesis, known as "two sides of the same coin", has been tested in terrestrial and aquatic environments for different taxonomic groups (fish, crustaceans, birds, reptiles and plants) (e.g., Murray et al. 2002; Marchetti et al. 2004; Blackburn and Jeschke 2009; Larson and Olden 2010; van Kleunen et al. 2010). The use of different definitions for invasive, non-invasive, threatened and rare species across articles, promoted the generation of contradictory evidence (van Kleunen and Richardson 2007; Blackburn and Jeschke 2009). Despite the methodological inconsistencies observed in the literature, it is still possible to draw some conclusions. The assumption that for all functional traits analyzed, threatened and successful species will always exhibit contrasting variants is an oversimplification (Tingley et al. 2016). The majority of the traits evaluated in the bibliography show small or no-difference among threatened and successful species (e.g., Jeschke and Strayer 2008; Tingley et al. 2016). It is important to highlight that the still fragmentary nature of the data for certain species could explain some of the obtained results (van Kleunen and Richardson 2007).

The current "absence" of trends in multiple-trait analyses questions the validity of the "two sides of the same coin" hypothesis (Jeschke and Strayer 2008; Blackburn and Jeschke 2009; Tingley et al. 2016). Available evidence makes it extremely difficult to speak about a set of traits that unequivocally predicts both extinction risk and species success, across environments and taxa. Nevertheless, results become more consistent if we just focus on extinctions (a process that has received much more attention in the last decades) and some specific traits. In particular, ecological and paleontological literature identified body mass as a major predictor of extinctions, i.e., large-bodied species are more likely to disappear. Body size tightly correlates with different life history traits and demographic characteristics determining the susceptibility of species to extinction-promoting drivers (e.g., Purvis et al. 2000; Springer et al. 2003; Barnosky 2008).[8] Important functional traits like trophic position, diet width, and productivity scale with body size. Thus, extinctions modify the size distribution of communities being able to alter the stability and functioning of ecosystems (Woodward et al. 2005). Observational and experimental examples have shown the consequences that the loss of "big" species has on ecosystem processes. Solan et al. (2004) showed that the loss of larger infaunal species reduces bioturbation and sediment oxygenation, altering the decomposition of organic matter and cycling of nutrients. Articles showing cascading effects of large predator's extinctions on overall ecosystems are probably those that better exemplify the impacts of body size changes. Estes et al. (2011) and Ripple et al. (2014) (and citations therein), reviewed the literature highlighting the relevance of top-down controls in ecosystems. Carbon uptake in freshwater and marine ecosystems, nutrients accumulation in soils and waters or primary production in coastal areas are just some examples of ecosystems processes affected by the extinction of apex consumers.

The question that still remains to be answered is whether the massive number of exotic species introduced worldwide will be able to functionally replace those that are lost (Fig. 1). Available data are insufficient to explain extinctions and introductions in terms of species traits and to determine the consequences of changes in those traits on ecosystem processes. Increasing research efforts on this topic are needed to accurately predict how ecosystems will respond under global change.

[8]The single consideration of mean adult body size (as has been done in most of the existing bibliography) in the mechanistic understanding of ecological and evolutionary processes could be misleading, since species usually show dramatic ontogenetic changes in body size (see Woodward et al. 2005 and Codron et al. 2012).

Tools for Analyzing Functioning: From Single Species to Functional Traits

Many studies that link biodiversity with ecosystem functioning have focused on different biodiversity metrics, multiple processes and ecological interactions (Reiss et al. 2009). The usage of experimental data and modeling has been discussed, since the combination of these approaches could allow the detection of early signs of functioning shifts due to predicted global change. In this section a set of tools for studying the functioning of ecosystems is proposed. First, we focus on dynamic energetic budget (DEB) for single-species analysis due to the importance of evaluating the contribution of each component of functional groups. Second, we illustrate the use of ecological network analysis (ENA) to study community-level interactions. Third, we suggest the use of loop analysis (LA) to investigate how external inputs affect ecosystems.

Species Level Analysis Using the Dynamic Energy Budget (DEB) Model

The first step for studying an ecosystem is to understand the contribution of each component since ecological processes can be related to multiple species and at the same time one species might be involved in multiple processes (Reiss et al. 2009). The DEB is an individual-based model proposed as a method to analyze the role of the individual into the functioning context.

Kooijman (2010) proposes the DEB model for analyzing energy fluxes within individuals (Fig. 2). The DEB theory is based on the first law of thermodynamics and assumes the conservation of energy and mass. The model focuses on three basic energy fluxes: assimilation, dissipation and growth. Assimilation is the inflow of energy that enters the reserve pool proportional to the surface area of the organism. It is represented by the feeding minus the material excreted via feces, in the case of heterotrophs. In photoautotrophs, assimilation refers to the acquisition of nutrients mainly by photosynthesis (Edmunds et al. 2011). The energy reserve is used by the organism for maintenance, growth and reproduction. Dissipation corresponds to maintenance processes that use part of the reserve, which will result in products released into the environment, i.e., respiration. Growth corresponds to the increase of body size. The model also includes energy from the reserve that is invested in reproduction.

The DEB model is ideal for integrating single-species experimental outcomes (Edmunds et al. 2011). The model connects data acquired from physiology and structure of the organisms, i.e., functional traits, to provide an overview of the species as a system. In addition, the DEB model is able to describe the impacts of disturbance, e.g., pollutants (Nisbet et al. 2000; van der Meer 2006). The model also allows the assessment of the organisms from larval to adult stage, e.g., Monaco et al. (2014) carried out experiments with the sea-star *Pisaster ochraceus* under different life-stages and determined the transitions according to body size. The empirical data was used to predict the responses (e.g., flow of energy from reserve, structure and gonads to biomass). However, to exploit the potential of DEB models, more experiments considering how the traits change under different environmental conditions (e.g., temperature) would be necessary. The analysis of energy fluxes under future environmental state may assist the prediction of how the species will respond to environmental shifts or even to the new regions where they can be introduced. Knowing how species will respond is crucial because the species may change (e.g., become more or less efficient in processing energy or even disappear), resulting in biodiversity reshuffling under the effect of global change drivers.

The software developed for the DEB model is called DEBtool[9] for Matlab. It enables the user to analyze eco-physiological data by calculating relationships between variables and check the model predictions.

Analyzing single-species systems corresponds to finding only one piece of the entire puzzle. Putting empirical data together using a DEB model has good potential for single species and population analysis but the usage for ecosystems is still not certain (Nisbet et al. 2000). Muller et al. (2009) used the DEB model for analyzing the flow of energy using carbon and nitrogen as currencies of an autotroph, a heterotroph, and the symbiotic interaction between them. However, for modeling the complex interactions of ecosystems we suggest ENA as a better approach.

Ecological Network Analysis (ENA)

In order to connect the species embedded into a system and their relationships with abiotic components, ENA is a useful tool. It increases the complexity of food web analysis by quantifying the flow of energy and including interactions with non-living compartments that are part of the ecosystem (Gaedke 1995; Fig. 2). Food web models depict topological webs, i.e., binary networks where the species are the nodes (compartments) and the connections between them are representations of "who eats whom". ENA analysis considers that the food webs are exchanging energy with and within non-living compartments as well (Magri et al. 2017). The analysis is considered weighted when it includes the infor-

[9]The software and additional information can be found here: http://www.bio.vu.nl/thb/deb/deblab/

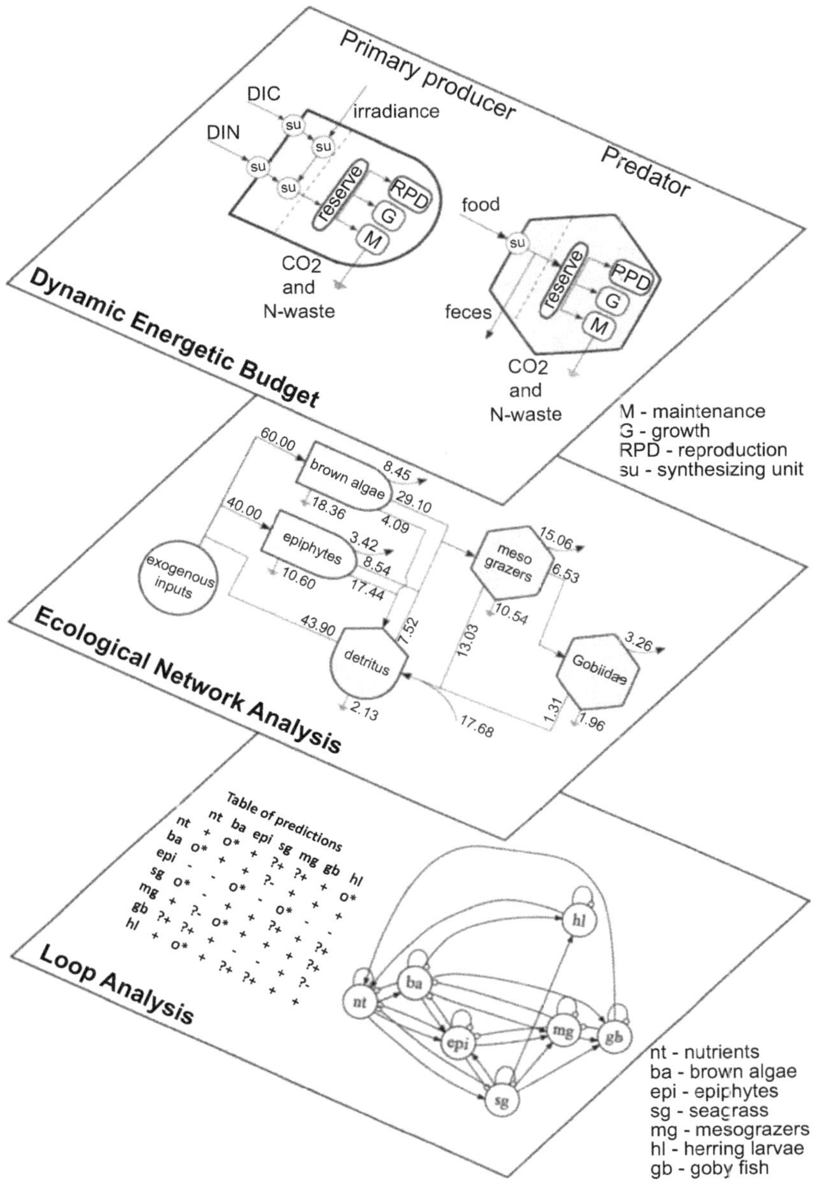

Fig. 2 Representation of the recommended tools for analyzing ecosystem functioning. The layers show that the analysis could target various organizational levels: single species (upper layer), ecological communities (in the middle), and interactions of ecological and human components, e.g., the increase of nutrient inputs in aquatic systems (lower layer).
The schematic representation of the upper layer refers to the dynamic energetic budget and illustrates the fate of energy flow in a primary producer and a predator. The (trophic) interactions between the species in the community are represented as ecological network analysis. The network traces carbon flows of a hypothetical coastal community of the Baltic Sea and the flows display matter circulation in terms of mg C m^{-2} day^{-1}. Finally, the loop analysis can bring together feeding interactions and other non-trophic relationships like symbiosis. In the hypothetical Baltic Sea community presented, the interactions of mesograzers and herring larvae with seagrass are related to habitat provision. Also, the brown algae and seagrass interactions with epiphytes are related to competition. The table of prediction for the community on the left side indicates the expected responses of column compartments following positive perturbations on the row compartments. The signed directed graph of the community on the right side of the loop analysis depicts positive interactions as arrows and negative interactions as empty circles

mation about feeding rates, that represent the strength of the connections (Ulanowicz 2004). ENA (like the DEB model) assumes conservation of energy and mass (i.e., all nodes must be at steady-state, with the same amount of energy exchanged by input and output processes). Therefore, ENA considers four types of energy flows: imports, exports, respirations (i.e., losses) and inter-compartmental exchanges. The energy flow can be expressed in the unit kcal and various mediums (currencies) such as carbon, nitrogen, phosphorus, and sulfur. The input of energy into the system usually is related to the gross primary productivity or even detritus aggregation that enters the system. The loss of energy corresponds to degraded material that might be represented by dissipation as heat (i.e., respiration), which is different from the export of usable energy to other systems (e.g., detritus that is flushed away from an eelgrass meadow). The inter-compartments corresponds to quantification of flows by energy transferred not only by the predator-prey interaction but also from living to non-living (and vice and versa) compartments (Kay et al. 1989). For example, this kind of analysis is useful for identifying cascade effects on the processes in an ecosystem. Indeed, ENA is able to connect information about the elements of the ecosystem to quantify how indirect effects spread along the system (Ulanowicz 2004). For example, ENA has been used for investigating changes due to eutrophication (Christian et al. 2009). One of the consequences detected was that eutrophication decreased the macrophyte biomass, lowering herbivory and causing impacts to the functioning of the overall system.

ENA is able to shed light on different aspects of ecosystem functioning. The algorithms of ENA provide indices that show how the systems respond to changes applied to them (Baird et al. 2004). Some output variables connected to the functioning of the systems are:

- The efficiency of the ecosystems in using the energy captured by primary producers can shift under different conditions (e.g., salinity gradients). The efficiency determines whether an ecosystem is more autotrophic or heterotrophic. The ENA provides the Lindeman spine, which is the representation of the complex network in terms of a linear food chain based on discrete trophic levels. It depicts the transfer of energy along compartments in a simplified way allowing the calculation of trophic efficiency (Baird and Ulanowicz 1993).
- Energy cycling can be a good indicator of stress (Ulanowicz 1995). Cycling refers to the recycling of the medium within the ecosystem, i.e., the ability of the nodes involved in the energy transfer to reuse the medium. In order to obtain a complete picture of the consequences of cycling it is important to analyze the number of cycles, length of the cycles (quantity of nodes involved) and species involved. The total amount of cycling is represented by the Finn cycling index (FCI). Mature ecosystems tend to have more cycles and increase the amount of energy circulating through them. However, eutrophication that represents a stress for ecosystems may also contribute to generate more cycles. The difference between mature and eutrophic systems is the length of these cycles. For example, mature ecosystems have longer cycles, while eutrophic systems present a high FCI but the cycles are shorter, so the energy does not reach higher trophic levels in the food web resulting in loss of functioning (Baird et al. 2004; Christian et al. 2005).
- Average residence time (ART) is related to the time that the medium is retained in the network. The residence time is not necessarily related to the aforementioned cycling since the intensity of the cycles (i.e., energy flowing within the cycles) can vary (Baird and Ulanowicz 1989). The ART is calculated by the ratio of the total system biomass and total output (Baird et al. 2004). The less time it spends in the system, the less efficient the system is in using energetic resources (Baird et al. 2004).
- Average path length expresses the quantity of compartments that the medium goes through before leaving the system. Shorter paths may be the response to stressful conditions in the ecosystem (Baird and Ulanowicz 1993).
- Total system throughput (TST) is related to the whole activity because it reports the amount of the medium flowing through the system. It is used to quantify ecosystems growth.
- Ascendency (A) corresponds to the organization (i.e., development) of the system considering the total activity (TST). It has also been suggested the use of "internal ascendency" (A_I) that considers only internal flows of the studied system. Ulanowicz (2004) suggests A_I for comparing growth and development of different ecosystems.
- Overhead takes into account the four types of flow while redundancy indicates the quantity of internal flows only. Both overhead and redundancy have been used to determine the resilience of the system. Increased values mean more resilient ecosystem according to Ulanowicz (2004).
- Development capacity is the upper limit of development that can be attained by ascendency. It is calculated as the sum of ascendency plus the overhead. It indicates the status of a system. Ascendency/development capacity ratios are good indicators of organization of the system (Ulanowicz 2004).

In order to use ENA for evaluating ecological processes and the impacts of environmental change, we have some recommendations. The first recommendation is to examine food webs throughout the seasons because the networks depict static snapshots of energy-matter flows in ecosystems. Traits of species such as body size, ontogeny and trophic interactions shift along the seasons (Warren 1989). Therefore, the simplification of the analysis (e.g., carrying out the ENA for the whole year) might lead to overlook patterns, e.g., cycling (Bondavalli et al. 2006). The analysis over the seasons is useful for studying temporal dynamics. Consequently, it helps to disentangle the changes driven by natural variability from stress, e.g., eutrophication (Bondavalli et al. 2006).

The second recommendation are the software tools for ENA, NETWRK 4.2 (Ulanowicz and Kay 1991) and Ecopath with Ecosim (Christensen and Pauly 1992). NETWRK 4.2 runs the ENA and the outputs include the indices and properties described above. It was written for DOS, however, there are Windows user-friendly versions like EcoNetwrk developed by NOAA Great Lakes Environmental Research Lab and WAND (Allesina and Bondavalli 2004). Ecopath is widely used for fishery management and includes intuitive functions to model incomplete dataset with algorithms that allow balancing the networks.

A final recommendation focuses on which data should be used to run the model, not only for ENA but also for DEB. Authors have used data from the literature and/or expert opinion only (Christian et al. 2009), but it could represent a limiting factor for the analysis. Although literature data is a valuable resource it is not possible to find updated data in many cases, which can alter the accuracy of the models. Therefore, we emphasize that generation of data broads the potential of the models. Experiments exposing organisms or even biological communities to environmental gradients or even testing the synergetic effects of possible stressors allow us to model the energetic flow and find optimal conditions for targeted organisms or ecosystems. Also the use of monitoring data is recommended in order to understand how the species or communities respond to seasonal or annual variability they go through. The use of experimental and monitoring data to feed the models enables us to understand the thresholds of tolerance range (plasticity) and make better predictions for future climatic changes and possible biological invasions.

Towards Functional Trait Assessment Using Loop Analysis (LA)

Even though ENA shows great potential for analyzing the functioning of ecosystems, there are some aspects that are not covered. The model is restricted to the application of only one type of currency to represent the interactions. When we refer to functional traits, the species may be grouped according to diverse characteristics depending on the function you are looking at. In this subsection, we aim to introduce the application of qualitative analysis as a tool to handle such complexity. In the same framework, it incorporates predator-prey, mutualistic and symbiotic relationships, while at the same time creating connections between human activities and ecosystems (Dee et al. 2017). Qualitative analysis is able to predict the response of the ecosystems to inputs (disturbances), e.g., biological invasions (Raymond et al. 2011) and overfishing (Rocchi et al. 2016).

LA is a holistic and qualitative analysis that is based on positive, negative, and absence of interactions between nodes (Levins 1974). It allows predicting how the impacts from perturbations that occur on target nodes may propagate through the interaction network, thus generating indirect effects on other nodes of the system. It has been used for many purposes: from explaining the interactions between organisms in a food web (Bodini et al. 1994) to modelling the effects that ecological processes have on society (Martone et al. 2017). A loop or circuit is defined as a pathway that crosses the nodes only once and finishes where it started, creating positive or negative feedbacks (Fig. 2). The pathways and feedbacks are determined based on the interactions described in the literature (Bodini 2000). For our purpose, the most interesting part in the analysis is calculating the sign of the feedbacks, since LA detects the cascade effects of the inputs on the functioning and predicts whether the nodes are going to increase, decrease or remain the same under the impact of different perturbations (Bodini 2000). Levins (1974) showed that the systems are stable when there are more negative feedbacks than positive ones. The predictions generated by LA are displayed in a matrix that presents the response of all nodes to the positive input of each variable (Martone et al. 2017; Fig. 2). Software solutions to run these models are available as pakages in R and GUI versions.[10] The software tools usually provide a matrix and a schematic figure (see Fig. 2) with the pathways and types of feedbacks that connect the nodes.

LA has proved to be a useful tool to bring together variables of different kind. Thus, as long as the type of interaction (positive, negative or neutral) is known, it can be a powerful tool to analyze the effect of functional traits independently on the functions used to define them. In addition, the traits can be connected to measure the efficiency of various management strategies, ecosystem functioning and services provided to society (Martone et al. 2017).

Conclusions

The functioning of ecosystems is modulated by the responses of different compartments (e.g., primary producers, herbivores), which determine how species interact. Thus, the horizontal analysis of single compartments using DEB models could help to understand the basis of ecosystems functioning. Nevertheless, a more holistic approach

[10]The software and additional information can be found here: https://www.alexisdinno.com/LoopAnalyst/

can be reached by integrating vertical analysis, i.e., how compartments influence each other by considering feeding preferences and the interaction with non-living elements in ENA. DEB and ENA are not necessarily meant to be used together, but they are complementary and using both of them may diminish uncertainties. Once we understood how the compartments of ecosystems behave, the LA might be the way to bring the discussion to another level. LA outputs can provide information about expected impacts of disturbances on the functioning and services provided by ecosystems. Literature attempting to ingrate the overall complexity of ecosystems and predict the expected consequences of global change drivers on their structure and functioning is still scarce. Hereby we suggest that this gap can be fulfilled based on rigorous algorithms and analytical methods.

Appendix

This article is related to the YOUMARES 8 conference session no. 6: "The Interplay Between Marine Biodiversity and Ecosystems Functioning: Patterns and Mechanisms in a Changing World". The original Call for Abstracts and the abstracts of the presentations within this session can be found in the appendix "Conference Sessions and Abstracts", chapter "11 The Interplay Between Marine Biodiversity and Ecosystems Functioning: Patterns and Mechanisms in a Changing World", of this book.

References

Allesina S, Bondavalli C (2004) WAND: an ecological network analysis user-friendly tool. Environ Model Softw 19:337–340. https://doi.org/10.1016/j.envsoft.2003.10.002

Baird D, Ulanowicz RE (1989) The seasonal dynamics of the Chesapeake Bay ecosystem. Ecol Monogr 59:329–364

Baird D, Ulanowicz R (1993) Comparative study on the trophic structure, cycling and ecosystem properties of four tidal estuaries. Mar Ecol Prog Ser 99:221–237. https://doi.org/10.3354/meps099221

Baird D, Christian RR, Peterson CH et al (2004) Consequences of hypoxia on estuarine ecosystem function: energy diversion from consumers to microbes. Ecol Appl 14:805–822. https://doi.org/10.1890/02-5094

Barnosky AD (2008) Megafauna biomass tradeoff as a driver of quaternary and future extinctions. Proc Natl Acad Sci U S A 105:11543–11548. https://doi.org/10.1073/pnas.0801918105

Barnosky AD, Matzke N, Tomiya S et al (2011) Has the Earth's sixth mass extinction already arrived? Nature 471:51–57. https://doi.org/10.1038/nature09678

Bengtsson J (1998) Which species? What kind of diversity? Which ecosystem function? Some problems in studies of relations between biodiversity and ecosystem function. Appl Soil Ecol 10:191–199. https://doi.org/10.1016/S0929-1393(98)00120-6

Blackburn TM, Jeschke JM (2009) Invasion success and threat status: two sides of a different coin? Ecography 32:83–88. https://doi.org/10.1111/j.1600-0587.2008.05661.x

Bodini A (2000) Reconstructing trophic interactions as a tool for understanding and managing ecosystems: application to a shallow eutrophic lake. Can J Fish Aquat Sci 57:1999–2009. https://doi.org/10.1139/f00-153

Bodini A, Giavelli G, Rossi O (1994) The qualitative analysis of community food webs: implications for wildlife management and conservation. J Environ Manag 41:49–65. https://doi.org/10.1006/jema.1994.1033

Bond WJ (1994) Keystone species. In: Schulze ED, Mooney HA (eds) Biodiversity and ecosystem function. Springer, Berlin/Heidelberg, pp 237–253

Bondavalli C, Bodini A, Rossetti G et al (2006) Detecting stress at the whole-ecosystem level: the case of a mountain lake (Lake Santo, Italy). Ecosystems 9(5):768–787. https://doi.org/10.1007/s10021-005-0065-y

Cardinale BJ, Duffy JE, Gonzalez A et al (2012) Biodiversity loss and its impact on humanity. Nature 486:59–67. https://doi.org/10.1038/nature11148

Carstensen DW, Lessard JP, Holt BG et al (2013) Introducing the biogeographic species pool. Ecography 36:1310–1318. https://doi.org/10.1111/j.1600-0587.2013.00329.x

Christensen V, Pauly D (1992) ECOPATH II — a software for balancing steady-state ecosystem models and calculating network characteristics. Ecol Model 61:169–185. https://doi.org/10.1016/0304-3800(92)90016-8

Christian RR, Baird D, Luczkovich J et al (2005) Role of network analysis in comparative ecosystem ecology of estuaries. In: Belgrano A, Scharler UM, Dunne J et al (eds) Aquatic food webs: an ecosystem approach. Oxford University Press, Oxford, pp 25–40

Christian RR, Brinson MM, Dame JK et al (2009) Ecological network analyses and their use for establishing reference domain in functional assessment of an estuary. Ecol Model 220:3113–3122. https://doi.org/10.1016/j.ecolmodel.2009.07.012

Codron D, Carbone C, Müller DWH et al (2012) Ontogenetic niche shifts in dinosaurs influenced size, diversity and extinction in terrestrial vertebrates. Biol Lett 8:620–623. https://doi.org/10.1098/rsbl.2012.0240

Cornell HV, Harrison SP (2014) What are species pools and when are they important? Annu Rev Ecol Evol Syst 45:45–67. https://doi.org/10.1146/annurev-ecolsys-120213-091759

Dee LE, Allesina S, Bonn A et al (2017) Operationalizing network theory for ecosystem service assessments. Trends Ecol Evol 32:118–130. https://doi.org/10.1016/j.tree.2016.10.011

Díaz S, Cabido M (2001) Vive la différence: plant functional diversity matters to ecosystem processes. Trends Ecol Evol 16:646–655. https://doi.org/10.1016/S0169-5347(01)02283-2

Díaz S, Fargione J, Chapin FS et al (2006) Biodiversity loss threatens human Well-being. PLoS Biol 4:e277. https://doi.org/10.1371/journal.pbio.0040277

Doney SC (2010) The growing human footprint on coastal and open-ocean biogeochemistry. Science 328:1512–1516. https://doi.org/10.1126/science.1185198

Edmunds PJ, Putnam HM, Nisbet RM et al (2011) Benchmarks in organism performance and their use in comparative analyses. Oecologia 167:379–390. https://doi.org/10.1007/s00442-011-2004-2

Elton CS (1958) The ecology of invasions by animals and plants, 1st edn. Chapman and Hall Ltd, London

Estes JA, Terborgh J, Brashares JS et al (2011) Trophic downgrading of planet earth. Science 333:301–306. https://doi.org/10.1126/science.1205106

Fargione J, Tilman D, Dybzinski R et al (2007) From selection to complementarity: shifts in the causes of biodiversity-productivity relationships in a long-term biodiversity experiment. Proc R Soc B 274:871–876. https://doi.org/10.1098/rspb.2006.0351

Gaedke U (1995) A comparison of whole-community and ecosystem approaches (biomass size distributions, food web analysis, network

analysis, simulation models) to study the structure, function and regulation of pelagic food webs. J Plankton Res 17(6):1273–1305. https://doi.org/10.1093/plankt/17.6.1273

Gilarranz LJ, Rayfield B, Liñán-Cembrano G et al (2017) Effects of network modularity on the spread of perturbation impact in experimental metapopulations. Science 357:199–201. https://doi.org/10.1126/science.aal4122

Gross K, Cardinale BJ, Fox JW et al (2014) Species richness and the temporal stability of biomass production: a new analysis of recent biodiversity experiments. Am Nat 183:1–12. https://doi.org/10.1086/673915

Gurevitch J, Padilla D (2004) Are invasive species a major cause of extinctions? Trends Ecol Evol 19:470–474. https://doi.org/10.1016/j.tree.2004.07.005

Hillebrand H, Blenckner T (2002) Regional and local impact on species diversity – from pattern to processes. Oecologia 132:479–491. https://doi.org/10.1007/s00442-002-0988-3

Hoegh-Guldberg O, Bruno JF (2010) The impact of climate change on the world's marine ecosystems. Science 328:1523–1528. https://doi.org/10.1126/science.1189930

Hull P (2015) Life in the aftermath of mass extinctions. Curr Biol 25:R941–R952. https://doi.org/10.1016/j.cub.2015.08.053

Isbell F, Gonzalez A, Loreau M et al (2017) Linking the influence and dependence of people on biodiversity across scales. Nature 546:65–72. https://doi.org/10.1038/nature22899

Jeschke JM, Strayer DL (2008) Are threat status and invasion success two sides of the same coin? Ecography 31:124–130. https://doi.org/10.1111/j.2007.0906-7590.05343.x

Kay JJ, Graham LA, Ulanowicz RE (1989) A detailed guide to network analysis. In: Wulff F, Field JG, Mann KH (eds) Network analysis in marine ecology. Springer, Berlin/Heidelberg, pp 15–61

Kooijman SALM (2010) Dynamic energy budget theory, 3rd edn. Cambridge University Press, Cambridge

Larson ER, Olden JD (2010) Latent extinction and invasion risk of crayfishes in the southeastern United States. Conserv Biol 24:1099–1110. https://doi.org/10.1111/j.1523-1739.2010.01462.x

Lehman CL, Tilman D (2000) Biodiversity, stability, and productivity in competitive communities. Am Nat 156:534–552. https://doi.org/10.1086/303402

Levins R (1974) Discussion paper: the qualitative analysis of partially specified systems. Ann N Y Acad Sci 231:123–138

Loreau M, Hector A (2001) Partitioning selection and complementarity in biodiversity experiments. Nature 412:72–76. https://doi.org/10.1038/35083573

Loreau M, Naeem S, Inchausti P et al (2001) Biodiversity and ecosystem functioning: current knowledge and future challenges. Science 294:804–808. https://doi.org/10.1126/science.1064088

Magri M, Benelli S, Bondavalli C et al (2017) Benthic N pathways in illuminated and bioturbated sediments studied with network analysis. Limnol Oceanogr 63:S68. https://doi.org/10.1002/lno.10724

Marchetti MP, Moyle PB, Levine R (2004) Invasive species profiling? Exploring the characteristics of non-native fishes across invasion stages in California. Freshw Biol 49:646–661. https://doi.org/10.1111/j.1365-2427.2004.01202.x

Martone RG, Bodini A, Micheli F (2017) Identifying potential consequences of natural perturbations and management decisions on a coastal fishery social-ecological system using qualitative loop analysis. Ecol Soc 22:34. https://doi.org/10.5751/ES-08825-220134

May RM (1971) Stability in multispecies community models. Math Biosci 12:59–79. https://doi.org/10.1016/0025-5564(71)90074-5

May RM (1972) Will a large complex system be stable? Nature 238:413–414. https://doi.org/10.1038/238413a0

May RM (1973) Stability and complexity in model ecosystems, 1st edn. Princeton University Press, Princeton

McCann KS (2000) The diversity–stability debate. Nature 405:228–233. https://doi.org/10.1038/35012234

McCann K, Hastings A, Huxel GR (1998) Weak trophic interactions and the balance of nature. Nature 395:794–798. https://doi.org/10.1038/27427

McKinney ML, Lockwood JL (1999) Biotic homogenization: a few winners replacing many losers in the next mass extinction. Trends Ecol Evol 14:450–453. https://doi.org/10.1016/S0169-5347(99)01679-1

McNaughton SJ (1977) Diversity and stability of ecological communities: a comment on the role of empiricism in ecology. Am Nat 111:515–525. https://doi.org/10.1086/283181

Monaco CJ, Wethey DS, Helmuth B (2014) A dynamic energy budget (DEB) model for the keystone predator *Pisaster ochraceus*. PLoS One 9(8):e104658. https://doi.org/10.1371/journal.pone.0104658

Montoya JM, Pimm SL, Solé RV (2006) Ecological networks and their fragility. Nature 442:259–264. https://doi.org/10.1038/nature04927

Muller EB, Kooijman SALM, Edmunds PJ et al (2009) Dynamic energy budgets in syntrophic symbiotic relationships between heterotrophic hosts and photoautotrophic symbionts. J Theor Biol 259:44–57. https://doi.org/10.1016/j.jtbi.2009.03.004

Murray BR, Thrall PH, Gill AM et al (2002) How plant life-history and ecological traits relate to species rarity and commonness at varying spatial scales. Austral Ecol 27:291–310. https://doi.org/10.1046/j.1442-9993.2002.01181.x

Naeem S (1996) Species redundancy and ecosystem reliability. Conserv Biol 12:39–45. https://doi.org/10.1111/j.1523-1739.1998.96379.x

Neutel A-M, Heesterbeek JAP, De Ruiter PC (2002) Stability in real food webs: weak links in long loops. Science 296:1120–1123. https://doi.org/10.1126/science.1068326

Nisbet RM, Muller EB, Lika K et al (2000) From molecules to ecosystems through dynamic energy budget models. J Anim Ecol 69:913–926. https://doi.org/10.1046/j.1365-2656.2000.00448.x

Odum EP (1953) Fundamentals of ecology, 1st edn. Saunders Co., Philadelphia

Olesen JM, Bascompte J, Dupont YL et al (2007) The modularity of pollination networks. Proc Natl Acad Sci U S A 104:19891–19896. https://doi.org/10.1073/pnas.0706375104

Paine RT (1992) Food-web analysis through field measurement of per capita interaction strength. Nature 355:73–75. https://doi.org/10.1038/355073a0

Pimm SL, Jenkins CN, Abell R et al (2014) The biodiversity of species and their rates of extinction, distribution, and protection. Science 344:1246752. https://doi.org/10.1126/science.1246752

Purvis A, Gittleman JL, Cowlishaw G et al (2000) Predicting extinction risk in declining species. Proc R Soc B 267:1947–1952. https://doi.org/10.1098/rspb.2000.1234

Raymond B, McInnes J, Dambacher JM et al (2011) Qualitative modelling of invasive species eradication on subantarctic Macquarie Island. J Appl Ecol 48:181–191. https://doi.org/10.1111/j.1365-2664.2010.01916.x

Reiss J, Bridle JR, Montoya JM et al (2009) Emerging horizons in biodiversity and ecosystem functioning research. Trends Ecol Evol 24:505–514. https://doi.org/10.1016/j.tree.2009.03.018

Reusch TBH, Ehlers A, Hammerli A et al (2005) Ecosystem recovery after climatic extremes enhanced by genotypic diversity. Proc Natl Acad Sci U S A 102:2826–2831. https://doi.org/10.1073/pnas.0500008102

Ripple WJ, Estes JA, Beschta RL et al (2014) Status and ecological effects of the world's largest carnivores. Science 343:1241484. https://doi.org/10.1126/science.1241484

Rocchi M, Scotti M, Micheli F et al (2016) Key species and impact of fishery through food web analysis: a case study from Baja California Sur, Mexico. J Mar Syst 165:92–102. https://doi.org/10.1016/j.jmarsys.2016.10.003

Rooney N, McCann KS (2012) Integrating food web diversity, structure and stability. Trends Ecol Evol 27:40–45. https://doi.org/10.1016/j.tree.2011.09.001

Sax DF, Gaines SD (2003) Species diversity: from global decreases to local increases. Trends Ecol Evol 18:561–566. https://doi.org/10.1016/S0169-5347(03)00224-6

Sax DF, Gaines SD (2008) Species invasions and extinction: the future of native biodiversity on islands. Proc Natl Acad Sci U S A 105:11490–11497. https://doi.org/10.1073/pnas.0802290105

Seehausen O (2004) Hybridization and adaptive radiation. Trends Ecol Evol 19:198–207. https://doi.org/10.1016/j.tree.2004.01.003

Solan M, Cardinale BJ, Downing AL et al (2004) Extinction and ecosystem function in the marine benthos. Science 306:1177–1180. https://doi.org/10.1126/science.1103960

Springer AM, Estes JA, van Vliet GB et al (2003) Sequential megafaunal collapse in the North Pacific Ocean: an ongoing legacy of industrial whaling? Proc Natl Acad Sci U S A 100:12223–12228. https://doi.org/10.1073/pnas.1635156100

Stachowicz JJ, Bruno JF, Duffy JE (2007) Understanding the effects of marine biodiversity on communities and ecosystems. Annu Rev Ecol Evol Syst 38:739–766. https://doi.org/10.1146/annurev.ecolsys.38.091206.095659

Thomas CD (2013a) The Anthropocene could raise biological diversity. Nature 502:7. https://doi.org/10.1038/502007a

Thomas CD (2013b) Local diversity stays about the same, regional diversity increases, and global diversity declines. Proc Natl Acad Sci U S A 110:19187–19188. https://doi.org/10.1073/pnas.1319304110

Tilman D, Isbell F, Cowles JM (2014) Biodiversity and ecosystem functioning. Annu Rev Ecol Evol Syst 45:471–493. https://doi.org/10.1146/annurev-ecolsys-120213-091917

Tingley R, Mahoney PJ, Durso AM et al (2016) Threatened and invasive reptiles are not two sides of the same coin. Glob Ecol Biogeogr 25:1050–1060. https://doi.org/10.1111/geb.12462

Ulanowicz RE (1995) Trophic flow networks as indicators of ecosystem stress. In: Polis GA, Winemiller KO (eds) Food webs: integration of patterns and dynamics. Chapman and Hall, New York, pp 358–368

Ulanowicz RE (2004) Quantitative methods for ecological network analysis. Comput Biol Chem 28:321–339. https://doi.org/10.1016/j.compbiolchem.2004.09.001

Ulanowicz RE, Kay JJ (1991) A package for the analysis of ecosystem flow networks. Environ Softw 6:131–142. https://doi.org/10.1016/0266-9838(91)90024-K

van der Meer J (2006) An introduction to dynamic energy budget (DEB) models with special emphasis on parameter estimation. J Sea Res 56:85–102. https://doi.org/10.1016/j.seares.2006.03.001

van Kleunen M, Richardson DM (2007) Invasion biology and conservation biology: time to join forces to explore the links between species traits and extinction risk and invasiveness. Prog Phys Geogr 31:447–450. https://doi.org/10.1177/0309133307081295

van Kleunen M, Weber E, Fischer M (2010) A meta-analysis of trait differences between invasive and non-invasive plant species. Ecol Lett 13:235–245. https://doi.org/10.1111/j.1461-0248.2009.01418.x

Vitousek PM, Mooney HA, Lubchenco J et al (1997) Human domination of earth' s ecosystems. Science 277:494–499. https://doi.org/10.1126/science.277.5325.494

Warren PH (1989) Spatial and temporal variation in the structure of a freshwater food web. Oikos 55:299–311. https://doi.org/10.2307/3565588

Woodward G, Ebenman B, Emmerson M et al (2005) Body size in ecological networks. Trends Ecol Evol 20:402–409. https://doi.org/10.1016/j.tree.2005.04.005

Worm B, Barbier EB, Beaumont N et al (2006) Impacts of biodiversity loss on ocean ecosystem services. Science 314:787–790. https://doi.org/10.1126/science.1132294

Yachi S, Loreau M (1999) Biodiversity and ecosystem productivity in a fluctuating environment: the insurance hypothesis. Proc Natl Acad Sci U S A 96:1463–1468. https://doi.org/10.1073/pnas.96.4.1463

Microplastics in Aquatic Systems – Monitoring Methods and Biological Consequences

Thea Hamm, Claudia Lorenz, and Sarah Piehl

Abstract

Microplastic research started at the turn of the millennium and is of growing interest, as microplastics have the potential to affect a whole range of organisms, from the base of the food web to top predators, including humans. To date, most studies are initial assessments of microplastic abundances for a certain area, thereby generally distinguishing three different sampling matrices: water, sediment and biota samples. Those descriptive studies are important to get a first impression of the extent of the problem, but for a proper risk assessment of ecosystems and their inhabitants, analytical studies of microplastic fluxes, sources, sinks, and transportation pathways are of utmost importance. Moreover, to gain insight into the effects microplastics might have on biota, it is crucial to identify realistic environmental concentrations of microplastics. Thus, profound knowledge about the effects of microplastics on biota is still scarce. Effects can vary regarding habitat, functional group of the organism, and polymer type for example, making it difficult to find quick answers to the many open questions. In addition, microplastic research is accompanied by many methodological challenges that need to be overcome first to assess the impact of microplastics on aquatic systems. Thereby, a development of standardized operational protocols (SOPs) is a pre-requisite for comparability among studies. Since SOPs are still lacking and new methods are developed or optimized very frequently, the aim of this chapter is to point out the most crucial challenges in microplastic research and to gather the most recent promising methods used to quantify environmental concentrations of microplastics and effect studies.

T. Hamm
GEOMAR Helmholtz Center for Ocean Research, Kiel, Germany
e-mail: thamm@geomar.de

C. Lorenz (✉)
Alfred Wegener Institute (AWI), Helmholtz Centre for Polar and Marine Research, Biologische Anstalt Helgoland, Helgoland, Germany
e-mail: claudia.lorenz@awi.de

S. Piehl
Department of Animal Ecology I and BayCEER, University of Bayreuth, Bayreuth, Germany
e-mail: sarah.piehl@uni-bayreuth.de

Introduction

Literature on microplastic (MP) abundance in aquatic environments and observed effects on biota has exponentially increased over the last 7 years (Connors et al. 2017). Within the current literature, MP sampling is imbalanced and studies are most often conducted on sandy beaches and the sea surface, followed by bottom sediment samples and water column samples (Duis and Coors 2016; Bergmann et al. 2017). Individual studies examining MP abundance, i.e., deep sea sediments (Van Cauwenberghe et al. 2013b; Woodall et al. 2014), sea ice (Obbard et al. 2014) or marine snow (Zhao et al. 2017) exist. Thereby, attempts to compare data taken from similar sampling matrices have been made in almost every study (Filella 2015), whereas for most studies this is often hampered by the various sampling methods applied (Hidalgo-Ruz et al. 2012; Filella 2015; Löder and Gerdts 2015; Costa and Duarte 2017). Hidalgo-Ruz et al. (2012) was the first article that showed the huge variety of different methods used for MP data collection and suggested the need for standardized operational protocols (SOPs). In the "Guidelines for Monitoring of marine litter" published by Hanke et al. (2013) the authors suggested methods based on the most often used techniques but also stressed that further standardization is needed. The NOAA made initial attempts of standardization in laboratory methods (Masura et al. 2015). Moreover, Löder and Gerdts (2015), as well as more recently Costa and Duarte (2017), took up the issue and critically assessed the different methods used for MP analysis. However, different environments can only be compared

S. Jungblut et al. (eds.), *YOUMARES 8 – Oceans Across Boundaries: Learning from each other*,
https://doi.org/10.1007/978-3-319-93284-2_13

to a certain extent, as the different sample matrices require different sampling methods. Moreover, as replication of samples is limited within a project, the high spatial and temporal variability of MPs in the various environments poses another major challenge in MP research (Goldstein et al. 2013; Moreira et al. 2016; Imhof et al. 2017). Whereas some recommendations for spatial replication have been made, no general consensus exists about temporal replication (Hanke et al. 2013). As a next step, the impact of the determined environmental concentrations of MP on biota is interesting. Parallel to monitoring studies, the toxicological implications for biota have been addressed in many studies. So far, we know that MPs are ingested by a wide range of organisms from the base of the food web up to top predators. As the environmental concentrations have not yet been sufficiently analyzed, exposure to MPs in laboratory studies are applying high concentrations to get first insights into possible effects following ingestion. This chapter aims to summarize the main results of the latest 3 years of research on sampling and monitoring methods as well as to give an overview about observed effects of MP exposure on biota.

Sampling Design

Previous research already addressed the problem of an appropriate sampling design (Browne et al. 2015; Löder and Gerdts 2015; Costa and Duarte 2017). A detailed review on the topic is given by Underwood et al. (2017). Over the last years, some studies focused on improving sampling design (Chae et al. 2015; Kang et al. 2015; Barrows et al. 2017) and aimed to investigate spatial and temporal patterns of MPs (Goldstein et al. 2013; Heo et al. 2013; Besley et al. 2017; Fisner et al. 2017; Imhof et al. 2017). Moreover, a few recommendations and protocols for sampling exist (Hanke et al. 2013; GESAMP 2016; Kovač Viršek et al. 2016). Potential factors which need to be considered when sampling beach sediments are summarized in Fig. 1. Some of the main issues are discussed in the following for both, water and sediment samples.

In each study, scientists should first determine the appropriate study area suitable for their research question. Thereby, factors such as, for example, proximity to potential sources (i.e., cities, harbors, industry), ocean currents and sampled sediment type need to be considered, as they can influence composition of MPs as well as the abundances (Hanvey et al. 2017). As a next step, a sampling design needs to be chosen, which suits the study question and is representative of the study area. Although most studies are initial assessments of MP concentrations, most often potential accumulation sites have been sampled (e.g., high tide line on beaches or ocean surface) (Filella 2015; Bergmann et al. 2017; Hanvey et al. 2017). Therefore, results cannot be extrapolated to the whole study area, as this kind of sampling is designed to find MP contamination. If the objective of the study is to assess the contamination level of the whole area, the sampling design could be improved by expanding the sampling to spots, which are not expected to have high amounts of MPs. Thus, random

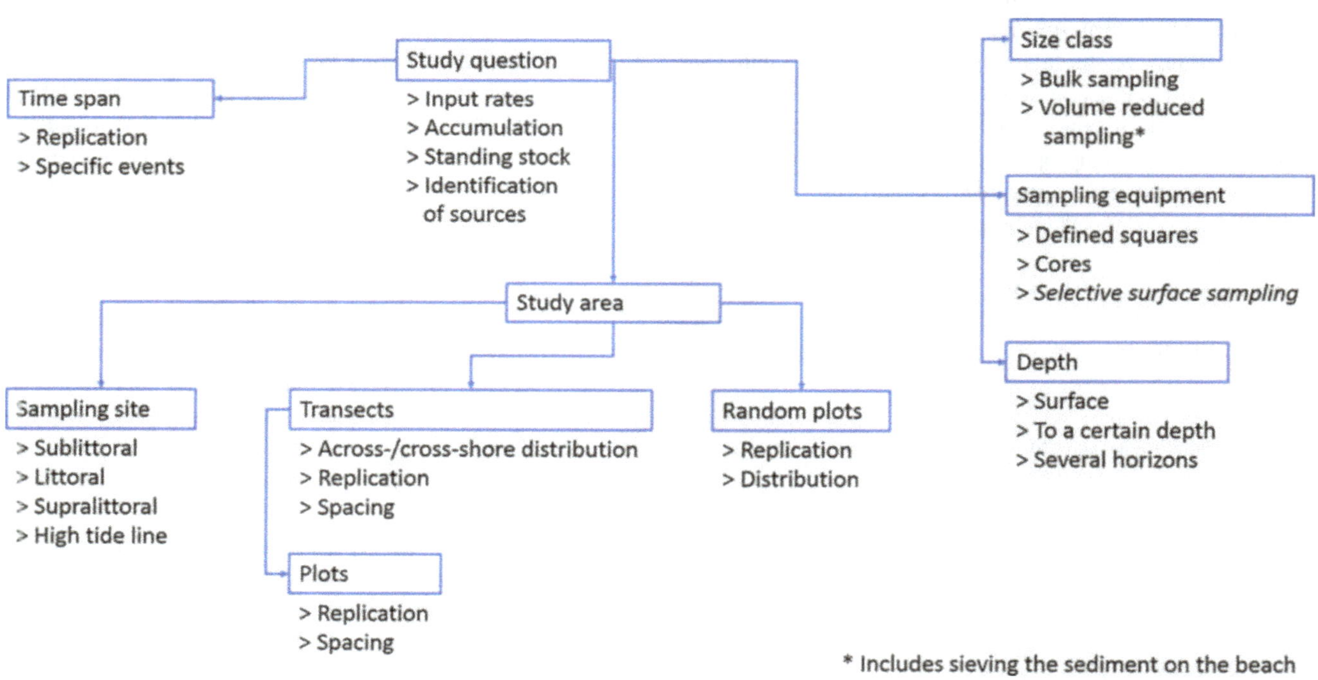

Fig. 1 Overview of factors, which need to be considered when planning a microplastics sampling campaign, exemplary for beach sediment samples

sampling, e.g., of a section of a beach, including the whole vertical and horizontal dimension, could be an option, although not yet conducted for MPs. In any case, care should be taken when formulating research questions, as this will set the framework for considerations regarding the sampling design.

Spatial and Temporal Replication

To get a representative sample, care needs to be taken with respect to appropriate replication as well as the amount of sample, which will be taken. If study areas of various sizes are compared, it needs to be considered, whether the number of replicates is kept the same or whether they are adjusted to the area (balanced vs. unbalanced sampling design). For beach sediment, Kim et al. (2015) adjusted sampling effort to beach size, whereas the majority of studies kept replicate numbers the same. In the current literature, replicate samples for one beach can range from one to 88 (Besley et al. 2017), whereas recommendations suggest a replication of at least five (Hanke et al. 2013). For beach sediments, Dekiff et al. (2014) found no significant variability in MP abundance within a 100 m transect, taking six replicate samples. Low spatial variability on a small scale (within tens of m) was further found in a recent study from Fisner et al. (2017) on plastic pellets (~ 1–6 mm; (Hidalgo-Ruz et al. 2012), whereas this study further found a high spatial variability on a large scale (within km). Contrary, Besley et al. (2017), including smaller MPs (300–5000 μm), found a high spatial variability among ten samples on a transect of 100 m. Confidence intervals around the mean in this study decreased rapidly after a replication of five, and 11 replicates would be needed to reach a 0.5 standard deviation at a confidence level of 90% (Besley et al. 2017). Those results are supported by a further study concentrating on large MPs (1–5 mm) on a 100 m transect on a tropical beach (six replicates; (Imhof et al. 2017). For surface water samples there is one study investigating spatial variability within the eastern North Pacific, off California (~ 20°–40°N, 120°–155°W; (Goldstein et al. 2013). They found that MP concentrations were highly variable over relatively small scales (tens of km) as well as for large scales (hundreds to thousands of km).

It is also stated that MP abundance varies over numerous temporal scales and detection of temporal trends are often hampered by the sampling design (Browne et al. 2015). Recent studies conducted on beaches found high daily variability due to tidal dynamics (Moreira et al. 2016; Imhof et al. 2017). One possibility to improve knowledge about temporal patterns could be through ice or sediment cores (Costa and Duarte 2017), by analyzing different layers separately. For the water surface, high inter-annual variability was found (Law et al. 2010; Doyle et al. 2011; Law et al. 2014), whereas Law et al. (2010), investigating a 22-year dataset of surface plankton net tows, found no strong temporal trends in MP concentrations within this data set. Nevertheless, the time span needed for a sampling campaign should be considered beforehand. For example for beach sediment sampling, sampling periods range over several hours to years (Browne et al. 2015).Whereas for some study questions, sampling over a certain period of time may not be a problem, for others it could lead to biased results. This might, for instance, apply to the sampling of various river mouths at a delta over several days. Strongly changing precipitation between sampling days could hamper comparability, as MP runoff could be enhanced during days of heavy rainfall, similar to what was hypothesized in a recent study comparing MP load of waste water treatment plants effluents on two different dates with differing participation events (Primpke et al. 2017a).

Sampling Depth

For both, sediments and water column, the optimal sampling depth remains another open question. Sediment sampling is recommended to a depth of at least 5 cm (Hanke et al. 2013; Besley et al. 2017), whereas studies report that a potential proportion can be lost if deeper sediment layers are not sampled (Carson et al. 2011; Claessens et al. 2011). Thus, it has already been stated that samples should be taken at a depth to 1 m, to get a more precise picture of MP abundances (Turra et al. 2014; Fisner et al. 2017). For the water column, only few studies exist where different depths were concurrently sampled (Lattin et al. 2004; Reisser et al. 2015). In one study, no significant differences were found between the sea surface, the water column (5 m depth), and above the bottom (Lattin et al. 2004), whereas the other found that MP concentrations decreased exponentially, with highest amounts within the first 0.5 m of the water column (Reisser et al. (2015). This is confirmed by Goldstein et al. (2013), detecting the highest concentrations of MPs during low wind conditions, when minimal mixing occurs between shallow and deeper water layers. The optimal sampling depth will finally be a compromise between increasing sampling surface and sampling depth and thus will also be determined by the research question.

Reporting of Data

Though different methods are necessary depending on the research question, researchers should aim for standardization, the most important one being size classes and reporting units. Regarding size classes the upper limit for MPs is 5 mm, whereas the lower limit will be defined by the sampling device, as well as the analytical method. Initial studies investigating size distribution found generally increasing abundances with decreasing size classes (Imhof et al. 2016). Even though the applied methodology will define the lower size limit, the post-sampling procedures will allow for classification into different size classes. Thereby, Hanke et al. (2013) recommended to allocate MP particles into size bins of 100 µm. Although this recommendation would provide high resolution datasets, in practice this is almost not feasible, as the preparation of microplastic samples is already very time consuming and, for instance, additional sieving steps would further increase analysis time. Further, depending on the research question different size categories are of importance. If, for example, pictures of the microplastic particles are taken during analysis, it is possible to obtain data on the size at a later time point in case the data would be requested for comparative analysis.

Standardization of reporting units is a further necessity to increase comparability among data sets. So far, different sampling strategies have led to various reporting units (e.g., m^2, m^3, ml, l, g, kg) (Hidalgo-Ruz et al. 2012; Löder and Gerdts 2015; Costa and Duarte 2017). For MPs in the environment (excluding biota samples) either bulk or volume reduced samples are taken. Thus, a volume measurement can always be obtained and should be the minimum information reported. Additional reporting of sampling depth as well as weight measurements for sediment samples will further increase data quality.

Finally, reporting of meta data like prevailing wind direction, sea state, beach morphology, rainfall, and so on would improve the interpretation of the data collected (Barrows et al. 2017). In the current literature, missing information range from unreported size ranges, replication, detected numbers of particles to sampling locations (Filella 2015; Besley et al. 2017). Comprehensive reporting of the applied methods is a crucial part and not only a requirement for reproducibility, but further gives the reader the ability to judge about the representativeness of the study, as well as the conclusions drawn from the results.

Sampling Equipment

Further considerations should be made on the sampling equipment, as this will define the size range of MPs in the study, as well as reporting units. For beach sediments, sampling equipment is well established (Hidalgo-Ruz et al. 2012; Hanvey et al. 2017), it only remains important to consider, whether to collect a bulk or a volume reduced sample. For the latter, a lower size limit is defined. For bottom sediments corers, Van Veen or Ekman grabs can be used, however, grabs disturb the surface layer of the sediment and corers do not only take the sediment but also the water layer above the sediment (Löder and Gerdts 2015).

For water samples, nets of various types have been used (Table 1 gives an overview of the used equipment found in the current literature). Most commonly, manta nets are the device of choice (Costa and Duarte 2017), where the reduced sample volume limits the lowest size class of investigated MPs mostly to 300–350 µm (Filella 2015). Thus, some researchers used bottles to take bulk samples of the water surface (Dubaish and Liebezeit 2013; Barrows et al. 2017), which, however, results in small sample volumes. Nevertheless, sampling lower size ranges, Barrows et al. (2017) found MP concentrations were several orders of magnitude higher in bottle samples than manta samples. To obtain larger sample volumes, others took several bottles or buckets of surface water and concentrated the material on filters with smaller mesh sizes on board (hand-nets; Chae et al. 2015; Kang et al. 2015). Moreover, contamination issues through high air exposure times during a manta trawl, as well as filtering samples on board, motivated researchers to further develop pumping systems (Desforges et al. 2014; Lusher et al. 2014; Enders et al. 2015). One of the first studies comparing different methodologies for the same size class (300–5000 µm) was conducted by Setälä et al. (2016) comparing their custom-made pump to manta trawls. Preliminary results from the pump (collecting surface water in a depth of 0–0.5 m) did not significantly differ from the results obtained by the manta net. Another interesting solution to decrease sampling effort has been published by Edson and Patterson (2015). They designed an automated sampling device (MantaRay), which automatically pumps sea surface water at a depth of 30 cm, while drifting through the water. Thereby, particles are concentrated on a filter and 28 successive samples can be taken. For the prototype, 500 µm stainless steel sieves were used. Such an instrument can decrease sampling effort and airborne contamination, which is often a challenge when conducting trawls. One drawback could be the autonomous operation of the MantaRay, which limits the control over the area sampled. Moreover, an optical sensor is implemented to ensure that only water containing particulate matter is filtered. Thereby, especially small MP particles could be overlooked so the influence on the obtained results must be further evaluated.

Independent of the applied method, decreasing mesh sizes will increase the content of organic and inorganic material, which could lead to smaller sample sizes as meshes will become clogged faster, but also to increased sample preparation time in the laboratory. In any case, negative controls should be run, as most of the used methods may contain polymer materials which are a further source for contamination.

Table 1 Comparison of various methods used to collect water samples for the analysis of microplastics (MP) in different compartments. Pro and contra are always relative with regard to the sampling devices used for the specific compartment

Sampled compartment	Most common used equipment	General description	Pro	Contra	References
Sea surface microlayer (SML)	Rotating drum sampler	Drum is towed over the water surface and SML is sampled under capillary force by the rotating drum and collected in glass containers	reduced contamination issues large sample volume	only a small part of SML is sampled (50-60 µm)* water adhering to the drum may dilute the sample device materials need to be considered	Ng and Obbard (2006)
	Screen sampler	Water surface is gently touched with a metal sieve with specific pore size; MP particles and SML water is trapped within the metal sieve mesh by surface tension	easy handling and transport larger part of SML is covered compared to rotating drum sampler	only a part of SML is sampled (150-400 µm)* variation can be caused by different operators contamination through higher air exposure times	Song et al. (2014)
Water surface	Manta or plankton/ neuston nets with flowmeter	Net is towed over the water surface to a certain depth (depending on mouth opening) and volume recorded with a flowmeter	large sample sizes exact for the water surface layer integrates a high area of sea surface	investigated size class limited (mesh size often ~300 µm) contamination through higher air exposure times and material of equipment plankton/neuston nets: opening obstructed by ropes for towing	Barrows et al. (2017) and Costa and Duarte (2017)
	Bulk sampling with bottles	Water samples are taken directly from water surface and bottles closed below surface to reduce contamination	whole size range of MPs can be sampled reduced contamination issues	small sample sizes may result in a high variability varying sampling depth	Dubaish and Liebezeit (2013) and Barrows et al. (2017)
	Bulk sampling with hand-net	Water sample is taken with a container and poured over stainless steel meshes on board	whole size range of MPs can be sampled pre-separation of size classes possible large sample sizes can be obtained	varying sampling depth contamination through higher air exposure times device materials need to be considered	Chae et al. (2015) and Kang et al. (2015)
	Pumping systems	Seawater is either collected via the intake of a ship, a hose or a submersible pump	whole size range of MPs can be sampledpre-separation of size classes possible large sample sizes can be obtained reduced contamination issues	varying sampling depth smaller mesh sizes lead to faster blocking of the filters device materials need to be considered	Desforges et al. (2014), Enders et al. (2015), Lusher et al. (2014) and Setälä et al. (2016)
Water column	Bongo nets	Paired zooplankton nets joined by a central axle	large sample sizes integrates a high area of water column unobstructed by towing ropes	investigated size class is limited through mesh size contamination through material of equipment	Lattin et al. (2004)
	Continuous plankton recorder (CPR)	A box for filtering particles at a depth between 5–10 m; material is concentrated on continuously moving bands of filter silk	low operation effort archived data records available	smaller MP particles, which cannot be hand-picked can probably not be recovered from the silk material	Reid et al. (2003) and Thompson et al. (2004)
	Epibenthic sled	A sled which is towed over the sea bottom with a net placed at a certain distance (20 cm) over the bottom such that no resuspended sediment is collected	large sample volumesintegrates a high area of water column	high operation effort obstacles on the ground could block the net or make the sample useless due to resuspended material investigated size class is limited through mesh size contamination through material of equipment	Lattin et al. (2004)

Sample Preparation

The environmental samples taken for MP analysis usually contain a high amount of biogenic material (biota and detritus) and inorganic material (clay, silicates). Therefore, extraction of MPs from the environmental matrix is crucial to facilitate the subsequent identification of MPs. Sometimes, sieving is used to remove larger particles (> 5 mm) from the samples as well as to divide them into distinct size fractions that might be further analyzed differently (Löder and Gerdts 2015). Especially for bulk sediment samples, MPs have to be extracted from the inorganic sediment matrix first while for water and biota samples the removal of the biogenic matrix is put first.

Extraction Techniques

Removing inorganic material from environmental samples is based on the fact that most MPs possess a considerably lower density (0.90–1.55 g cm^{-3}; Table 2) than the inorganic components of sediments like quartz sand or other silicates (2.65 g cm^{-3}) (Hidalgo-Ruz et al. 2012). The most prominent extraction techniques are density separation or fluidization/elutriation. According to Hanvey et al. (2017) density separation is by far the most prevalent one and is defined by the liquid used, the mixing time, the time for settling and the limits of subsequent size fractionation (Hanvey et al. 2017). The most common salt solution for separation is sodium chloride (NaCl) with a density of 1.2 g/cm^3 (Thompson et al. 2004; Hidalgo-Ruz et al. 2012; Hanvey et al. 2017). Due to being inexpensive and non-hazardous, the use of NaCl is also recommended by Hanke et al. (2013), despite its relatively low density. By raising the density of the separation fluid, mainly by using other salt solutions, a better density gradient can be obtained (Filella 2015). These solutions include zinc chloride ($ZnCl_2$) with a density of 1.5–1.7 g cm^{-3} (Imhof et al. 2012; Imhof et al. 2013; Imhof et al. 2016; Mintenig et al. 2017), sodium iodide (NaI) with a density of 1.6 g cm^{-3} (Van Cauwenberghe et al. 2013a; Van Cauwenberghe et al. 2013b; Dekiff et al. 2014; Nuelle et al. 2014; Fischer and Scholz-Böttcher 2017), sodium polytungstate with a density of 1.4–1.5 g cm^{-3} (Corcoran et al. 2009; Corcoran 2015), zinc bromide ($ZnBr_2$) with a density of 1.71 g cm^{-3} (Quinn et al. 2017) and calcium chloride ($CaCl_2$) with a density of 1.30–1.46 g cm^{-3} (Stolte et al. 2015; Courtene-Jones et al. 2017). Samples are added to the separation fluid and either stirred or shaken for a defined time to separate MPs from the sediment matrix (Hanvey et al. 2017). These periods vary considerably between studies if indicated at all (Hidalgo-Ruz et al. 2012; Filella 2015; Hanvey et al. 2017). This is also true for settling times after mixing (Besley et al. 2017; Hanvey et al. 2017) which vary between several minutes (Nuelle et al. 2014; Corcoran 2015) and hours (Stolte et al. 2015; Imhof et al. 2016; Mintenig et al. 2017). Since the aim is to allow for all the sediment particles to sink and all MPs to rise through the whole fluid column according to their respective density, Besley et al. (2017) suggested a minimum settling time of 5–8 hours. Especially for small sample amounts, density separation can be done simply in a

Table 2 Density, heat deflection temperature (HDT), and chemical resistance of common plastic types (Osswald et al. 2006; Bürkle GmbH 2015; Qiu et al. 2016)

Plastic type	Density ρ	HDT	Chemical resistance									
			HCl		H_2SO_4	HNO_3		NaOH		KOH	H_2O_2	NaClO
	g cm^{-3}	°C	5% 2 M	35% 11 M	40%	5%	66%	4% 1 M	30% 10 M	10%	30%	12.5% Cl
Acrylonitrile butadiene styrene (ABS)	1.04–1.06	95–105	–	–	–	–	–	–	–	–	–	–
High-density polyethylene (HDPE)	0.94–0.96	~50	1/1	1/1	1/1	1/1	2/4	1/1	1/1	1/1	1/1	2/3
Low-density polyethylene (LDPE)	0.91–0.92	~35	1/1	1/1	1/1	1/1	3/4	1/1	1/1	1/1	1/2	2/3
Polyamide (PA)	1.02–1.14	55–120	4/4	4/4	4/4	4/4	4/4	1/–	1/–	1/–	4/4	4/4
Polybutylene terephthalate (PBT)	1.31	60	–	–	–	–	–	–	–	–	–	–
Polycarbonate (PC)	1.20	125–135	1/1	4/4	2/–	1/2	4/4	3/4	4/4	4/4	1/1	2/3
Polyethylene terephthalate (PET)	1.37	80	2	4	4	2	4	3	4/4	4/4	1/–	3
Polymethyl methacrylate (PMMA)	1.17–1.20	75–105	–	–	–	–	–	–	–	–	–	–
Polyoxymethylene (POM)	1.41–1.42	100–160	4/4	4/4	4/4	4/4	4/4	1/1	1/3	1/1	4/4	4/4
Polypropylene (PP)	0.90–0.91	55–70	1/1	1/2	1/1	1/1	4/4	1/1	1/1	1/1	1/3	2/3
Polystyrene (PS)	1.05	65–85	1/1	3/3	2/–	2/4	4/4	2/2	1/–	–	1/2	1/3
Polysulfone (PSU)	1.24	170–175	1/1	1/1	3/–	1/3	4/4	1/1	1/–	–	1/1	1/1
Polytetrafluoroethylene (PTFE)	2.15–2.20	50–60	1/1	1/1	1/1	1/1	1/1	1/1	1/1	1/1	1/1	1/1
Polyurethane (PUR)	1.05	–	–	–	–	–	–	–	–	–	–	–
Polyvinyl chloride (PVC)	1.16–1.55	65–75	1/1	2/3	1/3	1/2	3/4	1/1	1/3	–	1/1	1/3
Styrene acrylonitrile (SAN)	1.08	95–100	1/3	1/3	1/1	1/3	–	–	–	–	1/–	1/1

Chemical resistances are listed for temperatures of +20 °C (left digit and color code) and + 50 °C (right digit): – = no data available, 1/green = resistant, 2/yellow = practically resistant, 3/orange = partially resistant, 4/red = not resistant beaker or flask where the supernatant is decanted or removed with a pipette or in a separatory funnel, where the inorganic material is removed via the bottom valve (Maes et al. 2017b; Mintenig et al. 2017; Zobkov and Esiukova 2017). Constructed devices like the Munich/MicroPlastic Sediment Separator (MPSS) by Imhof et al. (2012), designed for the extraction of MPs from large quantities of sediment (up to 6 kg), and the small-scale Sediment-Microplastic Isolation (SMI) unit by Coppock et al. (2017) usually achieve very good recovery rates (96%) even for small MPs (< 1 mm; (Imhof et al. 2012), when applied with $ZnCl_2$. According to Kedzierski et al. (2017), it is possible to extract 54% of the plastics produced in Europe with NaCl of 1.18 g cm^{-3} density while with a 1.8 g cm^{-3} solution (achievable with, e.g., NaI, polytungstate, $ZnCl_2$) the extraction of 93–98% is feasible. Therefore, achieved recovery rates are not only dependent on the device but mainly on the separation liquid used.

Another density based technique to separate MPs from sediment matrix is elutriation/fluidization, where water or air is pumped through the fluid column containing the sample and water or a salt solution (Claessens et al. 2013; Nuelle et al. 2014; Zhu 2015; Kedzierski et al. 2016). Recently, a non-density based extraction approach with canola oil has been developed by Crichton et al. (2017). The approach makes use of the oleophilic properties of MPs. So far it has only been tested with MPs larger than 500 µm, but showed high recovery rates of 96% (Crichton et al. 2017). When choosing one of the available methods, factors like sample volume or mass, time needed, costs, safety, toxicity, and extraction efficiency have to be considered.

For small amounts of sediment, approaches in flasks or funnels can be used or the novel developed SMI unit (Coppock et al. 2017; Maes et al. 2017b). If larger sediment volumes (1–6 L) are processed, elutriation systems or the MPSS would be a better choice (Imhof et al. 2012; Nuelle et al. 2014).

The time necessary for shaking should be adjusted to the sediment amount. The more sediment, the longer the mixing interval should be to assure that all MP particles are separated from the sediment particles. For settling, the span depends on the density gradient between MPs and liquid as well as the length of the fluidization column. Furthermore, the settling times have to be adjusted to the solutions used since particles rise and settle more slowly in more viscous solutions like $CaCl_2$ or $ZnCl_2$ (Crichton et al. 2017).

The most inexpensive approaches are simple setups with flasks and NaCl or oil. Zinc chloride is more expensive in relation to NaCl, especially when adjusted to higher densities but by far less expensive than NaI and polytungstate (Coppock et al. 2017). At best, an effective and cost efficient setup is used with a high density solution that can be refurbished and that allows for a proper mixing of the sediment as well as a proper settling time.

Concentrated NaCl solutions as well as canola oil do not pose any hazard to the environment. Other salt solutions are more hazardous to health and the environment in ascending order: NaI, $CaCl_2$, polytungstate, $ZnCl_2$. These solutions should therefore be recycled as far as possible due to financial and environmental reasons (Löder and Gerdts 2015). Kedzierski et al. (2017) showed that NaI can effectively be recycled without major density loss. Zinc chloride can be refurbished in large quantities quite easily via pressure filtration (Löder and Gerdts 2015). Miller et al. (2017) did an extensive comparison of different separation techniques on the basis of current literature and listed advantages and disadvantages. Based on this list, the authors recommended the use of $ZnBr_2$ (Miller et al. 2017). Nevertheless, $ZnBr_2$ has to date just been used by one study (Quinn et al. 2017) and $ZnCl_2$ is not included in the list although it is suitable for the same density range, less expensive ($ZnBr_2$: 165 € kg^{-1}, $ZnCl_2$: 92.50 € kg^{-1}, Merck Millipore, December 2017) and more widely used. Therefore, other authors have recommended the use of $ZnCl_2$ as well (Löder and Gerdts 2015; Ivleva et al. 2016; Primpke et al. 2017a).

Independent of the extraction method chosen the next step is to filter the residual fluid or the supernatant of the (density) separation containing MPs to remove the respective salt solution and to concentrate the sample to certain size fractions.

Sample Purification

Before the samples can be analyzed the biogenic matter has to be removed. Sediment samples after density separation contain usually a relatively low amount of biogenic matter (benthic diatoms, copepods, polychaetes, bivalves, etc.). In contrast, samples from the sea surface, mostly taken with plankton nets, are normally very rich in biogenic matter (phyto- and zooplankton) as well as biota samples. The main digesting agents used for the removal of biogenic matter are acids like hydrochloric acid (HCl), nitric acid (HNO_3) and sulphuric acid (H_2SO_4) (Claessens et al. 2013; De Witte et al. 2014; Klein et al. 2015), bases like sodium hydroxide (NaOH) and potassium hydroxide (KOH) (Foekema et al. 2013; Dehaut et al. 2016; Karami et al. 2017; Wagner et al. 2017), oxidative agents like sodium hypochlorite (NaClO) and hydrogen peroxide (H_2O_2) (Nuelle et al. 2014; Avio et al. 2015; Collard et al. 2015; Tagg et al. 2017) and enzymes (Cole et al. 2014; Löder and Gerdts 2015; Courtene-Jones et al. 2017; Fischer and Scholz-Böttcher 2017; Mintenig et al. 2017). Several studies showed the destructive effects, i.e., discoloration, embrittlement or a loss in surface area, of acids (e.g., HNO_3) and bases (e.g., NaOH) on MPs especially at high temperatures (Cole et al. 2014; Nuelle et al. 2014; Bürkle GmbH 2015; Karami et al. 2017). Heat deflection temperatures of some plastics are around 50–80 °C or even below for PE (Osswald et al. 2006; Qiu et al. 2016). Therefore, it is generally recommended to use temperatures of less than 50 °C.

For H_2O_2, negative effects on synthetic polymers have been shown by Nuelle et al. (2014), but just after a week-long treatment. The needed incubation time and effectiveness can be further improved by a new approach from Tagg et al. (2017) who used Fenton's reagent, a mixture of iron sulphate ($FeSO_4$) and H_2O_2. The digestion with enzymes is regarded to be non-destructive to MPs, targeting specifically proteins, polysaccharides and lipids. Cole et al. (2014) presented an approach with Proteinase-K and an up to 97.7% effective removal of biogenic matter. Courtene-Jones et al. (2017) digested mussel tissue with trypsin with an efficiency of 88%. The biggest disadvantage of these treatments is the high cost of these specific enzymes. The succession of several technical enzymes in combination with sodium dodecyl sulphate (SDS) and an oxidative agent (i.e., H_2O_2) seems to be an effective, inexpensive, and non-hazardous alternative (Löder and Gerdts 2015; Löder et al. 2015; Fischer and Scholz-Böttcher 2017; Mintenig et al. 2017; Primpke et al. 2017b).

When choosing the most suitable digestion method several factors have to be considered: time, cost, destructiveness, and effectiveness.

Purification can take several minutes (Tagg et al. 2017), several hours (Cole et al. 2014; Dehaut et al. 2016) or several days (Foekema et al. 2013; Löder and Gerdts 2015; Karami et al. 2017). Generally, longer incubation times improve the effectiveness but might also negatively impact MPs. For example, Nuelle et al. (2014) showed a negative effect of a week-long treatment with H_2O_2 while no significant effect has been shown for shorter application periods (Nuelle et al. 2014; Tagg et al. 2017). Application time should be reduced to the maximum time before causing negative effects and to the minimum time necessary to cause the highest possible effectiveness.

Specific enzymes like Proteinase-K and trypsin are very expensive. Technical enzymes, on the other hand, can be used as an inexpensive alternative (Löder and Gerdts 2015; Löder et al. 2017; Mintenig et al. 2017).

It is noticeable that methods using acids are more destructive, especially at higher temperatures, than other methods. Only at low concentrations and low temperatures (5%, 25 °C) HCl and HNO_3 are less destructive than non-acid based methods, although they are also less effective at low temperatures and concentrations. For the alkaline treatments, KOH is more effective than NaOH with the same level of destructiveness. When comparing two oxidative treatments most frequently used, H_2O_2 is more effective than NaClO and less destructive.

Next to the potential destructiveness, the effectiveness of the treatment has to be taken into account when considering the most suitable digesting agent (Fig. 2). For most treatments, an increase in temperature provokes an increase in effectiveness but often also an increase in destructiveness. Some treatments might be very effective but also relatively destructive to MPs like HNO_3 (69%) and HCl (37%) and other treatments are less destructive but also less effective like NaOH and NaClO (Karami et al. 2017). Enzymatic treatments represent the best choice in terms of being non-destructive to MPs. Several working groups have shown the high effectiveness of enzymatic digestion with different enzymes (Cole et al. 2014; Courtene-Jones et al. 2017; Karlsson et al. 2017; Löder et al. 2017; Mintenig et al. 2017).

Microplastics Identification

Once the environmental samples have been purified and concentrated by removing the biogenic and inorganic matter the MPs within the samples have to be identified. This identification is most easily performed by visual inspection either with the naked eye or with the use of a (stereo) microscope (Shim et al. 2017). The sorting is based on several criteria defined in a pilot-study by Norén (2007), which include having no visible cell-structure, homogenous coloration, and equal thickness for fibers (Enders et al. 2015). Nonetheless, Hidalgo-Ruz et al. (2012) stated that up to 70% of particles that potentially resembled MPs based on merely visual inspection could not be confirmed to be of synthetic origin. These limits of visual identification, even by experienced operators, have been shown by several studies (Eriksen et al. 2013; Dekiff et al. 2014; Lenz et al. 2015; Löder and Gerdts 2015; Song et al. 2015). Despite this high proneness to errors, many studies still rely on the visual identification of MPs. An overestimation can be avoided when a chemical characterization is subsequently performed to confirm plastics. If the chemical characterization is based on a prior visual sorting of potential MPs, an underestimation, especially of very small particles is still very likely (Song et al. 2015). Stains can be used to facilitate visual analysis, like Nile Red (Desforges et al. 2014; Shim et al. 2016; Erni-Cassola et al. 2017; Maes et al. 2017a) or rose bengal (Ivleva et al. 2016). Maes et al. (2017a) presented an approach using Nile Red that enabled for a reliable identification of MPs (96.6% recovery for MPs of a 100–500 μm size range). Nevertheless, this approach does not allow for a differentiation of distinct polymer types (Maes et al. 2017a), and may only be suitable for identification of MPs used in organism studies, where the specific polymer type is known. For environmental samples, chemical characterization is needed and can be achieved by spectroscopic analyses like Fourier transform infrared (FTIR), Raman and energy dispersive X-ray (EDX) spectroscopy or thermal analysis (Ivleva et al. 2016; Shim et al. 2017).

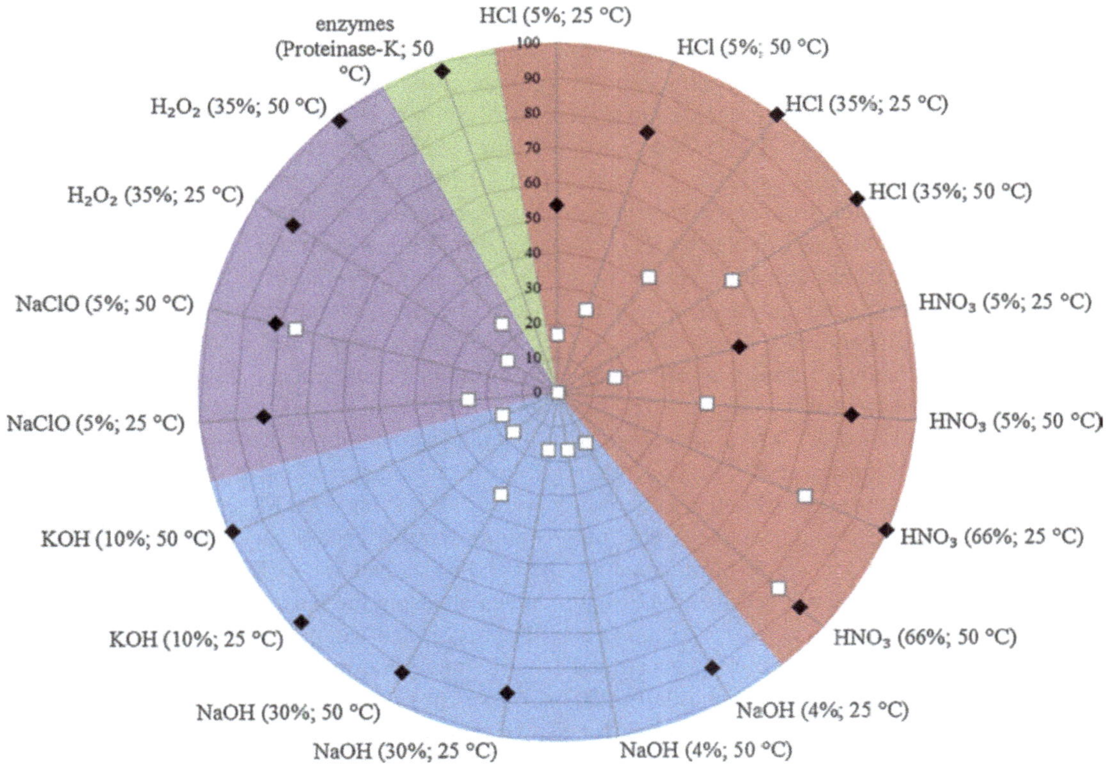

Fig. 2 Effectiveness of different digestion treatments (black symbols, in %) and maximum percentage of microplastics negatively affected by the treatments (white symbols, based on 12 polymers). Different colored sectors highlight the different treatments = red: acid, blue = alkaline, violet = oxidative, green = enzymatic (based on Cole et al. 2014; Bürkle GmbH 2015; Karami et al. 2017)

When combining EDX with scanning electron microscopy (SEM), this technique can provide information on the elemental composition of a particle and therefore distinguish plastics from inorganic materials (Eriksen et al. 2013; Vianello et al. 2013; Ivleva et al. 2016; Wagner et al. 2017; Wang et al. 2017). The identification of different plastic types is limited and therefore this method is recommended to be used for surface characterization and visualization additional to previous FTIR analysis (Vianello et al. 2013; Shim et al. 2017). FTIR analysis is a vibrational spectroscopic technique based on infrared radiation that excites molecular bonds resulting in vibrations that can be detected and transferred into characteristic absorbance spectra. These spectra can further be compared to a database of reference spectra allowing for the reliable identification of different polymer types. FTIR spectroscopy can be used in different modes, namely transmission (Löder et al. 2015; Käppler et al. 2016; Mintenig et al. 2017; Primpke et al. 2017b), reflection (Harrison et al. 2012; Vianello et al. 2013; Tagg et al. 2015) and attenuated total-reflectance (ATR) (Song et al. 2015; Käppler et al. 2016; Crichton et al. 2017; Imhof et al. 2017; Wagner et al. 2017). To measure very small particles FTIR spectroscopy can be coupled to microscopy (µFTIR) and be used in all three modes as well (Ivleva et al. 2016; Shim et al. 2017). All these modes have several advantages and limitations. While the transmission mode provides high quality spectra it is restricted to a certain thickness of material to allow infrared radiation to pass through the sample without being fully absorbed (Löder and Gerdts 2015; Ivleva et al. 2016). Reflectance mode on the other hand provides spectra of thick and opaque particles but does depend on the surface properties since uneven surfaces can cause scattering effects which cause refractive errors (Löder and Gerdts 2015; Shim et al. 2017). High quality spectra can be achieved by µATR-FTIR with the disadvantage of potentially damaging particles since a crystal has to be pressed on the sample (Ivleva et al. 2016; Shim et al. 2017). Another vibrational spectroscopy, that is complementary to FTIR, is Raman spectroscopy (Käppler et al. 2016). Monochromatic light, usually provided by a laser, irradiates the sample and vibrations are resulting in a Raman shift, which can be presented as substance characteristic spectra (Ivleva et al. 2016; Shim et al. 2017). Raman micro-spectrometry has successfully been used to identify MPs in environmental samples (Enders et al. 2015; Fischer et al. 2015; Frère et al. 2016; Imhof et al. 2016; Wagner et al. 2017). For thermal analysis, pyrolysis-gas chromatography-mass spectrometry (Pyr-GC-MS) and thermoextraction and desorption (TED) coupled with GC-MS are the most prevalent and promising ones (Fries et al. 2013; Dümichen et al. 2015; Fischer and Scholz-Böttcher 2017).

Both methods provide the chemical composition based on heating the sample and analyzing the decomposition products (Ivleva et al. 2016). Pyrograms or ion chromatograms are obtained that can be compared to references, equivalent to spectra of spectroscopic techniques (Löder and Gerdts 2015; Dümichen et al. 2017; Shim et al. 2017).

Most studies use these methods to analyze preselected particles. Recently Fischer and Scholz-Böttcher (2017) presented also for Pyr-GC-MS an approach independent of a prior visual sorting by analysing whole filters on which previously purified samples had been concentrated. That also TED-GC-MS can be used to analyze subsamples of environmental samples without pre-selection to identify MPs has been shown by Dümichen et al. (2017). An advantage of TED-GC-MS presented by Dümichen et al. (2017) is that a relatively high sample amount of up to 100 mg can be processed, which, depending on the condition of the environmental sample, obviates the need for sample purification.

Chemical imaging approaches developed for µFTIR and Raman spectroscopy eliminate the need for a visual pre-selection. Therefore, the purified samples are concentrated on filters that are directly scanned. The filter chosen for the analysis has to be compatible to the method by not interfering with the sample analysis (Käppler et al. 2015; Löder et al. 2015). For µFTIR the use of Focal plane array (FPA) detectors have substantially improved the time needed for the analysis of whole filter areas (Löder et al. 2015; Tagg et al. 2015; Käppler et al. 2016; Mintenig et al. 2017; Primpke et al. 2017b). Although the imaging using FPA is independent of a prior visual selection of potential MPs, the approach presented by Löder et al. (2015) still involves an operator-based selection of MPs based on their spectral signature. Therefore, advances are automated approaches independent of human bias like it has been recently presented by Primpke et al. (2017b).

Shim et al. (2017) recently reviewed the advantages and disadvantages of currently used methods for identification of MPs. Furthermore, Elert et al. (2017) added to the comparison a classification of the different techniques in terms of restrictions, requirements and the analytical information received.

The major advantage of thermal analysis is the simultaneous analysis of polymer and containing additives, while the major disadvantage is the destruction of the sample by combustion. While thermal analyses provide mass-related results only, spectroscopic analyses are normally non-destructive and provide particle-related results (Shim et al. 2017). The lower size limit for µFTIR is at 10 µm due to the diffraction limit (Löder and Gerdts 2015; Shim et al. 2017), whereas for Raman spectroscopy particles down to 1 µm size can be analyzed (Ivleva et al. 2016). Residual water hampers FTIR analysis while for Raman spectroscopy fluorescence of residues of the environmental matrix is a problem as well as the interference from pigments (Imhof et al. 2016; Käppler et al. 2016; Shim et al. 2017). Käppler et al. (2016) showed that Raman imaging provides a better identification of MPs < 20 µm when compared to using FPA-µFTIR in transmission mode but with the major drawback for Raman imaging that the measurement time was more than 100-times higher than the µFTIR analysis. Currently, µFTIR imaging of large filter areas is considerably faster than Raman imaging, even when reducing the resolution for Raman imaging, resulting in a comparable quality to FTIR imaging (Käppler et al. 2016).

All above mentioned methods share the commonality that to avoid misinterpretation of spectra and programs alike as well as identifying dyed MPs, efficient sample purification is of utmost importance (Löder and Gerdts 2015; Crichton et al. 2017; Fischer and Scholz-Böttcher 2017; Maes et al. 2017b). When choosing the most appropriate method: time demand, size range, and sample preparation have to be considered. Furthermore, thermal analysis should be used when a fast assessment of mass-related data is required, while spectroscopic analysis provides particle-related data but might take considerably longer. A holistic approach would involve FTIR-analysis of MPs down to 10 µm, Raman-analysis for MPs below 10 µm and a subsequent thermal analysis.

Biological Effects of Microplastics on Biota

Although research on MPs in aquatic systems regarding monitoring and abundance in animals has dramatically increased in the last years, profound knowledge about the effects of MPs on biota is still scarce (Ribeiro et al. 2017). Here, we give a short overview about investigated consequences of MP exposure, methods, and their effects on organisms.

Images circulating the media of sea turtles, dolphins or seals entangled in plastic bags and other macroplastics are well known, but what about the plastic we do not see? Microplastics can pose a danger to organisms, when they are ingested (Avio et al. 2017). Reasons for ingestion in the first place are either MPs being mistaken for food or prey due to similarities in size, shape or color (Wright et al. 2013) or because the organism is not selective with food particles, which is, for example, the case for most filter and deposit feeders (Van Cauwenberghe et al. 2015). Although filter feeders often possess some mechanisms to avoid particles that are too big or inedible, MPs are very similar to actually nutritious food and thus sorting mechanisms might not work (Ward and Shumway 2004).

The Risk of Exposure to Microplastics

When trying to assess the danger of MPs in the marine environment, various things have to be taken into consideration. The risk to be exposed to MPs varies a lot with compartments, usually divided into water surface, water column and sediments. Sediments are thought of as being the most affected compartment, because they function as a sink for MPs (Hidalgo-Ruz et al. 2012; Duis and Coors 2016). All compartments considerably vary spatially and temporally and distribution of MPs is, therefore, difficult to assess (Chubarenko et al. 2016).

Risk of exposure is different for the different types of polymers (buoyancy, fragmentation rate) and the habitat of the organism (surface layer, water column, sediment) (Andrady 2017). Polymers with low density tend to stay longer in the surface layer, possibly aggregating with phytoplankton in the euphotic zone (Long et al. 2017). They can also be overgrown by microbes and other fouling organisms and sink down in the water column together with MPs of neutral buoyancy. Higher density polymers such as polyvinyl chloride sink quickly and are readily available for benthic filter feeders or deposit feeders such as bivalves and polychaetes (Avio et al. 2017).

The hazard that MP poses for organisms also varies depending on the functional group such as the trophic level of the organism. So far, mainly low trophic levels such as filter feeders, deposit feeders and planktivorous fish have been found to be contaminated with MPs, but recently MP particles have also been detected in predatory pelagic fish such as tuna (Romeo et al. 2015) and even filter feeding mammals such as humpback whales (Besseling et al. 2015).

Studies have accumulated on examining fish guts for MPs and have found evidence of MPs among multiple species and life stages across different functional groups (Vendel et al. 2017). Transfer to higher trophic levels, such as fish preying on zooplankton that has ingested MPs, has been hypothesized but no clear evidence has been found yet (Santana et al. 2017). So far, studies give contradictory results with some claiming that MPs cascade to higher trophic levels (Setälä et al. 2014) while others disagree or argue that they travel to predators but do not persist in the gut (Santana et al. 2017).

Effects Due to the Specific Properties of Microplastic Particles

All types of MPs are hypothesized to cause gut blockage or a false sense of fullness, if not excreted within reasonable time span (Gall and Thompson 2015). Most MPs are so-called secondary MPs resulting from fragmentation of larger particles. Therefore, the shape of MPs can cause internal ruptures and injuries. Most studies have been conducted with primary MPs: spherical, highly defined microbeads not reflecting the situation in the environment, as the most commonly found types are fragments and fibers. This calls for the use of fragments or fibers in laboratory studies to enhance significance of the obtained results.

The effects of MPs on an organism depend a lot on its size. Seabirds often take up colorful plastic particles that fill up their stomach and can be too large for gut passage (van Franeker et al. 2011). Contrarily, very small particles (1–400 nm) (GESAMP 2015), called nanoplastics if <100 nm (Löder and Gerdts 2015), can potentially be implemented in body cells after ingestion as they are small enough to pass pores in membranes. Inside the cell, the particles can potentially disturb other tissues than the digestive system such as the liver or lymph system (von Moos et al. 2012).

Impacts on biota can vary depending on the polymer type of the encountered MPs. Some polymers such as silicone are sturdier and break down slower under the influence of temperature and wave action than others due to their chemical composition. They fragment slower and are also less likely to leach pollutants as leaching of additives is dependent on surface area which increases with decreasing particle size (Suhrhoff and Scholz-Böttcher 2016). Other polymers, however, are already toxic in themselves by leaching monomers or oligomers such as polyvinyl chloride (PVC) or polystyrene (PS). These monomers have been shown to act as endocrine disruptors (Espinosa et al. 2016).

When MPs are introduced into the environment they are free of microorganisms and have not yet been impacted by waves or UV light. With time, MPs weather, pollutants adsorb and leach, and microorganisms start growing on the particles. These processes lead to changed characteristics of the MPs. With growing or adhering organisms the buoyancy changes and low density polymers start to sink and become available for a different range of organisms. Furthermore, biofilm-coated particles might not be recognizable anymore as MPs or seem more palatable due to chemical cues emitted from the microorganisms and are ingested with higher probability. Bacterial assemblages on MPs have also been found to be different from other surfaces with yet unknown consequences (Kesy et al. 2016).

Considering all of the above, there are various things to be accounted for when working with MP in the laboratory. Glass containers or glass material should be used as much as possible to reduce contamination sources. As this is onl possible to a certain extent, negative controls should al account for plastic materials used within the experime set-up. To assess the effects of MPs in experiments, the centrations have to be determined, to which the orga are exposed. Using spherical beads, this can be cal via diameter, density and mass of the spheres. Irregul are more difficult to handle. Simple methods usual counting chambers (Syberg et al. 2015), light n

(Nobre et al. 2015) and quantitative filtering of the water samples. Other common methods are flow cytometer (Sussarellu et al. 2016) and the use of a coulter counter (Syberg et al. 2015). For characterization of the beads, FTIR (Lusher et al. 2017), Raman spectroscopy and electron microscopy (Murray and Cowie 2011) are the preferred methods. The next step is to determine the presence of MPs in the organism. This is usually achieved by dissecting to check for presence in gut systems or histological analysis of tissue samples (Farrell and Nelson 2013). Effects on the organism can be directly determined via deformations of larvae. Potentially, MPs can induce epigenetic effects, e.g., in copepods (Heindler et al. 2017). Epigenetics are usually viewed as a quick and advantageous mechanism for an F1 generation to adapt to a stressor to which the F0 generation was exposed. Microplastics can also cause a decrease in reproduction (Heindler et al. 2017) and are therefore directly affecting fitness. Sussarellu et al. (2016) also reported reductions in feeding activity, accumulation and inhibition of acetylcholinesterase activity in bivalves.

Microplastics as Vector for Pollutants

The effects of the combination of MPs with pollutants are ambivalently discussed. First, it is important to differentiate between pollutants adhering to plastics, which belong to the group of persistent organic pollutants (POPs), originating from the water, and between additives leaching from the MP particles or emittance of monomers or oligomers from the MPs themselves. The difference here is that pollutants that adhere to plastics are usually already widespread in the environment, while pollutants associated with plastic have only been around since the production of plastics, so roughly the 1950s (Hammer et al. 2012). Both groups of chemicals suggest a role of MPs as a vector to organisms. This is very debatable for pollutants already present in the environment ... some argue that other pathways such as food and water are ...al magnitudes higher than the intake via MPs, simply ...the fact that MPs are still not that abundant in the ... therefore, bioaccumulation of this POPs is not ... MPs yet. Additionally, it is also discussed if ...es from MPs are of major concern. Here, it is ...entiate between primary MPs and second- ...Ps are introduced already in the size ...secondary MPs are often introduced ...macroplastics that fragment over ... over time and it remains an ...es are still present within

... study on the toxicity ...) and the common ...P) that copepod nauplii are far more sensitive to exposure than adults. This stresses the need for assessing the toxicity of MPs at different life stages and focusing on juveniles or larvae, which are usually more sensitive to stressors than adults. Effects of MPs on younger life-stages can have knock-on effects on populations if for example mortality is significantly higher and fewer individuals reach sexual maturity and reproduce.

Regarding laboratory methods, again, glassware should be used where appropriate to make sure that no pollutant is adhered to the experimental container and, therefore, removed from the experiment. Toxin burdens can for example be assessed in different compartments (water, plastic, and biota) via high throughput liquid chromatography (HPLC) (Brennecke et al. 2015).

Conclusion

Although intensive research activities have already resolved some methodological issues in MP research, there are still some challenges, which need to be overcome before standardized operational protocols (SOPs) can be defined. Sampling effort (spatial and temporal replication, as well as sample volume) within a project is still limited by the high demand for personnel and physical resources as well as the long analysis time for MP samples. Thereby, an adequate sampling design should be chosen to answer pre-defined research questions as precisely as possible. It is obvious that different research questions require the use of different methods, which in turn will hamper complete standardization of methods. Nevertheless, a comprehensive and proper data recording, as well as gathering additional information, e.g., environmental data, will contribute to high quality datasets.

In addition, the extraction of MPs from environmental matrices is a crucial step, as inorganic and organic substances concurrently sampled with the potential MPs, can interfere with the subsequent analysis. Lately, many protocols have been proposed to remove inorganic or organic materials from samples. Thereby, developments were made to improve extraction efficiency, while not affecting fragile MPs, i.e., applying high density solutions for density separation of inorganic material or enzymatic purification of organic material. Finally, for a reliable identification of MPs a solely visible analysis is insufficient, and a chemical characterization is highly recommended. Spectroscopic methods like Raman- or FTIR-spectroscopy are state-of-the-art, providing particle related data (e.g., numbers, sizes) as well as thermal extraction methods like Pyr-GC-MS and TED-GC-MS, which provide mass related data and information about absorbed pollutants or contained additives.

Both methods will provide relevant information for further studies on the effects of MPs on organisms. There is still a huge lack of knowledge and besides evidence that MPs are ingested, either by mistake or because the organism is not selective in feeding, information about the effects is limited. The exposure to MPs will largely depend on the habitat of the organism (water surface, water column, sediment) and the feeding type (predatory, herbivory, planktivory). How the MPs affect the organism after ingestion is largely unknown but depends most likely very much on specific properties of the MPs (polymer type, size, shape) as well as life stage of the organism. The interaction of MPs with adsorbed pollutants seems to be negligible compared to already existing pathways (food, water), however, the effect of leaching additives is not yet determined. Thus, the effects of MPs on organisms still need intense research activities to come up with proper risk assessments for different life stages of different species to determine, who is most at risk and how to protect them.

Appendix

This article is related to the YOUMARES 8 conference session no. 13: "Microplastics in Aquatic Habitats – Environmental Concentrations and Consequences". The original Call for Abstracts and the abstracts of the presentations within this session can be found in the appendix "Conference Sessions and Abstracts", chapter "12 Microplastics in Aquatic Habitats – Environmental Concentrations and Consequences", of this book.

References

Andrady AL (2017) The plastic in microplastics: a review. Mar Pollut Bull 119:12–22. https://doi.org/10.1016/j.marpolbul.2017.01.082

Avio CG, Gorbi S, Regoli F (2015) Experimental development of a new protocol for extraction and characterization of microplastics in fish tissues: first observations in commercial species from Adriatic Sea. Mar Environ Res 111:18–26. https://doi.org/10.1016/j.marenvres.2015.06.014

Avio CG, Gorbi S, Regoli F (2017) Plastics and microplastics in the oceans: from emerging pollutants to emerged threat. Mar Environ Res 128:2–11. https://doi.org/10.1016/j.marenvres.2016.05.012

Barrows APW, Neumann CA, Berger ML et al (2017) Grab vs. neuston tow net: a microplastic sampling performance comparison and possible advances in the field. Anal Methods 9:1446–1453. https://doi.org/10.1039/C6AY02387H

Bergmann M, Tekman MB, Gutow L (2017) LITTERBASE: an online portal for marine litter and microplastics and their implications for marine life. In: Baztan J, Jorgensen B, Pahl S et al (eds) Fate and impact of microplastics in marine ecosystems. Elsevier, MICRO 2016, Amsterdam, pp 106–107. https://doi.org/10.1016/B978-0-12-812271-6.00104-6

Besley A, Vijver MG, Behrens P et al (2017) A standardized method for sampling and extraction methods for quantifying microplastics in beach sand. Mar Pollut Bull 114:77–83. https://doi.org/10.1016/j.marpolbul.2016.08.055

Besseling E, Foekema EM, Van Franeker JA et al (2015) Microplastic in a macro filter feeder: humpback whale *Megaptera novaeangliae*. Mar Pollut Bull 95:248–252. https://doi.org/10.1016/j.marpolbul.2015.04.007

Brennecke D, Ferreira EC, Costa TMM et al (2015) Ingested microplastics (>100 µm) are translocated to organs of the tropical fiddler crab *Uca rapax*. Mar Pollut Bull 96:491–495. https://doi.org/10.1016/j.marpolbul.2015.05.001

Browne MA, Chapman MG, Thompson RC et al (2015) Spatial and temporal patterns of stranded intertidal marine debris: is there a picture of global change? Environ Sci Technol 49:7082–7094. https://doi.org/10.1021/es5060572

Bürkle GmbH (2015) Chemische Beständigkeit von Kunststoffen. Bürkle GmbH. URL:www.buerkle.de/files_pdf/wissenswertes/chemical_resistance_en.pdf. Accessed 03 Aug 2017

Carson HS, Colbert SL, Kaylor MJ et al (2011) Small plastic debris changes water movement and heat transfer through beach sediments. Mar Pollut Bull 62:1708–1713. https://doi.org/10.1016/j.marpolbul.2011.05.032

Chae D-H, Kim I-S, Kim S-K et al (2015) Abundance and distribution characteristics of microplastics in surface seawaters of the Incheon/Kyeonggi coastal region. Arch Environ Contam Toxicol 69:269–278. https://doi.org/10.1007/s00244-015-0173-4

Chubarenko I, Bagaev A, Zobkov M et al (2016) On some physical and dynamical properties of microplastic particles in marine environment. Mar Pollut Bull 108:105–112. https://doi.org/10.1016/j.marpolbul.2016.04.048

Claessens M, Meester SD, Landuyt LV et al (2011) Occurrence and distribution of microplastics in marine sediments along the Belgian coast. Mar Pollut Bull 62:2199–2204. https://doi.org/10.1016/j.marpolbul.2011.06.030

Claessens M, Van Cauwenberghe L, Vandegehuchte MB et al (2013) New techniques for the detection of microplastics in sediments and field collected organisms. Mar Pollut Bull 70:227–233. https://doi.org/10.1016/j.marpolbul.2013.03.009

Cole M, Webb H, Lindeque PK et al (2014) Isolation of microplastics in biota-rich seawater samples and marine organisms. Sci Rep 4:4528. https://doi.org/10.1038/srep04528

Collard F, Gilbert B, Eppe G et al (2015) Detection of anthropogenic particles in fish stomachs: an isolation method adapted to identification by Raman spectroscopy. Arch Environ Contam Toxicol 69:331–339. https://doi.org/10.1007/s00244-015-0221-0

Connors KA, Dyer SD, Belanger SE (2017) Advancing the quality of environmental microplastic research. Environ Toxicol Chem 36:1697–1703. https://doi.org/10.1002/etc.3829

Coppock RL, Cole M, Lindeque PK et al (2017) A small-scale, portable method for extracting microplastics from marine sediments. Environ Pollut 230:829–837. https://doi.org/10.1016/j.envpol.2017.07.017

Corcoran PL (2015) Benthic plastic debris in marine and fresh water environments. Env Sci Process Impact 17:1363–1369. https://doi.org/10.1039/c5em00188a

Corcoran PL, Biesinger MC, Grifi M (2009) Plastics and beaches: a degrading relationship. Mar Pollut Bull 58:80–84. https://doi.org/10.1016/j.marpolbul.2008.08.022

Costa MF, Duarte AC (2017) Microplastics sampling and sample handling. Compr Anal Chem 75:25–47. https://doi.org/10.1016/bs.coac.2016.11.002

Courtene-Jones W, Quinn B, Murphy F et al (2017) Optimisation of enzymatic digestion and validation of specimen preservation methods for the analysis of ingested microplastics. Anal Methods 9:1437–1445. https://doi.org/10.1039/C6AY02343F

Crichton EM, Noel M, Gies EA et al (2017) A novel, density-independent and FTIR-compatible approach for the rapid extraction of microplastics from aquatic sediments. Anal Methods 9:1419–1428. https://doi.org/10.1039/C6AY02733D

De Witte B, Devriese L, Bekaert K et al (2014) Quality assessment of the blue mussel (*Mytilus edulis*): comparison between commercial and wild types. Mar Pollut Bull 85:146–155. https://doi.org/10.1016/j.marpolbul.2014.06.006

Dehaut A, Cassone A-L, Frère L et al (2016) Microplastics in seafood: benchmark protocol for their extraction and characterization. Environ Pollut 215:223–233. https://doi.org/10.1016/j.envpol.2016.05.018

Dekiff JH, Remy D, Klasmeier J et al (2014) Occurrence and spatial distribution of microplastics in sediments from Norderney. Environ Pollut 186:248–256. https://doi.org/10.1016/j.envpol.2013.11.019

Desforges J-PW, Galbraith M, Dangerfield N et al (2014) Widespread distribution of microplastics in subsurface seawater in the NE Pacific Ocean. Mar Pollut Bull 79:94–99. https://doi.org/10.1016/j.marpolbul.2013.12.035

Doyle MJ, Watson W, Bowlin NM et al (2011) Plastic particles in coastal pelagic ecosystems of the Northeast Pacific Ocean. Mar Environ Res 71:41–52. https://doi.org/10.1016/j.marenvres.2010.10.001

Dubaish F, Liebezeit G (2013) Suspended microplastics and black carbon particles in the jade system, southern North Sea. Water Air Soil Pollut 224:1352–1359. https://doi.org/10.1007/s11270-012-1352-9

Duis K, Coors A (2016) Microplastics in the aquatic and terrestrial environment: sources (with a specific focus on personal care products), fate and effects. Environ Sci Eur 28:2

Dümichen E, Barthel A-K, Braun U et al (2015) Analysis of polyethylene microplastics in environmental samples, using a thermal decomposition method. Water Res 85:451–457. https://doi.org/10.1016/j.watres.2015.09.002

Dümichen E, Eisentraut P, Bannick CG et al (2017) Fast identification of microplastics in complex environmental samples by a thermal degradation method. Chemosphere 174:572–584. https://doi.org/10.1016/j.chemosphere.2017.02.010

Edson EC, Patterson MR (2015) MantaRay: a novel autonomous sampling instrument for in situ measurements of environmental microplastic particle concentrations. Paper presented at the OCEANS 2015 – MTS/IEEE Washington, DC, pp 19–22 Oct 2015

Elert AM, Becker R, Duemichen E et al (2017) Comparison of different methods for MP detection: what can we learn from them, and why asking the right question before measurements matters? Environ Pollut 231:1256–1264. https://doi.org/10.1016/j.envpol.2017.08.074

Enders K, Lenz R, Stedmon CA et al (2015) Abundance, size and polymer composition of marine microplastics $\geq$ 10 μm in the Atlantic Ocean and their modelled vertical distribution. Mar Pollut Bull 100:70–81. https://doi.org/10.1016/j.marpolbul.2015.09.027

Eriksen M, Mason S, Wilson S et al (2013) Microplastic pollution in the surface waters of the Laurentian Great Lakes. Mar Pollut Bull 77:177–182. https://doi.org/10.1016/j.marpolbul.2013.10.007

Erni-Cassola G, Gibson MI, Thompson RC et al (2017) Lost, but found with Nile red: a novel method for detecting and quantifying small microplastics (1 mm to 20 μm) in environmental samples. Environ Sci Technol. https://doi.org/10.1021/acs.est.7b04512

Espinosa C, Esteban MÁ, Cuesta A (2016) Microplastics in aquatic environments and their toxicological implications for fish. In: Soloneski S, Larramendy ML (eds) Toxicology – new aspects to this scientific conundrum. InTech, Rijeka, pp 113–145. https://doi.org/10.5772/64815

Farrell P, Nelson K (2013) Trophic level transfer of microplastic: *Mytilus edulis* (L.) to *Carcinus maenas* (L.). Environ Pollut 177:1–3. https://doi.org/10.1016/j.envpol.2013.01.046

Filella M (2015) Questions of size and numbers in environmental research on microplastics: methodological and conceptual aspects. Environ Chem 12:527–538. https://doi.org/10.1071/EN15012

Fischer M, Scholz-Böttcher BM (2017) Simultaneous trace identification and quantification of common types of microplastics in environmental samples by pyrolysis-gas chromatography–mass spectrometry. Environ Sci Technol 51:5052–5060. https://doi.org/10.1021/acs.est.6b06362

Fischer D, Kaeppler A, Eichhorn K-J (2015) Identification of microplastics in the marine environment by Raman microspectroscopy and imaging. Am Lab 47:32–34

Fisner M, Majer AP, Balthazar-Silva D et al (2017) Quantifying microplastic pollution on sandy beaches: the conundrum of large sample variability and spatial heterogeneity. Environ Sci Pollut Res 24:13732–13740. https://doi.org/10.1007/s11356-017-8883-y

Foekema EM, De Gruijter C, Mergia MT et al (2013) Plastic in North Sea fish. Environ Sci Technol 47:8818–8824. https://doi.org/10.1021/es400931b

Frère L, Paul-Pont I, Moreau J et al (2016) A semi-automated Raman micro-spectroscopy method for morphological and chemical characterizations of microplastic litter. Mar Pollut Bull 113:461–468. https://doi.org/10.1016/j.marpolbul.2016.10.051

Fries E, Dekiff JH, Willmeyer J et al (2013) Identification of polymer types and additives in marine microplastic particles using pyrolysis-GC/MS and scanning electron microscopy. Environ Sci Processes Impact 15:1949–1956. https://doi.org/10.1039/C3EM00214D

Gall SC, Thompson RC (2015) The impact of debris on marine life. Mar Pollut Bull 92:170–179. https://doi.org/10.1016/j.marpolbul.2014.12.041

GESAMP (2015) Sources, fate and effects of microplastics in the marine environment: a global assessment. In: Kershaw PJ (ed) (IMO/FAO/UNESCO-IOC/UNIDO/WMO/IAEA/UN/UNEP/UNDP Joint Group of Experts on the Scientific Aspects of Marine Environmental Protection). Rep. Stud. GESAMP No. 90

GESAMP (2016) Sources, fate and effects of microplastics in the marine environment: part two of a global assessment. In: Kershaw PJ, Rochmann CM (eds) (IMO/FAO/UNESCO-IOC/UNIDO/WMO/IAEA/UN/UNEP/UNDP Joint Group of Experts on the Scientific Aspects of Marine Environmental Protection). Rep. Stud. GESAMP No. 93

Goldstein MC, Titmus AJ, Ford M (2013) Scales of spatial heterogeneity of plastic marine debris in the Northeast Pacific Ocean. PLoS One 8:e80020. https://doi.org/10.1371/journal.pone.0080020

Hammer J, Kraak MHS, Parsons JR (2012) Plastics in the marine environment: the dark side of a modern gift. In: Whitacre DM (ed) Reviews of environmental contamination and toxicology, vol 220. Springer, New York, pp 1–44. https://doi.org/10.1007/978-1-4614-3414-6_1

Hanke G, Galgani F, Werner S et al (2013) MSFD GES technical subgroup on marine litter. Guidance on Monitoring of marine litter in European Seas European Commission. https://doi.org/10.2788/99475

Hanvey JS, Lewis PJ, Lavers JL et al (2017) A review of analytical techniques for quantifying microplastics in sediments. Anal Methods 9:1369–1383. https://doi.org/10.1039/C6AY02707E

Harrison JP, Ojeda JJ, Romero-González ME (2012) The applicability of reflectance micro-Fourier-transform infrared spectroscopy for the detection of synthetic microplastics in marine sediments. Sci Total Environ 416:455–463. https://doi.org/10.1016/j.scitotenv.2011.11.078

Heindler FM, Alajmi F, Huerlimann R et al (2017) Toxic effects of polyethylene terephthalate microparticles and Di(2-ethylhexyl) phthalate on the calanoid copepod, *Parvocalanus crassirostris*. Ecotoxicol Environ Saf 141:298–305. https://doi.org/10.1016/j.ecoenv.2017.03.029

Heo N, Hong S, Han G et al (2013) Distribution of small plastic debris in cross-section and high strandline on Heungnam beach, South Korea. Ocean Sci J 48:225–233. https://doi.org/10.1007/s12601-013-0019-9

Hidalgo-Ruz V, Gutow L, Thompson RC, Thiel M (2012) Microplastics in the marine environment: a review of the methods used for identification and quantification. Environ Sci Technol 46:3060–3075. https://doi.org/10.1021/es2031505

Imhof HK, Schmid J, Niessner R et al (2012) A novel, highly efficient method for the separation and quantification of plastic particles in sediments of aquatic environments. Limnol Oceanogr Methods 10:524–537. https://doi.org/10.4319/lom.2012.10.524

Imhof HK, Ivleva NP, Schmid J et al (2013) Contamination of beach sediments of a subalpine lake with microplastic particles. Curr Biol 23:R867–R868

Imhof HK, Laforsch C, Wiesheu AC et al (2016) Pigments and plastic in limnetic ecosystems: a qualitative and quantitative study on microparticles of different size classes. Water Res 98:64–74. https://doi.org/10.1016/j.watres.2016.03.015

Imhof HK, Sigl R, Brauer E et al (2017) Spatial and temporal variation of macro-, meso- and microplastic abundance on a remote coral island of the Maldives, Indian Ocean. Mar Pollut Bull 116:340–347. https://doi.org/10.1016/j.marpolbul.2017.01.010

Ivleva NP, Wiesheu AC, Niessner R (2016) Microplastic in aquatic ecosystems. Angew Chem Int Ed 56:1720–1739. https://doi.org/10.1002/anie.201606957

Kang J-H, Kwon OY, Lee K-W et al (2015) Marine neustonic microplastics around the southeastern coast of Korea. Mar Pollut Bull 96:304–312. https://doi.org/10.1016/j.marpolbul.2015.04.054

Käppler A, Windrich F, Löder MJ et al (2015) Identification of microplastics by FTIR and Raman microscopy: a novel silicon filter substrate opens the important spectral range below 1300 cm^{-1} for FTIR transmission measurements. Anal Bioanal Chem 407:6791–6801. https://doi.org/10.1007/s00216-015-8850-8

Käppler A, Fischer D, Oberbeckmann S et al (2016) Analysis of environmental microplastics by vibrational microspectroscopy: FTIR, Raman or both? Anal Bioanal Chem 408:8377–8391. https://doi.org/10.1007/s00216-016-9956-3

Karami A, Golieskardi A, Choo CK et al (2017) A high-performance protocol for extraction of microplastics in fish. Sci Total Environ 578:485–494. https://doi.org/10.1016/j.scitotenv.2016.10.213

Karlsson TM, Vethaak AD, Almroth BC et al (2017) Screening for microplastics in sediment, water, marine invertebrates and fish: method development and microplastic accumulation. Mar Pollut Bull 122:403–408. https://doi.org/10.1016/j.marpolbul.2017.06.081

Kedzierski M, Le Tilly V, Bourseau P et al (2016) Microplastics elutriation from sandy sediments: a granulometric approach. Mar Pollut Bull 107:315–323. https://doi.org/10.1016/j.marpolbul.2016.03.041

Kedzierski M, Le Tilly V, César G et al (2017) Efficient microplastics extraction from sand. A cost effective methodology based on sodium iodide recycling. Mar Pollut Bull 115:120–129. https://doi.org/10.1016/j.marpolbul.2016.12.002

Kesy K, Oberbeckmann S, Müller F et al (2016) Polystyrene influences bacterial assemblages in *Arenicola marina*-populated aquatic environments in vitro. Environ Pollut 219:219–227. https://doi.org/10.1016/j.envpol.2016.10.032

Kim I-S, Chae D-H, Kim S-K et al (2015) Factors influencing the spatial variation of microplastics on high-tidal coastal beaches in Korea. Arch Environ Contam Toxicol 69:299–309. https://doi.org/10.1007/s00244-015-0155-6

Klein S, Worch E, Knepper TP (2015) Occurrence and spatial distribution of microplastics in river shore sediments of the Rhine-main area in Germany. Environ Sci Technol 49:6070–6076. https://doi.org/10.1021/acs.est.5b00492

Lattin GL, Moore CJ, Zellers AF et al (2004) A comparison of neustonic plastic and zooplankton at different depths near the southern California shore. Mar Pollut Bull 49:291–294. https://doi.org/10.1016/j.marpolbul.2004.01.020

Law KL, Morét-Ferguson S, Maximenko NA et al (2010) Plastic accumulation in the North Atlantic subtropical gyre. Science 329:1185–1188. https://doi.org/10.1126/science.1192321

Law KL, Morét-Ferguson SE, Goodwin DS et al (2014) Distribution of surface plastic debris in the eastern Pacific Ocean from an 11-year data set. Environ Sci Technol 48:4732–4738. https://doi.org/10.1021/es4053076

Lenz R, Enders K, Stedmon CA (2015) A critical assessment of visual identification of marine microplastic using Raman spectroscopy for analysis improvement. Mar Pollut Bull 100:82–91. https://doi.org/10.1016/j.marpolbul.2015.09.026

Löder MGJ, Gerdts G (2015) Methodology used for the detection and identification of microplastics—a critical appraisal. In: Bergmann M, Gutow L, Klages M (eds) Marine anthropogenic litter. Springer, Cham, pp 201–227. https://doi.org/10.1007/978-3-319-16510-3_8

Löder MGJ, Kuczera M, Mintenig S et al (2015) Focal plane array detector-based micro-Fourier-transform infrared imaging for the analysis of microplastics in environmental samples. Environ Chem 12:563–581. https://doi.org/10.1071/EN14205

Löder MGJ, Imhof HK, Ladehoff M et al (2017) Enzymatic purification of microplastics in environmental samples. Environ Sci Technol. https://doi.org/10.1021/acs.est.7b03055

Long M, Paul-Pont I, Hégaret H et al (2017) Interactions between polystyrene microplastics and marine phytoplankton lead to species-specific hetero-aggregation. Environ Pollut 228:454–463. https://doi.org/10.1016/j.envpol.2017.05.047

Lusher AL, Burke A, O'Connor I et al (2014) Microplastic pollution in the Northeast Atlantic Ocean: validated and opportunistic sampling. Mar Pollut Bull 88:325–333. https://doi.org/10.1016/j.marpolbul.2014.08.023

Lusher AL, Welden NA, Sobral P et al (2017) Sampling, isolating and identifying microplastics ingested by fish and invertebrates. Anal Methods 9:1346–1360. https://doi.org/10.1039/C6AY02415G

Maes T, Jessop R, Wellner N et al (2017a) A rapid-screening approach to detect and quantify microplastics based on fluorescent tagging with Nile Red. Sci Rep 7:44501. https://doi.org/10.1038/srep44501

Maes T, Van der Meulen MD, Devriese LI et al (2017b) Microplastics baseline surveys at the water surface and in sediments of the north-East Atlantic. Front Mar Sci 4:135. https://doi.org/10.3389/fmars.2017.00135

Masura J, Baker J, Foster G et al (2015) Laboratory methods for the analysis of microplastics in the marine environment: recommendations for quantifying synthetic particles in waters and sediments. NOAA Technical Memorandum, Silver Spring

Miller ME, Kroon FJ, Motti CA (2017) Recovering microplastics from marine samples: a review of current practices. Mar Pollut Bull 123:6–18. https://doi.org/10.1016/j.marpolbul.2017.08.058

Mintenig SM, Int-Veen I, Löder MGJ et al (2017) Identification of microplastic in effluents of waste water treatment plants using focal plane array-based micro-Fourier-transform infrared imaging. Water Res 108:365–372. https://doi.org/10.1016/j.watres.2016.11.015

Moreira FT, Prantoni AL, Martini B et al (2016) Small-scale temporal and spatial variability in the abundance of plastic pellets on sandy beaches: methodological considerations for estimating the input of microplastics. Mar Pollut Bull 102:114–121. https://doi.org/10.1016/j.marpolbul.2015.11.051

Murray F, Cowie PR (2011) Plastic contamination in the decapod crustacean *Nephrops norvegicus* (Linnaeus, 1758). Mar Pollut Bull 62:1207–1217. https://doi.org/10.1016/j.marpolbul.2011.03.032

Ng KL, Obbard JP (2006) Prevalence of microplastics in Singapore's coastal marine environment. Mar Pollut Bull 52:761–767. https://doi.org/10.1016/j.marpolbul.2005.11.017

Nobre CR, Santana MFM, Maluf A et al (2015) Assessment of microplastic toxicity to embryonic development of the sea urchin *Lytechinus variegatus* (Echinodermata: Echinoidea). Mar Pollut Bull 92:99–104. https://doi.org/10.1016/j.marpolbul.2014.12.050

Norén F (2007) Small plastic particles in Coastal Swedish waters. KIMO Sweden, Lysekil

Nuelle M-T, Dekiff JH, Remy D et al (2014) A new analytical approach for monitoring microplastics in marine sediments. Environ Pollut 184:161–169. https://doi.org/10.1016/j.envpol.2013.07.027

Obbard RW, Sadri S, Wong YQ et al (2014) Global warming releases microplastic legacy frozen in Arctic Sea ice. Earth's Future 2:315–320. https://doi.org/10.1002/2014EF000240

Osswald TA, Baur E, Brinkmann S et al (2006) International plastics handbook. Hanser, München

Primpke S, Imhof H, Piehl S et al (2017a) Mikroplastik in der Umwelt. Chem unserer Zeit 51:402–412. https://doi.org/10.1002/ciuz.201700821

Primpke S, Lorenz C, Rascher-Friesenhausen R et al (2017b) An automated approach for microplastics analysis using focal plane array (FPA) FTIR microscopy and image analysis. Anal Methods 9:1499–1511. https://doi.org/10.1039/C6AY02476A

Qiu Q, Tan Z, Wang J et al (2016) Extraction, enumeration and identification methods for monitoring microplastics in the environment. Estuar Coast Shelf Sci 176:102–109. https://doi.org/10.1016/j.ecss.2016.04.012

Quinn B, Murphy F, Ewins C (2017) Validation of density separation for the rapid recovery of microplastics from sediment. Anal Methods 9:1491–1498. https://doi.org/10.1039/C6AY02542K

Reid PC, Colebrook JM, Matthews JBL et al (2003) The Continuous Plankton Recorder: concepts and history, from Plankton Indicator to undulating recorders. Prog Oceanogr 58:117–173. https://doi.org/10.1016/j.pocean.2003.08.002

Reisser J, Slat B, Noble K et al (2015) The vertical distribution of buoyant plastics at sea: an observational study in the North Atlantic Gyre. Biogeosciences 12:1249–1256. https://doi.org/10.5194/bg-12-1249-2015

Ribeiro F, Garcia AR, Pereira BP et al (2017) Microplastics effects in *Scrobicularia plana*. Mar Pollut Bull 122:379–391. https://doi.org/10.1016/j.marpolbul.2017.06.078

Romeo T, Pietro B, Pedà C et al (2015) First evidence of presence of plastic debris in stomach of large pelagic fish in the Mediterranean Sea. Mar Pollut Bull 95:358–361. https://doi.org/10.1016/j.marpolbul.2015.04.048

Santana MFM, Moreira FT, Turra A (2017) Trophic transference of microplastics under a low exposure scenario: insights on the likelihood of particle cascading along marine food-webs. Mar Pollut Bull 121:154–159. https://doi.org/10.1016/j.marpolbul.2017.05.061

Setälä O, Fleming-Lehtinen V, Lehtiniemi M (2014) Ingestion and transfer of microplastics in the planktonic food web. Environ Pollut 185:77–83. https://doi.org/10.1016/j.envpol.2013.10.013

Setälä O, Magnusson K, Lehtiniemi M et al (2016) Distribution and abundance of surface water microlitter in the Baltic Sea: a comparison of two sampling methods. Mar Pollut Bull 110:177–183. https://doi.org/10.1016/j.marpolbul.2016.06.065

Shim WJ, Song YK, Hong SH et al (2016) Identification and quantification of microplastics using Nile Red staining. Mar Pollut Bull 113:469–476. https://doi.org/10.1016/j.marpolbul.2016.10.049

Shim WJ, Hong SH, Eo SE (2017) Identification methods in microplastic analysis: a review. Anal Methods 9:1384–1391. https://doi.org/10.1039/C6AY02558G

Song YK, Hong SH, Jang M et al (2014) Large accumulation of micro-sized synthetic polymer particles in the sea surface microlayer. Environ Sci Technol 48:9014–9021. https://doi.org/10.1021/es501757s

Song YK, Hong SH, Jang M et al (2015) A comparison of microscopic and spectroscopic identification methods for analysis of microplastics in environmental samples. Mar Pollut Bull 93:202–209. https://doi.org/10.1016/j.marpolbul.2015.01.015

Stolte A, Forster S, Gerdts G et al (2015) Microplastic concentrations in beach sediments along the German Baltic coast. Mar Pollut Bull 99:216–229. https://doi.org/10.1016/j.marpolbul.2015.07.022

Suhrhoff TJ, Scholz-Böttcher BM (2016) Qualitative impact of salinity, UV radiation and turbulence on leaching of organic plastic additives from four common plastics – a lab experiment. Mar Pollut Bull 102:84–94. https://doi.org/10.1016/j.marpolbul.2015.11.054

Sussarellu R, Suquet M, Thomas Y et al (2016) Oyster reproduction is affected by exposure to polystyrene microplastics. Proc Natl Acad Sci U S A 113:2430–2435. https://doi.org/10.1073/pnas.1519019113

Syberg K, Khan FR, Selck H et al (2015) Microplastics: addressing ecological risk through lessons learned. Environ Toxicol Chem 34:945–953. https://doi.org/10.1002/etc.2914

Tagg AS, Sapp M, Harrison JP et al (2015) Identification and quantification of microplastics in wastewater using focal plane array-based reflectance micro-FT-IR imaging. Anal Chem 87:6032–6040. https://doi.org/10.1021/acs.analchem.5b00495

Tagg AS, Harrison JP, Ju-Nam Y et al (2017) Fenton's reagent for the rapid and efficient isolation of microplastics from wastewater. Chem Commun 53:372–375. https://doi.org/10.1039/C6CC08798A

Thompson RC, Olsen Y, Mitchell RP et al (2004) Lost at sea: where is all the plastic? Science 304:838. https://doi.org/10.1126/science.1094559

Turra A, Manzano AB, Dias RJS et al (2014) Three-dimensional distribution of plastic pellets in sandy beaches: shifting paradigms. Sci Rep 4:4435. https://doi.org/10.1038/srep04435

Underwood AJ, Chapman MG, Browne MA (2017) Some problems and practicalities in design and interpretation of samples of microplastic waste. Anal Methods 9:1332–1345. https://doi.org/10.1039/C6AY02641A

Van Cauwenberghe L, Claessens M, Vandegehuchte MB et al (2013a) Assessment of marine debris on the Belgian Continental Shelf. Mar Pollut Bull 73:161–169. https://doi.org/10.1016/j.marpolbul.2013.05.026

Van Cauwenberghe L, Vanreusel A, Mees J et al (2013b) Microplastic pollution in deep-sea sediments. Environ Pollut 182:495–499. https://doi.org/10.1016/j.envpol.2013.08.013

Van Cauwenberghe L, Claessens M, Vandegehuchte MB et al (2015) Microplastics are taken up by mussels (*Mytilus edulis*) and lugworms (*Arenicola marina*) living in natural habitats. Environ Pollut 199:10–17. https://doi.org/10.1016/j.envpol.2015.01.008

van Franeker JA, Blaize C, Danielsen J et al (2011) Monitoring plastic ingestion by the northern fulmar Fulmarus glacialis in the North Sea. Environ Pollut 159:2609–2615. https://doi.org/10.1016/j.envpol.2011.06.008

Vendel AL, Bessa F, Alves VEN et al (2017) Widespread microplastic ingestion by fish assemblages in tropical estuaries subjected to anthropogenic pressures. Mar Pollut Bull 117:448–455. https://doi.org/10.1016/j.marpolbul.2017.01.081

Vianello A, Boldrin A, Guerriero P et al (2013) Microplastic particles in sediments of lagoon of Venice, Italy: first observations on occurrence, spatial patterns and identification. Estuar Coast Shelf Sci 130:54–61. https://doi.org/10.1016/j.ecss.2013.03.022

Viršek MK, Palatinus A, Koren Š et al (2016) Protocol for microplastics sampling on the sea surface and sample analysis. J Vis Exp 118:55161. https://doi.org/10.3791/55161

von Moos N, Burkhardt-Holm P, Kohler A (2012) Uptake and effects of microplastics on cells and tissue of the blue mussel *Mytilus edulis* L. after an experimental exposure. Environ Sci Technol 46:11327–11335. https://doi.org/10.1021/es302332w

Wagner J, Wang Z-M, Ghosal S et al (2017) Novel method for the extraction and identification of microplastics in ocean trawl and fish gut matrices. Anal Methods 9:1479–1490. https://doi.org/10.1039/C6AY02396G

Wang Z-M, Wagner J, Ghosal S et al (2017) SEM/EDS and optical microscopy analyses of microplastics in ocean trawl and fish guts. Sci Total Environ 603:616–626. https://doi.org/10.1016/j.scitotenv.2017.06.047

Ward EJ, Shumway SE (2004) Separating the grain from the chaff: particle selection in suspension- and deposit-feeding bivalves. J Exp Mar Biol Ecol 300:83–130. https://doi.org/10.1016/j.jembe.2004.03.002

Woodall LC, Sanchez-Vidal A, Canals M et al (2014) The deep sea is a major sink for microplastic debris. R Soc open Sci 1:140317. https://doi.org/10.1098/rsos.140317

Wright SL, Thompson RC, Galloway TS (2013) The physical impacts of microplastics on marine organisms: a review. Environ Pollut 178:483–492. https://doi.org/10.1016/j.envpol.2013.02.031

Zhao S, Danley M, Ward JE et al (2017) An approach for extraction, characterization and quantitation of microplastic in natural marine snow using Raman microscopy. Anal Methods 9:1470–1478. https://doi.org/10.1039/C6AY02302A

Zhu X (2015) Optimization of elutriation device for filtration of microplastic particles from sediment. Mar Pollut Bull 92:69–72. https://doi.org/10.1016/j.marpolbul.2014.12.054

Zobkov M, Esiukova E (2017) Microplastics in Baltic bottom sediments: quantification procedures and first results. Mar Pollut Bull 114:724–732. https://doi.org/10.1016/j.marpolbul.2016.10.060

Appendices

Appendix 1: List of Conference Participants

Last name	First name	Presentation type/Function
Afoncheva	Sofia	Oral Presentation
Amptmeijer	David	Oral Presentation
Andreasen	Laurits	Listener
Andskog	Mona	Session Host
Anschuetz	Anna-Adriana	Listener
Armelloni	Enrico Nicola	Poster Presentation
Baali	Ayoub	Oral Presentation
Barboza	Francisco	Session Host and Workshop
Barrett	Chris	Oral Presentation
Basse	Wiebke	Organization Team
Bazzicalupo	Enrico	Invited Session Speaker
Bluhm	Wally	Invited Artist
Bode	Maya	Organization Team
Böhm	Frederike	Organization Team
Böhmer	Astrid	Poster Presentation
Böök	Imke	Poster Presentation
Bourassi	Hajar	Oral Presentation
Bourreau	Mathilde	Oral Presentation
Brüwer	Jan D.	Organization Team, Session Host and Oral Presentation
Buck-Wiese	Hagen	Session Host and Oral Presentation
Cadier	Charles	Oral Presentation
Campen	Hanna	Workshop
Campos	Camila	Session Host
Carcedo	Cecilia	Listener
Carus	Jana	Session Host and Oral Presentation
Chen	Shaomin	Listener
Chi	Xupeng	Oral Presentation
Christiansen	Svenja	Oral Presentation
Cormon	Xochitl	Session Host and Oral Presentation
Creemers	Marie	Oral Presentation
Cziesielski Olschowsky	Maha Joana	Session Host and Oral Presentation
Dippe	Tina	Session Host
Dohr	Jessica	Listener
Doolittle-Llanos	Sara	Oral Presentation
Drinkorn	Catherine	Oral Presentation
Dudeck	Tim	Listener
Durucan	Furkan	Oral Presentation
Earp	Hannah	Session Host and Oral Presentation
Eich	Charlotte	Oral Presentation
Eich	Andreas	Organization Team
El Kassar	Jan Riad	Oral Presentation
Ezekiel	Jeoline	Organization Team
Fahning	Jana	Listener
Färber	Leonie	Oral Presentation
Folkers	Mainah	Oral Presentation
Ford	Amanda K.	Invited Session Speaker
Franz	Markus	Session Host and Workshop
Frey	Dmitry	Oral Presentation
Fricke	Enno	Oral Presentation
Friedland	Rene	Poster Presentation
Garthe	Stephan	Invited Session Speaker
Geburzi	Jonas	Session Host and Workshop
Gerdes	Rüdiger	Listener
Geuer	Jana K.	Session Host and Oral Presentation
Gluchowska	Marta	Invited Session Speaker
Goerres	Matthias	Session Host and Oral Presentation
Golikov	Alexey	Oral Presentation
Gruber-Vodicka	Harald	Invited Session Speaker
Guevara	Emily	Oral and Poster Presentation
Gustavs	Lydia	Workshop
Hamer	Jorin	Listener
Hamm	Thea	Session Host and Organization Team
Hanfland	Claudia	Invited Plenary Speaker
Hartmann	Daniel	Workshop
Heckmann	Saskia	Inivted Industry Representative
Heel	Lena	Organization Team
Heidenreich	Marie	Workshop
Hennigs	Laura	Poster Presentation
Hentschel	Lisa	Organization Team and Workshop
Herhoffer	Vanessa	Oral Presentation
Hewitt	Olivia	Listener
Heylen	Brigitte C.	Session Host and Oral Presentation
Hildebrandt	Lars	Listener
Hohensee	Dorothee	Organization Team
Horn	Myriel	Session Host

S. Jungblut et al. (eds.), *YOUMARES 8 – Oceans Across Boundaries: Learning from each other*,
https://doi.org/10.1007/978-3-319-93284-2

Last name	First name	Presentation type/Function
Hoving	Henk-Jan T.	Invited Session Speaker
Ito	Maysa	Session Host and Workshop
Jäger	Anne	Poster Presentation
Johnson	Mildred	Poster Presentation
Joon	Sudhir Kumar	Listener
Juhls	Bennet	Poster Presentation
Jungblut	Simon	Organization Team, Session Host and Poster Presentation
Kaiser	Patricia	Listener
Kalita	Sabrina	Oral Presentation
Käse	Laura	Session Host and Oral Presentation
Käß	Melissa	Oral Presentation
Keck	Therese	Session Host and Oral Presentation
Khosravi	Maral	Organization Team
Kittu	Leila	Listener
Kleinschmidt	Birgit	Oral Presentation
Klimpel	Veit	Workshop
Köberle	Cornelia	Listener
Korez	Špela	Oral Presentation
Kovacs	Tamas	Poster Presentation
Kuhl	Theresa	Listener
Lachs	Liam	Oral Presentation
Latif	Mojib	Invited Plenary Speaker
Launspach	Janna	Listener
Laurenz	Jan	Poster Presentation
Lazareva	Margarita	Poster Presentation
Letschert	Jonas	Listener
Liebich	Viola	Organization Team
Linck Rosenhaim	Ingrid	Oral Presentation
Lishchenko	Anastasiia	Oral Presentation
Lishchenko	Fedor	Session Host
Liu	Bi Lin	Poster Presentation
Liu	Yangyang	Session Host
Lorenz	Claudia	Session host
Lübbecke	Joke	Invited Session Speaker
Mańko	Maciej	Session Host
Mascarenhas	Veloisa	Session Host and Organization Team
Maureaud	Aurore	Listener
Mauvisseau	Quentin	Oral and Poster Presentation
McCarthy	Morgan Lee	Session Host and Oral Presentation
Melis	Eleni	Oral Presentation
Milinski	Sebastian	Oral Presentation
Mpinou	Evangelia Layla	Listener
Nachtsheim	Dominik A.	Session Host and Oral Presentation
Nieva	Joyce	Poster Presentation
Nour	Ola	Organization Team
Nugroho	Avianto	Listener
Olivier	Pierre	Oral Presentation
Olsen	Maria Winberg	Oral Presentation
Oltmanns	Marilena	Oral Presentation
Onate Casado	Javier	Listener
Paraskiv	Artem	Poster Presentation
Pauli	Nora-Charlotte	Oral Presentation
Paulweber	Kyra	Poster Presentation
Piehl	Sarah	Session Host
Poiesz	Suzanne	Oral Presentation
Pradisty	Novia Arinda	Oral Presentation
Prinz	Natalie	Session Host and Oral Presentation
Prume	Julia	Listener
Raczkowska	Anna	Oral Presentation
Rech	Sabine	Oral Presentation
Reichert	Jessica	Oral Presentation
Reif	Farina	Listener
Richter	Maren Elisabeth	Oral Presentation
Riesbeck	Sarah	Poster Presentation
Rist	Sinja	Oral Presentation
Rochner	Andrea	Listener
Rölfer	Lena	Poster Presentation
Rombouts	Titus	Poster Presentation
Roscher	Lisa	Poster Presentation
Rubio	Iratxe	Oral Presentation
Rümmler	Marie-Charlotte	Oral Presentation
Sahlin	Sara	Oral Presentation
Sazonova	Olga	Organization Team
Schellenberg	Lisa	Listener
Schiller	Jessica	Oral Presentation
Schmitz	Jana	Poster Presentation
Schneider	Matthias	Listener
Schwarz	Richard	Session Host
Schweikert	Frank	Workshop
Schwermer	Heike	Listener
Scotti	Marco	Invited Session Speaker
Senff	Paula	Oral Presentation
Sguotti	Camilla	Session Host, Oral and Poster Presentation
Shojaei	Mehdi	Oral Presentation
Siegfried	Lydia	Oral Presentation
Sievers	Imke	Oral Presentation
Slaby	Beate	Oral Presentation
Słomińska	Marta	Oral Presentation
Smilenova	Angelina	Oral Presentation
Smirnova	Maria	Oral Presentation
Sondej	Greta	Listener
Soundararajan	Vigneshwaran	Oral Presentation
Spangenberg	Ines	Listener
Starke	Claudia	Poster Presentation
Stippkugel	Angela	Oral Presentation
Storkenmaier	Nadine	Listener
Straub	Sandra	Oral Presentation
Stuthmann	Lara	Poster Presentation
Szewc	Karolina	Oral Presentation
Titova	Alena	Listener
Tompson	Timothy	Organization Team

Last name	First name	Presentation type/Function
van de Ven	Clea	Poster Presentation
van Kuijk	Tiedo	Listener
Visbeck	Martin	Workshop
Walczyńska	Katarzyna	Session Host
Waldrop Bergman	Lovisa	Listener
Weidung	Mara	Organization Team and Oral Presentation
Weinheimer	Alaina Rose	Oral Presentation
Wekerle	Claudia	Invited Session Speaker
Welden	Natalie	Invited Session Speaker
Willenbrink	Nils	Oral Presentation
Wilson-Green	Jack	Listener
Wölfelschneider	Mirco	Oral Presentation
Wrobel	Iwona	Oral Presentation
Wu	Yu-Chen	Listener
Xi	Hongyan	Invited Session Speaker
Zurhelle	Christian	Oral Presentation

Appendix 2: Conference Sessions and Abstracts

In the following appendix, the Calls for Abstracts and the abstracts of the oral and poster presentations of each session of YOUMARES 8 are listed. The appendix chapters here are ordered according to the corresponding proceedings chapters of the book, not according to the session numbers during the conference. While the appendix chapters 1 to 12 have a corresponding proceedings chapter, the chapters 13 to 15 do not.

1 Physical Processes in the Tropical and Subtropical Oceans: Variability, Impacts, and Connections to Other Components of the Climate System

Tina Dippe[1] and Martin Krebs[1]

[1]GEOMAR Helmholtz Centre for Ocean Research Kiel, Germany

1.1 Call for Abstracts

The tropical-subtropical oceans are key regions of the global climate system and affect livelihoods across the globe. We invite research that is concerned with understanding and predicting variability, or dealing with its regional and global impacts. Possible contributions include, but are not limited to, observational and modeling studies about the El Niño-Southern Oscillation, the Indian Ocean dipole, tropical Atlantic variability, equatorial and eastern boundary upwelling systems, tropical monsoons, and tropical cyclones. While we wish to focus the session on physical processes, we strongly encourage researchers from other disciplines to submit studies that link biological, chemical, and other research to the tropical oceans.

1.2 Abstracts of Oral Presentations

1.2.1 El Niño and Its Little Sister in the Tropical Atlantic: Similarities, Differences and Interaction

Joke F. Lübbecke[1*]

[1]GEOMAR Helmholtz Centre for Ocean Research Kiel, Düsternbrooker Weg 20, 24105 Kiel, Germany

*invited speaker, corresponding author: jluebbecke@geomar.de

In this talk, the main modes of climate variability in the equatorial Pacific and Atlantic Ocean, namely the El Niño – Southern Oscillation (ENSO) and the Atlantic Niño mode will be compared to each other and their interaction will be discussed. It will be shown that while both modes involve the Bjerknes feedback, they exhibit differences with respect to strength, timing and symmetry. As an example, the Atlantic Niño mode is more strongly damped than ENSO, mainly due to a weaker thermocline feedback that favors the growth of SST anomalies in response to wind stress forcing. Also, while ENSO is asymmetric for warm and cold events with respect to amplitude, spatial patterns and temporal evolution, in the equatorial Atlantic, cold events are effectively mirror images of warm events. This behavior is rationalized by means of the individual Bjerknes feedback components.

1.2.2 Detecting Changes in Internal Variability in Response to Global Warming

Sebastian Milinski[1,2*], Jürgen Bader[1,3], Johann H. Jungclaus[1], Jochem Marotzke[1]

[1]Max Planck Institute for Meteorology, Hamburg, Germany

[2]International Max Planck Research School on Earth System Modelling, Hamburg, Germany [3]Uni Climate, Uni Research and the Bjerknes Centre for Climate Research, Bergen, Norway

*corresponding author: sebastian.milinski@mpimet.mpg.de

Keywords: Internal variability, Detection and attribution, Tropical rainfall

The response of the climate system to past and possible future anthropogenic emissions has been one of the guiding questions for several decades. The previously investigated responses to anthropogenic climate change can be grouped into changes of the mean state and changes of variability. Both types of questions are difficult to answer, mainly because the large internal variability in the climate system might conceal small forced changes. Furthermore, the internal variability itself might also change under global warming. In our study, we analyze changes in the rainfall variability in the tropical Atlantic region. We use a 100-member ensemble of historical (1850-2005) model simulations with the Max Planck Institute for Meteorology Earth System Model (MPI-ESM) to separate the directly forced response and internal variability. To investigate the effects of global warming, we employ an additional ensemble of model simulations with stronger external forcing (1% CO_2-increase per year, same integration length as the historical simulations) with 68 ensemble members. We find changes in the internal variability of tropical rainfall in all ocean basins. However, the magnitude and sign of the change is regionally different. In the tropical Atlantic, the strongest internal variability is simulated on the southern flank of the Intertropical Convergence Zone (ITCZ). Under global warming the variability on the southern ITCZ flank increases and is located further north than under preindustrial conditions, coinciding with a narrowing and intensification of the mean rainfall in the Atlantic ITCZ.

1.2.3 Linking the Tropical and Subtropical Atlantic: A Benguela Upwelling Perspective

Lydia Siegfried[1*], Martin Schmidt[1], Volker Mohrholz[1]

[1]Leibniz Institute for Baltic Sea Research, Warnemünde, Seestraße 15, 18119 Rostock-Warnemünde, Germany

*corresponding author: lydia.siegfried@io-warnemuende.de

Keywords: SACW, Benguela, Regional model, Upwelling

Upwelling in the Benguela region is mainly fed by two different Central Water types. The interplay between the tropical South Atlantic Central Water (SACW) and the subtropical ESACW (Eastern SACW) determines the physical and chemical properties of the Benguela ecosystem. We use a regional circulation model (Modular Ocean Model) to analyze the source region of the upwelled water. Passive tracers are employed to study propagation pathways of Central Water Masses and their temporal and spatial variation. We show that water originating from the equatorial current system contributes substantially to the upwelling in the northern Benguela system. The Equatorial UnderCurrent (EUC) is one of the strongest currents in the South-east Atlantic and its water is transported to the Benguela system by two mechanisms: the trapping of water inside the equatorial and coastal wave guides and by a poleward bending of the EUC between 8°W and 0°E feeding the South Equatorial UnderCurrent (SEUC). Water loss from the EUC into the SEUC dominates in austral summer whereas the strength of transport in the wave guides is relatively constant throughout the year. The SE-transport in the open ocean might be established by a Sverdrup balance. A strong interannual transport anomaly was found for the Benguela Niño 2010/11. Transport from the EUC to the Benguela upwelling system was enhanced, maximal SEUC transport peaked earlier in the winter season and coastal upwelling was reduced. Once transported to the Benguela upwelling system, part of the surface and subsurface water contributes to coastal upwelling in the Kunene Cell and is advected westward due to Ekman transport. The remaining part of the water feeds the poleward undercurrent which is the main pathway for tropical SACW transport inside the Benguela system. Strength and location of the poleward undercurrent are determined, among other things, by wind forcing and shelf topography.

1.2.4 Bottom Currents in the Tropical Fractures of the Northern Mid-Atlantic Ridge

D. I. Frey[1*], E. H. Morozov[1], N. I. Makarenko[2]

[1]Shirshov Institute of Oceanology, Russian Academy of Sciences, Moscow, Russia

[2]Lavrentiev Institute of Hydrodynamics, Russian Academy of Sciences, Novosibirsk, Russia

*corresponding author: dima.frey@gmail.com

Keywords: Abyssal currents, LADCP measurements, Numerical modeling, Ocean circulation

The properties of bottom flows in the tropical part of the Atlantic were studied on the basis of hydrographic measurements and numerical modeling of the oceanic circulation. Antarctic Bottom Water (AABW, potential temperature $\theta<2$ °C) occupies the deepest layer of the major part of the Atlantic Ocean. This water propagates from the Weddell Sea to the north filling the lower part of the ocean basins. Interbasin exchange of bottom waters occurs through the depressions in the topographic obstacles. Mean velocities of the northward propagation of AABW are less than 1 cm/s; however, when AABW flows through narrow abyssal channels, the current strongly accelerates. Thermohaline properties and velocities of the currents in the region of the Mid-Atlantic Ridge were measured onboard the R/V "Akademik Sergey Vavilov" by the scientists of the Shirshov Institute of Oceanology in 2014–2016. A part of these hydrographic observations at several fracture zones was carried out for the first time. The numerical simulation was performed using the Institute of Numerical Mathematics Ocean Model (INMOM). The observations of velocities were used for verification of the numerical model. The simulated three-dimensional velocity fields with high spatial resolution in the lower layer allow us to study the bottom currents over their entire length. This research was supported by the Russian Science Foundation (project 16-17-10149).

2 The Physics of the Arctic and Subarctic Oceans in a Changing Climate

Camila Campos[1] and Myriel Horn[1]
[1]Alfred Wegener Institute (AWI), Helmholtz Centre for Polar and Marine Research, P.O. Box 120161, 27570 Bremerhaven, Germany

2.1 Call for Abstracts

The Arctic climate system has been drastically impacted by the changing global climate. While summer sea ice reduced dramatically/significantly, and the atmospheric warming is amplified over the Arctic, changes in the ocean are less obvious due to its higher inertia. However, Arctic and Subarctic Oceans play a crucial role in modulating mid-latitude and polar climate. Thus, investigating the pathways and timescales of oceanic and atmospheric dynamics is crucial to understand major aspects of current climate change. We invite presentations advancing our understanding of the relevance of Arctic and Subarctic Oceans.

2.2 Abstracts of Oral Presentations

2.2.1 Declining Convection in the North Atlantic Initiated Through Warming Summers

Marilena Oltmanns[1*], Johannes Karstensen[1], Jürgen Fischer[1]
[1]GEOMAR Helmholtz Centre for Ocean Research Kiel, Düsternbrooker Weg 20, 24105 Kiel, Germany
*corresponding author: moltmanns@geomar.de

Keywords: Atmosphere-ocean interactions, Subpolar North Atlantic, Ocean circulation, Atmospheric dynamics, Climate variability

A shutdown of ocean convection, triggered by enhanced melting over Greenland, is regarded as potential tipping element in future climate and recent model studies, reporting that rising meltwater fluxes from Greenland are already pervading the North Atlantic, have substantiated the emerging threat. Yet, from an observational viewpoint, the atmospheric forcing is deemed more important than melting as explicit evidence for an impending halt in convection is still lacking. Here we show, based upon a comprehensive set of in situ hydrographic observations, that warm summers – associated with salient fresh layers, capping the convective region, increased sea surface temperatures and enhanced melting – evoke reinforcing atmospheric responses, including heavier precipitation in fall and reduced ocean heat losses throughout winter, which contribute to maintaining a strong stratification and coactively shorten the period for convection. Considering that the summer warming over the last decades was accompanied by concurrent trends in the identified feedbacks and is expected to proceed further, we anticipate that the critical threshold for a convective shutdown will be reached faster than estimates, premised only on melt rates, suggest.

2.2.2 Upper Layer Circulation in the Nordic Seas – Spatial Distribution Variability of Heat, Freshwater and Potential Energy

Catherine Drinkorn[1,2*], Görkan Björk[1]
[1]Department of Marine Sciences, University of Gothenburg, Carl Skottsbergsgata 22B, 41319
Gothenburg, Sweden
[2]Faculty of Geosciences, University of Bremen, Klagenfurter Str. 2-4, 28359 Bremen, Germany
*corresponding author: gusdrica@student.gu.se

Keywords: Subarctic, Upper layer circulation, Climate change, Hydrography, Variability

The Nordic and Subarctic Seas constitute a major region for the exchange of warm and salty Atlantic water with cool and fresher Arctic water masses. This modification of Atlantic and Arctic water masses is the main driver for the northern deep-water formation branch of the global overturning circulation and hence an important climatic factor. In order to investigate its stability and robustness against climate warming, contemporary hydrographic data of great spatial and temporal resolution needs to be persistently examined and represented in an insightful manner. Therefore, the aim of this project is to pursue a previous study by Björk et al. (2001, Polar Res 20: 161–168), who illustrated the upper layer circulation in the Nordic Seas inferred from spatial distribution of freshwater, heat and potential energy based on data from numerous hydrographic stations until 1998. It will be investigated whether and to what extent seasonal and decadal cycles or trends occur in the whole picture by adding hydrographic data from after 1998 to the previous data set. Furthermore, correlation with known natural cycles, trends or properties, e.g., the North Atlantic Oscillation (NAO), Arctic sea ice extent or characteristic topography, may lead to a better understanding of the possible variability properties and their causes.

2.2.3 Fram Strait Recirculation and the East Greenland Current

Maren Richter[1,2*], Wilken-Jon von Appen[1]
[1]Alfred-Wegener-Institut, Helmholtz-Zentrum für Polar- und Meeresforschung, Am Handelshafen 12, 27570 Bremerhaven, Germany
[2]Universität Bremen, Bibliothekstraße 1, 28359 Bremen, Germany
*corresponding author: mrichter@awi.de

Keywords: Fram Strait, East Greenland Current, Atlantic Water, Recirculation, Transport

In Fram Strait (FS), between Greenland and Svalbard, heat is transported to, and sea-ice is exported from the Arctic Ocean. The oceanography in the area plays an important role in the stability of two of the Greenland ice-sheet's largest outlet glaciers and thus mass loss of the Greenland ice-sheet to the ocean. In Fram Strait, a source of North Atlantic Deep Water is in contact with the atmosphere last before contributing to the Atlantic Meridional Overturning Circulation. The East Greenland Current (EGC) in western FS flows southward. The West Spitsbergen Current (WSC) in eastern FS transports relatively warm Atlantic Water (AW), originating at lower latitudes, northward. However, not all AW entering FS from the south is transported to the Arctic Ocean. Rather a part of it "recirculates" in FS, it flows westward to join the EGC. Here, we use hydrographical and velocity data (yielding absolute geostrophic velocity) collected by RV *Polarstern* in FS in summer 2016 to investigate the meridional extent and spatial structure of this recirculation to determine at what latitudes AW joins the EGC and how that changes the structure of the EGC. Further, we investigate possible AW pathways in two troughs on the Greenland shelf, leading to the large outlet glaciers. Four sections cross the EGC between 77.8°N and 80.8°N; two are located at the mouths of the troughs, while one continues across the central FS and the WSC. Additionally, a meridional section at 0°EW spans the recirculation in the central FS. We use these sections to show the first estimate of absolute geostrophic transports of different water masses in the recirculation and the EGC north of 79°N and their spatial structure at an appropriate spatial resolution. Based on these results we update the circulation scheme in FS and to the outlet glaciers and elucidate transport mechanisms.

2.2.4 Measuring Temperature Variability of Deep Atlantic Water in the Arctic

Sara E. Sahlin[1*]

[1]Department of Marine Sciences, University of Gothenburg, Box 115, 405 30 Gothenburg, Sweden

*corresponding author: gussahlisa@student.gu.se

Keywords: Physical Oceanography, Greenland, Temperature, Variability, Arctic, NADW

Warm waters melt the polar ice shelves from below. For the marine terminating glaciers of Greenland and Antarctica, melting not only adds freshwater to a rising sea level, but these ice tongues may also act as buffers between the ocean and the large land-based glaciers. However, very little is known of the variability of the temperature of these warm water masses. In August 2015, ten prototype autonomous expendable bottom temperature sensors were deployed in Petermann Fjord and adjacent Nares Strait in the northwest of Greenland. The instrument measures bottom temperature over a relatively long time (up to 10 years) and at high frequency (every half hour). After several months measuring temperature, the sensors floated to the surface, sent the recorded data by satellite, and then acted as drifters. Part of this project is to analyze the temperature data from these Lotus buoys and relate it to other measurements in the area, primarily CTD casts. Using historical observations from CTD casts and moorings, the depth and location of deployment of a second batch of sensors was defined for the Barents Sea shelf break during expedition PS106.2 of the icebreaking research vessel *Polarstern* in July 2017. The warm core of the Atlantic Water can be identified just as it enters the Arctic, and its temperature variability can be monitored with an accuracy of at least 0.05 °C using these sensors on the sea bed. The aim of the project is to determine the utility and advantages of these buoys, and to come up with a strategy of how to put them to best use in future deployments. Their potential as a cost-effective monitoring device is of great importance to the increasing demand of temporally high resolution large-range hydrographic data in the field of climate change related research effort.

2.2.5 Characteristic of the Arctic Climate as a Response to Air-Sea Interaction

Iwona Wrobel[1,2*], J. Piskozub[1], P. Makuch[1], V. Drozdowska[1], P. Markuszewski[1], T. Petelski[1], T. Zielinski[1]

[1]Institute of Oceanology Polish Academy of Sciences Sopot, Poland

[2]IOPAS – Centre for Polar Studies KNOW (Leading National Research Centre), Sopot, Poland

*corresponding author: iwrobel@iopan.gda.pl

Keywords: Arctic Ocean, Air-sea interaction, Carbon cycle, Climate change

There are three active reservoirs for carbon dioxide (CO_2): the atmosphere, the oceans, and the terrestrial system. The atmosphere is the link with the other reservoirs, and the ocean plays a major part in determining the atmosphere's concentration of CO_2 through physical, chemical, and biological processes. It is well known that the Arctic Ocean (AO) is an overall sink for CO_2 throughout the year even though continental shelves can either be regional or seasonal sinks or sources of atmospheric CO_2. Relevant knowledge of air-sea CO_2 fluxes and their spatial and temporal variability is essential to gain the necessary understanding of the global carbon cycle and to fully resolve the ocean's role in climate variability. At present, the net air-sea CO_2 fluxes in the AO have been estimated at -0.12 ± 0.06 Pg C year^{-1} with net global ocean CO_2 uptake at -2.2 ± 0.5 Pg C year^{-1}. The direction and rates of net air-sea CO_2 exchange are determined by the product of the difference in values between pCO_2 in seawater and the atmosphere, and also by the rate of k. Net air-sea CO_2 fluxes were calculated using the FluxEngine toolset, which was created as a part of the

European Space Agency funded OceanFlux Greenhouse Gases project. Results from my study indicate that the variability in wind speed and, hence, the gas transfer velocity, generally play a major role in determining the temporal variability of CO_2 uptake, while variability in monthly ΔpCO_2 plays a major role spatially, with some exception.

2.2.6 The Winter Stratification on the East Siberian Shelf

Imke Sievers[1*], Göran Björk[1], Karen Assmann[1]

[1]Department of Marine Sciences, University of Gothenburg, Carl Skottsbergsgata 22B, 41319 Gothenburg, Sweden

*corresponding author: imke-s@gmx.de

Keywords: East Siberian Shelf, Mixing, 1-D modeling, Methane transport, Sub-sea permafrost

One of the climate relevant processes in the Arctic is the release of methane from the thawing sub-sea permafrost and sediments in the shelf regions. The escape of methane to the atmosphere is very much dependent on the density stratification on the shelf. The stratification is controlled by freshwater supply from rivers, exchange of water between the shelf and the outside deep ocean, ice production, ice export and wind forcing. The focus of this study was to make computations, using a 1D model of the development of winter stratification from wind data, air temp, river water supply and ice export in order to quantify how far out from the coast the shelf can be well mixed during winter and under which conditions. By answering this it also aimed to contribute to our understanding of the methane release from shelf regions.

2.2.7 Eddy-Resolving Simulation of the Atlantic Water Circulation in the Fram Strait

Claudia Wekerle[1*], Qiang Wang[1], Wilken-Jon von Appen[1], Sergey Danilov[1], Vibe Schourup-Kristensen[1], Thomas Jung[1]

[1]Alfred Wegener Institute, Helmholtz Centre for Polar and Marine Research, Bremerhaven, Germany

*invited speaker, corresponding author: claudia.wekerle@awi.de

Keywords: Ocean modeling, Unstructured meshes, Fram Strait, Arctic Ocean

The Fram Strait is the deepest and widest gateway that connects the Arctic Ocean with the Nordic Seas and thereby the Atlantic Ocean, thus being the major place for heat and water mass exchange. The two main currents responsible for the exchange are the West Spitsbergen Current (WSC) on the eastern side of Fram Strait transporting warm and salty Atlantic Water (AW) northward, and the East Greenland Current on the western side carrying cold and fresh Polar Water southward. Eddy driven recirculation of AW in the Fram Strait modifies the amount of heat that reaches the Arctic Ocean, but is difficult to constrain in ocean models due to very small Rossby radius there. Here we explore the effect of resolved eddies on the AW circulation in a locally eddy-resolving simulation of the global Finite-Element-Sea ice-Ocean-Model (FESOM). An eddy-permitting simulation serves as a control run. Our results suggest that resolving local eddy dynamics is critical to realistically simulate ocean dynamics in the Fram Strait. Strong eddy activity simulated by the eddy-resolving model, with peak in winter and lower values in summer, is comparable in magnitude and seasonal cycle to observations from a long-term mooring array, whereas the eddy-permitting simulation underestimates the observed magnitude. Furthermore, a strong cold bias in the central Fram Strait present in the eddy-permitting simulation is reduced due to resolved eddy dynamics, and AW transport into the Arctic Ocean is increased with possible implications for the Arctic Ocean heat budget. Given the good agreement between the eddy-resolving model and measurements, it can help filling gaps that point-wise observations inevitably leave. For example, the path of the WSC offshore branch, measured in the winter months by the mooring array, is shown to continue cyclonically around the Molloy Deep in the model, representing the major AW recirculation branch in this season.

2.3 Abstracts of Poster Presentations

2.3.1 Wind Forcing of the Arctic and North Atlantic Freshwater System

Tamas Kovacs[1*], Rüdiger Gerdes[1,2]

[1]Alfred-Wegener-Institut, Helmholtz-Zentrum für Polar- und Meeresforschung, Am Handelshafen 12, 27570 Bremerhaven, Germany

[2]Jacobs University Bremen gGmbH, Bremen, Germany

*corresponding author: tamas.kovacs@awi.de

Keywords: Arctic Ocean, Modeling, Freshwater, Atmospheric forcing

Oceanic processes in the Arctic and in the North Atlantic that play a key role in the global ocean circulation are often sensitive to density stratification of water, which is greatly shaped by salinity, or in another measure, by freshwater content. The freshwater budgets of these oceans are connected by currents that convey large volumes of water of different characteristics between one another. However, these budgets show spatial and temporal variations, and the fluxes between them cannot be considered constant either. The freshwater system of the Arctic linked to the North Atlantic is dynamic with changes and anomalies on different time scales, and the changes of this joint system seem to be in correlation with the evolution of atmospheric forcing patterns. Previous studies suggest the importance of wind stress forcing over key regions such as the Beaufort Sea or the Greenland Sea in influencing the distribution of freshwater. In this study we examine the sensitivity of freshwater distribution and fluxes between the Arctic and the North Atlantic oceans to wind stress forcing through numerical experiments. The tool for this is the Modini-system, a partial coupling technique that allows flexible experiments with prescribed wind stress fields for the ocean in the otherwise fully coupled Earth System Model of the Max Planck Institute. In this work we present the first results in investigating the role of atmospheric forcing in shaping freshwater reservoirs and exchanges between different oceanic subregions by comparing our model results using external wind stress forcing with the Modini-system, and fully coupled runs.

3 Ocean Optics and Ocean Color Remote Sensing

Veloisa Mascarenhas[1], Yangyang Liu[2], and Therese Keck[3]

[1]Institut für Chemie und Biologie des Meeres (ICBM), Universität Oldenburg, Schleusenstrasse 1, 26382 Wilhelmshaven, Germany

[2]Alfred Wegener Institute (AWI), Helmholtz Centre for Polar and Marine Research, P.O. Box 120161, 27570 Bremerhaven, Germany

[3]Institute for Space Sciences, Freie Universität Berlin, Carl-Heinrich-Becker-Weg 6-10, 12165 Berlin, Germany

3.1 Call for Abstracts

Ocean color remote sensing (OCRS) supports many research fields such as ocean bio-geo-chemistry, physical oceanography, ocean-system modeling and other climate change studies with its unique capability of providing synoptic view of the aquatic ecosystem. This session invites in-situ and satellite studies of marine bio-optics and OCRS such as hyperspectral radiometric observations, light interactions with optically active constituents (phytoplankton, colored dissolved organic matter, total suspended matter), inherent and apparent optical properties, algorithm development & validation, atmospheric correction algorithms, time series analysis, products and applications using multiple platforms and coupled models.

3.2 Abstracts of Oral Presentations

3.2.1 Dominant Phytoplankton Group Identification by Using Simulated and in situ Hyperspectral Remote Sensing Reflectance

Hongyan Xi[1*]

[1]Institute of Coastal Research, Helmholtz-Zentrum Geesthacht (HZG), Max-Planck-Str. 1, 21502 Geesthacht, Germany

*invited speaker, corresponding author: hongyan.xi@hzg.de

The Environmental Mapping and Analysis Program (EnMAP) is a German hyperspectral satellite mission that aims at monitoring and characterizing the Earth's environment on a global scale. One of its applications focuses on the aquatic ecosystems. With advanced spectral resolution of EnMAP hyperspectral imager, it provides much potential to identify phytoplankton taxonomic groups and improve mapping of phytoplankton community composition both for global oceans and regional waters. Given that the commonly used parameter obtained directly from hyperspectral earth observation sensors is the remote sensing reflectance (Rrs), this study focused on identification of dominant phytoplankton groups by using Rrs spectra directly. Based on five standard absorption spectra representing five different phytoplankton spectral groups, a simulated database of Rrs (C2X database, compiled within the ESA SEOM C2X Project) that includes 100,000 different water optical conditions was built with HydroLight. A test dataset was constructed using absorption spectra of 128 individual cultures from different taxonomic phytoplankton groups by simulating Rrs for 120 different water conditions. An identification approach is proposed to determine phytoplankton groups with the use of simulated C2X data and the test data; the skill of the identification is tested by investigating how and to what extend water optical constituents (Chl, NAP, and CDOM) impact the accuracy of this identification. The applicability of this approach in natural waters was also tested with the use of in situ measurements from different regions.

3.2.2 PlanktonID – Combining *in situ* Imaging, Deep Learning and Citizen Science for Global Plankton Research

Svenja Christiansen[1*], Rainer Kiko[1], Simon-Martin Schröder[2], Reinhard Koch[2], Lars Stemmann[3]

[1]GEOMAR, Helmholtz Centre for Ocean Research Kiel, Hohenbergstraße 2, 24105 Kiel, Germany

[2]Department of Computer Science, Christian-Albrechts University Kiel, Olshausenstr. 4, 24148 Kiel, Germany

[3]Sorbonne Universités, UPMC Univ Paris 06, UMR 7093, France and LOV, Observatoire Océanologique, 06230 Villefranche/mer, France

*corresponding author: schristiansen@geomar.de

Keywords: Zooplankton, Imaging, Citizen science, Deep learning, Rhizaria

Recent publications revealed the global importance of single-celled zooplankton, belonging to the super group Rhizaria and highlighted the need of in-situ imaging to study these fragile organisms. The advance of *in situ* plankton imaging techniques leads to increasing amounts of image data sets that require identification to different taxonomic levels. Automatic classification by computer algorithms provides the means for fast data availability, however the accuracy of those algorithms still requires manual identification by humans. We combined state of the art automatic image classification by convolutional neural networks (deep learning) with a citizen science project to classify a large dataset of ~3 million images from an Underwater Vision Profiler 5 (UVP5). On our website https://planktonid.geomar.de, citizen scientists can confirm or reject the automatic assignment of UVP5 images to different plankton categories in a memory-like game. Inbuilt quality controls and multiple validations per image enable scientific analysis of the citizen science data. We will present data on citizen scientist engagement, data quality and the distribution analysis of large protists (Rhizaria) in the Mauretanian, Benguela and Humboldt Current upwelling systems.

3.2.3 Study Phytoplankton Dynamics in the Arctic Using MERIS Sun-Induced Fluorescence

J. R. El Kassar[1*], R. Preusker[1], J. Fischer[1]

[1]Institute of Space Sciences, Carl-Heinrich-Becker-Weg 6-10, 12165 Berlin, Germany

*corresponding author: jan.elkassar@mail.met.fu-berlin.de

Keywords: Phytoplankton, Fluorescence, Chlorophyll, MERIS, Arctic

Analyzing the physiology of phytoplankton with sun-induced fluorescence (SIF) is a crucial aspect of ocean color. In this study, however, SIF and chlorophyll content have been used to study the phytoplankton dynamics in the Arctic. Chlorophyll content is taken from MERIS' "Algal Pigment Index 1" (AL1) dataset, derived from a blue-green ratio. SIF is estimated with the fluorescence line height (FLH) which is processed from MERIS reflectances in the red spectrum. We created a monthly climatology for the years 2003–2011 over the Arctic (60°N–90°N) and conducted a regression analysis between FLH and AL1. Monthly averages show that AL1 peaks in summer. FLH, however, peaks in spring and fall. The regression analysis shows a unimodal relationship between FLH and AL1 with high slopes in spring and autumn, whereas the relationship splits into two separate modes in summer. The upper mode shows high slopes similar to these in spring, whereas the lower mode shows very low slopes indicating weak or no correlation between AL1 and FLH. These two regimes are also visible in the spatial distribution of the ratio FLH/AL1. The results from the regression analysis correspond with the Arctic annual surface chlorophyll cycle. Spatial patterns of FLH/AL1 also align with currents exporting fresh water from the Arctic and currents advecting warm, salty water from the Atlantic which influence the stratification. This suggests that the results are related to seasonal and regional stratification processes and the vertical distribution of phytoplankton. The hypothesis is that FLH and AL1 contain information about phytoplankton from different depths due to different signal depths in the red (FLH) and blue-green (AL1) spectrum. Thus a combination of both parameters might be suitable to analyze not only the physiology but also the vertical distribution of phytoplankton.

3.2.4 Hyperspectral Simulation of Chl-a Fluorescence in Optically Complex Waters

Therese Keck[1*], Lena Kritten[1], René Preusker[1], Jürgen Fischer[1]

[1]Institute for Space Sciences, Freie Universität Berlin, Carl-Heinrich-Becker-Weg 6-10, 12165 Berlin, Germany

*corresponding author: therese.keck@wew.fu-berlin.de

Keywords: Ocean color, Fluorescence, Remote sensing

Phytoplankton is one of the main constituents in oceans, coastal and inland waters. Observing phytoplankton from space, mainly the pigments called chlorophyll are detected due to a very characteristic spectral properties. After correcting the remote sensing signals for the atmosphere, several techniques can retrieve the chlorophyll concentration. During photosynthesis, the pigments convert a part of the incoming visible light to photochemical energy for living. The other part dissipates as heat and is emitted as chlorophyll-a fluorescence. Optically complex waters contain various constituents like colored dissolved organic matter (cDOM) or sediments which can change the remotely sensed signal. The chlorophyll-a fluorescence located close to 682 nm is found to be relatively insensitive to other constituents. For phytoplankton populations close to the surface the fluorescence line height (FLH) gives good results for chlorophyll-a concentrations. The simulation of radiance in and above water enable us to understand how phytoplankton stratification and additional constituents influence .the fluorescence peak. In future there will be hyperspectral satellite sensors available (e.g., EnMAP) which may be used for novel hyperspectral fluorescence algorithms.

3.2.5 Characterization of CDOM and FDOM in the Nordic Seas

Anna Raczkowska[1,2*], Piotr Kowlaczuk[1], Sławomir Sagan[1], Monika Zabłocka[1], Mats A. Granskog[3], Alexey K. Pavlov[3], Colin Stedmon[4]

[1]Institute of Oceanology, Polish Academy of Sciences, ul. Powstańców Warszawy 55, 81-712 Sopot, Poland

[2]Centre for Polar Studies, Leading National Research Centre, 60 Będzińska Street, 41-200 Sosnowiec, Poland

[3]Norwegian Polar Institute, Fram Centre, 9296 Tromsø, Norway

[4]National Institute for Aquatic Resources, Technical University of Denmark, 2920 Charlottenlund, Denmark

*corresponding author: anraczkowska@gmail.com

Keywords: Absorption, Fluorescence, DOM, Polar regions

The aim of this study is to characterize optical properties of Chromophoric Dissolved Organic Matter (CDOM) and Fluorescent Dissolved Organic Matter (FDOM) in Nordic Seas. The experimental material has been collected during three summer seasons (2013–2015) onboard Polish r/v Oceania and Norwegian r/v Lance in the western and northern Spitsbergen Shelf, Norwegian Sea, Barents Sea and in the Fram Strait. Spatial distribution of DOM absorption and fluorescence from water samples is presented. A three channel WET Labs WET Star fluorometer was deployed, with channels for humic- and protein-like DOM and used to assess distribution of different FDOM fractions. A relationship between fluorescence intensity of the protein-like fraction of FDOM and chlorophyll a fluorescence was found and indicated the importance of phytoplankton biomass in West Spitsbergen Current waters as a significant source of protein-like FDOM. East Greenland Current waters has low concentration of chlorophyll a, and were characterized by high humic-like FDOM fluorescence. An empirical relationship between humic-like FDOM fluorescence intensity and CDOM absorption was derived and confirms the dominance of terrigenous like CDOM on the composition of DOM in the East Greenland Current. These high resolution profile data offer a simple approach to fractionate the contribution of these two DOM source to DOM across the Fram Strait and may help refine estimates of DOC fluxes in and out of the Arctic through this region.

3.3 Abstracts of Poster Presentations

3.3.1 100 Years of North Sea Clarity and Color Changes

J. Schmitz[1*], M. Wernand[2], J. Wollschläger[1], O. Zielinski[1]

[1]University of Oldenburg, Institute for Chemistry and Biology of the Marine Environment, Schleusenstraße 1, 26382 Wilhelmshaven, Germany

[2]NIOZ Royal Netherlands Institute for Sea Research, PO Box 59, 1790 AB Den Burg (Texel), The Netherlands

*corresponding author: jana.schmitz@uni-oldenburg.de

Keywords: Light, Water clarity, Water color, North Sea, Time series analysis

Thermal stratification and generation of currents, degradation of dissolved organic material and production of biomass by photosynthesis – all those processes are driven by light energy introduced into the seas and oceans. For visual predators and organisms depending on a circadian rhythm, light is furthermore essential for survival. Underwater light quality and penetration depth are largely determined by water constituents present (phytoplankton, colored dissolved organic matter and suspended sediments) and can be inferred from observations of water clarity and water color. Simple instruments that provide information about these parameters, the Secchi disc and the Forel-Ule scale, have been introduced more than 100 years ago. Originally, they were used by sailors for navigation purposes. Nowadays, in combination with modern instruments like submersible and above-water radiometers, their application provides a link between historical observations and modern light field measurements. The project Coastal Ocean Darkening aims to examine the changes of the light climate in coastal seas to get a better understanding of this highly sensitive and utilized ecosystem supporting its protection and sustainable use. The analysis of a large historic dataset of Secchi disc and Forel-Ule scale measurements of the past century is one core topic. Trends of darkening coastal seas are examined in detail with a focus on the North Sea. By making use of hydrographic parameters like temperature, salinity, and depth, a natural classification of the North Sea area into regions will be established. Clarity and color changes are analyzed with respect to these regions in order to investigate spatial variability in the large-scale trend of decreasing light availability. The results of this analysis will be used together with data of current measurements and lab experiments for modeling future scenarios within the Coastal Ocean Darkening project.

3.3.2 Land-Ocean Interactions in Arctic Coastal Waters – Ocean Color Remote Sensing in Optical Complex Waters

Bennet Juhls[1*], Birgit Heim[2], P. Paul Overduin[2], Jürgen Fischer[1]

[1]Institute of Space Sciences, Freie Universität Berlin, Carl-Heinrich-Becker-Weg 6-10, D-12165 Berlin, Germany

[2]Alfred Wegener Institute, Telegrafenberg A43, D-14473 Potsdam, Germany

*corresponding author: bjuhls@wew.fu-berlin.de

Keywords: Arctic, Ocean Color, Laptev Sea, Biogeochemical, OLCI

Arctic coastal waters and shelf regions such as the Laptev Sea are rapidly changing environments influenced by climate warming and consequent permafrost thaw and sea ice reduction, causing a longer open-water period. Biogeochemical interactions between land and ocean, e.g., river discharge and coastal erosion, control the ecosystem of these regions. During the short ice-free season, extremely dynamic and chaotic processes such as river and sea-ice break-up strongly influence the aquatic ecosystem for the rest of the year. In this season, large amounts of suspended sediments and dissolved organic matter discharged into the Arctic Ocean by the Lena River and along the coast control the transparency of the water. Therefore, it is highly important to obtain a synoptic view to complement spatially or temporally limited *in situ* data. Ocean Color Remote Sensing (OCRS) is a tool which provides high temporal resolution and synoptic information for biogeochemical water constituents in surface waters. Times series for up to 17 years can be used to detect interannual and long-term trends. The use of several satellites is increasing temporal data coverage. Besides the OLCI (Ocean and Land Colour Instrument) sensor on-board the

Sentinel 3 satellite, other satellite sensors as Landsat 7 ETM+, Landsat 8 OLI and Sentinel 2 MSI with higher spatial resolutions of up to 10 m are used. High spatial resolution helps resolve small-scale processes in coastal waters even though the spectral resolutions of those satellites are not as high as OLCI's. For the evaluation of Ocean Color satellite products, *in situ* biogeochemical samples and optical measurements of the water and its surface are needed. However, remote regions such as the Laptev Sea shelf are strongly undersampled. This study presents unique biogeochemical and hyperspectral radiometric measurements of several expeditions over the last decade and first match-up results of *in situ* vs. satellite data.

4 Phytoplankton in a Changing Environment – Adaptation Mechanisms and Ecological Surveys

Jana K. Geuer[1] and Laura Käse[2]

[1]Alfred Wegener Institute (AWI), Helmholtz Centre for Polar and Marine Research, P.O. Box 120161, 27570 Bremerhaven, Germany

[2]Alfred Wegener Institute (AWI), Helmholtz Centre for Polar and Marine Research, Biologische Anstalt Helgoland, Postbox 180, 27483 Helgoland, Germany

4.1 Call for Abstracts

The enormous importance of oceanic primary production is not only limited to the marine environment but has an equally great impact on the global atmosphere. Our changing climate affects the composition of phytoplankton species composition and requires the organisms to adapt to this changing environment, influencing micronutrient bioavailability and other biogeochemical parameters. Studies to monitor phytoplankton species composition are as important as are molecular and genetic approaches on understanding adaptation mechanisms of phytoplankton in regard to changing oceanic conditions. Studies dealing with different approaches on monitoring changes in phytoplankton, their impact on the microbial loop and adapting mechanisms are welcome.

4.2 Abstracts of Oral Presentations

4.2.1 Under Surveillance – Monitoring Phytoplankton and Hunting Their Traces

Laura Käse[1*], Jana K. Geuer[2*], Bernd Krock[2], Katja Metfies[2], Boris Koch[2,3], Alexandra Kraberg[1]

[1]Biologische Anstalt Helgoland, Alfred-Wegener-Institute Helmholtz Centre for Polar and Marine Research, Kurpromenade 201, 27498 Helgoland, Germany

[2]Alfred-Wegener-Institute Helmholtz Centre for Polar and Marine Research, Am Handelshafen 12, 27570 Bremerhaven, Germany

[3]Hochschule Bremerhaven, An der Karlstadt 8, 27568 Bremerhaven, Germany

*corresponding authors: laura.kaese@awi.de, jana.geuer@awi.de

Keywords: Phytoplankton, Microscopy, Marine ligands, Domoic acid, Next-generation sequencing

Marine primary production is not only crucial for regulating the atmosphere. Phytoplankton is also an important base for oceanic food webs. Their species composition depends on seasonal conditions, predators and micronutrient availability. (1) An existing monitoring system, the so called "Helgoland Roads time-series", should be improved by incorporating different monitoring methods to include so far underreported pico- and nanoplankton. (2) As molecular tracer the bioactive ligand domoic acid should be quantified in seawater from the Atlantic Ocean. Phytoplankton samples were taken two times per week and cells were counted optically via microscopy. The same samples were used for next-generation sequencing. For domoic acid quantification, seawater was filtered via combusted glass fiber filters (0.7 µm) and subsequently concentrated via solid phase extraction. Molecular characterisation was done with Fourier-transform ion cyclotron resonance mass spectrometry. Quantification was performed using high-pressure liquid chromatography coupled to mass spectrometry. Microscopic Utermöhl counting resulted in high abundances of different small-sized flagellates and cryptophytes. Smaller organisms included in the picoplankton group could not be distinguished further using this method. Bigger organisms like the diatom *Pseudonitzschia* sp. are hardly distinguishable on species level using this method. Therefore, comparison of these counts and the regular Helgoland Roads counts with sequencing data is necessary. Additionally the sequencing data will give lots of information about the different species of cryptophytes that were not further distinguished via microscope. In comparison to that a flow cytometer is going to be used for automated counting and cell sorting. Domoic acid's molecular formula could be detected in most samples. Additionally, it could be quantified in surface water. Higher concentrations were measured in the northern hemisphere (up to 172 pmol L^{-1}) compared to the southern hemisphere (up to 15 pmol L^{-1}). This method could potentially be transferred to detect and quantify other relevant target ligands.

4.2.2 NMR-Spectroscopic Study of Dissolved Organic Matter During a Microalgal Spring Bloom

Christian Zurhelle[1*], Julian Mönnich[1], Jan Tebben[2] and Tilmann Harder[1,2]

[1]Faculty of Biology and Chemistry, University of Bremen, Leobener Straße UFT, D-28359 Bremen, Germany

[2]Alfred Wegener Institute, Helmholtz Centre for Polar and Marine Research, Am Handelshafen 12, D-27570 Bremerhaven, Germany

*corresponding author: zurhelle@uni-bremen.de

Keywords: DOM, Metabolite profiling, Spring bloom Helgoland

Remineralization of algae biomass by heterotrophic bacteria plays an important role in carbon cycling. Teeling and co-workers observed the secondary bloom of planktonic bacteria succeeding a phytoplankton bloom in the German Bight (Teeling et al. 2012, *Science* 336: 608–611). These bacterioplankta have specialized carbohydrate-active enzymes that allow them to decompose organic matter released by microalgae (Teeling et al. 2012) resulting in recurring bacterioplankton succession patterns largely governed by substrate-induced forcing (Teeling et al. 2016, *eLife* 5: e11888). However, qualitative and quantitative shifts of organic matter (i.e., bacterial substrate) during and after an algal bloom are still poorly understood therefore limiting our understanding of community succession and biodiversity patterns. Here, I will present a study of how the dissolved organic matter (DOM) composition changed over a phytoplankton bloom in spring 2016 at Helgoland in the German Bight. Utilizing nuclear magnetic resonance spectroscopy (NMR), my data shows clear shifts in DOM constituents as well as quantitative bursts of DOM depended on abundance shifts of microalgae. My findings are the first step to linking the chemical specificity of phytoplankton-derived DOM with the subsequent bloom of substrate-specialized bacteria.

4.3 Abstracts of Poster Presentations

4.3.1 Identification of Microalgal Cyclic Imines and Their Potential Ecological Roles

Joyce Nieva[1*], Boris Koch[1], Jan Tebben[1], Bernd Krock[1], Urban Tillman[1], Ulf Bickmeyer[1]

[1]Alfred Wegener Institute, Am Handelshafen 12, 27570 Bremerhaven, Germany

*corresponding author: joyce.nieva@awi.de

Keywords: Cyclic imines, Chemical elucidation, Ecological role, Dissolved organic matter

Cyclic imines constitute the growing group of toxins produced by dinoflagellates. To date, of the two (2) main groups of cyclic imines, only 11 spirolides and 4 gymnodimines are structurally elucidated. With a high number of cyclic imines that are to be identified, this study aims to determine the other congeners and elucidate its structure, function to the organism and its ecological impact. Cyclic imines from *Alexandrium ostenfeldii* strains isolated from the Netherlands, Norway and Argentina will be chemically elucidated through combined use of LC-MS/MS and 1D- and 2D NMR techniques. The potency and structure activity relationship of cyclic imines on acetylcholine receptors (AChR) will be determined through cell line assays. For its role as an allelophatic compound, trace metal binding studies and grazing experiments will be conducted to determine the role of cyclic imines as a ligand for micronutrient uptake and as a deterrent compound, respectively. Further, this study will investigate the presence of the cyclic imines in the dissolved organic matter (DOM) by analysis of DOM samples from different oceanographic expeditions: MSM65 (Arctic/Greenland), HE 492 (Svalbard) and HE 514 (British Channel, Celtic Sea, North Sea). Expected results of this study include the determination of the novel cyclic imines, mechanism of action and their possible ecological role.

5 Reading the Book of Life – Omics as a Universal Tool Across Disciplines

Jan D. Brüwer[1,2] and Hagen Buck-Wiese[2]

[1]Red Sea Research Center, Division of Biological and Environmental Science and Engineering (BESE), King Abdullah University of Science and Technology (KAUST), Thuwal 23955-6900, Saudi Arabia

[2]Faculty of Biology and Chemistry, University of Bremen, P.O. Box 330440, 28334 Bremen, Germany

5.1 Call for Abstracts

From the poles to the tropics and from shallow waters to the deep sea: The marine environment is the greatest and most diverse system on the planet. As diverse as the habitats are the disciplines of marine sciences. Understanding the universal molecular languages, has revolutionized the information available to almost all of them. These advances in omics generate ever more data, demanding data mining and allow specific research question. We invite researchers across disciplines to submit their research to introduce cutting-edge methodologies, reveal insights achievable thanks to omics, and inspire approaches across the board.

5.2 Abstracts of Oral Presentations

5.2.1 PhyloFlash – Metagenomics, Microbial Molecular Ecology and a New Age of Discovery

Harald Gruber-Vodicka[1*], Elmar Pruesse[2], and Brandon K. B. Seah[1]

[1]Max Planck Institute for Marine Microbiology, Bremen, Germany

[2]University of Colorado Denver, Colorado, U.S.A

*invited speaker, corresponding author: hgruber@mpi-bremen.de

The classification of microbial life from species to higher taxa is largely based on the small subunit ribosomal RNA (SSU rRNA) gene. Microbial molecular ecology employs the SSU rRNA gene for cultivation-independent studies of communities where it serves both as a phylogenetic marker in sequence based analysis, and at the organismal level as a target for molecular probes through fluorescence in situ hybridization (FISH). With the advent of high-throughput next-generation sequencing (NGS), shotgun metagenomes and transcriptomes of full communities that sample from all domains of life can replace PCR-based SSU rRNA amplicon libraries as the basis for the 'full cycle rRNA approach' in molecular ecology. We have developed phyloFlash – a software tool to rapidly assess the phylogenetic composition of metagenomic or transcriptomic libraries, without assuming extensive bioinformatics expertise from its users. It reconstructs long and exact (>1200 bp) SSU rRNA sequences that are suitable for high-resolution phylogenetics and probe design, from short-read NGS libraries. The software builds upon a curated version of the SILVA SSU rRNA database and generates a taxonomic community profile, taxonomically-annotated full-length SSU rRNA sequences for bacteria, archaea and eukaryotes, and interactive plots. Multiple libraries can also be compared on the basis of their taxonomic composition. We have employed phyloFlash-based community composition analysis on diverse marine samples ranging from single-celled protists, minute worms, giant deep-sea clams and stratified sediments. The phyloFlash approach quickly unveiled unique patterns of diversity and symbiotic associations, such as a novel phylum of intracellular bacteria in animals and the plasticity of chemosynthetic symbiont communities in a single species of small gutless worms. The combined improvements in library preparation methods for NGS 'omics' and in software tools such as phyloFlash open a new age of discovery for marine environments where most of the biodiversity has to be investigated with cultivation-independent methods.

5.2.2 Challenges of Amplifying Degraded Nuclear and Mitochondrial DNA from Historical Dugong Skulls

Morgan L. McCarthy[1,2*], Kieren J. Mitchell[3], Jennifer M. Seddon[2], Janet M. Lanyon[1]

[1]The University of Queensland, School of Biological Sciences, St. Lucia, Queensland 4072, Australia

[2]The University of Queensland, School of Veterinary Science, Gatton, Queensland 4343, Australia

[3]Australian Centre for Ancient DNA, School of Earth and Environmental Sciences, University of Adelaide, North Terrace Campus, South Australia 5005, Australia

*corresponding author: m.l.mccarthy@uq.net.au

Keywords: Dugong, Ancient DNA, Population genetics, *Dugong dugon*, Queensland

The use of ancient DNA (aDNA), including DNA extracted from museum samples has been an instrumental tool in the study of historical ecology. Sequencing nuclear DNA for microsatellites and mitochondrial DNA (mtDNA) provides information on genetic diversity and estimates of historical populations, giving our modern calculations perspective. The nature of using museum samples is that DNA degrades overtime and that extraction, sequencing and amplification is inhibited by sequence fragmentation. As the densest bone in the mammalian body, the periotic bone of the petrous portion of the temporal lobe is reported to yield the least degraded DNA. Dentine from human teeth and enamel from elephant tusks have both yielded aDNA as well. This study tests the amplification of microsatellite loci and mitochondrial control regions in DNA extracted from periotic, cheek teeth and permanent incisors (tusks) of dugongs to determine the best portion of the skull to use when extracting DNA from museum samples. Such data can be used to calculate a historical effective population size and compare it with contemporary estimation methods based on shark-netting bycatch hind casting models in Queensland, Australia. Hence, its results will lay the framework to investigate whether a decline in dugong populations from extreme storm events and a government sponsored shark-netting program from the 1960s is reflected as a loss in present day genetic diversity when compared with present day DNA samples.

5.2.3 On the Way for Detecting and Quantifying Elusive Species in the Sea: The *Octopus vulgaris* Case Study

Quentin Mauvisseau[1*], Marina Parrondo[1], María del Pino Fernández[2], Lucía García[2], Jose Luis Martínez[3], Eva García-Vázquez[1], Yaisel Juan Borrell[1]

[1]Department of Functional Biology, University of Oviedo, Calle Julián Clavería S/N, 33006, Spain

[2]Centro de Experimentación Pesquera, Dirección de Pesca Marítima, Gobierno Del Principado De Asturias, Gijon, Spain

[3]Sequencing Unit, Edificio Severo Ochoa, C/Julian Claveria S/N, University of Oviedo, Oviedo 33006, Spain

*corresponding author: Q.Mauvisseau@derby.ac.uk

Keywords: Environmental DNA, Detection, Fisheries, *Octopus vulgaris*, Quantification

Environmental DNA (eDNA) can be a powerful method for assessing the presence and the distribution of aquatic species. We used this tool in order to detect and quantify eDNA from the elusive species *Octopus vulgaris*, using qPCRs (SybrGreen protocol). We designed species-specific primers, and set up an experimental aquarium approach to validate the new molecular tool in different controlled conditions. Field validation was conducted from sea water samples taken from 8 locations within an octopus fishery area in the Cantabrian Sea during February–March 2016. A significant positive correlation between the total biomass (g of *O. vulgaris* within thanks) and the amount of *O. vulgaris* eDNA detected (p-value = 0.01261) was found in aquarium experiments. The species was also detected by PCR in 7 of the 8 water samples taken at sea, and successfully quantified by qPCR in 5 samples. This preliminary study and innovative method opens very promising perspectives for developing quick and cheap tools for the assessment of *O. vulgaris* distribution and abundance in the sea. The method could help in a close future for quantifying unseen and elusive marine species, thus contributing to establish sustainable fisheries.

5.2.4 Single-Cell Transcriptomics: Gotta Catch 'em all!

Sabrina N. Kalita[1*], Uwe John[1], Nancy Kuehne[1]

[1]Alfred Wegener Institute Helmholtz Centre for Polar and Marine Research, Am Handelshafen 12, 27570 Bremerhaven DE

*corresponding author: Sabrina.Kalita@awi.de

Keywords: Single-cell, Transcriptomics, Sequencing, Data analysis

Eukaryotic genomes are difficult to sequence and moreover to assemble and annotate as they contain large intergenic regions, introns and repetitive DNA chunks. This means, that even when the complete sequence of the genome is known, it is often difficult to spot particular genes in the assembly. One approach to conquer this problem is to examine all the messenger RNA molecules transcribed from the genome as they directly connect to a gene function. So called transcriptomic sequencing makes it possible to profile organisms without detours as their coding regions are straightforwardly accessible and hereby revealing gene expression dynamics. More importantly, transcriptomic approaches lead to reference databases consisting of coding regions for further modeling. Unfortunately, some of the cellular properties are masked due to bulk and population-averaged samples retrieved from pure cultures. But in recent years, low-input RNA-sequencing methods have been adapted to work with single cells. Thus, making it possible to study single cell states, quantify intrapopulation heterogeneity potentially revealing cell subtypes or gene expression dynamics that were previously masked in bulk measurements. Even more, the possibility to work with one single cell eliminates the need of cell cultivation and enables the opportunity to process a cell directly from the environment. Although now robust and economically practical, it is still a challenge to develop sensitive, precise, and reliable protocols that lead to whole transcriptome sequencing from single cells. Here, we will report on the process of establishing the whole work-flow regarding single-cell transcriptomics in the laboratory, leading to an analytical evaluation of methods and processes for single-cell sequencing. This will include a feasibility study whether the method is suitable for on board analysis on research vessels (RV HEINCKE; Spitsbergen, NO; August '17), as well as data analysis, like gene annotations after Illumina sequencing.

5.2.5 Defining Gene Cluster Families from Globally-Distributed Seawater Samples Using Community Detection Methods

Alaina Weinheimer[1*], Jorge C. Navarro-Muñoz[2], Frank Oliver Glöckner[1,3], Marnix Medema[2], Antonio Fernandez-Guerra[1,3,4]

[1]Microbial Genomics and Bioinformatics Research Group, Max Planck Institute for Marine Microbiology, Bremen, Germany

[2]Bioinformatics Group, Wageningen University, 6708PB Wageningen, Netherlands.

[3]Jacobs University Bremen gGmbH, Bremen, Germany

[4]Oxford e-Research Centre (OeRC), University of Oxford, Oxford, UK

*corresponding author: alainarw94@gmail.com

Keywords: Natural products, Network analyses, Seawater metagenomics

The pharmaceutical and agricultural industries often rely on natural products synthesized by bacteria, plants, and other organisms. Once isolated from nature, these compounds or enzymes can be mass produced through cultivation. Alternatively, some industries genetically engineer organisms for the synthesis of such products. However, the process and results of doing so are typically laborious and/or unpredictable. Thus, finding useful enzymes and metabolites in nature is still an effective means of natural product discovery. Though many ecosystems have been investigated extensively for these compounds and enzymes, the ocean remains widely unexplored. The biosynthetic potential of organisms was examined in seawater samples collected from oceans around the globe by the TARA Oceans expedition. Genomes were then assembled from the metagenomic seawater samples. These genomes were searched for the presence of biosynthetic gene clusters (BGCs) using the program antiSMASH, identifying 1384 BGCs within 93 of the metagenome-assembled genomes (MAGs). The aim of this study was to investigate the relatedness of these BGCs and define families of closely related BGCs in marine samples. Each BGC was compared to each other using the program BiG-SCAPE, which generates a similarity-based network. Within this network, communities of BGCs were identified by employing various community detection algorithms, such as HDBSCAN and Louvain. Based on several metrics, such as entropy, the most informative community detection algorithm was affinity propagation. This algorithm was then applied to individual networks of BGCs belonging to different classes of enzymes (i.e., terpene-producers) to further understand the relatedness and novelty of BGCs in the world's oceans.

5.2.6 Marine Sponge Symbionts: United in Defense But Specialized in Metabolism

Beate M. Slaby[1,2*], Thomas Hackl[3], Hannes Horn[1,2], Kristina Bayer[1], Ute Hentschel[1,4]

[1]RD3 Marine Microbiology, GEOMAR Helmholtz Centre for Ocean Research Kiel, Germany

[2]Dept. of Botany II, Julius-von-Sachs Institute for Biological Science, University of Würzburg, Germany

[3]Dept. of Civil and Environmental Engineering, Massachusetts Institute of Technology, Cambridge, MA, USA

[4]Christian-Albrechts University of Kiel, Germany

*corresponding author: bslaby@geomar.de

Keywords: Metagenomics, PacBio, Illumina HiSeq, Hybrid assembly, Differential coverage binning

Marine sponges are ancient metazoans that are populated by distinct and highly diverse microbial communities. In order to obtain deeper insights into the functional gene repertoire of the Mediterranean sponge *Aplysina aerophoba*, we combined Illumina short-read and PacBio long-read sequencing followed by un-targeted metagenomic binning (Slaby et al., ISME J in press). We identified a total of 37 high-quality bins from 11 bacterial phyla and 2 candidate phyla, which are representative of the sponge symbiont consortium known from previous studies. For comparison to closely related bacteria from non-sponge environments, we selected reference genomes based on genome completeness, phylogenetic similarity, and habitat (marine preferred over other habitats). Statistical group-wise comparison of symbiont and reference genomes by Welch's t-test based on clusters of orthologous groups (COGs) revealed a significant enrichment of genes related to bacterial defense (restriction-modification systems, toxin-antitoxin systems) as well as genes involved in host colonization and extracellular matrix utilization in sponge symbionts. A within-symbionts genome comparison by principle component analysis (PCA) revealed a nutritional specialization of at least two symbiont guilds, where one appears to metabolize carnitine and the other sulfated polysaccharides, both of which are abundant molecules in the sponge extracellular matrix. A third guild of symbionts may be viewed as nutritional generalists that perform largely the same metabolic pathways but lack such extraordinary numbers of the relevant genes. This study characterizes the genomic repertoire of sponge symbionts at an unprecedented resolution and it provides greater insights into the molecular mechanisms underlying microbial-sponge symbiosis.

5.2.7 Investigating Transcriptome and Proteome Heat Stress Response of the Cnidarian Model Organism *Exaiptasia pallida*

Maha J. Cziesielski[1*], Yi Jin Liew[1], Sebastian Schmidt-Roach[1], Guoxin Cui[1], Sara Campana[1], Claudius Marondedze[2], Manuel Aranda[1*]

[1]King Abdullah University of Science and Technology (KAUST), Red Sea Research Center (RSRC), Biological and Environmental Sciences & Engineering Division (BESE), Thuwal, Saudi Arabia

[2]Cambridge Center for Proteomics, Department of Biochemistry, University of Cambridge, Cambridge, United Kingdom

*corresponding authors: maha.olschowsky@kaust.edu.sa, manuel.aranda@kaust.edu.sa

Keywords: Transcriptomics, Proteomics, Heat stress, *Exaiptasia*, Biomarkers

Corals, and their endosymbiotic dinoflagellates of the genus *Symbiodinium*, are key building blocks of the coral reef ecosystem. This symbiotic relationship is fragile and breaks down under heat stress, which leads to bleaching of the corals. Transcriptomic approaches to investigate potential mechanisms of acclimatization and adaptation have become increasingly popular with growing application of next generation sequencing (NGS) technology, in particular RNA- Seq. While significant information regarding coral and larvae stress response has been obtained on a transcriptomic level, proteomics has remained mostly unaddressed. Proteins ultimately dictate fitness, but studies on other model organisms (i.e., mice and humans) have continuously reported low correlations between mRNA and protein. Using the small anemone *Exaiptasia pallida,* we investigated transcriptome- and proteome-wide heat stress responses in a cnidarian. Anemones from North Carolina (CC7), Hawaii (H2) and the Red Sea (RS) were heat stressed for 24 h. Comparison across genotype transcriptomes showed a number of shared pathways previously suggested to represent a core cnidarian thermal stress response, however little commonality was observed on a proteomic basis. We report consistently low correlation between mRNA and protein, which reduced further when focusing on fold changes. In order to assess the heat stress response capacity of the genotypes, we tested previously suggested biomarkers. Using the multi-omics data obtained, we further investigated the stability of these biomarkers across layers and strains. We managed to validate 12 biomarkers and suggest new ones to be considered based on their transcriptomic consistency. We suggest important parameters to be kept under consideration during biomarker development and inter-strain response comparisons. Overall, our study highlights core cnidarian heat stress mechanisms and the importance of inter-disciplinary omics approaches.

5.2.8 Let's Go Viral on the Model Metaorganism Aiptasia

Jan D. Brüwer[1*], Christian C. Voolstra[1]

[1]Red Sea Research Center, Division of Biological and Environmental Science and Engineering (BESE), King Abdullah University of Science and Technology (KAUST), 23955 Thuwal, Saudi Arabia

*corresponding author: bruewer_j@gmx.de

Keywords: Coral reefs, Model system, Virus, Metaorganism, Symbiosis

Anthozoans, including stony corals and sea anemones, are associated with a variety of bacteria, archaea, and viruses and, thus, form metaorganisms or holobionts. They receive special attention, due to their endosymbiosis with zooxanthellae algae of the genus *Symbiodinium* that, in the case of corals, provide the foundation for the ecologically and economically important reef ecosystems. While the bacterial community is the main focus, comparatively few studies have investigated the viral component of metaorganisms. In order to study the viral community and potential ecological functions of viruses in the cnidarian-algae symbiosis, we re-analyzed a previously published RNA-Seq dataset of the model metaorganism Aiptasia (*sensu Exaiptasia pallida*) featuring three different symbiotic states with *Symbiodinium* (aposymbiotic, partially populated, and fully symbiotic). Our bioinformatic approach included the removal of host and endosymbiont sequences (Aiptasia and *Symbiodinium*) prior to viral sequence characterization. Aiptasia seem to harbor a diverse and relatively complex viral community, dominated by viruses of the families *Herpesviridae*, *Partitiviridae*, and *Picornaviridae.* Some distinct members of the viral community change significantly in relative abundance across altered symbiotic states of Aiptasia with *Symbiodinium*. Additionally, we assembled the Aiptasia core virome comprised of viruses from 11 viral families that were present across all specimens. Our study provides a first insight into the viral community of the cnidarian model system Aiptasia and a bioinformatics pipeline to tease out viral signatures from existing organismal RNA-Seq data. Aiptasia forms a dynamic assemblage with a variety of viruses in which presence and absence of *Symbiodinium* aligns with viral community differences.

5.3 Abstracts of Poster Presentations

5.3.1 Alternative Methods for Assessing Habitat Quality in River Systems

Quentin Mauvisseau[1*], Andrew Ramsey[1], Alix Blockley[1], Jim Campbell[2], Rein Brys[3], Michael Sweet[1]

[1]Aquatic Research Facility, Environmental Sustainability Research Centre, College of Life and Natural Sciences, University of Derby, Derby, UK

[2]SureScreen Scientifics Ltd, Morley Retreat Church Lane, Morley, Derbyshire, DE7 6DE, UK

[3]Research Institute for Forest and Nature, B-1070 Brussels, Belgium

*corresponding author: Q.Mauvisseau@derby.ac.uk

Keywords: Environmental DNA (eDNA), Detection, Freshwater ecosystems, Invasive species, Endangered species

Environmental DNA (or eDNA) refers to the traces of DNA (originating from skin, gametes or mucus, for example) which are left by any given organism or group of organisms in any given ecosystem. Using eDNA, it is possible to therefore assess the presence or absence of either a specific species (a more targeted approach) or the whole community (a metagenomic approach). Recent studies have also suggested that quantification of biomass can also be retrieved from eDNA data. Monitoring biodiversity is a cornerstone for the evaluation of ecosystem health. In freshwater ecosystems, the assessment of water quality relies heavily on biological monitoring and the detection of specific key indicator species, endangered species or those which may be harmful to the ecosystem or invasive in nature. The aim of this PhD is therefore to develop and assess various methods focused around the eDNA concept and explore and improve various aspects associated with this non-invasive methodology in order to assess habitat quality of freshwater systems. Throughout the various experiments associated with the PhD, new eDNA methods will be mapped against more traditional survey methods which are more routinely utilized to date; electrofishing and hand searching, for example. Each aspect of the study involves a variety of different 'project partners' who will insure the new techniques developed during the PhD move from principle into practice and start to influence policy through the UK and the rest of mainland Europe. Furthermore, working with our commercial project partners ensures each technique can be developed into a fully validated and commercially available product available to a wider end user group.

6 Sentinels of the Sea: Ecology and Conservation of Marine Top Predators

Dominik A. Nachtsheim[1,2] and Brigitte C. Heylen[3,4]

[1]Institute for Terrestrial and Aquatic Wildlife Research, University of Veterinary Medicine Hannover, Werftstrasse 6, 25761 Büsum, Germany

[2]BreMarE – Bremen Marine Ecology, Marine Zoology, University of Bremen, P.O. Box 330440, 28334 Bremen, Germany

[3]Behavioural Ecology and Ecophysiology, University of Antwerp, Universiteitsplein 1, 2610 Antwerp, Belgium

[4]Terrestrial Ecology Unit, Ghent University, K.L. Ledeganckstraat 35, 9000 Ghent, Belgium

6.1 Call for Abstracts

Marine top predators, such as cetaceans, seals, seabirds and elasmobranchs, represent an essential part of marine ecosystems. They are generally regarded as sentinels of the sea since their presence can, for instance, reflect high biological productivity and stabilize marine food webs. Besides their ecological role, these top predators are also socio-economically important. Many populations, however, experience dramatic declines attributed to various human induced threats (e.g., pollution, climate change), which highlight the need for effective conservation. This session invites contributions covering a broad range from fundamental ecological studies over modeling approaches to conservation issues.

6.2 Abstracts of Oral Presentations

6.2.1 Possible Impacts of Human Pressures on Seabirds: Insights from Recent Studies in the North and Baltic Seas

Stefan Garthe[1*]

[1]Research and Technology Centre (FTZ), Kiel University, Hafentörn 1, 25761 Büsum, Germany

*invited speaker, corresponding author: garthe@ftz-west.uni-kiel.de

Keywords: Human Pressure, North Sea, Baltic Sea, Seabird, Behavior

Humans have made use of sea products since hundreds of years. But only since the middle of the last century these activities have reached a dimension that impacts natural dynamics in northwestern European seas, mainly through the industrialization of fisheries and the exploitation of gas and oil resources. Within the last decade, humans developed many more uses of the seas, with offshore wind farms possibly being the most obvious one. Marine animals are thus facing increasing human pressures. In this talk I will first of all highlight how effects on seabirds may be investigated. Methods range from well-established seabirds at sea counts to latest developments in microelectronics. In such cases, tags are attached to animals that collect geographic position and often measure various other parameters from which animal behavior can be deduced. Examples for effects of human activities include offshore wind farms, fisheries, plastic pollution and habitat alteration. Case studies are presented for the offshore areas and coastal zones of the North and Baltic Seas which range among the most intensively used sea areas world-wide.

6.2.2 Drones as a New Tool for Antarctic Wildlife Monitoring

Marie-Charlott Rümmler[1*], Jan Esefeld[1], Osama Mustafa[2], Hans-Ulrich Peter[2]

[1]Friedrich Schiller University Jena, Institute of Ecology, Dornburger Str. 159, 07743 Jena, Germany

[2]ThINK – Thüringer Institut für Nachhaltigkeit und Klimaschutz, Leutragraben 1, 07743 Jena, Germany

*corresponding author: marie-charlott.ruemmler@uni-jena.de

Keywords: UAV, Disturbance, Penguins, Seals, Birds

Drones or UAVs (unmanned aerial vehicles) are a relatively new and rapidly expanding tool for environmental research, particularly in the Antarctic, where huge datasets of population censuses are often hard to accomplish due to the harsh Antarctic conditions. Colonies of breeding birds or congregations of terrestrial mammals can be difficult to estimate by traditional methods often involving a high level of disturbance. Drones promise to provide a solution here for various reasons, but the question of their impact on different species still remains largely unanswered. During two Antarctic summer seasons, we accomplished a series of disturbance experiments on various species to gather information on how they react to fly-overs by drones. Analyses of the first experiments on Adélie and gentoo penguins (*Pygoscelis adeliae/P. papua*) revealed a noticeable change in behavior with approaching UAVs: at about 40 m, gentoo penguins start to show first reactions to the drone. For Adélie penguins this distance is further than 50 m. For both species, a strong increase in impact seems to be caused by a drone flying at distances of about 20 m, where the disturbance has to be considered severe. We also conducted experiments on the distance of the take-off of our drone and found it to be sufficiently distant at 30 m for gentoo penguins, but farther for Adélie penguins. In the second season we extended the experiments to physiological measurements to get a better comparison between different disturbance situations by using heart-beat-measurement devices. Additionally, we added more species to the experiments, namely skuas (*Catharacta maccormicki/C. antarctica lonnbergii*), kelp gulls (*Larus dominicanus*), southern giant petrels (*Macronectes giganteus*) and Antarctic fur seals (*Arctocephalus gazella*). An overview over the results found in both seasons and the implications of these findings for monitoring and conservation of Antarctic wildlife will be presented.

6.2.3 Under-Shelf Ice Foraging of Weddell Seals

Dominik A. Nachtsheim[1,2,3*], Christoph Held[1], Nils Owsianowski[1], Joachim Plötz[1], Richard Steinmetz[1], Yasuhiko Naito[4], Horst Bornemann[1]

[1]Alfred-Wegener-Institut Helmholtz-Zentrum für Polar- und Meeresforschung, Am Handelshafen 12, 27570 Bremerhaven, Germany

[2]BreMarE – Bremen Marine Ecology, Marine Zoology, University of Bremen, P.O. Box 330440, 28334 Bremen, Germany

[3]present Address: Institute for Terrestrial and Aquatic Wildlife Research (ITAW), University of Veterinary Medicine Hannover, Foundation, Werftstr. 6, 25761 Büsum, Germany

[4]National Institute of Polar Research, 10-3, Midori-cho, Tachikawa-shi, Tokyo 190-8518, Japan

*corresponding author: Dominik.Nachtsheim@tiho-hannover.de

Keywords: Weddell seal, *Leptonychotes weddellii*, Shelf ice, Foraging ecology, Bio-logging

The Weddell seal (*Leptonychotes weddellii*) inhabits the Antarctic coastal ecosystem and aggregates in areas characterized by a stable fast ice layer. Due to their extreme diving capabilities, they are able to exploit both pelagic and benthic prey resources. They mainly feed on fishes but occasionally also take cephalopods and crustaceans. Weddell seals instrumented with still-picture camera loggers in the Drescher Inlet, eastern Weddell Sea, detected an unknown cryo-benthic community underneath the floating ice shelf. Images show dense aggregations of invertebrates that likely represent an attractive food horizon for Weddell seals. In this context, we conducted a retrospective analysis of dive profiles collected in the Drescher Inlet to identify favored hunting depths of Weddell seals and correlate those to the local physical and biological environment. A total of 34 adult Weddell seals were instrumented with dive loggers in the course of six summer field campaigns between 1990 and 2016. An automated broken stick algorithm was used to separate each dive profile into different segments. Segments with a high sinuosity were considered to indicate hunting. Segments characterized by a straight dive trajectories (low sinuosity), were assumed to be transit phases with no hunting activities. A tri-modal distribution of mean hunting depths suggests that Weddell seals concentrated their foraging activities in three depth strata. A peak in hunting depths below 370 m corresponds to the sea floor of the Drescher Inlet, indicating demersal foraging. A second peak between 110 and 160 m matches with the depth of the underside of the floating ice shelf, which suggests shelf ice associated foraging. The third peak probably represents hunting in the pelagic realm. Our investigation highlights the importance of the shelf ice underside as an attractive food horizon for Weddell seals suggesting a re-evaluation of trophic interactions and bentho-pelagic processes in the coastal Antarctic ecosystem.

6.2.4 Diet of Red-Throated Divers Within an Important Wintering Site, the German Bight, Analyzed with Molecular Tools

Birgit Kleinschmidt[1,2]*, Monika Dorsch[2], Stefan Heinänen[3], Julius Morkūnas[4], Ramūnas Žydelis[3], Claudia Burger[2], Georg Nehls[2], Rosemary Moorhouse Gann[5], William O. C. Symondson[5], Petra Quillfeldt[1]

[1]Department of Animal Ecology and Systematics, Justus Liebig University Giessen, Heinrich Buff Ring 38, 35392 Giessen, Germany

[2]BioConsult SH, Schobüller Str. 36, Husum 25813, Germany

[3]DHI, Agern Allé 5, Hørsholm 2970, Denmark

[4]Klaipeda University, Vilhelmo Berbomo g.10, 92221 Lithuania

[5]Cardiff School of Biosciences, Cardiff University, Sir Martin Evans Building, Museum Ave, Cardiff CF10 3AX, UK

*corresponding author: Birgit.Kleinschmidt@bio.uni-giessen.de; b.kleinschmidt@bioconsult-sh.de

Keywords: Red-throated diver, Diet, North Sea, Piscivorous, Next generation sequencing

The red-throated diver (*Gavia stellata*) is a strictly protected piscivorous migratory species, breeding primarily in the arctic regions and wintering in temperate coastal waters of the northern hemisphere. With high diver abundances in late winter and spring, the North Sea, especially the German Bight, represents one of the major resting sites for West Palearctic birds within the wintering season. The North Sea also represents a changing habitat with an increasing offshore windfarm development, and as red-throated divers are known to be sensitive with anthropogenic disturbance, this development can lead to conflicts. For habitat choice, resource availability, such as prey accessibility, might be essential, but the overall knowledge about diet choice of red-throated divers at the North Sea is rare. However, for conservation, a better understanding of dietary composition of red-throated divers within this important wintering and staging habitat is crucial to understand habitat choice and distribution of red-throated divers. We used standard molecular tools and Next generation sequencing to analyze fecal samples of red throated divers captured within the German Bight in 2015 and 2016 in winter and spring to obtain information of diet composition. Comparing the results of 28 samples we found a similar pattern between years. At all we found a wide ranged spectrum with 16 prey species of 11 families including flatfish, sand eels, and gobies but mackerel, gadoids, and clupeids as most abundant prey species. The results for the North Sea about dietary composition contribute existing knowledge that red-throated divers are in general opportunistic feeders with a preference for high energetic key species, such as for this area Atlantic herring and Atlantic sprat, and therefore indicate the importance of the occurrence of these prey species within utilized areas of red-throated divers, and extent the results known so far.

6.2.5 Protecting African Penguin Through Adaptive and Dynamic Management

B. C. Heylen[1,2*], L. Pichegru[3]

[1]EMBC+ International Master in Marine Biodiversity & Conservation, Ghent University, Belgium

[2]present Address: Avian Ecology and Evolution Group, Terrestrial Ecology Unit (TEREC), Ghent University, K.L. Ledeganckstraat 35, 9000 Ghent, Belgium

[3]Coastal and Marine Research Institute, Nelson Mandela Metropolitan University, Port Elizabeth, South Africa

*corresponding author: brigitte.heylen@ugent.be

Keywords: *Spheniscus demersus*, GPS tracking, Marine protected areas, Foraging behavior

Human disturbances are altering the functioning and provisioning of services in every ocean, and as such, affect marine wildlife profoundly. One of the problems is resource competition between marine predators and fisheries. Marine protected areas or MPAs can be a powerful tool for attenuating this threat, especially when designed with climate change in mind. However, additional management strategies are required. A case study of African penguin conservation in South Africa was used to illustrate the benefits of embracing adaptive and dynamic management in the marine environment. For this, future conservation plans were designed by using the results of a 7-year experiment, in which purse-seine fisheries were closed around penguin colonies in Algoa Bay. Those results indicated that the newly proposed MPA would provide a legal improvement to the current situation, but is not sufficient to increase numbers of African penguin populations. Larger no-take zones are necessary when prey availability is low. As an immediate measure, ongoing acoustic surveys could provide recommendations on prey availability to design flexible boundaries and increase the benefits for all stakeholders involved. In general, more experience and knowledge will always be necessary to obtain the ideal management plan for a specific area, but learning from the outcome is one of the aspects of adaptive management, and as such, it is never too early to implement it.

6.3 Abstracts of Poster Presentations

6.3.1 Hydrography-Driven Distribution of Seabirds and Cetaceans in the Temperate and Tropical Atlantic Ocean

Simon Jungblut[1,2,3*] Dominik A. Nachtsheim[1,2,3,4], Karin Boos[2,5], Claude R. Joiris[1,6]

[1]Laboratory for Polar Ecology (PolE), 1367 Ramillies, Belgium

[2]Bremen Marine Ecology (BreMarE), Marine Zoology, University of Bremen, Postbox 330440, 28334 Bremen, Germany

[3]Alfred Wegener Institute, Helmholtz Centre for Polar and Marine Research, Postbox 120161, 27570 Bremerhaven, Germany

[4]present address: Institute for Terrestrial and Aquatic Wildlife Research (ITAW), University of Veterinary Medicine Hannover, Foundation, Werftstr. 6, 25761 Büsum, Germany

[5]MARUM – Center for Marine Environmental Sciences, University of Bremen, Leobener Strasse, 28359 Bremen, Germany

[6]Conservation Biology Unit, Royal Belgian Institute for Natural Sciences, 1000 Brussels, Belgium

*corresponding author: jungblut@uni-bremen.de

Keywords: Seabirds, Marine mammals, Distributional patterns, Biogeography, Water masses, Atlantic Ocean

Water masses influence the distribution, abundance and species assemblage of top predators – seabirds and cetaceans. In polar regions, for instance, boundaries between water masses often induce the accumulation of prey organisms, leading to high abundances of top predators. Information on the structuring effects of water masses in temperate and tropical oceans and particularly in the Atlantic are relatively limited. Here we (1) provide baseline distributional data that may function as basis for future comparisons, e.g., in the course of climate change, and (2) test whether water masses and boundaries between them affect distributional patterns of top predators in the temperate and tropical Atlantic Ocean. Between 2011 and 2014, four trans-equatorial expeditions of RV *Polarstern* were used to conduct continuous half-hour transect counts from the vessels bridge without width limitation at traveling speed and daylight conditions. Temperature and salinity were automatically recorded by the navigational system of *Polarstern*. The hydrographical parameters mostly gradually changed from one water mass to the next. Compared to polar regions, counts of seabirds and cetaceans were generally low during all expeditions. Statistical analysis of the eight most abundant seabird species and the pooled cetaceans revealed the numbers to be dependent on water masses and seasons. However, no distinct changes or aggregations were identified in correlation to the borders between water masses. Numbers of seabirds were correlated to water masses but not to borders between them. This may mainly be due to only gradual changes in water masses which for instance fail to accumulate prey organisms. Other mechanisms like distance to productive areas (upwelling), competition effects and prey-accumulating sub-surface predators may be similarly important as hydrography in shaping Atlantic top predator distribution. Continuous monitoring of the at-sea distribution of seabirds and cetaceans is essential to understand the effects of climate change on their geographical distribution.

7 How Do They Do It? – Understanding the Success of Marine Invasive Species

Jonas Geburzi[1,2] and Morgan L. McCarthy[3,4]

[1]Kiel University, Zoological Institute and Museum, Hegewischstr. 3, 24118 Kiel, Germany

[2]Alfred-Wegener-Institute, Helmholtz-Centre for Polar and Marine Research, Wadden Sea Station, Hafenstr. 43, 25992 List/Sylt, Germany

[3]The University of Queensland, School of Biological Sciences, St. Lucia, Qld 4072, Australia

[4]Vrije Universiteit Brussel (VUB), Marine Biology, Pleinlaan 2, 1050 Brussels, Belgium

7.1 Call for Abstracts

From the depths of the oceans to the shallow estuaries and wetlands of our coasts, organisms of the marine environment are teeming with unique adaptations to cope with a multitude of varying environmental conditions. With millions of years and a vast volume of water to call their home, they have become quite adept at developing specialized and unique techniques for survival and – given increasing human mediated transport – biological invasions. The theme for this session calls for abstracts on marine invasions and how their life history strategies and specialized adaptations have made certain organisms particularly successful at invading new habitats.

7.2 Abstracts of Oral Presentations

7.2.1 Tracking Down the Origin of a Mediterranean Alien: *Bursatella leachii* (de Blainville, 1817)

Enrico Bazzicalupo[1,2*], Fabio Crocetta[3], Ángel Valdés[1]

[1]Department of Biological Sciences, California State Polytechnic University, 3801 Temple Avenue, Pomona, California 91768, USA

[2]present Address: Department of Biology, University of Florence, via Romana 17, 50125, Florence, Italy

[3]Hellenic Centre for Marine Research, Institute of Marine Biological Resources and Inland Waters, GR-19013 Anavyssos, Greece

*invited speaker, corresponding author: enricobazzical@gmail.com

Keywords: Mollusca, Genetics, Mediterranean, *Bursatella leachii*

Colonization of the Mediterranean Sea by alien species is one of the major threats to local biodiversity and human economies. Tropical species, in particular, appear to become established and thrive in the Mediterranean waters favored by human impacts on the environment, such as climatic changes due to global warming and the opening of the Suez Canal. The mollusc *Bursatella leachii* (de Blainville 1817) (Heterobranchia: Anaspidea: Aplysiidae) is a pantropical sea slug that colonized the Mediterranean Sea since around half of the twentieth century. In this study, we investigate the origin of the Mediterranean population of *B. leachii* by examining the population structure of this species and assessing relatedness levels of different regional populations. Sequence data from the cytochrome oxidase I (COI) and the 16S rRNA mitochondrial genes, and the Histone H3 nuclear gene, were used to construct a phylogenetic tree using samples coming from across the globe. With the use of population genetics analyses conducted on the COI gene, a haplotype network was constructed and fixation indexes between the Caribbean, Mediterranean, and Indo-pacific populations were calculated. The phylogenetic analysis suggests that *B. leachii*, as currently conceived, is a species complex with one species inhabiting the Indo-pacific region and another with a pantropical distribution. All the Mediterranean samples belong to the pantropical species, and the population genetics analyses show that the Caribbean and Mediterranean populations are not differentiated, while the Indo-pacific population is significantly different from the other two. These results suggest that *B. leachii* arrived into the Mediterranean Sea from the Atlantic Ocean, but the way this migration occurred is up to interpretation and debate.

7.2.2 First Record of the Invasive Kelp *Undaria pinnatifida* in Germany

Jessica Schiller[1*], Dagmar Lackschewitz[2], Christian Buschbaum[2], Kai Bischof[1]

[1]Bremen Marine Ecology (BreMarE), Marine Botany, University of Bremen, Leobener Str. 3, 28359 Bremen, Germany

[2]Alfred-Wegener-Institute, Helmholtz-Centre for Polar and Marine Research, Wadden Sea Station Sylt, Hafenstr. 43, 25992 List, Germany

*corresponding author: j.schiller@uni-bremen.de

Keywords: Invasive species, Neobiota, Laminariales, Schleswig-Holstein Wadden Sea National Park, Pacific oyster reef

The kelp *Undaria pinnatifida* is native to the Pacific coasts of China, Japan, Korea and southeastern Russia and has been introduced to many shores worldwide. It was first brought to Europe in 1971 (French Mediterranean) and 1983 to Brittany, from where it spread to other northwest European coastlines. Since 2016, thalli of *U. pinnatifida* are regularly found washed ashore at the island of Sylt (south-eastern North Sea, German Bight), which constitutes the northernmost site worldwide so far. Since its first detection in 2016, large numbers of thalli were found, many of them more than 1 m in length and fertile. In June 2017, field investigations were performed to locate and characterize populations of *U. pinnatifida* near Sylt. Sporophytes were found attached to an oyster/mussel reef in the shallow subtidal zone. A total of 91 attached thalli were collected and the attachment substratum was identified. Total length and midrib width were measured, the maturity state was determined and habitat properties assessed. The majority of the thalli (91%) were attached to Pacific oysters *Magallana gigas* (former *Crassostrea gigas*), while 7.7% were growing on *Mytilus edulis*. The mean actual length was 21.5 cm (full mean length estimated to 30.6 cm, due to damaged tips) and the longest thallus was estimated to be about 70 cm. Most individuals showed distinctive sporophylls and 88.9 % of them released spores. In laboratory studies a new generation of sporophytes was successfully grown from spores, released from a sporophyte collected in October 2016. All these findings prove that a fertile and self-sustaining population has already established in the northern Wadden Sea area. We will conduct further monitoring and physiological studies as well as genetic analysis to predict its potential further spread. Additionally, investigations on interactions with native species will provide insights on the effects on the native community.

7.2.3 Plastic as a Vector for Marine Invasive Species in the Galapagos Archipelago, Ecuador

S. Doolittle-Llanos[1*]

[1]University of Groningen, Nijenborgh 7, 9747 AG Groningen, Netherlands

*corresponding author: s.doolittle.llanos@student.rug.nl, saradoolittlellanos@gmail.com

Keywords: Plastic, Marine invasive species, Vector, Marine litter, Galapagos

On top of the well-known hazardous effects plastic litter has for ecosystems all over the globe, it may also act as a habitat and vector for marine fouling organisms to new environments, where they may spread and become invasive. Fouling of plastic litter varies significantly with latitude, with higher percentages of colonized items found closer to the equator. This study attempts to determine how the Galapagos Archipelago, Ecuador (marine reserve and UNESCO World Heritage Site), is affected by this issue, since it is hit by the Panama, Humboldt, and Cromwell currents, and is vulnerable to being reached by plastic debris from the north, south, and west Pacific Ocean. During a beach cleanup trip in collaboration with the Charles Darwin Research Station, the Galapagos National Park, and Conservation International, more than 1000 items of plastic debris were collected at 7 sites across 4 islands, in which 108 species of five different taxa (bryozoa, barnacles, molluscs, polychaetes, and crabs) were found and are to be identified by colleagues at the Smithsonian Institution, USA. We provide the first description of observed fauna fouling on plastic debris in the Galapagos Archipelago, an analysis of species richness and total coverage of said fauna, and correlations of species or taxonomic groups with certain characteristics of the plastic. Additionally, the results show a difference in abundance of colonized plastic among sites, suggesting some may be naturally more exposed to invasions than others. We also provide what is to the best of our knowledge, the first record of two invasive species in other points of the world, the barnacle *Balanus trigonus*, and the bryozoa *Bugula neritina*, on plastic debris in the field, showing that plastic colonization is indeed possible and might need to be considered as a route in which these organisms are dispersing.

7.2.4 Alien Invasive Species Rafting on Floating Plastic Litter

Sabine Rech[1*], Yaisel J. Borrell[1], Eva Garcia-Vazquez[1]

[1]Department of Functional Biology, Calle Julián Clavería s/n, 33006 Oviedo, Spain

*corresponding author: rechsabine@uniovi.es

Keywords: Invasive species, Rafting biota, Marine plastic pollution

Floating marine debris carrying fouling communities is a transport vector for (possibly invasive) species. Since the massive production and consumption of plastic products in the last decades has led to ever increasing quantities of discarded plastics in the global marine environment, the availability of floating vectors for rafting biota has strongly augmented. In the past, floating litter has usually not been included in investigations of sources and pathways of alien invasive species (AIS), but increasing evidence suggests that it may be an important underlying factor of introductions, for example, via canals and of secondary spread from areas where AIS have been introduced by other means, like aquaculture areas or ports. Although AIS rafting on marine plastics is commonly reported, there is no broad understanding of the dimension and underlying processes of this phenomenon, such as high-risk source and sink areas and anthropogenic activities, or high-risk vector items and materials. Here we present the investigations carried out and results gained on the phenomenon during the last 1.5 years of a PhD thesis within the ongoing project AQUAINVAD-ED (Marie Curie 2014 ITN H2020 AQUAINVAD-ED; Grant agreement no 642197). We concentrate on aquaculture areas as sources of rafting AIS on plastic litter. Rafting animals and the fouled vectors were analyzed on beaches at three sites of bivalve aquaculture in Italy and Portugal, including the Venice lagoon, which is a hotspot of AIS. We found that the fouled vectors (mostly plastics) carried a range of invasive species. This shows that aquaculture areas are indeed high-risk areas for the dispersal of invasive species via floating plastics and should be included in future regulations and policies.

7.2.5 Modeling of Ballast Water Discharge in the Arctic Ocean

Ingrid Linck Rosenhaim[1*], Kathrin Riemann-Campe[1], Andreas Herber[1], Rüdiger Gerdes[1]

[1]Alfred-Wegener Institute, Bussestraße 24, 27570 Bremerhaven, Germany

*corresponding author: Ingrid.linck-rosenhaim@awi.de

Keywords: Ballast water, Ocean-sea ice model, Invasive species, Arctic Ocean, Shipping

Sea-ice concentration in the Arctic Ocean decreased 9% per decade since 1978. With the decline of the sea-ice in the

Arctic Ocean, more vessels are navigating through, enlarging the risk of alien species to enter the Arctic Ocean through ballast water. Vessels depend on it for stability and maintaining structural integrity. However, it contains unintended aquatic species that are carried in the tanks. It is possible that the transferred species survive to establish a reproductive population in the host environment. I implemented a ballast water tracer (BW-tracer) into a high-resolution version of NAOSIM (North Atlantic/Arctic Ocean-Sea Ice Model). The focus of my work is to identify the flow of the BW discharged in the Arctic and areas of accumulation with potential for invasive species to survive. In 2013, 17.407 vessels navigated the Arctic waters, of those, 202 vessels were of Destination Traffic making 731 port calls in 37 different Arctic ports. The estimated amount of discharged BW by these vessels is about 13 million cubic meters. In the model, the BW-tracer is released in those areas where we know ships have been in 2013. My first results show that the seasonal cycle of the ocean mixing affects the BW-tracer distribution. In winter, the tracer reach depths of around 35 m deep in the Barents Sea, where the maximum depth is about 80 m, and 750 m deep near the west coast of Spitsbergen where the maximum depth is about 3000 m. In spring, the sea ice starts to melt creating stratification in the upper level of the ocean, and the BW-tracer remains within the first 15 m deep in both regions. Moreover, strong surface velocities spread the BW-tracer during spring and summer, and the availability of sun and nutrients lead to more favorable living conditions for non-indigenous species.

8 Tropical Aquatic Ecosystems Across Time, Space and Disciplines

Mona Andskog[1,2], Hannah Earp[1,2,3], Natalie Prinz[1,2], and Maha Joana Cziesielski Olschowsky[4]

[1]Faculty of Biology and Chemistry, University of Bremen, P.O. Box 330440, 28334 Bremen, Germany

[2]Leibniz Centre for Tropical Marine Research (ZMT), Fahrenheitstraße 6, 28359 Bremen, Germany

[3]current address: School of Ocean Sciences, Bangor University, Menai Bridge, LL59 5AB, Wales

[4]Red Sea Research Centre, King Abdullah University of Science and Technology, Thuwal 23955-6900, Kingdom of Saudi Arabia

8.1 Call for Abstracts

Coral reefs, mangroves and seagrasses are among the most diverse, productive and complex ecosystems on the planet. They are also among the most vulnerable and are declining at unprecedented rates. The high complexity of these environments means they are difficult to study from a single perspective or on single scales, thus making it of critical importance to study them across time, space, and scientific disciplines. Only then will we fully understand their functioning and how to successfully manage and preserve them for future generations. This session will explore the lessons being learned through current interdisciplinary and comparative tropical marine research.

8.2 Abstracts of Oral Presentations

8.2.1 Coral Reef Ecosystems in Times of Global Change

Amanda K Ford[1*]

[1]Leibniz Center for Tropical Marine Research (ZMT), Fahrenheitstrasse 6, Bremen 28359, Germany

*invited speaker, corresponding author: amanda.ford@leibniz-zmt.de

Keywords: Climate change, Local stressors, Coral reef functioning, Resilience, Social-ecological systems

Coral reefs represent one of the most diverse and productive ecosystems on earth. Despite having thrived for millennia, a multitude of local and climate change-related stressors are threatening the continued prevalence of this ecosystem within the current era of the Anthropocene. Rapid and strong human-driven changes in climate, terrestrial and marine systems are increasingly facilitating regime shifts from coral-dominated systems to those dominated by alternative organisms such as fleshy algae or sponges. These regime shifts are often associated with a loss of critical ecosystem services provided by coral reefs such as coastal protection and fisheries. To maximize the future prevalence of coral reefs, efforts towards reducing local stressors (e.g., fisheries, sewage pollution) should both reduce the likelihood of regime shifts and facilitate maximum ecosystem resilience (e.g., ability to resist and recover) to climate change-related disturbances. While the effects of climate change on coral reef ecosystems are relatively well understood, the role of various local human impacts on system resilience remains under debate. My work investigates how various levels and types of local impacts can influence coral reef communities and ecosystem functioning, with a primary focus on the Pacific Island region. By understanding the impacts of different local stressors, I investigate how they can affect future responses to increasing climate change-related stressors. This has involved an evaluation of the monitoring metrics that have the capacity to determine processes determining resilience. Where regime shifts have occurred, I identify self-reinforcing positive feedbacks that promote reef degradation, with a primary focus of my work being on shifts towards benthic cyanobacterial mats. A critical component of coral reef futures is the changing behaviors and perceptions of coastal human societies, who in turn are directly affected by reef condition. Thus, formulating effective management practices requires an understanding of feedbacks within social-ecological systems.

8.2.2 Spatial Distribution of Surface Water Quality Parameters in Menjangan Island

Novia Arinda Pradisty[1,2*], Eghbert Elvan Ampou[1], Rizki Hanintyo[1,3], Mardatilah[1], I. Nyoman Surana[1]

[1]Institute for Marine Research and Observation, Ministry of Marine Affairs and Fisheries – Republic of Indonesia, Jalan Baru Perancak, 82251 Bali, Indonesia

[2]Faculty of Biology and Chemistry (FB2), University of Bremen, Bibliothekstrasse 1, 28359 Bremen, Germany

[3]Faculty of Geo-Information Science and Earth Observation (ITC), Department of Water Resources, University of Twente, 7500 AE Enschede, the Netherlands

*corresponding author: novia.arinda@kkp.go.id

Keywords: Physicochemical parameters, Phytoplanktons, Anthropogenic disturbances, Bali, Indonesia

Tropical aquatic ecosystems are currently facing severe threats from climate change and anthropogenic disturbances. Menjangan Island is located inside West Bali National Park, Indonesia, which is home for various species of mammals, birds, seagrasses, mangroves, corals and fishes. The island is also recognized internationally for its marine ecotourism, especially the diving and snorkeling sites. Currently, the island is polluted by increasing nutrient loads and marine litters, which may lead to eutrophication and loss of biodiversity. From this observation, the aim of our study is to examine the spatial distribution of surface water quality parameters and to evaluate water quality status in the coastal area of Menjangan Island. In this study, various surface water quality parameters were collected in February 2017 at 10 sampling locations around Menjangan and northwestern Bali Island. The parameters examined in this study are temperature, dissolved oxygen, pH, salinity, turbidity, total alkalinity, total dissolved solids, total suspended solids, dissolved inorganic nutrients (nitrate, nitrite, ammonia, silicate, orthophosphate), chlorophyll *a* and phytoplankton. The results show that the value of turbidity, total suspended solids and nitrate on several sampling sites are beyond local quality standards for marine biota, which shows the evidence of high nutrient concentrations and marine litters in the coastal water. Phytoplankton of the island are generally classified as diatoms and dinoflagellates. *Chaetoceros* sp., which is known as the inducer of fish mortality also exist, although not in blooming condition. Hopefully these results will encourage further initiatives to protect both the aquatic ecosystem and ecotourism development of Menjangan Island.

8.2.3 Holobiont at the Vet – Pro- and Eukaryotic Components in an *Acropora* White Syndrome Case

Jan D. Brüwer[1,2], Markus T. Lasut[1], Davide Poli[3], Ajinkya Kulkarni[4], Indri Manembu[5], Elvy Like Ginting[5], Stenly Wullur[5], Wilmy Pelle[5], Marco Serge Reinach[6], Karl-Heinz Blotevogel[4], Michael W. Friedrich[4], Sebastian Ferse[7], Hagen Buck-Wiese[1,2*]

[1]Aquatic Science Study Program, Faculty of Fisheries and Marine Sciences, Sam Ratulangi University, 95115 Manado, Indonesia

[2]Faculty of Biology and Chemistry, University of Bremen, 28359 Bremen, Germany

[3]School of Biological Science, University of Queensland, 4072 Brisbane, Australia

[4]Microbial Ecophysiology, Faculty of Biology and Chemistry, University of Bremen, 28359 Bremen, Germany

[5]Marine Science Study Program, Faculty of Fisheries and Marine Sciences, Sam Ratulangi University, 95115 Manado, Indonesia

[6]Coral Eye Bangka, 95375 Lihunu, Bangka Island, North Sulawesi, Indonesia

[7]Centre for Marine Tropical Sciences, Fahrenheitsstrasse 6, 28359 Bremen, Germany

*corresponding author: h.buckwiese@gmail.com

Keywords: Coral disease, Vibrio, Ciliates, Pseudoalteromonas, Coral Triangle

The striking diversity and plasticity of microorganisms is an important part of the coral holobiont. To elucidate its ecology in the phase of vast anthropogenic impact, observations of stress induced microbial community shifts within the holobiont need to be complemented with an in-depth understanding of their respective functional roles. Here, we studied White Syndrome pathogenesis and cessation in *Acropora indonesia* using *in situ* observations. In addition, we conducted a comprehensive set of experiments to elucidate the disease etiology, including transmission trials and predatory protist infection experiments. Furthermore, we cultivated several microbes from diseased coral tissue and inoculated healthy coral fragments with candidate strains. Among them, we identified *Vibrio* spp., which have previously been described as pathogens in other corals and observed disease symptoms in *A. indonesia* subsequent to inoculation. One of the isolates, a *Vibrio owensii* strain, was not able to induce disease symptoms in *A. indonesia* in co-infection with a *Pseudoalteromonas* sp., suggesting the occurrence of antagonistic effects. Inoculations with ciliates isolated from diseased fragments did not provoke symptoms; however, we observed ciliates as secondary scavengers of coral tissue after bacterial infections. Our results suggest a complex etiology, in which bacterial agents trigger tissue necrosis, while opportunistic ciliates lead to typical symptoms. We have some indicators that *Pseudoalteromonas* spp. might play a role in the immunity of corals, which requires further investigations. This evidence complements recent findings on White Syndrome agents and allows a glimpse into functional roles of microbial components in coral holobiont ecology.

8.2.4 Attached or Hanging: Does Maintenance Method Affect the Response of *Pocillopora damicornis* to Thermal Stress?

Sofia Afoncheva[1*], Pia Kegler[1,2], Andreas Kunzmann[1]

[1]Leibniz Centre for Tropical Marine Research, Fahrenheitstraße 6, 28359 Bremen, Germany

[2]CORESea, Chaloklam 94 Moo 7 Koh Phangan, 84280 Surat Thani, Thailand

*corresponding author: sofiyaaf@gmail.com

Keywords: Thermal stress, Coral aquaculture, Coral fragments

Scleractinian corals are in high demand for marine ornamental trade, reef restoration and bioprospecting. These activities require regular supply with live corals. In order to reveal the most efficient method to grow corals in captivity two different maintenance methods (attached vs. hanging coral fragments) for fragments of the hermatypic coral *Pocillopora damicornis* were compared in response to thermal stress and with regard to the feeding status (fed vs. non-fed). The results showed that respiration was affected by the maintenance method. Oxygen consumption in attached corals under 30 °C was 26% higher than in hanging corals whereas hanging fragments did not show acute response to temperature stress. Physiological parameters (photosynthesis, quantum yield, Chl *a*, zooxanthellae density, protein content) were not influenced by maintenance method but were affected by availability of food. Fed fragments maintained a higher Chl *a* concentration, symbiont density and protein content compared to the non-fed ones. Concentration of Chl *a* was inversely correlated with fragments' weight due to the higher skeleton to tissue ratio in the bigger fragments. Growth rates of the smaller fragments tended to be higher than in the bigger ones due to higher metabolic rates and lower absolute energy demands. Despite of the benefits of the hanging method for coral growth in the field nurseries (higher growth rates, resistance to diseases), in closed systems coral growth does not change significantly regardless which maintenance method is applied. When choosing an appropriate maintenance method for coral culture in closed systems several things should be considered: the size of facilities, number of fragments and individual preferences of the coral grower between easy-to-handle options or higher ability to tissue recovery and resistance of corals to biofouling.

8.2.5 Underwater Vision: How a Coral Reef Fish (*Rhinecanthus aculeatus*) Discriminates Color

Emily F. Guevara[1 2,3*], Naomi F. Green[1], Andreas Kunzmann[3], N. Justin Marshall[4] Karen L. Cheney[1]

[1]School of Biological Sciences, University of Queensland, Brisbane, QLD 4072, Australia

[2]Universität Bremen, Bibliothekstr. 1, 28359, Bremen, Bremen, Germany

[3]Leibniz-Zentrum für Marine Tropenforschung (ZMT), Fahrenheitstr. 6, 28359 Bremen, Germany

[4]Queensland Brain Institute, University of Queensland, Brisbane, QLD 4072, Australia

*corresponding author: Emily.guevara1@gmail.com

Keywords: Color thresholds, Visual modeling, Receptor noise, Neurobiology

Healthy coral reefs are extremely colorful, chaotic, and dappled environments, cast under a constant flicker of incident sunlight that poses a particular challenge for visual navigation – a challenge that has been brilliantly solved using color vision. Color vision is finely tuned and indispensable to identify conspecifics, attract mates, deter or elude predators, and to find food and shelter. Many shallow reef fish are trichromats, creating color images much like humans, by combining light information from three types of color-sensing 'cone' cells in downstream neural opponency mechanisms. The ability to detect and discriminate colors is therefore determined by a combination of light availability, receptor sensitivity, and neural processing capability. We use visual modeling to investigate the latter; to understand the processes that underlie color vision. However, these models rely on key assumptions about animal visual systems that must be calibrated with behavioral testing – a rarity in the literature. Our aim is to understand how animals distinguish color, by examining behavioral threshold differences: when are two colors different enough to be considered independent? This study is part of the most comprehensive color vision test that has ever been done on a non-human vertebrate, and uses a new and exciting way to test color vision in animals that is based on the Ishihara method for human color blindness tests. Picasso triggerfish show an outstanding ability to generalize across colors, performing remarkably well in odd-one-out tasks. We therefore trained them to identify a target spot that differs from a background of spots in either saturation or hue. Using the Receptor Noise Limited model, we found that for the tested area of color space, the Picasso triggerfish exhibits behavioral thresholds that significantly deviate from the assumptions of the model. This is a novel finding that could not have been predicted based on modeling alone.

8.2.6 Chemical Cycles Disrupted by White Mangrove Logging in North-Brazil?

Mirko Wölfelschneider[1*], Véronique Helfer[2], Martin Zimmer[2], Moirah Paula Machado De Menzes[3], Ulf Mehlig[3]

[1]Universität Bremen, Bibliothekstraße 1, 28359 Bremen, Germany

[2]Leibniz Centre for Tropical Marine Sciences, Fahrenheitstraße 6, 28359 Bremen, Germany

[3]Federal University of Pará – Campus Bragança, Bragança – PA, 68600-000, Brazil

*corresponding author: mirco.woelfelschneider@gmx.de

Keywords: Mangrove-logging, Sustainable resource-use, Chemical cycle, Mangroves

Mangroves provide an array of ecosystem services to coastal communities, including coastal protection and livelihood. Yet, these fragile environments are among the most threatened ecosystems on the planet. As human populations, particularly in developing nations continue to grow, the number of people dependent on mangroves for their survival has increased. 'No-take' policies are no longer a feasible solution to protect and preserve mangroves for future generations, meaning there is an urgent need for well-conceived management plans. Communities along the coastline of North Brazil practice a traditional form of selective mangrove-logging involving the White mangrove (*Laguncularia racemosa*). The felled trees of this species have the ability to re-grow, thus this activity is often considered as sustainable resource-use. However, its subsequent ecological impacts have yet to be investigated. This study investigates how the re-growth process of the White mangrove influences plant chemistry and subsequent element-cycling within the ecosystem. Leaf and sediment samples obtained directly from the trees at four different stages of re-growth and their surrounding areas were analyzed for carbon and nitrogen content to gain an overview of the systems dynamics. Pyrolysis-gas chromatography-mass spectrometry (Py-GC/MS) was then employed to gain a more detailed insight into the chemical composition of both mangrove leaves and the organic matter of the sediment. The results show significant changes in the nitrogen content of leaf tissues during the different stages of the re-growth process. For the carbon level no such significant changes were revealed. Sediment sample analyses revealed trends in changes of organic matter, carbon or nitrogen content, however these could not be related to the changes within the leaf chemistry. The results from this study support the current assumption that selective logging is a sustainable resource-use with a very low impact on the surrounding ecosystem.

8.2.7 Long Term Impact of Experimental Harvesting on a Tropical Seagrass Meadow in Gazi Bay, Kenya

Charles Cadier[1,2,3*], Anna-Maria Frouws[1,2]

[1]School of Applied Sciences, Edinburgh Napier University, EH11 4BN, Edinburgh, UK

[2]Kenya Marine and Fisheries Research Institute, P. O. Box 81651, Mombasa, Kenya

[3]University of the Basque Country, Areatza z/g 48620 Plentzia-Bizkaia, Spain

*corresponding author: charlescadier@hotmail.fr

Keywords: Canopy removal, Tropical seagrass meadow, Recovery

Seagrass meadows provide a wealth of ecosystem services, among which the provision of habitats for many taxonomic groups and significant contribution to sediment carbon storage. Along the African coast research on seagrasses is still limited. This study examined the recovery capacity of four experimentally harvested (monthly removal of leaves during 1.5 years) plots (3*2 m) in a seagrass meadow in Gazi Bay, Kenya, and compared these with four control plots. Recovery was measured 6 months after disturbance ended. Seagrass health was determined through leaf puncturing techniques and leaf counts. Macrofauna was sampled through dropsamples and cores. Soil organic carbon was measured through Loss on Ignition. The removal phase showed a loss in fauna abundance (252.8 ± 73.3 and 1541.8 ± 606.4 (mean abundance ± SD) respectively) and % organic matter in sediment (1.5% ± 0.07 and 2.7 ± 0.70 (mean % organic matter ± SD) respectively) within removal plots compared with control plots. After 6 months of recovery, results displayed a tendency for lower growth rate for recovering plots compared to control plots (4.6 ± 2.1 and 7.0 ± 3.2 mm d^{-1} (mean growth rate ±SD) respectively). The % organic carbon in sediment displayed a similar trend (0.8% ± 0.4 and 3.0 ± 0.9% (mean % organic carbon ± SD) respectively). Fauna has been sampled within recovering plots and shows clear signs of disturbance, results are expected in July. These preliminary results indicate that half a year after the end of the disturbance, experimentally harvested seagrass meadows do not show signs of recovery.

8.2.8 Using a Social-Ecological System (SES) Framework as a Tool for Stakeholder Deliberation of Sustainability Challenges in the Gulf of Nicoya, Costa Rica

Vigneshwaran Soundararajan[1*], Stefan Partelow[1]

[1]Leibniz Center for Tropical Marine Research (ZMT), Fahrenheitsraße 6, 28359 Bremen, Germany

*corresponding author: vigneshwaran.soundar@gmail.com

Keywords: Social-ecological systems, Sustainability science, Participatory management, Small-scale fisheries, Costa Rica

This project attempts to operationalize a social-ecological system (SES) framework (McGinnis, M., & Ostrom, E. (2014). Social-ecological system framework: initial changes and continuing challenges. Ecology and Society, 19(2)) through experimental research in the case of small-scale artisanal fisheries in the Gulf of Nicoya, Costa Rica. The SESF provides a common structure for academia and practitioners to identify and integrate factors; and in this case factors that are crucial to appropriately manage common pool resources that have greater collective action challenges for management. This project used the SESF as a methodological tool beyond academia, and utilized it for practical deliberation. In doing so, the project aims to produce novel concepts to mitigate ongoing issues in small scale fisheries co-management. Focus groups were conducted in two communities in the Gulf of Nicoya (Chira and Tambor), and participants were given entry surveys prior to the focus group session, and exit surveys thereafter. The SESF was implemented as an illustration to facilitate focus group discussions. Multiple images were used to represent distinct first tier variables within a subsystem, intending to bring up conflicting topics during the group deliberation. This functions to bridge knowledge from academia (i.e., systems knowledge) to non-academic stakeholders (i.e., small-scale fishermen) and incorporate heterogeneous ideas. The role of deliberation in focus groups also has the potential to induce changes in socio-psychological attributes (value, belief, norm, behavioral intentions) of the participants. Focus group deliberation data was quantified qualitatively using the software MAXQDA to measure depth and complexity of focus group discussions, and quantitatively with R-Statistics to analyze entry and exit surveys for socio-psychological changes. Through promoting and fostering accountability and compliance among small-scale fishermen, this project aims to provide a tool to increase the efficacy of iterative co-management processes and develop effective mechanisms for stakeholder participation in decision making.

8.2.9 Research Networks in Tropical Marine Science – Where, on what and with whom?

Paula Senff[1*], Stefan Partelow[1,2], Achim Schlüter[1,2]

[1]Leibniz Centre for Tropical Marine Research (ZMT), Fahrenheitstr. 6, 28359 Bremen, Germany

[2]Jacobs University, Campus Ring Road 1, 28759 Bremen, Germany

*corresponding author: paula.senff@leibniz-zmt.de

Keywords: Social network analysis, Path dependency, Sustainability, Authorship, Collaboration

The body of research on tropical marine ecosystems has increased tremendously in the past decades and these areas now also face unprecedented pressures from growing human impacts. In Partelow et al. (2017, Conserv Lett, doi:10.1111/conl.12351), we identified several shortcomings in the published literature related to advancing sustainability contributions from science, including a lack of clear problem orientation, a dissimilarity of knowledge generation and communication between disciplines, and an uneven focus on certain geographic regions and ecosystems. This study further analyzed the literature for author affiliation and location, conducting an in-depth network analysis. Looking at country level authorship networks, strong regional trends as well as prioritization on coral reef ecosystems can be seen. Author collaborations favor domestic cooperation and developing countries in the tropics are heavily underrepresented. Furthermore, where empirical research is conducted, the focus of research and who publishes together result in part from the numerous path dependencies driving academic institutions. Institutional path dependencies can take multiple forms such as the availability of field labs and infrastructure, the degree of knowledge of field locations or topics, established collaborative relationships, or in a more discursive form, current funding and 'hot topics' in a scientific field. To further advance the discourse for a sustainability agenda in tropical marine science, we propose a stronger focus on international research in a variety of disciplines and subject areas.

8.3 Abstracts of Poster Presentations

8.3.1 A Common Coral-Algal Interaction Under the Influence of Climate Change and Ocean Acidification

Lena Rölfer[1,2*], Dorothea Bender-Champ[3], Hauke Reuter[1], Sebastian Ferse[1], Andreas Kubicek[3], Sophie Dove[3], Ove Hoegh-Guldberg[3]

[1]Leibniz Centre for Tropical Marine Research, Fahrenheitsstraße 6, 28359 Bremen, Germany

[2]Faculty of Biology and Chemistry, University of Bremen, D-28359 Bremen, Germany

[3]School of Biological Sciences, University of Queensland, QLD 4072, Australia

*corresponding author: lena@roelfer.de

Keywords: Coral-algal interaction, Climate change, Ocean acidification, *Porites lobata, Chlorodesmis fastigiata*

Competition between corals and algae is frequently observed on reefs and the outcome of such interactions affects the relative abundance of reef organisms and therefore reef health. Anthropogenic activities have resulted in increased in atmospheric CO_2 levels and a subsequent rise in ocean temperatures. Besides temperature increase, elevated CO_2 levels are leading to a decrease in oceanic pH resulting in ocean acidification. These two changes have the potential to alter ecological processes within the oceans including the outcome of coral-algal interactions. In this study, the combined effect of temperature increase and ocean acidification on the competition between the coral *Porites lobata* and the alga *Chlorodesmis fastigiata* was assessed. A temperature increase of +2 °C above preindustrial temperatures and CO_2 level of 450 ppm were used to simulate a RCP2.6 emission scenario for the mid- to late twenty-first century. Results show negative effects of both, algal contact and future conditions on coral metabolism. We observed decreased net photosynthesis of *P. lobata* in contact with *C. fastigiata* and increased respiration under RCP scenario. Dark calcification rates of corals under RCP conditions were negative and significantly decreased compared to ambient conditions. Light calcification rates were negatively affected by the interaction of algal contact and RCP scenario, leading to decreased calcification under RCP conditions irrespective of algal contact. Results indicate that the combined effects of temperature and CO_2 as well as the interaction with *C. fastigiata* may alter the growth and metabolism of *P. lobata* in the future.

8.3.2 Hide-and-Seek: How Prey Patterning Alters Detectability to a Marine Predator (*Rhinecanthus aculeatus*)

Emily F. Guevara[1,2,3*], Andreas Kunzmann[3], N. Justin Marshall[4], Karen L. Cheney[1]

[1]School of Biological Sciences, University of Queensland, Brisbane, QLD 4072, Australia

[2]Universität Bremen, Bibliothekstr. 1, 28359, Bremen, Bremen, Germany

[3]Leibniz-Zentrum für Marine Tropenforschung (ZMT), Fahrenheitstr. 6, 28359 Bremen, Germany

[4]Queensland Brain Institute, University of Queensland, Brisbane, QLD 4072, Australia

*corresponding author: emily.guevara1@gmail.com

Keywords: Color vision, Signal salience, Predator-prey interactions, Neurobiology

It is natural to presume that all other humans and creatures perceive the world in color in much the same way. However, color *per se* does not actually exist in the environment. What we perceive as color is simply the brain's way of representing a narrow range of wavelengths of the electromagnetic (EM) spectrum that have been harnessed to give contrast information where simple, monochromatic vision would be less helpful. Many reef animals rely heavily on color vision to attract mates, find food and shelter, signal their unpalatability or toxicity (i.e., aposematism), and avoid predators, among others. Clever exploitation of different ranges of the visual spectrum can result in beneficial communication to selected signal receivers, via detection and discrimination. There is a trend towards blue and yellow coloration in reef fish; blue light transmits furthest through the water column, and yellow is its complimentary (therefore highly contrasting) color; the same applies for green and red in greenish, eutrophied coastal waters. Both the wavelength and pattern arrangement of color is important to animals yet research is largely lacking on how animals *process* highly contrasting colors. 'Signal theory' states that a higher signal to background contrast increases the salience (a.k.a. conspicuousness) of that signal. Enhanced borders do exactly this, playing directly on the viewer's cognitive mechanism of edge detection by increasing this contrast. The present study expands on this idea, using a model coral reef predator, the Picasso triggerfish *Rhinecanthus aculeatus*, to behaviorally assess detectability. Using predator foraging biases on printed "prey" stimuli, it is possible to break down signal salience based on the presence of contrasting internal and/or bordering edges. Picasso triggerfish avoided aposematic signals with enhanced yellow borders faster than if signals contained only red or blue internal patterns, which aligns with previous theory.

8.3.3 Comparing Food Webs at Both Sides of the Isthmus of Panama: The Relative Importance of Mangrove Food Sources for Fish Communities

Lara Stuthmann[1*]

[1]Leibniz Centre for Tropical Marine Research, Fahrenheitstraße 6, 28359 Bremen, Germany

*corresponding author: lara-stuthmann@web.de

Keywords: Mangroves, Stable Isotope Analysis, Fish, Food web

Mangrove ecosystems are well known for playing an important role in coastal systems. The importance of mangroves as a primary food source for higher vertebrates is still in discussion, especially in comparison with other adjacent productive systems, like coral reefs and seagrass beds. Panama provides an extraordinary experimental ground to examine the role of mangroves as feeding area for fish. The rise of the Central American Isthmus closed the seaway between the Pacific and Atlantic Ocean around 3 Ma. It formed two coasts with contrasting environmental features: The Caribbean coast is shaped by a micro-tidal regime (ca. 0.5 m) with adjacent coral reefs and seagrass beds; while the Pacific coast is characterized by a macro-tidal regime (ca. 4 m) and the absence of seagrass beds and coral reefs but is in general more productive than the Caribbean Sea. Mangroves are present at both coasts. The aim of the study is to use this contrasting environmental set-up to reveal the food web structures and examine the relative importance of mangroves food sources for fish communities of two mangrove systems in Panama in the Caribbean ("Bocas del Toro") and the Pacific ("Gulf of Montijo"). Stable isotope analyses ($\delta^{15}N$ and $\delta^{13}C$) of the sampled prey items, primary producers, organic matter and fish will give insights in the food web structure under the different environmental circumstances. The data will be analyzed with the R statistical environment packages SIAR (Stable Isotope Analysis in R) and SIBER (Stable Isotope Bayesian Ellipses in R) to estimate the percentage of mangrove carbon in the fishes diets and to compare the isotopic niche width of the fish communities. The results will be available in September.

8.3.4 Using Perspectives to Understand Collective Action in Small-Scale Fisheries in the Gulf of Nicoya, Costa Rica

Anne Jäger[1,2*], Stephan Partelow[2], Achim Schlüter[2,3]

[1]Department of Biology and Chemistry, University Bremen, Bremen, Germany

[2]Department of Social Sciences, Leibniz Centre for Tropical Marine Research, Bremen, Germany

[3]Department of Economics, Jacobs University, Campus Ring 1, 28759 Bremen, Germany

*corresponding author: anne.jaeger@leibniz-zmt.de

Keywords: Stakeholders' perspectives, Collective action, SES- framework, Small-scale fisheries, Gulf of Nicoya

Overexploitation of many common-pool resources worldwide such as fisheries lead to adverse social and ecological consequences. Because of the complexity of social-ecological systems, effective management often poses great challenges. These can be addressed by cooperation among resource users which was often found to crucially improve long-term socio-economic benefits of the community and maintenance ecological sustainability. As a consequence of fisheries depletion in the Gulf of Nicoya, fishing communities have been formed and have requested the establishment of Marine Areas for Responsible Fishing in order to protect rich fishing points and allow for their sustainable use. Today, these areas are managed by several governmental and non-governmental stakeholders such as fishing organizations, INCOPESCA, coast guards, universities and NGO's. This case study aims to identify some key variables that affect stakeholders' perspectives by applying the SES- framework developed by E. Ostrom, and investigate the role of these perceptions on collective action in resource management of the Marine Areas for Responsible Fishing. The study takes place on Isla Venado, Gulf of Nicoya. I gained qualitative data in the form of participant observation, focus groups, and open-ended and semi-structured interviews with key informants and fishers in February and March 2017. In order to find entry points into the social network, interviewees were randomly sampled, followed by purposive snowball sampling and selection of key informants. I applied Elinor Ostrom's social-ecological systems (SES) framework to identify related variables affecting stakeholder's perspectives and hence collective action, and I used the SES- framework to structure data analysis with MAXQDA. Findings show the importance of socio-economic factors (e.g., heterogeneity, trust, social networks) and local perspectives on social-ecological outcomes of conservation for rules compliance and cooperation.

9 Polar Ecosystems in the Age of Climate Change

Maciej Mańko[1] and Katarzyna Walczyńska[1]

[1]Department of Marine Plankton Research, Institute of Oceanography, University of Gdańsk, Al. Marszałka Piłsudskiego 46, 81-378 Gdynia

9.1 Call for Abstracts

Polar regions are particularly susceptible to climate change. Even smallest modification of any of the physical characteristics of water will cause cascade of dramatic effects influencing all levels of biological organization. Already harsh environment will become even more challenging, thus affecting species reproduction, dispersal, and various other aspects of organismal biology and ecology. Complexity of polar ecosystems, together with logistical constraint in conducting research there are hampering proper recognition of their current state, and as such further predictions of possible impact of climate changes. This session encourages all polar-related contributions investigating these vulnerable ecosystems in the age of changing climate.

9.2 Abstracts of Oral Presentations

9.2.1 Zooplankton Variability Within Atlantic Water Flow to the Arctic

Marta Gluchowska[1*], Slawomir Kwasniewski[1], et al.[2,3,4]

[1]Institute of Oceanology Polish Academy of Sciences (IOPAN), Powstancow Warszawy 55, 81-712 Sopot, Poland

[2]Institute of Marine Research (IMR), P.O. Box 1870 Nordnes, 5817 Bergen, Norway

[3]Norwegian Polar Institute (NPI), Fram Centre, 9296 Tromsø, Norway

[4]Department of Vertebrate Ecology and Zoology, University of Gdansk (UG), PL-80-441 Gdansk, Legionow 9, Poland

*invited speaker, corresponding author: mgluchowska@iopan.pl

Keywords: Atlantic water, Barents Sea, *Calanus*, West Spitsbergen Current, Zooplankton

The northward incursion of warm Atlantic water (AW) not only affects thermal conditions and sea ice cover in the Arctic Ocean but it also allows plankton species to expand their distributions. Despite the fact that zooplankton plays multiple roles in ecosystem functioning, information on zooplankton variability the Arctic waters, are scare and highly fragmented in space and time. Here we present observations on the zooplankton taxonomical and functional structure at the border to the Arctic Ocean, in relation to physical properties of the water masses. The presented data were collected within zooplankton long term research programs, conducted since 2001 in cooperation between IOPAN, NPI, IMR and UG. Spatially the data include information from shelf areas of Spitsbergen and the Barents Sea, and the Greenland Sea with Fram Strait. During the study period, the AW temperature in the main pathways of the flow showed a noticeable increase, however in an oscillating manner. This gradual change was manifested in increasing proportion of the biomass contributed by boreal taxa such as *Calanus finmarchicus*. At the time of the warm event 2011, probably the second generation of the *C. finmarchicus* was observed for the first time in this region. Meanwhile, the zooplankton composition in the west Spitsbergen fjords was not changing considerably, however the community structure was undergoing noticeable modifications, especially in some fjords considered as of arctic character, most likely as a result of increased influx of AW. Through examination of available results on variability in zooplankton and in local and regional environmental factors, possibly influencing the zooplankton, allowed to propose a conceptual scenario of development of the present day situation towards different climate state. The 'warmer Arctic' scenario predicts a modification in Arctic marine food webs, and consequently, foresees changes in food resources available to the top Arctic predators, such as seabirds.

9.2.2 Community Structure of Macrobenthos in the Deep Fram Strait, Arctic Ocean

Melissa Käß[1,2,3*], Andrey Vedenin[4], Angelika Brandt[2,5], Thomas Soltwedel[1]

[1]Alfred Wegener Institute, Helmholtz Centre for Polar and Marine Research, Am Handelshafen 12, 27570 Bremerhaven, Germany

[2]University of Hamburg Centre of Natural History, Germany

[3]present Address: Stuttgart State Museum of Natural History, Rosenstein 1, 70191 Stuttgart, Germany

[4]P.P. Shirshov Institute of Oceanology, Laboratory of Ocean Bottom Fauna, 36, Nahimovski prospect, Moscow, Russia

[5]present Address: Senckenberg, Research Institute and Natural History Museum, Sencken-berganlage 25, 60325 Frankfurt, Germany

*corresponding author: melissa.kaess@smns-bw.de

Keywords: Macrofauna, Deep-sea, Arctic, Community structure, HAUSGARTEN

The eastern side of the Fram Strait is significantly influenced by the northern-bound comparably warm West Spitsbergen Current, whereas the western side of the strait is affected by the cold and less saline East Greenland Current flowing in a southerly direction. The current regime is the major factor in determining patterns in ice coverage in this area. In turn, distribution patterns of the sea-ice play an important role in determining the flux of potential food to the deep seafloor. The objective of this study was to compare the macrofaunal community structure in eastern and western parts of the Fram Strait along two bathymetrical transects (1000–2500 m water depth) at the LTER (Long-Term Ecological Research) observatory HAUSGARTEN. Material was collected during RV *Polarstern* expedition PS99.2 in June/July 2016 using an USNEL box corer with a sampling area of 0.25 m^2 deployed at a total of eight stations. Samples were processed through a 0.5-mm sieve. Results showed a generally higher macrofaunal density at the stations located in the eastern Fram Strait. On both sides of the strait, species richness, biomass and biodiversity showed a trend to decrease with increasing station depth. An exception was observed at a station at 2500 m water depth off Greenland, which was located in the marginal ice zone. At this site, macrofaunal densities and species diversity were higher than at the adjacent shallower sampling sites. In general, polychaetes were the most abundant taxon, followed by crustaceans and molluscs. Species composition along the bathymetrical transects on both sides of the strait changed with increasing depth. Sea-ice coverage and water depth, with the associated variables food quality and quantity at the seafloor, seem to be crucial factors driving the macrofaunal community patterns in the study area. However, more samples are necessary to support these first results.

9.2.3 Arctic Cephalopods Distribution During the Recent Climate Changes

Alexey V. Golikov[1*], Rushan M. Sabirov[1], Martin M. Blicher[2], Gudmundur Gudmundsson[3], Lis L. Jørgensen[4], Denis V. Zakharov[5], Olga L. Zimina[6], José C. Xavier[7,8]

[1]Kazan Federal University, Kremlyovskaya Street 18, 420008 Kazan, Russia

[2]Greenland Institute of Natural History, Kivioq Street 2, 3900 Nuuk, Greenland

[3]Icelandic Institute of Natural History, Urridaholtsstraeti 6-8, 212 Gardabaer, Iceland

[4]Institute of Marine Research, Sykehusveien Street 23, 9019 Tromsø, Norway

[5]Polar Research Institute of Marine Fisheries and Oceanography, Knipovitcha Street 6, 183038 Murmansk, Russia

[6]Murmansk Marine Biological Institute, Vladimirskaya Street 17, 183010 Murmansk, Russia

[7]Marine and Environmental Sciences Centre (MARE-UC), Department of Life Sciences, University of Coimbra, 3004-517 Coimbra, Portugal

[8]British Antarctic Survey, NERC, Madingley Road, CB3 0ET Cambridge, United Kingdom

*corresponding author: golikov_ksu@mail.ru

Keywords: Cephalopoda, Arctic, Climate Change

Recent climate change (mostly by warming) is impacting the Arctic region significantly. In terms of marine life, many fish and invertebrates from boreal and subtropical Atlantic are expanding into the Arctic. Here, we review the changes in the distribution of cephalopods in the Arctic due to the recent warming. During 1992–2016 extensive sampling has been carried out in the Barents Sea (and adjacent areas), around Iceland, in the Western and Eastern Greenland (Barents Sea Ecosystem Survey, BIOICE and INAMON programs respectively). There are 31 species of cephalopods in the Arctic, including those rarely appearing on the borders of the arctic region. Only 10 species constantly live in the Arctic throughout their lifecycle: *Gonatus fabricii*, *Rossia palpebrosa*, *R. moelleri*, *Cirroteuthis muelleri*, *Bathypolypus arcticus*, *B. bairdii*, *B. pugniger*, *Muusoctopus* sp., *M. sibiricus*, *M. leioderma*. The distribution range of *Gonatus fabricii* has been increasing eastward since 2006 to the shelf of the eastern Barents Sea and the Kara Sea. Three boreo-subtropical squids (*Todaropsis eblanae*, *Teuthowenia megalops*, *Galiteuthis armata*) and one bobtail squid (*Sepietta oweniana*) has been recorded in the Arctic since 2006. *Todarodes sagittatus* appeared in the Barents Sea in 2010, for the first time since 1983. Being known to appear in the Arctic as a foraging migrant, this record has therefore no relation to the ongoing climate changes. Thus, all the reported records of boreo-subtropical cephalopods in the Arctic are of two types: (1) regular foraging migrations within unknown period of time and without any connection with climate warming and (2) northward range expansions of boreal species into the Arctic during the last decade, induced by an ongoing warming of Arctic waters. Quantitative distribution of the Arctic cephalopods was studied only for the Barents Sea so far, therefore any recent climate change impacts on cephalopods elsewhere around the Arctic are not known.

9.2.4 Impact of Climate Change on Protist Communities in Isfjorden – Svalbard

Mathilde Bourreau[1*], Janne E. Søreide[2], Tove M. Gabrielsen[2]

[1]University Centre in Svalbard, P.O. Box 394 N-9171, Longyearbyen, Norway

[2]University Centre in Svalbard, P.O. Box 156, Longyearbyen, Norway

*corresponding author: mathilde.bourreau@etu.upmc.fr

Keywords: Protists, Climate change, Svalbard, 18S rRNA gene, Seasonality

Isfjorden on the western coast of Svalbard is seasonally ice-covered in the innermost parts. The Atlantic Water Current, strongly influencing western Spitsbergen fjords, has been warming 0.3 °C per decade over the last 50 years. This is altering Isfjorden sea water temperature, stratification and community composition. A change in the microbial community could have a major impact on the higher trophic levels and thus the arctic marine ecosystem. This research project is part of the IMOS project (Isfjorden Marine Observatory Svalbard) studying long-term effects of the changing Arctic. The aim of this project is to study the impact of climate change on the protist communities in Isfjorden, by considering if changes in the temperature, salinity and stratification are impacting their seasonality and diversity. Three stations from warm Atlantic water to more Arctic climate, along Isfjorden in Svalbard, were sampled regularly during 3 years. Water samples were taken combined with CTD measurements (salinity, fluorescence and temperature) and nutrient samples. Using Illumina sequencing, two size fractions (10 µm and 0.45–10 µm) of the water are analyzed and the species are identified by metabarcoding of the V4 region of the 18S rRNA gene. We will observe the modifications in community composition. We will compare the 3 years for inter-annual variation and long-term trends. Finally we will discuss the importance of factors such as light (is the bottleneck effect because of polar night remaining?), temperature, salinity and nutrients on the community diversity.

9.2.5 Sedimentation of Persistent Organic Pollutants (POPs) in Ryder Bay, Antarctica

David Amptmeijer[1*], Artem Krasnobaev[2], Nico van den Brink[2]

[1]Rijksuniversiteit Groningen (RUG), department of marine biology, Nijenborgh 7, 9747 AG Groningen, Nederland

[2]Wageningen university and Research (WUR), department of toxicology, Stippeneng 4, Helix, building 124, 6708 WE Wageningen, The Netherlands

*corresponding author: davidamptmeijer@gmail.com

Keywords: POPs, Antarctica, Modeling, Fugacity, NPZD

Strong lipophilic and halogenated POPs are accumulating in the Antarctic marine ecosystem. Although the main pathway of POPs into the Antarctic system is via atmospheric deposition, there is a higher bioaccumulation in benthic than pelagic fish. This suggests that there is a strong flux of POPs from surface water levels to the deeper water. In order to study the flux of POPs we linked the flux of organic matter to a chemical fugacity model. The flux of organic matter is calculated by a Nutrient, Phytoplankton, Zooplankton and Detritus (NPZD) model based on Chlorophyll and photosynthetic active radiation data collected by the Rothera Time Series (RaTS) program run by the British Antarctic Survey between 2012 and 2016. The measurements are taken biweekly in summer and weekly in winter using CTD at a 520 m deep sampling site located at: "67.570°S, 68.225°W" in Ryder Bay, Antarctica. Ryder bay is known for having a large spring diatom bloom after the sea ice melt, followed by sedimentation of a large fraction of these diatoms. We identified the association of POPs to phytoplankton during the phytoplankton bloom and subsequent co-sedimentation as the main pathway of transport of POPs from the pelagic to the benthic system. Subsequently the lipophilicity of the POP is the main contributor to the fractional flux of POPs from the pelagic to the benthic system.

10 Ecosystems Dynamics in a Changing World: Regime Shifts and Resilience in Marine Communities

Camilla Sguotti[1] and Xochitl Cormon[1]

[1]Institute for Hydrobiology and Fisheries Science, Centre for Earth System Research and Sustainability (CEN), University of Hamburg, Germany

10.1 Call for Abstracts

Over the last decades many marine ecosystems and fish populations all over the world have undergone drastic changes, termed regime shifts, because of the additive effects of anthropogenic and natural stressors. We invite you to present: (1) Studies related to marine ecosystems structure and population dynamics (i.e., food-webs, spatial distribution, regulating processes, top-down and/or bottom-up), as well as potential drivers of regime shifts. (2) Examples of shifts and resilience of ecosystems and studies comparing regimes across different scales (temporal or spatial). (3) Studies investigating impacts of regime shifts on ecosystem services and potential implications for management and stakeholders.

10.2 Abstracts of Oral Presentations

10.2.1 Regime Shift, a Global Challenge for the Sustainable Use of our Marine Resources

Camilla Sguotti[1*], Xochitl Cormon[1*]

[1]Institute for Hydrobiology and Fisheries Science, Centre for Earth System Research and Sustainability (CEN), University of Hamburg, Germany

*corresponding authors: camilla.sguotti@uni-hamburg.de, xochitl.cormon@uni-hamburg.de

Over the last decades many marine ecosystems and fish populations have undergone drastic changes often resulting in new, ecologically poorer and/or economically less valuable states. In particular, the additive effects of anthropogenic (e.g., fishing, climate change) and natural stressors (e.g., natural variability) seem to play a fundamental role in causing unexpected and sudden shifts between ecosystem states, generally termed regime shifts. Recently, many examples of regime shifts have been documented world-wide and their mechanisms and consequences have been vigorously discussed. Understanding causes and mechanisms of regime shifts is of great importance for the sustainable use of natural resources and their management, especially in marine ecosystems. In this session titled “Ecosystem dynamics in a changing world, regime shifts and resilience in marine communities” during the 8th YOUMARES conference (Kiel, 13–15th September 2017) we will present regime shifts and associated concepts to a broad range of marine scientists (not only biologists and/or ecologists) and highlight their importance for the marine ecosystems worldwide.

10.2.2 Regime Shift Analysis in Atlantic Herring Stocks

Leonie Färber[1*], Camilla Sguotti[2], Saskia A. Otto[2], Christian Möllmann[2]

[1]Centre for Ecological and Evolutionary Synthesis (CEES), Department of Biosciences, University of Oslo, PO Box 1066 Blindern, N-0316 Oslo, Norway

[2]Institute for Hydrobiology and Fisheries Science, University of Hamburg, Grosse Elbstrasse 133, 22767 Hamburg, Germany

*corresponding author: l.a.farber@ibv.uio.no

Keywords: Atlantic herring, *Clupea harengus*, Regime shift

External drivers such as temperature, nutrient loadings, habitat exploitation or fragmentation have profound effects on marine ecosystem dynamics playing a major role in observed large-scale changes. Those changes can happen gradually and linearly or abruptly and discontinuously, as observed in ecological regime shifts. For instance, many fish stocks experienced a strong depletion over time and some stocks might have undergone one or several abrupt shifts. In this work, 16 herring (*Clupea harengus)* stocks from the Atlantic, the North and Baltic Sea were analyzed for the detection of possible regime shifts. Herring is a commercially important species and the understanding of its dynamics is essential for management purposes. In specific, a trend change analysis was conducted, combined with a Bayesian change point analysis (bcp) in order to find possible sudden collapses in the spawning stock biomass (SSB) and recruitment (R). Only a few of the stocks showed a prolonged depletion in their SSB and hardly any changes in the recruitment during the time series. Most of the dynamics can be considered as natural fluctuations. Multiple statistical analyses were conducted in order to find the underlying drivers causing interannual fluctuations and, if existing, the collapse. Indices of large-scale climatic fluctuations, i.e., the North Atlantic Oscillation (NAO), the Atlantic Multidecadal Oscillation (AMO) index, the Sea Surface Temperature (SST) and fishery mortality were used as explanatory variables for the dependent variables SSB and R. Linear and additive models were used to find changes in the functional relationship. The analyses indicated that the herring stocks experienced changes over the course of the time series, but only few underwent regime shifts with fishing pressure being the major driver of these dynamics. Those results will be compared to dynamics of cod (*Gadus morhua*) stocks from nearby regions and possible differences will be investigated.

10.2.3 Effects of Wounding on Detritus Production by Caribbean Reef Sponges

M. L. Folkers[1,2*], K. C. Tomson[1,2], B. Mueller[1,2], J. M. de Goeij[1,2]

[1]Caribbean Marine Biological Institute (CARMABI), Piscaderabaai, Willemstad, Curaçao

[2]present Address: Department of Freshwater and Marine Ecology (FAME), Institute for Biodiversity and Ecosystem Dynamics (IBED), University of Amsterdam (UvA), Science Park 904, 1098 XH Amsterdam, Netherlands

*corresponding author: mainahfolkers@gmail.com

Keywords: Reef sponges, Detritus production, Wounding

Coral reefs are the most biodiverse and productive ecosystems. To thrive in oligotrophic waters efficient energy and nutrient recycling is essential. Despite being the largest energy and nutrient source on coral reefs, dissolved organic matter (DOM) is primarily consumed by microbes and sponges. Encrusting cavity sponges use most of the energy and nutrients obtained from filtering DOM to rejuvenate their filtration apparatus. Old filter cells are frequently shed and expelled as detritus. Sponge detritus production is increasingly recognized to be an important link to pass energy and nutrients stored in DOM through sponges to detritivores and subsequently higher trophic levels. However, little is known about how environmental factors, such as increased stress, influence sponge detritus production rate. This study examined the effects of wounding to simulate stress on the detritus production rate of four encrusting Caribbean reef sponge species (*Scopalina ruetzleri*, *Haliclona vansoesti*, *Halisarca caerulea* and *Chondrilla caribensis*). Based on previous research, detritus production rate is expected to decrease in wounded sponges compared to healthy sponges. Detritus was collected on six consecutive days from healthy and wounded sponges. A linear mixed effects analysis showed that mean detritus production rate of wounded *S. ruetzleri* individuals was significantly affected by wounding ($X^2(1) = 8.8274$, $p = 0.0030$). Wounding lowered the mean detritus production rate of healthy *S. ruetzleri* individuals ($\mu = 0.0204$ mg/cm^2/h $\pm$ 0.0039 (SE)) by 46%. In contrast, no effect of wounding on detritus production rates was found in *H. vansoesti*, *H. caerulea* and *C. caribensis*. This suggests that the effects of wounding are not similar among different Caribbean reef sponge species. Energy and nutrients taken up by sponges are estimated to be in the same order of magnitude as gross primary production on coral reefs. Reduced detritus production by some sponges due to increased stress may contributed to ecosystem regime shifts and reduce ecosystem resilience.

10.2.4 The Nutrient Limitation Status of Benthic Reef Macroalgae (Phaeophyceae) in Moorea, French Polynesia

Hannah S. Earp[1*], Mona Andskog[1], Joost den Haan[2,3], Mirta Teichberg[4], Deron Burkepile[5], Maggy M. Nugues[2,6]

[1]Universität Bremen, Bibliothekstraße 1, 28359 Bremen, Germany

[2]EPHE, PSL Research University, UPVD-CNRS, USR3278 CRIOBE, F-66860 Perpignan, France

[3]Max Planck Institute for Marine Microbiology, Celsiusstraße 1, 28359 Bremen, Germany

[4]Leibniz Centre for Tropical Marine Sciences, Fahrenheitstraße 6, 28359 Bremen, Germany

[5]University of California – Santa Barbara, Santa Barbara, CA 93106, USA

[6]Labex Corail, CRIOBE, 98729 Moorea, French Polynesia

*corresponding author: hannahsearp@hotmail.com

Keywords: Nutrient limitation, Macroalgae, Coral reef, NIFT experiments, Growth

A central paradigm of coral reef ecosystems is that algal productivity is limited by nutrient availability. Nutrient enrichment can relieve this limitation, allowing algae to proliferate at the expense of live coral. It is widely accepted that remote Pacific island reefs are more resilient to phase-shifts in part because they do not receive periodic influxes of nutrient-rich aeolian dust. To investigate the role of nutrients on algal proliferation in the Pacific, we compared the response of brown macroalgae to iron, nitrogen and phosphorous enrichment in Moorea, French Polynesia. The nutrient-limited productivity of three macroalgae (*Dictyota* spp., *Sargassum pacificum*, *Turbinaria ornata*), collected from sites along a human population gradient, was examined using Nutrient Induced Fluorescence Transient (NIFT) experiments. The nutrient-limited growth of two macroalgae (*S.pacificum*, *T.ornata*) was determined using a 14-day *in-situ* fertilization experiment. It was hypothesized that iron would be the primary nutrient limiting both productivity and growth. Our NIFT results found macroalgae in the vicinity of highly-populated areas to exhibit greater nutrient-limited productivity, compared to macroalgae nearby less-populated areas. No significant difference in productivity between nutrient treatments or algal species was observed. The *in-situ* fertilization results found no significant biomass increase of the two macroalgae species in response to the different nutrient treatments. Our results suggest that nutrient limitation is not homogeneous across Moorea's reefs and exhibits strong environmental dependence. It could be hypothesized that the observed disparity is a result of nutrient inputs from human populations which may offset nutrient-limited productivity, thus increasing the risk of an algal dominated phase-shift.

10.2.5 Resilience of Canopy Formers to Different Profiled Marine Heatwaves

Sandra C. Straub[1*], Thomas Wernberg[1]

[1]UWA Oceans Institute & School of Biological Sciences, The University of Western Australia, 39 Fairway, Crawley 6009 WA, Australia

*corresponding author: Sandra.straub@research.uwa.edu.au

Keywords: Positive temperature anomaly, Extreme climatic event, Kelp forests, Temperature threshold, *Ecklonia radiata*

Kelp forests, and the ecosystem services they provide, are increasingly challenged by ocean warming and climate extremes such as marine heatwaves. Marine heatwaves are anomalously warm-water temperatures over a prolonged period of time, and in recent years a number of marine heatwaves have occurred across the globe with devastating biological and socioeconomic impacts. In 2011, a marine heatwave off Western Australia had a strong impact on the main canopy-formers *Ecklonia radiata* and *Scytothalia dorycarpa*, which retracted their range southwards and decreased drastically in abundance in several locations, leading to a regime shift from a three-dimensional kelp forest to a turf-dominated state in several locations. Marine heatwaves can have different 'profiles', varying in intensity, duration and return frequency, but little is known about the relative importance of these characteristics in assessing their impacts. We determined how vulnerable the main canopy-former *Ecklonia radiata* is to different marine heatwaves by exposing the kelp to a variety of heatwaves in the laboratory, with varying duration (2, 4, 6 or 8 weeks), maximum intensity (25 °C and 26.5 °C) and with/without a plateau phase at maximum intensity. We measured photosynthetic efficiency, growth, biomass, surface area, mortality and health throughout the duration of the experiment. The kelp showed clear differences in response to the different profiles, with longer heatwaves having a strong detrimental impact, and varying maximum intensity showing trends that 1.5°C temperature difference can have a strong impact over extended periods of time. We saw a strong significant effect of duration on mortality, health status, surface area and growth rate, and a delayed effect of temperature exposure on mortality indicating short-term resilience of the kelp. These results will help to understand how *Ecklonia radiata* and other kelps can deal with future heatwaves, how likely survival is under different scenarios and if recovery of impacted ecosystems is possible.

10.2.6 Temperature and Food Conditions Affect Fitness of *Aurelia aurita* Polyps

Xupeng Chi[1*], Doerthe Mueller-Navarra[2], Samuel Hylander[3], Ulrich Sommer[1], Jamileh Javidpour[1]

[1]Experimental ecology, GEOMAR Helmholtz Centre for Ocean Research, Duesternbrooker Weg 20, Kiel 24105, Germany

[2]Centre for Marine and Climate Research, Institute for Hydrobiology and Fisheries Research, University of Hamburg, Olbers Weg 24, Hamburg D-22767, Germany

[3]Centre for Ecology and Evolution in Microbial Model Systems-EEMiS, Faculty of Health and Life Sciences, Linnaeus University, Kalmar, Sweden

*corresponding author: xchi@geomar.de

Keywords: *Aurelia aurita*, Polyp, Tolerance curve, Food quality, Asexual reproduction

Summer blooms of some scyphomedusae cause high stress on marine ecosystem structure by competing for fish ecological niche, or directly predate fish eggs and larvae. They are known to be dependent to the proliferation of sessile-overwintered polyps (strobilation). So far temperature was defined as the main abiotic factor driving presence/absence of some jellyfish outbreaks. However, little is known about combined effects of biotic and abiotic factors on the fitness of polyp stage. To investigate the survival, growth and phase transition ecology of *Aurelia aurita* polyps, we designed a factorial experiment manipulating food quality (*Artemia salina* and two manipulated *Acartia tonsa* fed with different prey), food quantity (20 µg C, 5 µg C and 1.5 µg C polyp^{-1} d^{-1}) and temperature (13 °C, 20 °C and 27 °C) representing warming scenarios of the North Sea in the late autumn, early spring and summer seasons. As expected, temperature was the key factor to determine the phase transition of polyps. Interestingly, while high food concentration and better food quality could promote the organisms' production in each life cycle mode (buds, ephyrae, podocysts). Besides, food quality seems to compensate the physiological stress caused by temperature and food concentration. Based on the findings in this experiment and previous studies, we put forward the polyp's temperature tolerance curve, which may help us to understand that the variation of medusae population regarding the overlay and shifting of different environmental factors' tolerance curve.

10.2.7 Catch Distribution Shifts in Tropical Tuna Fisheries

Iratxe Rubio[1,2*], Elena Ojea[2]

[1]Basque Centre for Climate Change BC3, Sede Building 1, 1st floor. Scientific Campus of the University of the Basque Country 48940 Leioa, Spain

[2]Future Oceans Lab, University of Vigo, Campus Lagoas Marcosende, 36310 Vigo, Spain

*corresponding author: iratxe.rubio@bc3research.org

Keywords: Climate change, Tropical tunas, Transboundary stocks, Industrial fisheries, Atlantic Ocean

There is broad evidence of climate change causing shifts in fish distribution worldwide. This evidence concentrates mostly in the North Atlantic Ocean, lacking observations in tropical regions. According to the Intergovernmental Panel on Climate Change, the strongest ocean warming will be in tropical and Northern Hemisphere subtropical regions. Therefore, we see a necessity of investigating impacts of climate change in tropical areas. Moreover, even if research has documented shifts in fish distribution due to warming climates, less is known about the response of fisheries to these shifts. For that purpose, fleets fishing tropical tunas have been selected as a case study since tunas are highly migratory species with extensive international trade, tuna fisheries being among the most valuable in the world. We use catch data of Skipjack, Bigeye and Yellowfin tunas in the Atlantic from the public database of the International Commission for the Conservation of Atlantic Tunas. Data from longliners, purse seiners and bait boats operating in the Atlantic from 1970 to 2015 is explored in order to investigate past trends of fleet distribution. A spatiotemporal analysis is conducted creating first a three dimensional (spatial and temporal) data matrix of catches covering the study to detect catch latitudinal and longitudinal shifts. We expect statistically significant catch distribution changes over time and will investigate how these trends are related to environmental conditions, using climate and biological satellite data (SST, chlorophyll, etc.). We also expect to derive from our analysis the preferred temperature of the catch for the three tuna species. We will then discuss the implications of changes in distribution in terms of Economic Exclusive Zones and institutional arrangements focusing on the Sustainable Fisheries Partnership Agreements of the European Union. As a result of this study we can derive policy implications for tropical tuna catches distribution shifts and international agreements.

10.2.8 Island Resort Runoff Threatens Reef Ecosystems: An Isotopic Assessment of the Extent and Impact of Sewage-Derived Nitrogen Across Redang Island, Terengganu, Peninsular Malaysia

Liam Lachs[1,2*]

[1]Marine Biology, Vrije Universiteit Brussel (VUB), 1050 Brussels, Belgium

[2]Institute of Oceanography and Environment, Universiti Malaysia Terengganu, 21030 Kuala Terengganu, Terengganu, Malaysia

*corresponding author: liamlachs@gmail.com

Coral reef resilience is widely accepted to be driven by a site-specific combination of controls, the bottom-up control of nutrient availability and the top-down control of herbivory, which both directly affect algal growth and competition with coral. The herbivorous fish community on Redang is assumed to be less affected by fishing, the primary source of variation to herbivory, due to national marine park fishing restrictions. Development of island tourism since 1995 in marine parks of Terengganu has resulted in an increase in the number of visitors from 22,725 to over 244,762, contributing to enhancement of pollution and sewage runoff which may present an ecosystem scale threat to the coral reefs across the islands. The aim of this in situ study is to assess nitrogen uptake by coral reef organisms from around Pulau Redang in order to discern the ecological extent and impact of sewage-derived nitrogen. The expected trend is of a higher isotopic δ15N value close to sewage pollution. Differential rates of isotopic fractionation between algae (short-term, days), herbivorous gastropods or bivalves (long-term, weeks or months) allow both estimation of the pollution impact on coral food web structure as well as determination of the most suitable functional group for use as a bio-indicator. Enrichment of δ15N values are expected in anthropogenically influenced fringing reefs on the eastern side of Redang compared to more pristine reefs on the uninhabited northern side. These northern sites may provide an isotopic baseline for monitoring where new resort developments are planned. The results of this study may contribute to the elucidation of an ecosystem threshold of resort pollution, above which coral reef communities become degraded, which would be of great use as a tool for the sustainable management of these vulnerable ecosystems.

10.3 Abstracts of Poster Presentations

10.3.1 The Failed Recovery of Atlantic Cod Stocks

Camilla Sguotti[1*], Romain Frelat[1], Saskia Otto[1], Marie Plambech[2], Martin Lindegren[2], Christian Möllmann[1]

[1]University of Hamburg, Institute for Hydrobiology and Fisheries Science, Center for Earth

System Research and Sustainability (CEN), Hamburg, Germany

[2]Centre for Ocean Life, National Institute of Aquatic Resources (DTU-Aqua), Technical

University of Denmark, Charlottenlund Castle, 2920 Charlottenlund, Denmark

*corresponding author: camilla.sguotti@uni-hamburg.de

Atlantic cod populations have been exploited for centuries and have experienced dramatic collapses over the last decades. In order to facilitate the recovery of this ecologically and economically important species, drastic and unpopular management measures have been implemented. Here we conducted a meta-analysis on cod recovery based on assessment data of 19 stocks over its distribution area in the North Atlantic. Based on a set of recovery criteria, such as achievement of management targets, restoration of former population structures and levels of fishing mortality, we classified only 3 stocks out of the 19 as recovering. We further investigated the reasons for the lack of recovery applying regime shift theory. We conducted multiple analyses to detect the presence of stable states (i.e., trend analyses, detection of hysteresis), and also examined the relationship between spawning stock biomass, environmental conditions (sea surface temperature) and fishing pressure to highlight the drivers that could have played a fundamental role in the decline of these stocks. Our analyses revealed that cod recovery is hindered by discontinuous behaviors, and the additive effect of fishing pressure and climate change. These conditions delay the recovery process or in the worst cases make it even impossible. Our results are relevant because they show how additive human and natural drivers, resulting in discontinuous systems´ responses, can prevent the recovery of such an iconic species such as cod.

10.3.2 A Deep-Sea Sponge-Loop? Tracing Carbon and Nitrogen from DOM to Sponges and Detritivores

Clea van de Ven[1*], Titus Rombouts[1*], Martijn Bart[1], Benjamin Mueller[1], Jasper de Goeij[1]

[1]Institute for Biodiversity and Ecosystem Dynamics, University of Amsterdam, Science Park P.O. Box 94248, 1090 GE Amsterdam, The Netherlands

*corresponding authors: clea.vandeven@student.auc.nl, titusrombouts@gmail.com

Keywords: Deep-sea sponge-loop, DOM uptake, Detritivores

Deep-sea sponges form complex ecosystems on the seafloor known as sponge grounds. Their ecological importance and biotechnological potential are estimated to be similar to, or even greater than that of other deep-sea ecosystems such as hydrothermal vents. Sponges are considered to be key ecosystem drivers on shallow tropical reefs. They take up carbon and nitrogen from dissolved organic matter (DOM) which is assimilated in sponge cells, shed as detritus, and subsequently consumed by detritivores. These steps facilitate the transfer of nutrients from DOM to higher trophic levels by a process termed "the sponge-loop". Contrastingly, the role of sponges in deep-sea food webs is largely unknown. Recently, ex-situ experiments demonstrated DOM uptake by deep-sea sponges and subsequent shedding of detritus. This provided the first evidence for the existence of a sponge-loop pathway in the deep-sea. However, the final consumption of sponge-derived detritus by detritivores has yet to be confirmed. With this poster, we present our approach to uncover unique evidence for carbon and nitrogen transfer in the form of DOM to sponges into detritus and eventually detritivores. Isotopically labeled DOM and particulate food (bacteria) are fed to the sponges to trace the bulk assimilation and respiration rates of ^{13}C- and ^{15}N-enriched food. Second, we collect labeled sponge-derived detritus and determine the uptake and respiration by deep-sea detritivores (brittle stars). Corroborative time-lapse observation with detritivores is used to confirm the consumption of sponge-detritus by detritivores. Preliminary experiments and results will be presented during the conference. Based on these experiments we expect to provide the first complete assessment of sponge-mediated carbon and nitrogen cycling in the deep-sea. The potential presence of a sponge-loop in the deep-sea could demonstrate how sponge grounds function in their energy-limited environment and will aid in the conservation and sustainable exploitation of deep-sea sponge grounds.

10.3.3 A Stressful Environment: Macrobenthic Bioturbation Under Hypoxia and Heat Waves

A. Böhmer[1,2*], A. M. Queirós[2], D. Wain[1], L. D. Bryant[1]

[1]University of Bath, Claverton Down, Bath, BA27AY, United Kingdom

[2]Plymouth Marine Laboratory, Prospect Place, The Hoe, Plymouth, PL13DH, United Kingdom

*corresponding author: A.Boehmer@bath.ac.uk

Keywords: Habitat thresholds, Biological turnover, Macrobenthic community bioturbation, Biogeochemical cycling, Climate stressors

Oxygen (O_2) and temperature are key environmental parameters that govern water quality and biogeochemical processes and define important habitat thresholds for aquatic life. The activity and biological turnover of benthic organisms are strongly influenced by variations in these parameters. In turn, benthic organisms have significant influence on aquatic O_2 and carbon cycling via bioturbation (i.e., rework of sediment by biota) and respiration (i.e., biological O_2 uptake) and thus play a highly complex and sensitive role in biogeochemical cycling in natural waters. Evaluation of benthic O_2 dynamics and other biogeochemical fluxes within the water column and at the sediment-water interface is vital to understanding O_2 and carbon budgets and overall aquatic ecosystem health. Thus, our hypotheses are: (1) Bioturbation contributes to carbon and O_2 uptake by the sediment; (2) Environmental stressors (hypoxia & heat waves) impact the ability of benthic fauna to transport carbon and O_2 into the sediment. In this study, we are investigating the effects of bioturbation on benthic (biological) turnover and corresponding O_2 and carbon transport under varying O_2 and temperature conditions. From June to August 2017 laboratory experiments will be conducted to measure macrobenthic bioturbation and respiration within specially designed mesocosms and respiration chambers. To give realism to experiments and to do something novel, we will (1) study the macrobenthic response to the combined effect of concurrent climate stressors (i.e., O_2 and temperature), (2) unify established work on benthic-water-column interactions, and 3) draw a more integrated picture representing the functional role of macrobenthic community bioturbation in ecosystem functioning. In the frame of the 2017 YOUMARES conference in Kiel, I will present sampling and experimental methods. The main focus will be on the experimental set up in the mesocosm and macrobenthic community respiration measurements. Further I will present some initial results, showing the comparison in macrobenthic respiratory response to varying environmental stressors (hypoxia & heat waves).

10.4 Session Report

The session was attended by 30–40 participants and chaired by Camilla Sguotti and Xochitl Cormon.

First, a general introduction about regime shift theory was given by Sguotti and Cormon. Afterwards, there were 2 h of in total seven oral presentations, three poster presentations and a session wrap-up conducted interactively with the audience.

Most of the studies presented were experimental studies investigating potential drivers (parameters and state variables; Folkers, Earp, Straub, Chi, van de Ven & Rombouts and Böhmer). Interestingly, most studies were related to bottom-up control (nutrients and temperature). Coastal systems dynamics were most studied (Folkers, Earp, Straub, Lachs, and Böhmer); followed by pelagic systems (Färber, Chi, Rubio, and Sguotti) and the deep-sea (van de Ven & Rombouts). Pelagic studies were the only ones exploring time series (Färber, Rubio, and Sguotti) and focusing on either commercial fish stocks (herring and cod) or fishing effort (tuna fleets) to detect potential regime shifts. No studies on management applications were presented but four studies (Rubio, Lachs, Sguotti, and Färber) were directly linked to management in their objectives. Most of the current and known regime shift related issues were highlighted in the session:

- Difficulty in regime shift detection

All the three studies aiming to understand non-linear dynamics of fish stocks (Färber, Rubio, and Sguotti) highlighted the difficulty to recognized regime shifts. For example, Färber highlighted the difficulty to disentangle abrupt changes and natural oscillations using three different models to analyze herring spawning stock biomass and recruitment dynamics, each model providing a different vision of these dynamics.

- Hysteresis and feedback mechanisms

Hysteresis phenomena were evidenced in the study of Straub, where after a heatwave, kelp always ended up bleaching if not even dying. Sguotti showed that most of Atlantic cod stock never recovered even after management measures, while van de Ven & Rombouts evidenced a feedback loop in deep sea sponges and DOM structuring the system (as known in coral reefs).

- Requirement of large amounts of data

All the studies highlighted the need of "big data" provided by long and extensive monitoring and/or experiments. For example, the studies of Sguotti, Färber, and Earp were constrained by the availability of life-history trait information, which may be key to detect/understand the regime shifts: e.g., age-at-maturation, growth rate, etc., while Rubio's study of tune fishery effort spatial distribution was limited to purse-seine fleet analysis, which resolution was fine enough compared to the other gears data.

- Requirement of high technical skills and equipment

Challenges related to statistical analysis (Färber and Sguotti), high-tech material (van de Ven & Rombouts), or chemical compose manipulation (Earp) were highlighted as limitation to the study of various systems/drivers.

- Scale issues

Straub highlighted how temporal scales matters in the reaction of kelp forest to heatwaves showing an important lag response as well as the importance of heatwave duration. The studies of Böhmer and Folkers were limited in their experimental time and might have shown different results with longer experiment durations. Spatial scales issues were highlighted by the studies of Lachs and Rubio.

- Multiple drivers, interactive and cumulative effects

Most of the experimental studies generally focused of two drivers or more (Folkers, Earp, Sandra, Chi, and Böhmer) and generally encountered challenges trying to disentangle the various drivers potentially impacting their system of interest.

- Stakeholder conflicts and governance issues

The studies of Earp and Lachs both highlighted potential conflicts concerning marine space and resource use (heavy vs eco-tourism, conservation, fishery, etc.) and the necessity of cooperation between the different involved stakeholders and the different scales of governance organization.

To conclude, this session managed to directly highlight the current issues related to the topic through the content of the presented studies. Indirectly, it also highlighted the current needs concerning future science within the Ecosystem-Based Management (EBM) framework. For example, the only domain of science represented in this session was marine ecology. No physics, social, economic, governance, etc., abstracts were submitted at all, while we know how important each component of social-ecological systems (and their interactions) are for the understanding of the systems as well as for the implications to management. Similarly, the lack of management applications presented here may suggest that young marine scientists tend to develop knowledge and understanding of processes rather than apply current findings to current global problematics.

11 The Interplay Between Marine Biodiversity and Ecosystems Functioning: Patterns and Mechanisms in a Changing World

Francisco R. Barboza[1], Maysa Ito[1], and Markus Franz[1]

[1] GEOMAR Helmholtz Centre for Ocean Research Kiel, Düsternbrooker Weg 20, 24105 Kiel, Germany

11.1 Call for Abstracts

Existing knowledge on the role of biodiversity in determining ecosystems' characteristics and their capacity to cope with human impacts has been mainly generated by terrestrial ecologists, leaving the information about marine environments restricted to a small number of publications. Thus, a bigger number of observational and experimental works are required, in an effort to generate valuable data that allow effective conservation strategies in the sea. The scientific session welcomes contributions on the interplay between biodiversity and structure-functioning in marine ecosystems, over all spatial and temporal scales. Theoretical and applied researchers, with a purely ecological or interdisciplinary background are welcome to share their work and ideas.

11.2 Abstracts of Oral Presentations

11.2.1 Phenology-Mediated Changes in the Structure of a Benthic Food Web

Marco Scotti[1*], Agnes Mittermayr[1,2], Ulrich Sommer[1]

[1]GEOMAR Helmholtz Centre for Ocean Research Kiel, Düsternbrooker Weg 20, 24105 Kiel, Germany

[2]Center for Coastal Studies, 5 Holway Avenue Provincetown, MA 02657, USA

*invited speaker, corresponding author: mscotti@geomar.de

Keywords: Biodiversity, Body size, Food web topology, Functional traits, Network analysis

Food webs are binary representations of who eats whom in ecosystems. The arrangement of feeding interactions can be predicted from species body size, and food web structure provides clues on functioning (e.g., pathways responsible for carbon recycling). Previous studies investigated how anthropic pressures impact biodiversity and percolate their effects on food web structure. Most food web analyses rely on average annual trophic behavior, thus neglecting the relevance of seasonal changes. Such simplification impairs the chance of modeling the role of phenology (e.g., size-frequency distribution along the life cycle) on food web structure. We present the food web analysis of a benthic community in the Baltic Sea. Fifteen food webs were constructed using triple isotope analysis (δ13C, δ15N and δ34S) of samples collected every 2 weeks, from March to September 2011. Details concerning species body size and environmental conditions (e.g., temperature) were recorded. We found that during summer: (1) species are less omnivore and feed at lower trophic levels than in spring; (2) trophic specialization is associated with low trophic similarity; (3) carbon is recycled through short cycles. Such changes are coherent with the size-frequency distribution trends observed for some species of polychaetes and amphipods. Species with annual life cycle attain larger body size in spring, while the drastic reduction of average body size occurs in summer following reproduction. These trends explain the decrease of average trophic level, the decline of omnivory, and structure of cycling from spring to summer. With the construction of time-explicit food webs we showed the link between size-frequency distribution and food web structure. Size-frequency distribution is regulated by the phenology and its effects would have remained hidden with an average annual food web. This connection might turn to be particularly useful to investigate the consequences of climate change on phenology and, indirectly, on food web functioning.

11.2.2 Food Webs: A Central Piece in Biodiversity-Ecosystem Functioning Theory

P. Olivier[1*], M. C. Nordström[1]

[1]Åbo Akademi University, Environmental and Marine Biology, BioCity, Artillerigatan 6, FI-20520 Åbo, Finland

*corresponding author: pierre.olivier@abo.fi

Keywords: Food webs, Traits, Biodiversity, Ecosystem functioning

Due to human and environmental pressures on marine ecosystems (e.g., fishing and climate change), marine biodiversity is decreasing at an unprecedented rate. It is vital to understand how this change impacts ecosystems and their ability to maintain functioning. A food web topology describes the diversity of species and of their trophic interactions, i.e., who eats whom. We know that species interactions play a crucial role in how species contribute to ecosystem functions. However, little is known about how the food web structure and the functional diversity of species influence each other. Traits constitute a link between food webs and ecosystem processes. On one hand, trophic interactions are only possible when traits of the prey match with traits of the predator (e.g., body size ratio between predator and prey). On the other hand, functional traits influence species contributions to ecosystem functions (e.g., high diversity of foraging traits of predators increases their ability to exploit prey resources and subsequently increases productivity of those predators). We developed a new ecological network, the trait web, which shows links between traits based on species trophic interactions. Using the Baltic Sea food web as a starting point, the structural analysis of both, the food web and the trait web, helps us to understand which species are important for the stability of the ecosystem and which of their traits are most likely to influence ecosystem functions. Using this dual approach, we will be able to investigate how environmental drivers (e.g., environmental gradients) alter the food web structure by removing or allowing interactions (i.e., local extinction or invasion) and how these changes may affect the relationship between food webs and ecosystem functions.

11.2.3 Functional Diversity of Puck Bay – The Role of Macrofauna in Ecosystem Functioning

Marta Słomińska[1*], Urszula Janas[1], Halina Kendzierska[1]

[1]Institute of Oceanography, Department of Experimental Ecology of Marine Organisms, Aleja Marszałka Piłsudskiego 46, 81-378 Gdynia, Poland

*corresponding author: m-slominska@wp.pl

Keywords: Functional diversity, BTA, Nutrient fluxes, Puck Bay

Puck Bay, thanks for its geological characteristics and nearness of urban agglomeration is a very interesting basin in terms of the assessment of the anthropogenic impact on ecosystem health. Very important is to understand what are the relations between the role of benthic macroinvertebrates and ecosystem functioning. With this knowledge we are able to create basis to conservation of valuable and sensitive habitats of this animals. Functional diversity is a tool allowing us to study the relationships of organisms with elementary factors in ecosystems. In our study we concentrate on the role of macrozoobenthos on organic matter cycles and oxygen and nutrient fluxes on the sediment-water interface. To realize this aim we compared 6 different coastal biotopes of Puck Bay, including *Mytilus* beds, *Zostera* meadows and different types of sediments, for structural and functional diversity using inter alia Biological Trait Analyses (BTA). This approach depends on the use of faunal characteristics to assess the benthic functional structure. A substantial work was to select existing and also to create new functional traits to describe the real impact of organisms on studied factors. Preliminary results shows that among all analyzed biotopes the most different according to functional diversity was *Mytilus* bed. This study was realized within BONUS-COCOA project.

11.3 Abstracts of Poster Presentations

11.3.1 High Functional Redundancy in the North Sea Benthic System

Mehdi Gh Shojaei[1*], Lars Gutow[2], Jennifer Dannheim[2], Thomas Brey[2]

[1]Department of Marine Biology, Faculty of Marine Science, Tarbiat Modares University, Noor, Iran

[2]Alfred Wegener Institute Helmholtz Centre for Polar and Marine Research, Am Handelshafen 12, 27570 Bremerhaven, Germany

*corresponding author: shojaeimgh@gmail.com

Keywords: Biodiversity, Functional diversity, Functional traits, Redundancy, North Sea

The relationship between species biodiversity and ecosystem functioning (BEF) has recently become an important subject in marine ecosystem studies. BEF relationships have usually been studied by creating random species assemblages or by experimentally manipulating species richness. Whereas biological trait analysis have emerged as a promising way for addressing ecological functioning based on traits exhibited by members of biological assemblages. We used the functional diversity (FD) as a surrogate of actual ecosystem functioning. The relationship between species richness and FD of the North Sea benthos was best explained by a positive power function. The model predicts that at low species numbers, variation in taxonomic diversity induces substantial changes in FD. In contrast, in species-rich assemblages, a change in taxonomic diversity would have only minor effects on the functionality, indicating a high functional redundancy of the benthic assemblage. Functionally redundant ecosystems are assumed to be particularly resilience to environmental disturbance as ecosystem functioning is buffered against species loss by mutual compensation of functionally similar species.

12 Microplastics in Aquatic Habitats – Environmental Concentrations and Consequences

Thea Hamm[1], Claudia Lorenz[2], and Sarah Piehl[3]

[1]GEOMAR Helmholtz Center for Ocean Research, Kiel, Germany

[2]Alfred Wegener Institute (AWI), Helmholtz Centre for Polar and Marine Research, Biologische Anstalt Helgoland, Postbox 180, 27483 Helgoland, Germany

[3]University of Bayreuth, Department of Animal Ecology I and BayCEER, Bayreuth, Germany

12.1 Call for Abstracts

Microplastics (< 5 mm) have recently become a topic of great societal concern and a widely studied field. However, there are still many knowledge gaps concerning abundances, sources, sinks and transportation pathways, as well as their impact on biota. Reasons for this are the challenging analytical methods and demanding experimental set-ups for organism studies as well as the lack of standardized operational procedures. To close these gaps and generate comparable data in the future, we invite young scientists to present innovative methodologies along with studies assessing the concentration of microplastics in the environment or their impact on aquatic life.

12.2 Abstracts of Oral Presentations

12.2.1 Micro-Methods to Macro-Results: From a Novel Pollutant to an International Environmental Issue

Natalie A. Welden[1*]

[1]Faculty of Science, Technology, Engineering and Maths, Open University, Milton Keynes, United Kingdom

*invited speaker, corresponding author: natalie.welden@open.ac.uk

Recognition of the increasing proportion of plastics under 5 mm in size has driven global analysis of microplastics in marine biota, sediments and the water column. Over the past 10 years there has been almost exponential growth in the number of studies of microplastic pollution. From initial observations of the presence of microplastic, studies have expanded to encompass numerous impacts of microplastics, including the ecological, morphological, and ecotoxicological effects. In addition to the diversification of the field, much more is required in terms of the number of samples, geographic area and general scope of studies for publication in high impact journals. The quest for standardization is ongoing, and researchers face the challenges of plastic recovery, identification and statistical analysis. Recently, methods to quantify microplastics in the environment have come from the citizen science community. They use a combination of achievable activities and subtle education to gather information and incite behavioral change. Notable successes include the Great Nurdle Hunt, which gathers global sightings of stranded nurdles. By including the public in research, we generate meaningful contact with stakeholders and an increased understanding of the project. Such engagement is vital. The massive global action to reduce plastic pollution (micro- and macro-) has not come from researchers. It is from the investment of conservation charities and the concern of the public, the scale of pressure put on policy makers and manufacturers. Globally it has been the public voice that has translated the issues identified in peer reviewed literature into measurable change. By stepping outside our laboratories we are often able to reach more people in a day than our publications ever will; by speaking, person to person, we can communicate the way we feel about and issue, rather than just its statistical significance; and by inspiring one person, we can reach many more.

12.2.2 Behind the Scenes: Who Is the Polluter of Slovenian Coast?

Špela Korez[1*], Lars Gutow[1], Reinhard Saborowski[1]

[1]Alfred Wegener Institute, Helmholtz Centre for Polar and Marine Research, Am Handelshafen 12, 27570 Bremerhaven, Germany

*corresponding author: spela.korez@awi.de

Keywords: Adriatic Sea, Slovenia, Sediment samples, Density separation with NaCl, Microplastics

Plastic litter in the marine environment is receiving increasing attention. Namely, through rivers, ship traffic and recreation is anthropogenically produced litter introduced into the marine environment. Sun radiation and mechanical impact degrade plastic items into smaller pieces, i.e., secondary microplastics. Alternatively, primary microplastics from cosmetic products can enter the environment through waste water drains. The number of microplastics at beaches can vary substantially. In this study, the occurrence and the quantity of microplastics in the Slovenian beach sediments was investigated. Samples were taken in March 2017. Microplastic particles were isolated by density separation in an aqueous solution of common salt (NaCl). The floating particles were separated and filtered over a 100-µm metal filter. Putative microplastics were photographed and their characteristics were noted (i.e., source, form, color) and stored in glass vials with a Teflon lid. Potential contamination during the processing of the samples was controlled with a white filter in petri dish exposed to air. The reliability of the separation method was confirmed by a recovery experiment which yielded 86–90% recovery of intentionally added microplastics. Secondary microplastics were present as fragments, films, and fibers. However, the present study showed lower concentrations of microplastic at the Slovenian beaches than previously reported by Laglbauer et al. (2014, Marine Pollution Bulletin 89: 356–366). The differences in the concentrations are probably due to seasonal fluctuations of microplastic concentration. Moreover, contamination was not considered in the former study. Potential pollutants of the Slovenian coast are tourism, industry, sea- and road traffic, wastewater treatment plant, port, and agriculture. Since Laglbauer et al. (2014) excluded tourism as the contributor of the microplastic pollution the real contributors need to be identified. Polluted beaches are not attractive and may repel tourists and, thus, impair economy.

12.2.3 Microplastics in Atmospheric Deposition to an Urbanized Coastal Zone – Preliminary Results

Karolina Szewc[1*], Bożena Graca[1], Katarzyna Grochowska[2], Danuta Zakrzewska[1], Gerard Śliwiński[2], Anita Lewandowska[1]

[1]Institute of Oceanography, University of Gdańsk, Marszałka Piłsudskiego Av. 46, 81-378 Gdynia, Poland

[2]The Szewalski Institute of Fluid-Flow Machinery, Polish Academy Of Sciences, Fiszera 14, 80-231 Gdańsk, Poland

*corresponding author: karolina.szewc@phdstud.ug.edu.pl

Keywords: Microplastics, Atmosphere, Raman spectroscopy

Due to increasing production and low biodegradability, plastic pollution has become a serious problem in the marine environment. One of possible major sources of microplastics (< 5 mm) in the marine environment could be atmospheric deposition. It probably has a great importance in highly urbanized areas where factories, traffic, construction sites, extensive urban infrastructure and numerous households can emit microplastics into the atmosphere. However, likewise an impact of microplastics on terrestrial organisms, these issues have been poorly studied. Dry and mixed deposition samples were collected in the urbanized coastal zone in Gdynia. The samples were collected in a weekly cycle into glass beakers. Samples' temperature, pH and conductivity were measured. Precipitation samples were filtered through 25 mm Whatman GF/A filters. Dry deposition samples were diluted with 100 mL of deionized water prior to filtration. Microplastics collected on filters were counted and measured under a microscope and then polymers were identified using Raman spectroscopy. Additionally, in order to determine air masses origin for sampling periods, air masses trajectories (HYSPLIT model) were designated, and ionic composition of deposition samples were examined. Obtained results indicate that a major source of microplastics in the urban atmosphere can be traffic (car tires rubbed in contact with asphalt) but optical methods and Raman spectroscopy are not sufficient methods for distinguishing polymers included in tires from other carbon-rich particles. To confirm the obtained results, further research using Fourier transform infrared spectroscopy and extension of the study area to non-urbanized area are planned.

12.2.4 Fouling and Degradation of Plastic Bags – An *in situ* Experiment

Nora-Charlotte Pauli[1,2*], Jana S. Petermann[1,3], Christian Lott[4], Miriam Weber[4]

[1]Institute of Biology, Freie Universität Berlin, Königin-Luise-Str. 1-3, 14195 Berlin, Germany

[2]present Address: GEOMAR Helmholtz Centre for Ocean Research Kiel, Wischhofstr. 1-3, 24148 Kiel, Germany

[3]present Address: Department of Ecology and Evolution, University of Salzburg, Hellbrunnerstrasse 34, 5020 Salzburg, Austria

[4]HYDRA Institute for Marine Sciences, Elba Field Station, Via del Forno 80, 57034 Campo nell'Elba (LI), Italy

*corresponding author: npauli@geomar.de

Keywords: Biodegradable plastic, Polyethylene (PE), Tensile properties, Biodiversity, Oxygen production

The growing amount of plastic debris poses an increasing threat for the marine environment, simultaneously providing a new substrate for fouling organisms. Those fouling communities on plastic have not received much scientific attention. We present a first comprehensive analysis of their community composition, their primary production and the polymer degradation comparing conventional polyethylene (PE) and a biodegradable starch-based plastic blend in the Mediterranean Sea. Samples of the two polymers were exposed to a sedimentary sublittoral and a pelagic coastal habitat over the duration of 1 year. Biodiversity and oxygen production of the fouling community were investigated to assess its possible environmental and ecological role. Moreover, we tested the degradability of the two polymer types as changes of tensile properties and loss of surface area. The biomass of the fouling layer increased significantly over time and each sample became heavy enough to sink to the seafloor. The fouling communities, consisting of 21 families, were distinct between habitats, but not between polymer types. In contrast to the benthic habitat, positive oxygen production was measured only in the pelagic habitat, suggesting that large accumulations of floating plastic could pose a source of oxygen for local ecosystems, as well as a carbon sink. The biodegradable plastic showed a significant loss of tensile strength and surface area after 9.5 months of exposure in both habitats. In contrast, the polyethylene polymer showed no signs of degradation. These results indicate that in the marine environment biodegradable polymers disintegrate at higher rates than conventional polymers. This should be considered for the development of new materials, environmental risk assessment and waste management strategies, especially for applications where an introduction into nature is unavoidable or likely.

12.2.5 Aggregation of Microplastics with Marine Biogenic Particles

Angela Stippkugel[1*], Jan Michels[1,2]

[1]GEOMAR Helmholtz Centre for Ocean Research Kiel, Düsternbrooker Weg 20, 24105 Kiel, Germany

[2]present Address: Department of Functional Morphology and Biomechanics, Institute of Zoology, Christian-Albrechts-Universität zu Kiel, Am Botanischen Garten 1–9, D-24118 Kiel, Germany

*corresponding author: astippkugel@geomar.de

Keywords: Coagulation, Marine plastic pollution, Marine biogenic particles, Microplastics

Although plastic pollution of the oceans has resulted in a dramatic accumulation of microplastics (i.e., plastic particles smaller than five millimeters) in the marine environment, knowledge of the impact of microplastics on biological processes in marine ecosystems is still very scarce. Today, microplastics are present everywhere in the oceans, including even very remote habitats such as deep-sea sediments and sea ice. However, the concentration of microplastics in the surface ocean is considerably lower than expected given the ongoing replenishment of microplastics and the tendency of many plastic types to float. It has been hypothesized that microplastics leave the upper ocean by aggregation and subsequent sedimentation. In the present study, we tested this hypothesis by investigating the interactions of microplastics with marine biogenic particles collected in the Bay of Kiel. Our laboratory experiments performed with roller tanks revealed a large potential of microplastics to rapidly coagulate with biogenic particles. In control experiments with microplastics and either filtered seawater or artificial seawater, only single microplastics spontaneously attached to each other. Together with the biogenic particles, the microplastics formed relatively large aggregates within a few days. In the presence of microplastics, the particle aggregation was faster and more pronounced compared with that observed between biogenic particles only. The results of our study suggest that microplastics become involved in the natural particle aggregation processes taking place in the marine water column. We assume that the aggregation of microplastics with marine biogenic particles facilitates the export of microplastics from the surface ocean and plays a significant role in the redistribution of microplastics in the oceans.

12.2.6 Quantitative Investigation of Microplastic Interaction with Blue Mussel Adults and Larvae

Sinja Rist[1*], Ida Mathilde Steensgaard[1], Olgac Guven[2], Lene Friis Møller[2], Torkel Gissel Nielsen[2], Lene Hartmann Jensen[2], Anders Baun[1], Nanna B. Hartmann[1]

[1]Department of Environmental Engineering, Technical University of Denmark, Bygningstorvet, Building 115, Kgs. Lyngby, Denmark

[2]National Institute for Aquatic Resources, Technical University of Denmark, Kemitorvet, Building 202, Kgs. Lyngby, Denmark

*corresponding author: siri@env.dtu.dk

Keywords: Microplastics, *Mytilus edulis*, Uptake, Quantification

An increasing number of studies are showing that marine invertebrates ingest microplastics (< 5 mm). One of the species studied in this context is the blue mussel (*Mytilus edulis*). As a filter feeder it is considered especially susceptible to microplastics uptake and it is a key species of benthic ecosystems. Although it is known that blue mussels ingest microplastics in a wide range of sizes, and the particles have been found to impair their physiology, the dynamics of ingestion and egestion of microplastics are only poorly understood and rarely quantified. Furthermore, all studies have so far been carried out with adult mussels and it is unclear how blue mussel larvae are affected. The aim of this research was therefore to investigate the fate of microplastics during an exposure of blue mussels and to quantify particle ingestion and egestion in adult mussels as well as mussel larvae. In the first stage methodologies for particle quantification in the whole animals were developed. These included enzymatic digestion protocols and fluorescence measurements. Subsequently, mussels were exposed to different exposure scenarios (varying microplastic sizes, concentrations, food availability and exposure time) and the fate of the particles was examined in the animals, in their feces and in the water. The extent to which the animals egest the taken up particles was examined by transferring them to clean water after exposure. This study found that not only adult blue mussels but also the larvae ingest a wide range of particle sizes. Furthermore, we investigated a wide range of particle sizes and how the ingested quantities relate to particle size, concentration, time and food availability. This work contributes to enhancing our knowledge on the mechanisms of microplastics – blue mussel interaction, which is decisive for understanding and evaluating the potential hazard of microplastics to this species.

12.2.7 Does Heat Stress Amplify the Negative Effects of Microplastics on Two Bivalve Species?

Vanessa Herhoffer[1*], Mone Ota[2], Masahiro Nakoka[2], Mark Lenz[3]

[1]Department of Economics, Faculty of Business, Economics and Social Sciences, Kiel University, Wilhelm–Seelig-Platz 1, 24118 Kiel, Germany

[2]Akkeshi Marine Station, Hokkaido University, Aikappu 1, Akkeshi, Hokkaido 088-1113, Japan

[3]GEOMAR Helmholtz Centre for Ocean Research Kiel, Düsternbrooker Weg 20, 24105 Kiel, Germany

*corresponding author: vanessa.herhoffer@web.de

Keywords: Microplastics, Global warming, Filter feeders, Interactive effects, Respiration rate

Small plastics particles (< 5 mm), so called microplastics, are now ubiquitous in the marine environment and can mistakenly be taken up by organisms like benthic filter feeders – what can affect them negatively. Furthermore, in the past 50 years, the average surface temperature of the world's ocean has increased and will continue to rise in the future. The two stressors therefore currently act in concert in many coastal ecosystems worldwide. In an experimental approach, we addressed the question whether elevated temperatures can amplify the negative effects of microplastics on marine benthic filter feeders. We investigated this in the blue mussel *Mytilus trossulus* (N = 120) and in the Pacific oyster *Crassostrea gigas* (N = 120) by performing a laboratory experiment in which the test animals were subjected to four different microplastics concentrations (0, 2, 20 and 200 mg/L) and three elevated temperature levels (17, 20 and 23 °C), simultaneously. The animals were collected near Akkeshi in Japan, where 17 °C is the ambient summer sea surface temperature. After 82 days of exposure, respiration rates of both test organisms were found to decline with increasing microplastics concentrations and increasing water temperatures. However, elevated water temperatures did not amplify the negative effects of microplastics on the performance of the bivalve species we tested. We discuss these results in the context of the increasing awareness of microplastics as a further component of marine global change.

12.2.8 Impact of Synthetic and Natural Microparticles in the Shrimp *Palaemon varians*

Mara Weidung[1*], Reinhard Saborowski[1], Lars Gutow[1]

[1]Alfred Wegener Institute, Am Handelshafen 12, 27570 Bremerhaven, Germany

*corresponding author: mara.weidung@awi.de

Keywords: Microplastics, Titanium dioxide, Oxidative stress

Microplastics (< 5 mm) have become ubiquitous in waters. The smaller they are the easier they can be taken up by aquatic organisms. Once ingested they can cause various harmful effects. This study investigates the effects of size of artificial and natural particles on the induction of cellular stress in the common ditch shrimp (*Palaemon varians*). The study includes feeding experiments with different sizes of fluorescent microplastic particles, nanosized titanium dioxide particles and silica powder of diatoms as a reference for natural particles. The uptake and distribution of particles in the digestive organs was observed by fluorescence microscopy. As marker for oxidative stress we measured the activities of the antioxidant enzymes catalase and superoxide dismutase in extracts of the midgut glands of animals which were fed with particles from 2 to 48 h. The larger particles (2 µm and 10 µm) remained in the stomach and in the lumen of the gut. The smaller particles (0.1 µm) were translocated into the surrounding tissues and entered the cells of the midgut gland. Crustaceans have a stomach with fine-meshed filter structures which prevent the uptake of particles > 1 µm into the digestive gland. Superoxide dismutase (SOD) activity was rapidly induced when the animals were exposed to 0.1 µm plastic particles. The activity increased within 2 h after microplastic ingestion and remained high after 48 h. Slight difference appeared between natural and synthetic particles. The diatom powder also induced SOD activity which, however, continuously decreased with time. It can be assumed that any particles < 1 µm enter the cells of the midgut gland and induce oxidative stress. Histological analysis of cryosections and scanning electron microscopy will help to clear up how far the different particles penetrate the cells of the digestive organs.

12.2.9 Differential Effects of Microplastics on Growth and Survival of Corals

Jessica Reichert[1*], Angelina Arnold[1], Patrick Schubert[1], Thomas Wilke[1]

[1]Department of Animal Ecology & Systematics, Justus Liebig University Giessen, Heinrich-Buff-Ring 26-32 (IFZ), D-35392 Giessen, Germany

*corresponding author: Jessica.Reichert@allzool.bio.uni-giessen.de

Keywords: 3D scanning, Growth rates, Microplastic, Scleractinian corals, Survival

Microplastics (i.e., plastic fragments <5 mm) gained recent attention in public and science as they are considered to be a major threat to marine ecosystems. As these plastic particles are mainly of terrestrial origin, coastal ecosystems such as coral reefs are particularly threatened. Previously it has been shown that scleractinian corals ingest microplastic particles. However, little is known about other responses and the subsequent effects on health and survival of the coral holobiont under realistic long-term conditions. Thus, the influence of low microplastic concentrations (polyethylene, size 35–650 µm, concentration 200 particles L^{-1}) on scleractinian corals (*Acropora*, *Pocillopora*, and *Porites* spp.) was examined in a 9-month fully controlled lab experiment. In particular, we studied growth and survival rates of the corals by utilizing 3D scanning and 3D model analyses. Our preliminary results indicate that different species responded differently to microplastic exposure. Both increases and decreases in growth and survival rates might be associated with the exposure to microplastics. This is an important baseline study of scleractinian corals from three genera showing the diverse impacts of microplastic exposure, calling for further investigations of the effects of microplastics on the integrity of the coral holobiont.

12.3 Abstracts of Poster Presentations

12.3.1 Does Heat Stress Amplify the Negative Effects of Microplastics on the Blue Mussel *Mytilus edulis*?

Kyra Paulweber[1*], Nadine Yvonne Müller[2], Mark Lenz[3]

[1]Department of Natural Resource Conservation, Faculty of Agricultural und Nutritional Sciences, Kiel University, Olshausenstraße 40, House I, 24118 Kiel, Germany

[2]Department of Biosciences, Faculty of Marine Sciences, Rostock University, Albert Einstein Straße 3, 18059 Rostock

[3]GEOMAR Helmholtz Centre for Ocean Research Kiel, Düsternbrooker Weg 20, 24105 Kiel, Germany

*corresponding author: kyra.paul@icloud.com

Keywords: Microplastics, Global warming, Filter feeders, *Mytilus edulis*, Interactive effects, BCI

Nowadays microplastic particles are ubiquitous in the marine environment. Because of their small size (< 5 mm), positive buoyancy and their low degradation rates, they are transported quickly across great distances and accumulate even in remote ecosystems. They can negatively affect marine benthic filter feeders, like the blue mussel *Mytilus edulis*, which ingest micro-sized plastic particles together with their food. In addition to this pollution, sea surface temperatures worldwide rise as a consequence of global warming, putting a further pressure on mussels in shallow water habitats. While there is good knowledge about the effects of heat stress and microplastic pollution separately, we know little about possible interactive effects of these two stressors. Therefore, we tested whether elevated temperatures can amplify the negative effects of microplastics on blue mussels in a laboratory experiment. For this, mussels from the Menai Strait, Wales, where 14 °C is the ambient summer sea surface temperature, were exposed to three temperature levels (14 °C, 17 °C, 20 °C) and four microplastic concentrations (0, 2, 20, 200 mg/L), simultaneously. After 79 days of exposure, the Body Condition Index (BCI) of the mussels was found to decrease with increasing water temperature and with increasing microplastic concentration. However, the latter effect was less pronounced at 17 °C and 20 °C than at 14 °C. This was due to the fact that the BCI at the two elevated temperature levels was generally lower. Although no interactions between the two factors emerged, the elevated temperatures obviously had a stronger impact on *Mytilus edulis* than the applied microplastic concentrations. We discuss these results in the light of the increasing public and scientific awareness of microplastics as an additional stressor in marine ecosystems.

12.3.2 Microplastic Occurrence in North Sea Surface Waters

Lisa Roscher[1*], Claudia Lorenz[1], Sebastian Primpke[1], Gunnar Gertds[1]

[1]Alfred-Wegener-Institut Helmholtz Zentrum für Polar- und Meeresforschung, Biologische Anstalt Helgoland, Kurpromenade 201, 27498 Helgoland, Germany

*corresponding author: lisa.roscher@awi.de

Keywords: Microplastics, North Sea, Infrared Spectroscopy

The global plastic production is increasing steadily. More and more studies focus on the occurrence of microplastics, i.e., synthetic organic polymers with a size < 5 mm, in environmental samples. These pollutants are omnipresent and hardly degradable. They are easily ingested by a wide range of animals throughout the food web and may act as a vector for persistent organic pollutants (POPs). For a valid evaluation of microplastic pollution in marine ecosystems appropriate assessment strategies are crucial. By now, no standardized sampling and analysis techniques are available, which is urgently needed in order to generate solid and comparable data bases. In this work, state-of-the-art methods were used for the identification and quantification of microplastics in North Sea neuston samples. The samples were split into two size fractions, on which two different methodological approaches were applied. Microplastics > 500 µm were extracted using a stereomicroscope, followed by polymer identification via Attenuated Total Reflection based Fourier Transform Infrared spectroscopy (FTIR-ATR). For the size fraction < 500 µm, more complex methodologies were employed: a recently developed enzymatic purification protocol was used in order to extract microplastics from the sample matrix. This was conducted in a novel filtration system (Microplastic-Reactor), followed by spectroscopic analysis via focal plane array based µ-Fourier-Transform Infrared spectroscopy (µFTIR-FPA). A subsequent automated analysis provided detailed information on particle number and sizes as well as chemical composition. Microplastic concentrations ranged from 0 to 2.5 m^{-3} (0–2.7 × 10^5 km^{-2}) in the size fraction > 500 µm and from 16.1 to 393.1 m^{-3} (1.3 × 10^6 to 4.3 × 10^7 km^{-2}) in the size fraction < 500 µm. Small-sized particles clearly dominated in both fractions. In total, 17 different synthetic polymers were detected with comparably high abundances of polyethylene, polypropylene, varnish and rubber, possibly originating from land-based sources or shipping activities.

12.3.3 Does Microplastic Induce Oxidative Stress in Marine Invertebrates?

Sarah Riesbeck[1,2*], Lars Gutow[2], Reinhard Saborowski[2]

[1]Technische Universität Darmstadt, Karolinenplatz 5, 64289 Darmstadt, Germany

[2]Alfred Wegener Institute for Polar and Marine Research, Am Handelshafen 12, 27570 Bremerhaven, Germany

*corresponding author: sarah.riesbeck@awi.de

Keywords: Crustacea, Toxicity, Cellular stress, ROS, Fluorescence Microscopy

In the last decades the production of plastic increased continuously. Simultaneously, environmental pollution by plastic became a rising issue. Marine litter can have adverse effects on animals. Some species may get trapped in lost fishing nets or they may starve to death upon ingestion of plastic which may clog their digestive tracts. Degradation of plastic items generates a continuously increasing number of smaller-sized particles. Microplastic, finally ranging in the µm-size classes can have adverse effects on marine invertebrates upon ingestion. Most of these effects can be attributed to the cellular level. How can particles in the microscale harm organisms? In this study the ingestion of microplastic by marine invertebrates and, moreover, the possible transfer into cells of the digestive tract will be examined. As model species we chose the Atlantic ditch shrimp (*Palaemon varians*). This species inhabits coastal regions, estuaries, and brackish water systems which are most affected by anthropogenic pollution. Effects will be determined in the cells of the midgut gland of *P. varians*. Measuring the formation of reactive oxygen species (ROS) is a suitable method to detect cellular stress. Quantification of ROS-formation will be done by confocal laser scanning microscopy and the aid of the fluorogenic substrates Dihydroethidium (DHE) and 2′, 7′ – Dichlorodihydrofluorescin diacetate (DCFDA). The results will help to identify cellular reactions after exposure to microparticles and indicate the toxicological impact on cells and whole organisms.

13 Cephalopods: Life Histories of Evolution and Adaptations

Fedor Lishchenko[1] and Richard Schwarz[2]

[1]Russian Federal Research Institute of Fisheries and Oceanography (VNIRO), Laboratory of Commercial Invertebrates and Algae, Moscow, Russia

[2]GEOMAR Helmholtz Centre for Ocean Research Kiel, Germany

This session was no. 4 of the YOUMARES 8 conference. It does not have a corresponding proceedings article.

13.1 Call for Abstracts

More than 400 million years ago the first cephalopods started inhabiting the World Oceans. Today their modern representatives spread around the world. Impetuous squids, intelligent octopuses and other cephalopods can be found everywhere from epipelagic layer to abyssal depths, from the coast to the open seas in almost every latitude from the tropics to Polar Regions. Living in such different habitats and ecological niches demanded from cephalopods development of several morphological, biological and behavioral adaptations. We warmly welcome biologists, ecologists, paleontologists, neurobiologists and specialists in fisheries management to present results of their studies at our session and to discuss these adaptations.

13.2 Abstracts of Oral Presentations

13.2.1 In situ Observations Reveal the Diversity of Life History Strategies in Oceanic Cephalopods

Henk-Jan T. Hoving[1*]

[1]GEOMAR Helmholtz Centre for Ocean Research Kiel, Düsternbrooker Weg 20, 24105 Kiel, Germany

*invited speaker, corresponding author: hhoving@geomar.de

Cephalopods are typically considered as active, carnivorous, short lived, mass spawning and dying invertebrates. However, this view is mostly based on neritic species. Biological knowledge on many oceanic and deep-sea cephalopods is still absent, but where available it shows a more diverse array of feeding and reproductive strategies than what is known from shallow water relatives. Such new knowledge is particularly coming to light through the application of in situ observational technologies, which allow the study of cephalopods and their behavior in the natural environment of the open ocean. In this seminar an overview of recent advancements in cephalopod behavior and ecology as well as life history strategies will be presented, and how the use of deep-sea observational technology has advanced our traditional view.

13.2.2 *Alloteuthis subulata* Aging: From Methods to Implications

Chris Barrett[1*], Chris Firmin[1], Rosana Ourens[1], and Vladimir Laptikhovsky[1]

[1]Centre For Environment, Fisheries & Aquaculture Science (Cefas), Pakefield, Lowestoft, Suffolk, NR33 0HT, England

*corresponding author: Christopher.barrett@cefas.co.uk

Keywords: *Alloteuthis subulata*, Ageing, Conservation, Biology, Squid

The European 'common' squid, *Alloteuthis subulata* are a relatively small, fast-growing and short-lived species, reaching up to 14 cm mantle length and living between 6 and 12 months. Furthermore, males and females have similar length/weight relationships until 7 cm ML, where females become heavier than same-sized males though they do not reach the same size. Squid were worth £6.4m to the U.K. in 2015, with *A. subulata* contributing highly to these landings, along with loligids *Loligo forbesii* and *Loligo vulgaris.*

Despite their high value, there are currently no management plans in place for *A. subulata* in U.K. waters, whilst squid are strictly managed to differing degrees in other countries. There is therefore a need to better understand the biology of U.K. specimens; ascertaining spatiotemporal differences in age-at-maturity could help determine whether enforcing byelaws for squid management could be beneficial for conserving stocks. Samples of *A. subulata* were collected from surveys on the RV Cefas Endeavour, as well as from commercial catches from the English Channel. Specimen length, weight, maturity stage (using Lipinski's 5-stage scale) were recorded, and ageing tools were extracted. Ageing tools (statoliths, beaks and gladii) were also extracted, prepared, read and analyzed to determine specimen growth rate and age-at-maturity, as well as determining, for future work, which tool is the most feasible and reliable for ageing research. Preliminary results of biological analyses and age readings are discussed, along with implications for fishery management.

13.2.3 The Influence of Environmental Factors on *Berryteuthis magister* (Berry, 1913) Aggregations Density in the Area of the Northern Kuril Islands

A. Lishchenko[1*], K. Kivva[1]

[1]Russian Federal Research Institute of Fisheries and Oceanography (VNIRO), 107140, Russia, Moscow, Verkhnaya Krasnoselskaya 17

*corresponding author: gvajta@ya.ru

Keywords: *Berryteuthis magister*, Cephalopoda, Environmental factors, Aggregations density

The schoolmaster gonate squid *Berryteuthis magister* (Berry 1913) is the most exploited cephalopod species in Russian waters. More than 100000 tones is caught by the trawling fleet in the area adjacent to the Northern Kuril Islands annually. The main aim of this work is to study the influence of environmental factors on the density of *B. magister* aggregations on the fishing grounds in this area. The commercial statistics data, provided by large-capacity fleet, was obtained from the fishery database. This data set included: date, time, coordinates, tow speed and depth, volume of catch for each trawling, and total daily catch for the fishery seasons from 2009 to 2016. Additionally we used the data set provided by scientific observers which contained detailed information on temperature and aggregations density collected during surveys on *B. magister* fishery. Presented work is the continuation of the study begun in 2013 which showed significant relationship between bottom temperature, presence of the bottom relief anomalies and squid aggregation density. Our study confirmed periodical short-term changes both in the volume of catches and the density of small-scale *B. magister* aggregations. However unlike previous studies we haven't confirmed correlations between bottom relief and aggregations density. Between studied characteristics only bottom temperature showed strong and constant correlation with the density of squid aggregations. According to this finding we discuss the role of the bottom temperature as the factor determining the formation of aggregations, or just an indicator of environmental processes significant for *Berryteuthis magister*.

13.2.4 Reproductive Strategies of Bobtail Squids in the Arctic (Cephalopoda: Sepiolida)

Alexey V. Golikov[1*], Rushan M. Sabirov[1]

[1]Kazan Federal University, Kremlyovskaya Street 18, 420008 Kazan, Russia

*corresponding author: golikov_ksu@mail.ru

Keywords: Cephalopoda, Sepiolida, Arctic, Reproductive Biology, Reproductive Strategy

Morphology of the reproductive system and reproductive strategies are very important while understanding the life history of the marine organisms. These characters are especially useful in the organisms living in the critical environmental conditions, such as the Arctic. Reproductive biology and ecology were studied in arctic-boreal bobtail squid *Rossia palpebrosa* Owen, 1834 (326 specimens) and high-arctic *R. moelleri* Steenstrup, 1856 (30 specimens) collected in the Barents Sea and adjacent waters. Females have unpaired ovary, left oviduct with oviducal gland, paired nidamental glands and accessory nidamental glands. Fecundity in mature females is 120–274 (191 ± 8.84) oocytes in *R. palpebrosa* and 310–531 (396 ± 48.32) oocytes in *R. moelleri* with mature oocyte diameter 6.1–11.4 (8.7 ± 0.9) mm, 15.00–37.73 (21.97 ± 3.18) % of mantle length and 8.0–13.0 (11.1 ± 1.1) mm, 11.94–19.40 (16.24 ± 1.23) % accordingly. Early stages of oocytes development are absent earlier during ontogenesis in comparison to tropical and temperate cephalopods. Males have unpaired testis occupying asymmetric position and spermatophoric complex with loop-like coiled basal part of the spermatophoric sac. Spermatophore numbers are 13–62 (31 ± 2) with length 8.9–19.0 (13.5 ± 0.08) mm, 30.00–54.21 (41.50 ± 0.26) % in *R. palpebrosa* and 84–141 (109 ± 6) with length 17.5–21.7 (19.7 ± 0.07) mm, 40.00–53.33 (44.56 ± 0.18) % in *R. moelleri*. Sizes of reproductive products in both species are bigger than in tropical/temperate Rossiinae, but their number is lower. *Rossia* thus shows tendency to increase reproductive *K*-strategy features while moving northward to the Arctic with secondary increase of already enlarged oocytes and spermatophores in *R. moelleri*. The same time morphology of the reproductive system clearly bear ancestral features of all Sepiolida. So the morphology of the reproductive system is phylogenetically explained the same time its functioning being adaptation to the critical environmental conditions of the Arctic.

13.3 Abstracts of Poster Presentations

13.3.1 Determination of Squid Age Using Upper Beak Rostrum Sections: Technique Improvement and Comparison with Statolith

Bi Lin Liu[1,2*], Xin Jun Chen[1] , Yong Chen[2], Guan Yu Hu[1]

[1]College of Marine Sciences, Shanghai Ocean University, 999 Hucheng Ring Road, Lingang New City, Shanghai, China, 201306

[2]School of Marine Sciences, University of Maine, Orono, Maine 04469, USA

*corresponding author: bl-liu@shou.edu.cn

Keywords: Upper beak, Rostrum sagittal sections, Age validation, *Dosidicus gigas*, *Ommastrephes bartramii*, *Illex argentinus*, *Sthenoteuthis oualaniensis*

Analysis of growth increments in beak rostrum sagittal sections (RSS) has been increasingly used for estimating octopus age. In this study we develop an effective method to process and read the RSS of 4 oceanic ommastrephid squid (*Dosidicus gigas*, *Ommastrephes bartramii*, *Illex argentinus* and *Sthenoteuthis oualaniensis*) and validate the daily deposition of the increments by comparing to corresponding statolith-determined ages. The proposed method of processing yielded readable rates ranging from 42.9% to 71.7% for samples of different species. The high precision of the increment readings with low independent counting coefficient of variation (CV) indicates that the processing and counting methods used are reliable. This study suggests that the RSS of the upper beak is an appropriate tool for estimating the age of *D. gigas*, *O. bartramii* and perhaps *S. oualaniensis*, although possible erosions of the rostral region may result in an underestimation of squid ages.

13.3.2 Ontogeny of Upper Beak in *Octopus vulgaris* Cuvier, 1797

E. N. Armelloni[1,2*], M. J. Lago-Rouco[1], A. Bartolome[1], E. Almansa[1], G. Scarcella[2,3], C. Perales-Raya[1]

[1]Instituto Español de Oceanografía. Centro Oceanográfico de Canarias, Vía Espaldón Dársena Pesquera PCL 8, 38180, Sta. Cruz de Tenerife, Spain

[2]Ms.C. of Marine biology. School of Science. University of Bologna. Ravenna Campus, via S. Alberto 163, 48123 Ravenna, Italy

[3]Institute of Marine Science (ISMAR), National Research Council (CNR), L.go Fiera della Pesca, 60125 Ancona, Italy

*corresponding author: enrico.e.armelloni@gmail.com

Keywords: Octopus, Beak, Embryo, Age, Growth increments

Octopus vulgaris (Cuvier 1797) is a candidate for aquaculture diversification, but a large mortality rate at early stages is a bottleneck for the commercial production. Comparison of wild and cultured paralarvae of similar ages is of great interest to establish requirements for culture conditions. The current methodologies for ageing octopus paralarvae use daily incrementents in Rostrum surface or in Lateral Walls of upper beak. Ontogeny of the beak microstructure would provide information to assess the presence of pre-hatching increments, nonetheless it is still unexplored in cephalopods. We provide a morphological description of upper beak ontogeny in *Octopus vulgaris*, addressing the onset of microstructural features and assessing the presence of any pre-hatching increments. We have used seven stages to divide late phase of ontogeny. From each stage, an upper beak was extracted and photographed wet under a coverslip using transmitted light with Differential Interference Contrast (DIC-Nomarski). Our preliminary results indicate that upper beak at a very early stage is created from two layers and that one of those already shows teeth outline. Soon the layers overlap in the front creating an overlapping area named Core. Afterwards, a third layer appears over teeth outline. It grows to outcomes the beak surface and creates Shoulder and Hood. The row of teeth is the apical part of the Rostrum and it seems to arise from a sheath just before hatching. This process leaves a hatching mark in the Rostrum surface, which corresponds to the first increment. On the other hand, the increments in Lateral Walls soon appear in embryonic development. These increments create a pattern which continues without interruptions up to paralarval stage, thus hindering the identification of any hatching mark in Lateral Walls.

14 Coastal Ecosystem Restoration – Innovations for a Better Tomorrow

Jana Carus[1] and Matthias Goerres[1]

[1]TU Braunschweig, Institute of Geoecology, Landscape Ecology and Environmental System Analysis, Langer Kamp 19c, 38106 Braunschweig, Germany

This session was no. 12 of the YOUMARES 8 conference. It does not have a corresponding proceedings article.

14.1 Call for Abstracts

Coastal ecosystems provide a variety of services. Due to increasing anthropogenic pressures, such as large-scale shipping, overfishing and eutrophication, the degradation and loss of suitable habitat in the past decades has led to numerous – yet more failed than successful – efforts of ecosystem restoration. This evokes a necessity for innovative approaches and alternative solutions. This session will comprise of the assessment of coastal ecosystem integrity, the identification of suitable restoration sites as well as the design of restoration measures and products. Studies covering these issues in salt marshes, seagrass/macrophyte beds, mussel banks, mangroves or coral reefs are very welcome.

14.2 Abstracts of Oral Presentations

14.2.1 Coastal Ecosystem Restoration – Innovations for a Better Tomorrow

Jana Carus[1*], Matthias Goerres[1*], Maike Paul[1]

[1]TU Braunschweig, Institute of Geoecology, Landscape Ecology and Environmental System Analysis, Langer Kamp 19c, 38106 Braunschweig, Germany

*corresponding authors: j.carus@tu-bs.de, m.goerres@tu-bs.de

Coastal ecosystems provide a variety of services. Due to increasing anthropogenic pressures, such as large-scale shipping, overfishing and eutrophication, the degradation and loss of suitable habitat in the past decades has led to numerous – yet more failed than successful – efforts of ecosystem restoration. This evokes a necessity for innovative approaches and alternative solutions. For instance, seagrass ecosystems are very dynamic and under constant change. Once vanished, seagrass meadows are difficult to re-establish because enhanced hydrodynamic energy and turbidity levels hinder resettlement. The project SeaArt aims to enhance seagrass restoration success by developing biodegradable artificial seagrass that improves these environmental conditions. Field measurements will help to quantify the dynamics of existing seagrass meadows and the effect of real seagrass on hydrodynamics and turbidity. Mesocosm experiments will be conducted to shed light on required establishment conditions. Within the scope of this project, we want to gain information on the most appropriate design and the required effect of the artificial seagrass. Furthermore, we aim to quantify the effect of a successful restoration on wave attenuation and thus coastal protection.

14.2.2 Coral Transplantation – Scientific Method or PR-Instrument?

Lena Rölfer[1*], Margaux Y. Hein[2,3], Sebastian Ferse[1]

[1]Leibniz Center for Tropical Marine Research, Fahrenheitstraße 6, 28359 Bremen, Germany

[2]College of Science and Engineering, James Cook University, Townsville, Queensland, 4811, Australia

[3]Australian Research Council (ARC) Centre of Excellence for Coral Reef Studies, Townsville, Queensland, 4811, Australia

*corresponding author: lena@roelfer.de

Keywords: Coral transplantation, Reef restoration, Survey, Transplantation techniques

Coral reefs are under threat of local factors such as chemical pollution, ship groundings, dynamite fishing and tourist damage and under threat of climate change on a global scale. Coral transplantation is a form of active reef restoration that is usually designed to assist natural recovery processes of a reef, but the method seems to be used in a broader context without prior assessment of necessity and effectiveness. To design a successful and sustainable transplantation project, many factors have to be taken into account, such as the receiving area, source of transplants, species selection, and monitoring and maintenance of transplants. To gauge the extent to which these principles are taken into account, we designed a survey to assess whether transplantation techniques are used in a scientific way, or if the term coral transplantation is rather used as a PR-method. Organizations conducting coral transplantation projects worldwide were contacted and asked to participate in the survey. Preliminary results show that only in 65% of the projects the cause of degradation of the reef was assessed. However, 76% assessed the environmental conditions prior to transplantation and 93% conducted monitoring after transplantation. All non-monitored projects were run by businesses. Main objectives of transplantation efforts differed among types of organizations. While research institutions, private organizations and NGOs concentrate clearly on habitat restoration, businesses follow various objectives such as relocation of threatened species, creation of tourist attraction, creation of new habitat and habitat restoration. In total 34% of all respondents indicated that the aim of the project is the creation of a tourist attraction. Results show that coral transplantation projects are used for various aims, not only the initial one of habitat recovery. Voluntary comments at the end of the survey indicate that more scientific knowledge and better monitoring is necessary to improve future transplantation projects.

14.3 Abstracts of Poster Presentations

14.3.1 Can Nutrient Mitigation Measures (Mussel Farms) Help to Restore Submerged Macrophytes?

R. Friedland[1*], A.-L. Buer[1], S. Dahlke[2], S. Paysen[1], G. Schernewski[1,3]

[1]Leibniz-Institute for Baltic Sea Research Warnemünde, Germany

[2]University Greifswald, Germany

[3]Klaipeda University, Marine Science and Technology Center, Lithuania

*corresponding author: rene.friedland@io-warnemuende.de

Many coastal waters are struggling with their heavily eutrophied state, resulting in high amounts of phytoplankton and low Secchi Depth which led to a strong decline of submerged macrophytes. For example, in Szczecin Lagoon (Oder Lagoon) dense macrophyte stocks were reported until the 1970ties, while nowadays only sparse spots are left. This was accompanied by a drop of Secchi Depth from up to 2 m to 60 cm over the last century, indicating that light availability may be the key limiting factor. On the hand, filter feeders like *Mytilus* or *Dreissena spp.* are known to reduce the phytoplankton densities substantially, resulting in an increased Secchi Depth. Hence, mitigation measures, like mussel farms, are not only an option to enhance the nutrient retention within coastal waters, but also to support the restoration of submerged macrophytes, if their growth is light limited.

14.3.2 Hydrological, Sedimentation and Biotic Ways of Seawater Self-Purification of Sevastopol Bay Waters from Plutonium Alfa-Radionuclides

A. A. Paraskiv[1*], N. N. Tereshchenko[1], V. Y. Proskurnin[1]

[1]The A.O. Kovalevsky Institute of Marine Biological Research of RAS, Sevastopol, Russian Federation

*corresponding author: artem.paraskiv@mail.ru

Keywords: Black Sea, Sevastopol Bay, $^{239+240}$Pu, Vertical and spatial distribution, Ways of $^{239+240}$Pu elimination from the water masses of the bay

Since the second half of the twentieth century, with the beginning of the use of nuclear energy for military and peaceful purposes, many artificial radionuclides entered natural ecosystems as a result of the normative work of nuclear enterprises as well as accidents involving nuclear technologies. The Black Sea has a huge catchment basin. Along with a large number of pollution sources on land, from which pollution comes from surface and sewage, the largest rivers of Europe flow into the Sea, as continuously operating suppliers of a wide variety of pollutants. Nowadays, in the post-Chernobyl period the main dose-forming technogenic radioisotopes are ^{137}Cs, ^{90}Sr, and ^{239,240}Pu as well as ^{243}Am alpha-isotopes in the Black Sea ecosystems. This fact determined the importance of studying the radioecological migration regularities of plutonium in the Black Sea in off shore and near shore areas. Half-lives of ^{239,240}Pu are thousands and tens of thousands of years, so they are a long-term technogenic radioecological factor with a radiation-toxic effect on human and biota. Sevastopol Bay is one of the largest and most widely used bays in the Black sea. Therefore, it is important to study radioecological parameters of distribution of plutonium radionuclides. It were studied the levels of plutonium contamination in the bottom sediments, seawater and some groups of hydrobionts. As is known, bottom sediments serve as a long-term plutonium depot. So the main attention was paid to the study of plutonium distribution in bottom sediments. In the paper are presented the data of spatial and vertical distribution of $^{239+240}$Pu and ^{238}Pu in the Bay sediments. Radiotracer technologies helped to estimate sedimentation rates and evaluation of the plutonium biogeochemical migration parameters (such as fluxes of plutonium) and the contribution of different ways of $^{239+240}$Pu elimination to self-purification of bay waters from plutonium contamination.

14.3.3 Approaches to Terrestrial Coastal Ecosystems' Restoration of the Leningrad region

Margarita Lazareva[1,2*]

[1]V.V. Dokuchaev Central Soil Museum, Birzhevoy proezd 6, Saint-Petersburg, Russia

[2]V.V. Dokuchaev Soil Science Institute, Pyzhyovskiy lane 7, building 2, Moscow, Russia

*corresponding author: margoflams@mail.ru

Keywords: Terrestrial ecosystems, Coastal ecosystems, Anthropogenically disturbed soils

When constructing of buildings, roads, product pipelines the natural ecosystems are completely or partially destroyed. In this case negative consequences of direct anthropogenic impact are indirectly manifested also in the adjacent territories (changes in their hydrological regime, pollution). Therefore, in order to protect the natural environment, it is extremely necessary to carry out preventive and rehabilitation measures for disturbed territories, as well as territories with natural environment that are at risk of extinction. Soil is the basis part of environment. All organisms of marine and terrestrial ecosystems are related with soil. Soil is the habitat for different organisms, the source of nutrients and the store for spores and seeds, the substrate for roots, the reservoir of water for plants, and is also the important link of the matter and energy cycles' system. Arenosols are the wide spread soils of the terrestrial coastal ecosystems of the Leningrad region. These soils are formed on the sandy soil forming materials, have the low fertility and are extremely unstable to anthropogenic pressure. Destruction of these soils as an important link of the system will inevitably lead to the destruction of the entire ecosystem. When analyzing a digital medium-scale soil map made in the V.V. Dokuchaev Central Soil Museum (E.Yu. Sukhacheva, B.F. Aparin, T.A. Andreeva, E.E. Kazakov, M.A. Lazareva, 2016) we identified significant changes in the soil cover of the Leningrad region. Virtually in all landscapes, we found a large number (>50%) of soil cover structures, the components of which, along with natural soils, are anthropogenically disturbed soils, anthropogenic soils and non-soil formations. In order to preserve the terrestrial coastal ecosystems, increase the resistance to anthropogenic impact, the following measures are recommended: (1) Reduction of the anthropogenic pressure; (2) Implementation of measures to protect soil from erosion; (3) Thickening of the high humus layer.

15 Open Session

Simon Jungblut[1,2]

[1]BreMarE – Bremen Marine Ecology, Marine Zoology, University of Bremen, P.O. Box
330440, 28334 Bremen, Germany

[2]Alfred Wegener Institute, Helmholtz Centre for Polar and Marine Research, P.O. Box 120161, 27570 Bremerhaven, Germany

This session was no. 15 of the YOUMARES 8 conference. It does not have a corresponding proceedings article.

15.1 Call for Abstracts

The Marine Sciences are a vast and divers field of research and barely any conference is able to represent all topics with a separate session. The Open Session aims to summarize contributions of young marine scientists from all research fields which do not feel to fit into one of the other sessions.

15.2 Abstracts of Oral Presentations

15.2.1 On the Hydrography of Water Masses in the Southern Rockall Trough – A Synoptic View

Angelina Smilenova[1,2*], Kieran Lyons[2], Glenn Nolan[2,3], Martin White[1]

[1]Earth and Ocean Sciences, School of Natural Sciences, National University of Ireland, Galway (NUIG), University Road, Galway, Ireland

[2]Oceanographic Services, Ocean Sciences and Information Services (OSIS), Rinville, Co. Galway, Ireland, H91 R673

[3]European Global Ocean Observing System (EuroGOOS), Avenue Louise 231, 1050 Brussels, Belgium

*corresponding author: Angelina.Smilenova@Marine.ie

Keywords: Water masses, Rockall Trough, Northeast Atlantic, Subpolar Gyre/Subtropical Gyre interplay

Full depth wintertime CTD transect data, spanning an 8 year period (2006–2013) and acquired along the 53°N–55°N and 54.5°N–56°N lines are assessed in relation to inter-annual variability of Subarctic Intermediate Water (SAIW), Mediterranean Outflow Water (MOW) and Labrador Sea Water (LSW) water masses in the southern entrance of Rockall Trough. The region is one of significant mesoscale variability, where mixing between SAIW and MOW upper intermediate water masses hinders inter-annual variability quantifications. Water column structure and water masses present in the vicinity of the hydrographic transect are examined by property-property diagrams, where temperature and salinity, as well as potential vorticity are used as water masses tracers. Freshening of the upper and intermediate water column south of 55°N in the Rockall Trough region during 2012 and 2013 is observed. General increase in wintertime salinity, 2009 onwards, is detected close to Porcupine Bank continental margin at the MOW depth range (1000 m), which could be representative of a westward extension of the MOW salinity plume into the northeast North Atlantic. At lower intermediate depths, 800–1500 m, cabbeling could be a potential mechanism for inter-annual water masses modifications. Following a deep (1250 dbar isobar) convection event in the Labrador Sea in the early 2000s, strong signals of LSW at depths below 1500 m were detected in 2006, peaking in 2009. Consideration is given to geostrophic current velocities and volume transports. Geostrophic current velocities calculations, based on mean temperature-salinity relationships derived dynamic height, appear reasonable, allowing for preliminary regional volume transport estimates. The southern Rockall Trough is uniquely positioned on the boundary between the Subpolar and Subtropical Gyres in the northeast North Atlantic, making the region a key area of inter-gyre exchange. Therefore, existing hydrographic data from the region could provide some support to the recently suggested north-eastward expansion of the Subpolar Gyre.

15.2.2 Halacarid (Prostigmata: Acari) Species from the Mediterranean Coast of Turkey, with a Checklist of Halacarids from the Coast of Turkey

Furkan Durucan[1*]

[1]Işıklar Caddesi, 07100 Antalya, Turkey

*corresponding author: f_durucan@hotmail.com

Keywords: Halacaridae, Systematic, Mediterranean Sea, Marine biodiversity, Meiobenthos

Halacarid mites are relatively small benthic animals, the adult body length is less than 1 mm. Little is known about the halacarid species in Turkey. This study focused on biodiversity and distribution of halacarid species which were collected in Antalya. For this purpose, samples were collected from 10 stations between September and October 2015-August, September and October 2016 along West Coast of Antalya (Lara-Kalkan). In this study, a total of 714 individuals belonging to 16 genus, 37 species were determined. According to the species numbers, *Copidognathus* ranked first with 10 species. *Copidognathus* was respectively followed by *Agauopsis*, *Rhombognathus* with 5 species, by *Acaromantis*, *Agaue*, *Scaptognathus*, *Simognathus* with 2 species, by *Acarochelopodia*, *Actacarus*, *Anomalohalacarus*, *Atelopsalis*, *Halacaropsis*, *Halacarus*, *Lohmannella*, *Maracarus* and *Thalassarachna* with 1 species. In summary, this study has contributed to an understanding of the halacarid diversity and ecology of halacarid species in Antalya. In addition, a check-list of the halacarid species that have been reported from the coasts of Turkey to date is provided.

15.2.3 The Effect of Temperature and Pyrene Exposure on *Calanus finmarchicus* Males and the Consequences for Population Sex Ratios

Maria Winberg Olsen[1*], Khuong Van Dinh[2], Torkel Gissel Nielsen[2]

[1]Institute of Biology, Ole Maaløesvej 5, 2200 Copenhagen N, Denmark

[2]National Institute of Aquatic Ressources, Anker Engelunds Vej 1, Building 202, 2800 Kgs. Lyngby, Denmark

*corresponding author: kalixvuc@gmail.com

Keywords: Climate change, Oil pollution, Zooplankton, Copepod, Sex ratio

Future warmer oceans will most likely increase ship traffic in the northern ocean environment. This study investigates the sensitivity to higher temperature (10 °C and 14 °C) and pyrene exposure (Water control, Acetone control, 1 nM, 10 nM, 100 nM and 100+ nM) on lab-cultured male *Calanus finmarchicus*. The males are generally smaller than the females, have significantly less lipid and produce less biomass as fecal pellets. Male survival decreased rapidly (86.1% to 28.9% at ended experiment) due to temperature and decreased 50% due to pyrene exposure. It was found that male *Calanus finmarchicus* produce less than 10% of female production given the same conditions, but both sexes decreased Specific Pellet Production (SPP) when exposed to pyrene treatments. Due to the higher sensitivity of the male copepod the sex ratio of the population will be skewed towards the female. The population response to so few males in *Calanus finmarchicus* is not well known and further studies will be needed.

15.2.4 Population Dynamics of Crustacean Cirriped *Pollicipes pollicipes* in the Moroccan Atlantic Coast

Hajar Bourassi[1,2*], Hakima Zidane[2], Ayoub Baali[1,2], Mohamed I. Malouli[2], Imane Haddi[3,2], Ahmed Yahyaoui[1]

[1]Zoology and general biology laboratory, Sciences Faculty of Rabat, Morocco

[2]Department of Fisheries Resources, National Institute of fishery Research, Casablanca, Morocco

[3]Earth sciences department, Faculty of Sciences, University Hassan II, Casablanca, Morocco

*corresponding author: hajar.bourassi@gmail.com

Keywords: *Pollicipes pollicipes*, Density, TAC, Intertidal biodiversity, Population dynamic

The fisheries environment has been subject of increasing pressure of the industrial and human activities. Many species that are considered as biological indicators of value and information are suffering the consequences, such as crustacean cirripeds: *Pollicipes pollicipes* (goose barnacle). Those represent important coastal resources for population livelihoods and coastal ecosystems. Yet, they are informally exploited despite the ministerial decrees that regulate their exploitation. To support the implementation of a management plan, various scientific studies are conducted on this species. Accordingly, considering the current concern for the conservation of the resources, we carried out a monthly monitoring program, within our larger scale study on the population dynamic of *P. pollicipes*, during the year 2016–2017 at two exploitable areas: Mansouria and Souiria Kdima. Those are characterized by the important deposits of the species and the high rate of its exploitation, but also vary by biotope features. This work studied goose barnacle population structure, density and biomass of different populations. The results obtained show that goose barnacle's abundance and biomass significantly differ between the seasons and from one site to another ($P < 0.05$) due to the biotope features: the density of Mansouria's population is over 50% higher than Souiria Kdima's. We also compared the RC size (corresponding to the distance between the two calcareous plates: Rostrum and Carina) of the goose barnacles in function of the area. The results showed that Mansouria's individuals are significantly larger than those from Souiria Kdima ($P < 0.05$). This may be due to the low rate of exploitation and to the low shore level where they grow, so a longer duration of immersion. The structure of both populations showed that the minimum size fixed by the ministry of fisheries (RC > 25 mm) is too low. This work together with the ongoing histological study will provide correct information to adjust the size regulation and more of the management plan of this resource.

15.2.5 First Characterization of Deep-Sea Sponges on Seamounts in the Galapagos Marine Reserve

Marie F. Creemers[1,2*], Salomé Buglass[1], Henry Reiswig[3], Sarah M. Griffiths[4], Evelyn Taylor-Cox[4], Robert Winch[4], Etienne Rastoin[1], Pelayo Salinas de León[1,5], Amanda Bates[4], Patricia Martí Puig[1]

[1]Charles Darwin Foundation, Av Charles Darwin s/n, Puerto Ayora, Santa Cruz, Galapagos Islands, Ecuador

[2]Ocean & Earth Sciences Department, University of Southampton, Southampton, United Kingdom

[3]University of Victoria and Royal B.C. Museum, Victoria, B. C., Canada

[4]School of Science and the Environment, Manchester Metropolitan University, Manchester, United Kingdom

[5]Pristine Seas, National Geographic Society, Washington, D.C., USA

*corresponding author: mcreemers@alumni.ulg.ac.be

The Galapagos Marine Reserve (GMR) harbors approximately 300 seamounts providing ecosystem services to the local community, yet, little is known about their habitats and biodiversity. Recognized as priority habitats by the United Nations, seamounts have recently become the focus of rising interest with regards to sustainable management and conservation. With the revised zonation of the GMR, increasing our current knowledge of Galapagos seamount fauna has arisen as an important priority. Deep-sea sponges are a habitat-forming taxon of key ecological importance and great economic interest. Baseline information is therefore necessary to improve the management of sponge-associated systems. The Charles Darwin Foundation initiated the project "Seamounts of the GMR" in partnership with the Galapagos National Park. In 2015, the Ocean Exploration Trust E/V *Nautilus* explored three seamounts in the northern part of the GMR using remotely operated vehicles (ROVs), providing over 40 h of high definition video transects, 39 biological samples and *in situ* records of environmental parameters. Nine species demosponges and four of hexactinellids were morphologically identified and molecularly characterized at the mitochondrial COI gene. Eleven species, all from different families, are potentially new to science. One intriguing potentially new demosponge is a carnivorous Cladorhizidae, while another presents a unique, sensational, kebab-like branching shape and might belong to a new genus of Phellodermidae. One gigantic branched hexactinellid could be a member of a new genus of Farreidae. Morphological results for both potentially new genera are supported by genetics. Other potentially new glass sponges include a species of *Vitrollula*, a genus only previously observed in Japan, and a species of *Tretodictyum* with a coral-like arborescent body. These results encourage our team to further investigate the mysterious sponge biodiversity of the Galapagos seamounts, with the aim to characterize sponge communities and assess their value in order to improve seamount conservation and sustainable management.

15.2.6 Influence of an External Magnetic Field on a Planktonic Foraminifer

Charlotte Eich[1,2*], Nina Keul[1], Stefanie Ismar[2]

[1]Christian-Albrechts-Universität – Institut für Geowissenschaften, Ludewig-Meyn Straße 10, 24118 Kiel, Germany

[2]Experimental Ecology – Food Webs, GEOMAR Helmholtz Institute for Ocean Research Kiel, Düsternbrooker Weg 20, 24105 Kiel

*corresponding author: eich.charlotte@gmx.de

Keywords: *Globigerinoides ruber,* Gametogenesis, External triggers, Lunar cycle, Spinose planktonic foraminifera

Many planktonic species, as, for example, foraminifera, conduct a simultaneous gametogenesis, which seems to be coupled to the lunar cycle. The change of the Earth's magnetic field during the lunar cycle, especially around full moon, has been suggested as one potential external trigger. We experimentally tested, whether an external magnetic field of the strength of the Earth's magnetic field (50 μT) triggers gametogenesis in the planktonic foraminifera *Globigerinoides ruber* by quantifying the survival and gametogenesis rates and assessing floating behaviour. All experiments were conducted during a cruise (MSM58, Reykjavík (Iceland) to Ponta Delgarda (Azores)) aboard the *Maria S. Merian.* Understanding the trigger behind simultaneous gamete release in *G. ruber* might be a useful model for understanding other big spawning events of planktonic species and help us to further understand both the physiology and ecology of planktonic foraminifera.

15.2.7 Applications of Stereo-Camera Systems in Marine Conservation

Nils T. Willenbrink[1*], Barbara Horta E. Costa[1]

[1]Centre of marine sciences (CCMAR), University of the Algarve, Campus de Gambelas, 8005-139 Faro, Portugal

*corresponding author: n.willenbrink@hotmail.de

Keywords: Stereo-cameras, Sampling methods, Ichthyofauna, Marine reserve

The utilization of underwater cameras has brought numerous advantages into the field of marine biology. In situ recordings are non-extractive in comparison to fisheries methods and less demanding on personal and logistics than scuba methods. Technological advances considerably enhanced video and information quality but almost always at high material cost. We present a low-cost stereo-camera dropdown system that outperforms single-camera setups by obtaining precise measurements through the principle of triangulation. Tested in a coastal marine reserve in SW Portugal, this system enables scientists to assess the areas performance in protecting ichthyofauna. We will discuss possibilities and drawbacks of the method and how modifications can be made depending on research interest and budget.

15.2.8 Feeding Plants to Predators: Lupin as a Feed Supplement for the European Sea Bass (*Dicentrarchus labrax)*

Enno Fricke[1*], Reinhard Saborowski[1], Monika Weiß[1], Christina Hörterer[1], Sinem Zeytin[1], Matthew James Slater[1]

[1]Alfred Wegener Institute, Helmholtz Centre for Polar and Marine Research, Am Handelshafen 12, 27570 Bremerhaven, Germany

*corresponding author: cef@hotmail.de

Keywords: Feed supplements, Plant protein, Digestibility, Enzymes, Growth

Lupin presents a promising alternative protein source for aquaculture diets, particularly for locally sourced or GMO-free use feeding high value finfish species. In the current study a feeding trial with juvenile European sea bass (*Dicentrarchus labrax*, initial weight ca. 15 g) was conducted to evaluate two experimental diets containing 50% dry matter untreated and enzymatically fermented dehulled lupin meal (*Lupinus angustifolius*). Weight gain of *D. labrax* fed the lupin based diets was about 20% lower than those fed a fishmeal control diet. Fish fed the untreated lupin meal diet showed a significantly reduced condition factor, whereas fish feeding on the fermented lupin meal diet had a significantly lower hepatosomatic index, both compared to the fishmeal control. Apparent digestibility coefficients of dry matter and carbon differed significantly between all diets with untreated lupin meal exhibiting the lowest, fermented lupin meal intermediate and the fish meal diet the highest values. The highest protein digestibility coefficients were measured in the fermented lupin meal diet, exceeding the values of the fishmeal control significantly. The fermentation of lupin meal also significantly increased the apparent phosphorus digestibility. Both lupin based diets impair chymotrypsin activities in the intestine and the pyloric caeca but did not significantly alter alkaline phosphatase activities. Lipase activities were significantly higher in the pyloric caeca of fish receiving the fishmeal control treatment. Results suggest that the very early juvenile stages of European sea bass do not cope well with the high lupin inclusion rates due in part to enzyme impairment and poor nutrient utilization. However, fermentation positively affected the digestibility of lupin and an appropriately adjusted supplementation rate could diminish the negative effects on growth and the digestive physiology.

15.2.9 Pacific Cod Population Structure in the Southern Part of Species' Range

M. A. Smirnova[1*], S. Y. Orlova[1], P. V. Kalchugin[2], M. I. Bojko[2], J.-H. Park[3], A. M. Orlov[1,4]

[1]Russian Federal Research Institute of Fisheries and Oceanography (VNIRO), 107140, Moscow, Russia

[2]Pacific Scientific Research Fisheries Center (TINRO-Center), 690091, Vladivostok, Russia

[3]National Institute of Fisheries Research (NIFS), 46083, Busan, South Korea

[4]A.N. Severtsov Institute of Ecology and Evolution, 119071, Moscow, Russia; Dagestan State University, 367000, Makhachkala, Dagestan; Tomsk State University, 634050, Tomsk, Russia

*corresponding author: masmirnova209@gmail.com

Keywords: Pacific cod, Microsatellite loci, D-loop, Genetic variation, Population structure

Pacific cod *Gadus microcephalus* is widely distributed in the coastal waters of the Northern Pacific and is one of the most important commercial fishery species in the Far Eastern waters. Research data indicate structural heterogeneity of Pacific cod population and existence of its several groups. Results of the molecular genetic study of Pacific cod population structure in the southern part of its range and some adjacent regions are presented. Samples were collected in 8 different areas of the Yellow Sea, East/Japan Sea, Sea of Okhotsk and Pacific Ocean off the South Kuril Islands. Control region of mtDNA and microsatellite loci were used as genetic markers. Pairwise genetic differentiation evaluation revealed heterogeneity of Pacific cod populations in the southern part of the range. According to the obtained F_{st} values Pacific cod from the Republic of Korea waters (Yellow Sea side) and north-western part of the Sea of Okhotsk significantly differ from all other studied regions. Significant differentiation was also revealed between samples from the waters of Tatar Strait and all other regions except for South Kuril Pacific cod (both Sea of Okhotsk and Pacific Ocean sides). These two latter sample collections were similar to each other as well. Low level of differentiation was also shown for the Peter the Great Bay and the East/Japan Sea waters of Republic of Korea. Results of the assignment test carried out in the "Structure" software for all regions including north-western Sea of Okhotsk showed clear division of all Pacific cod populations of the studied area into 3 clusters. One of the clusters corresponds to population from the north-western part of the Sea of Okhotsk, the second one – from the Peter the Great Bay. All other populations were presented as a combination of Pacific cod from these two clusters and the third one with different proportion.

15.2.10 Epibenthic Feeding Juvenile Flounder *Platichthys flesus* do not Suffer from Summer Growth Reduction as Benthic Feeders Suffer from

Henk W. van der Veer[1], Joana F. M. F. Cardoso[2], Vânia Freitas[2], Suzanne S. H. Poiesz[1*], Johannes I. J. Witte[1]

[1]Royal Netherlands Institute for Sea Research, P.O. Box 59, 1790 AB Den Burg Texel, The Netherlands

[2]CIIMAR/CIIMAR, Centro Interdisciplinar de Investigação Marinha e Ambiental, Universidade do Porto, Rua dos Bragas 289, 4050-123 Porto, Portugal

*corresponding author: suzannepoiesz@gmail.com

Keywords: 0-group flounder, Dutch Wadden Sea, Dynamic Energy Budget model, Growth, Otolith, Settlement

Summer growth reduction (observed versus maximum possible growth) has been described for various juvenile flatfish species over a large latitudinal gradient. It was hypothesized that the underlying mechanism was a lower activity of macrozoobenthcs after the spring phytoplankton bloom, resulting in reducing benthic prey availability and ultimately a reduction in food intake and hence in growth. This hypothesis was tested for juvenile plaice *Pleuronectes platessa* L., a flatfish species mainly feeding on benthos. Otolith microstructure analysis validated the observed summer growth reduction and showed that it coincided with a decrease in stomach content. In this paper, the test was extended to juvenile flounder *Plathichthys flesus* L., another flatfish species, whereby much lower summer growth reduction was expected, since flounder relied more on abundant epibenthic prey (crustaceans). Growth performance of 0-group flounder was analyzed at the Balgzand intertidal by means of otolith microstructure analysis for three different years, a relatively cold (1996), an average (2000) and a relatively warm (1995) year. In all 3 years, summer growth reduction in 0-group flounder was much lower than observed in 0-group plaice at the same areas, confirming expectations that summer growth reduction is at least partly caused by a lower activity of macrozoobenthos after the spring phytoplankton bloom.

15.2.11 Lipid Storage of the North Sea Shrimp *Crangon crangon*

Eleni Melis[1*], Diana Martínez-Alarcón[1,2], Reinhard Saborowski[2], Wilhelm Hagen[1]

[1]Bremen Marine Ecology (BreMarE), Marine Zoology, University of Bremen, P.O. Box 330440, 28334 Bremen, Germany

[2]Alfred Wegener Institute, Helmholtz Centre for Polar and Marine Research, P.O. Box 120161, 27570 Bremerhaven, Germany

*corresponding author: eleni.melis@uni-bremen.de

Keywords: Brown shrimp, Fatty acid composition, Midgut gland and eggs, Lipid transfer, Capital and income breeding

The brown shrimp *Crangon crangon* inhabits European coastal areas within a large latitudinal range. It also inhabits the North Sea, a habitat with pronounced seasonal fluctuations. In spite of its high ecological and economical value, our knowledge about adaptive processes and strategies of *C. crangon* to cope with these changing conditions is scarce. In marine decapods, lipids are mainly stored in the midgut gland and used as major energy source for numerous metabolic processes, including reproduction. Many decapods are capital breeders, using lipid deposits from the midgut gland to fuel egg production. During oogenesis in *C. crangon*, a significant share of the lipid stores is suggested to be transferred from the midgut gland to the developing eggs. This would indicate that *C. crangon* also follows the strategy of capital breeding. This study aims at elucidating lipid transfer and breeding strategy of *C. crangon*. Total lipid contents and fatty acid (FA) compositions of midgut glands and early developmental eggs were determined and compared. Gas chromatography and principal component analysis (PCA) were used to determine differences in FA composition between midgut glands and eggs. Comparison of total lipid levels did not reveal significant differences between the two tissues. However, PCA based on FA composition clearly separated two clusters, grouping samples of the same tissues. Hence, *C. crangon* may either follow the income breeding strategy, with FAs in the eggs originating from recently ingested food items, or capital breeding, with a transfer of deposited FAs from the midgut gland to the eggs. The capacity to utilize the benefits of either reproductive strategy may offer a selection advantage in variable habitats.

15.2.12 Biology and Population Dynamics of *Sardinella aurita* in South Morocco

A. Baali[1*], H. Bourassi[1], N. Charouki[2], K. Manchih[2], K. Amenzoui[2], A. Yahyaoui[1]

[1]Zoology and General Biology Laboratory, Faculty of Science, Mohammed V University, Rabat, Morocco

[2]Fishery Research National Institute, Casablanca, Morocco

*corresponding author: ayoubbaali22@gmail.com

Keywords: *Sardinella aurita*, Reproduction biology, Population dynamics, Gonadosomatic index, Sea Surface Temperature, South of Morocco

Round sardinella (*Sardinella aurita*) has gained much attention lately because of its biomass decrease, which might be the result of climatic changes occurring across the Atlantic Sea. Little information is known about the biology of this species in the Moroccan Atlantic area. This lack of study is rather surprising given the considerable catches of the round sardinella and the strong contributions in the total catch in last years in particular in the southern Moroccan Atlantic. The objective of this study was to explore some aspects of the reproductive biology of *Sardinella aurita* and to understand recent biomass decrease using monthly landed catch and Sea Surface Temperature (SST) data from 2009 to 2016. To fill the knowledge gaps in round sardinella biology, our study investigated this species population dynamics in the South of Morocco. Using a Chi test, we first investigated the overall female to male ratio, which was not statistically different, although it varied seasonally and according to the length of the fish. The monthly changes in the gonadosomatic index and the macroscopic characteristics of gonads showed that round sardinella have spawning peak on April in this area. They were no differences in size at maturation between sexes ($p > 0.05$) while females reach sexual maturity at a smaller size than males (26.17 and 26.78 cm respectively). Regarding the SST and biomass of the round sardinella we observed that we have a high correlation coefficient between the two parameters especially in the 22° of latitude, which allows us to say that the temperature influences the presence of this species in the present area and to conclude that the global warming impact Moroccan fisheries.

15.2.13 To Feed or not to Feed? Artificial Feeding Affects Reef Fish Functions

Natalie Prinz[1,2*], Sebastian C. A. Ferse[2], Sonia Bejarano[2]

[1]University of Bremen, Bibliothekstraße 1, 28359 Bremen, Germany

[2]Leibniz Center for Tropical Marine Research (ZMT) Fahrenheitstraße 6, 28359 Bremen, Germany

*corresponding author: nprinz@uni-bremen.de

Keywords: Coral reef tourism, Supplementary feeding, Fish feeding rates, Functional groups, South-Pacific

Over centuries humans have fed wild animals, driven by the desire to provoke close contact. Artificial feeding has become a regular habit in ecotourism activities worldwide, with poorly known consequences for ecosystem function. This study quantifies for the first time (1) how effective is artificial feeding at attracting reef fishes, (2) which feeding guilds are most attracted, and (3) how are natural levels of corallivory and grazing-detritivory affected. Data were collected in sites where fish are regularly fed bread by snorkelers, and adjacent control sites, where bread was only provided for this study, within the Aitutaki lagoon (Cook Islands). The fish community was censused and feeding rates (bread versus natural food) of three model species (*Chaetodon auriga, Ctenochaetus striatus, Lutjanus fulvus*) were quantified. 24% of all species observed across sites were effectively attracted by bread. Fish biomass was significantly higher in feeding, than control sites. Taxonomic richness decreased during bread feeding, compared to 1 h before and after across sites. Carnivores and omnivores dominated the community, suggesting concentrated predation pressure. The effect of artificial feeding on natural foraging rates varied between species. *C. auriga* fed significantly more on bread in feeding sites versus control sites, *C. striatus* fed less on the benthos during feeding, compared to before and after, suggesting an effect on foraging behavior. Stakeholder interviews revealed differences in perceptions on the issue of fish feeding between natives and tourists. Paradoxically, natives are strongly in favor to improve tourist satisfaction, whereas tourists appreciate snorkeling regardless of whether fish are artificially attracted. Finding ways for humans to appreciate wildlife closely while causing minimal disruptions is crucial to balance awareness raising and conservation. Future research on fish metabolism and cascading effects on the reef benthos may reveal further negative impacts of artificial feeding.

15.3 Abstracts of Poster Presentations

15.3.1 The Effect of Elevated and Fluctuating *p*CO_2 Concentrations on the Growth of Calcifying Marine Epibionts

M. Johnson[1,2#*], L. Hennigs[2,3#*], C. Pansch[2], M. Wall[2]

[1]Christian-Albrechts Universität zu Kiel, Christian-Albrechts-Platz 4, 24188 Kiel, Germany

[2]GEOMAR Helmholtz-Zentrum für Ozeanforschung Kiel, Düsternbrooker Weg 20, 24105 Kiel, Germany

[3]Carl-von-Ossietzky Universität Oldenburg, Ammerländer Heerstrasse 114-118, 26129 Oldenburg, Germany

#shared first authorship

*corresponding authors: mildredjjohnson@gmail.com, laura.hennigs@uni-oldenburg.de

Keywords: Macroalgae, Environmental fluctuations, Boundary layer, Epibiont calcification

The projected reduction in surface seawater pH by 2100 has led to a significant increase in ocean acidification research. These studies have mostly focused on constant conditions which produced valuable results but provide limited information on the important role of naturally occurring environmental fluctuations within and on habitat-forming organisms. In temperate regions, marine seaweeds dominate rocky shores and via their metabolism, photosynthesis and respiration, induce diurnal fluctuations of pH and oxygen within their surroundings. Dependent on hydrodynamic conditions, these processes act at the larger scale (slow flow conditions) shifting environmental conditions within the entire seaweed bed or at smaller scale at the macroalgae-seawater interface and create unique microenvironments, the so-called boundary layer (BL). In particular, epibionts that live at the macroalgae-seawater interface are constantly exposed to fluctuating conditions. This raises the question whether such diurnal variations may amplify or buffer the responses of organisms to future environmental pressures. To determine the effect of the small scale and large-scale BL on the development of epibionts, the growth rates of *Electra pilosa* and *Balanus improvisus* where compared on 2 substrates namely *Fucus serratus* and acrylic glass slides, under 4 different *p*CO_2 conditions and 2 temperature regimes. A multifactorial experiment was conducted at the Kiel Indoor Mesocosms programmed with 2 constant (400 and 1250 µatm) and 2 fluctuating *p*CO_2 concentrations (400/2400 µatm and 100/1250 µatm) testing the present and future role of small and large-scale BL fluctuations, respectively. These treatments were conducted at 2 different temperatures (10 and 15 °C) simulating present day and future average temperatures in May. Preliminary observations indicate that higher growth rates occur on *Fucus* substrates and at higher temperatures. Whether the fluctuating boundary layers of *Fucus* meadows provide short term refuge from low pH and O_2 conditions and buffer the responses to future ocean acidification still warrants resolve.

15.3.2 Impact of Submarine Groundwater Discharge on Fish, Plankton and Biofouling

C. Starke[1*], N. Moosdorf[2], W. Ekau[2]

[1]Institut of Hydrobiology and Fishery Science (IHF), Olbersweg 24, 22767 Hamburg, Germany

[2]Leibniz Centre for Tropical Marine Research (ZMT), Fahrenheitstr. 6, 28359 Bremen, Germany

*corresponding author: cls@zmt-bremen.de

Keywords: Groundwater, Nutrient input, Marine organisms

Submarine groundwater discharge (SGD) into the sea is widely spread and increasingly studied. However, the impact of SGD on marine organisms is still unclear. For better understanding that effect on fish and plankton abundance as well as on the early stage of biofouling processes, we investigated submarine wells in a coastal lagoon of Tahiti, French Polynesia with three different sampling methods. The occurrence and behavior of fish around a wellspring was investigated by means of underwater photos using two GoPro Hero 4 cameras fixed at a freshwater spring and a control site. Plankton samples were taken at a catchment area of around six meters around the same spring and the control site to test if the spring has influence on abundance and composition of micro- and mesozooplankton. For a biofouling experiment, settlement panels were installed along a transect through a freshwater spring. The study is based on the assumption that enhanced biological production occurs due to increased nutrient inflow caused by terrestrial freshwater supply. Our results suggest slightly higher plankton abundance outside the spring and higher settlement in the spring in our biofouling experiments. Fishes, however, as found in the underwater photos seem to avoid direct contact with the low salinities in the freshwater wells. In conclusion, due to higher nutrient concentrations in the freshwater there might be an increased primary production leading to an increase of primary consumers and in theory also in secondary consumers near the spring, which have been noted by fishermen around the world. Further investigations with optimized methodology is necessary for a better understanding of the subject.

15.3.3 The Urgent Need of Scientific Divers in Ecological Research on the Example of Investigations in the Comau Fjord, Chile

Jan Laurenz[1*], Jürgen Laudien[2]

[1]Christian-Albrechts-University Kiel, Zoological Institute, Am botanischen Garten 5-9, 24118 Kiel, Germany

[2]Alfred Wegener Institute Helmholtz Center for Polar and Marine Research, Bremerhaven, Germany

*corresponding author: j.laurenz@gmx.de

Technological developments allow performing scientific work underwater by remotely operated devices and make dangerous work under excessive pressure unnecessary in many situations. Nevertheless, some specific research tasks, in particular concerning ecological issues, can currently not be performed by any devices available on the market. Therefore, the work of scientific divers is essential, which can be demonstrated using the example of the Comau Fjord, Chile. The structure of the fjord itself, the water depths and the complexity of the data generation as well as the installation of the experimental setup makes it impossible to operate a remotely vehicle (ROV). This area is in the focus of several research projects of the Alfred Wegener Institute, Bremerhaven. Topics includes population analyses of cold-water corrals, planktonic observations, growths parameters, colonization, biodiversity, sedimentation and long term monitoring of population diversity. Many tasks concerning these research projects can only be performed by scientific divers, highlighting the importance to employ scientific divers. Examples include pushnet sampling, sediment sampling, long-term monitoring, underwater drilling, sampling organisms, colonization monitoring, underwater documentation and measurements of growth parameters. These examples underline the urgent need of scientific divers in modern marine science with their ability of interpreting what they see, the sensitivity of their perception and their ability to react spontaneously to new and unexpected situations. Ultimately the human mind and their handicraft skills cannot be replaced completely by modern technology.

15.3.4 Extracellular Enzymes of Invertebrate Origin

Imke Böök[1,2*], Reinhard Saborowski[2]

[1]University of Bremen, Leobener Straße, NW2, 28359 Bremen, Germany

[2]Alfred-Wegener-Institut für Polar und Meeresforschung, Am Handelshafen 12, 27570 Bremerhaven, Germany

*corresponding author: imke.boeoek@awi.de

Keywords: Biocatalysts, Organic matter, Remineralization, Nutrient cycles, Fluorophores

Extracellular enzymes are key drivers in the remineralization of organic matter in marine systems. According to the widespread view such enzymes derive mainly from bacteria. However, a large number of extracellular enzymes are released into the water by invertebrates through "sloppy feeding", molting, and excretion. These enzymes have the potential to degrade organic matter and boost subsequent microbial growth. The aim of this study is, therefore, to investigate the extracellular enzyme activity in molts and egesta of different marine invertebrate species with sensitive fluorometric assays. Visualization of enzymes leaking from molts and fecal pellets will be achieved by using agarose plates incubated with fluorogenic substrates. Several 4-Methylumbilliferone (MUF) derivatives will be used to detect enzymatic activity of selected enzyme classes: MUF-Phosphate for phosphatase, MUF-Butyrate for esterase, MUF-N-acetyl-beta-D-glucosaminide for exochitinase and MUF-beta-D-Glucoside for glucosidase). Molts and feces will be placed directly on agar plates and enzymatic activity will result in a measurable fluorescence signal. First results results show phosphatase, esterase and glucosidase activity in fecal pellets of isopods (***Idotea baltica*** and ***Idotea emarginata***). Furthermore, phosphatase activity was verified in feces of the decapod shrimp (***Palaemon*** sp.) and the gastropods (***Littorina littorea***). High chitinolytic activity was found in molts of ***I. baltica*** but no chitinolytic activity was detected in the egesta of the isopods. These results support the hypothesized important role of extracellular enzymes from marine invertebrates in remineralization processes. Further investigation will focus on the quantification and detailed characterization of these proteins to distinguish them from microbial enzymes.

www.ingramcontent.com/pod-product-compliance
Lightning Source LLC
LaVergne TN
LVHW082124200926
848484LV00008B/2003
* 9 7 8 1 0 1 3 2 7 6 1 1 8 *